License to Drive
in Illinois

JOIN US ON THE WEB: www.licensetodrive.com

"Your Online Driver Education Resource"

or at
www.delmardriversed.com

Visit www.thomson.com for information on 35 Thomson publishers and more than 25,000 products! or email: findit@kiosk.thomson.com

License to Drive
in Illinois

Alliance for Safe Driving
Santa Monica, California

Africa • Australia • Canada • Denmark • Japan • Mexico • New Zealand • Philippines
Puerto Rico • Singapore • Spain • United Kingdom • United States

NOTICE TO THE READER

Publisher and author do not warrant or guarantee any of the products described herein. Publisher and author do not perform any independent analysis in connection with any of the product information contained herein. Publisher and author do not assume, and expressly disclaim, any obligation to obtain and include information other than that provided to it by the manufacturer.

The reader is expressly warned to consider and adopt all safety precautions that might be indicated by the activities herein and to avoid all potential hazards. By following the instructions contained herein, the reader willingly assumes all risks in connection with such instructions.

The publisher and author make no representations or warranties of any kind, whether express or implied, including but not limited to, the warranties of fitness for particular purpose or merchantability, nor are any such representations implied with respect to the material set forth herein, its reliability, completeness, correctness, accuracy, legality, practicality or operativeness and the publisher and author do not take any responsibility with respect to such material. The publisher and author shall not be liable for any special, consequential, or exemplary damages resulting, in whole or part, from the readers' use of, interpretation of, application of or reliance upon, this material. The material contained herein does not in any way constitute and is not intended to be legal advice, or any promise, guarantee or binding statement of the state of the law.

Delmar Staff:
Business Unit Director: Susan L. Simpfenderfer
Executive Editor: Marlene McHugh Pratt
Acquisitions Editor: Zina M. Lawrence
Developmental Editor: Andrea Edwards Myers
Editorial Assistant: Elizabeth Gallagher
Executive Marketing Manager: Donna J. Lewis
Channel Manager: Nigar Hale
Executive Production Manager: Wendy A. Troeger
Production Manager: Carolyn Miller
Cover Image: Photo courtesy of Alliance for Safe Driving

COPYRIGHT © 2002
By Alliance for Safe Driving

Delmar is a division of Thomson Learning. The Thomson Learning logo is a registered trademark used herein under license.

Printed in the United States of America
1 2 3 4 5 6 7 8 9 10 XXX 05 04 03 02 01 00

For more information contact:
Delmar, 3 Columbia Circle, Box 15015, Albany, NY 12212-5015;
or find us on the World Wide Web at http://www.delmar.com

All rights reserved Thomson Learning 2002. The text of this publication, or any part thereof, may not be reproduced or transmitted in any form or by any means, electronics or mechanical, including photocopying, recording, storage in an information retrieval system, or otherwise, without prior permission of the publisher and/or the author.

You can request permission to use material from this text through the following phone and fax numbers. Phone: 1-800-730-2214; Fax 1-800-730-2215; or visit our Web site at http://www.thomsonrights.com

Library of Congress Cataloging-in-Publication Data

License to drive in Illinois / Alliance for Safe Driving, Santa Monica, California.
    p.   cm.
  Includes index.
  ISBN 0-7668-2370-9
   1. Automobile driving—Illinois. 2. Drivers' licenses—Illinois.
  I. Alliance for Safe Driving (U.S.)

Figure 21-5 copyright by and courtesy of GeoSystems Global Corporation

This book is dedicated to the memory of Rick Price,
whose life was tragically cut short at the young age of 23 by a drunk driver.
He never had a chance to fulfill his dreams.

About the Authors

The Alliance for Safe Driving is a driving and traffic safety organization dedicated to improving the education of drivers in all jurisdictions. Through many years of educating drivers on behalf of departments of motor vehicles, municipal courts, and traffic courts across North America, the Alliance for Safe Driving has developed a successful, reality-based, interactive learning system that comprises textbooks, workbooks, videotapes, and computerized educational products, including www.licensetodrive.com. The Alliance for Safe Driving's driver safety and traffic education professionals have all contributed to the development of this unique and innovative *License to Drive in Illinois* driver education course of study.

Contents

Special Boxes .... *xii*
Foreword .... *xvii*
Preface .... *xix*
Acknowledgments .... *xxi*
How to Use License to Drive .... *xxiv*

Chapter 1
Introduction to Driving 1
   1–1   Driver Education 3
   1–2   Driving Is a Privilege 4
   1–3   The Highway Transportation System 9

UNIT 1
THE DRIVER AND THE DRIVING TASK 19

Chapter 2
Defensive Driving 21
   2–1   Defensive-Driving Skills 23
   2–2   Managing Time, Space,
          and Visibility 26
   2–3   The "SAFE" Method 33

Chapter 3
The Condition of the Driver 45
   3–1   The Physical Condition
          of the Driver 47
   3–2   The Mental Condition of the Driver 54

UNIT 2
DRIVING FUNDAMENTALS 67

Chapter 4
Preparing to Drive 69
   4–1   Vehicle Instrumentation 71
   4–2   Operating Devices 74
   4–3   Vehicle Controls 81

Chapter 5
Vehicle Operation Basics 91
- 5-1 Starting the Engine and Engaging the Transmission 93
- 5-2 Steering 98
- 5-3 Backing up 102
- 5-4 Parking 104

Chapter 6
The Speed Factor: Negotiating Curves and Braking 119
- 6-1 Physical Laws Affecting Driving 121
- 6-2 Negotiating Curves 125
- 6-3 Braking 129
- 6-4 Speed Limits 134

UNIT 3
WELCOME TO THE ROAD 141

Chapter 7
Signs, Signals, and Roadway Markings 143
- 7-1 Traffic Signs 145
- 7-2 Traffic Signals 152
- 7-3 Roadway Markings 157

Chapter 8
Intersections and Right-of-Way 169
- 8-1 Intersections 171
- 8-2 Right-of-Way 180
- 8-3 Railroad Crossings 187

Chapter 9
Turning 195
- 9-1 Turning Basics 197
- 9-2 Left Turns 199
- 9-3 Right Turns 207
- 9-4 Reversing Your Direction 211

Contents ◆ ix

UNIT 4
THE DRIVING ENVIRONMENT 223

Chapter 10
City Driving 225
- 10–1 The City Driving Environment 227
- 10–2 Lane Positioning 231
- 10–3 Passing 240

Chapter 11
Highway and Rural Driving 247
- 11–1 The Rural Driving Environment 249
- 11–2 Defensive Driving on Highways 256
- 11–3 Other Dangers of Highway and Rural Driving 260

Chapter 12
Freeway Driving 269
- 12–1 The Freeway Driving Environment 271
- 12–2 Entering Freeways 278
- 12–3 Exiting Freeways 283
- 12–4 Other Dangers of Freeway Driving 286

Chapter 13
Sharing the Road 295
- 13–1 Pedestrians 297
- 13–2 Bicycles 302
- 13–3 Motorcycles 306
- 13–4 Trucks and Buses 313

UNIT 5
CHALLENGING DRIVING 323

Chapter 14
Challenging Driving Conditions 325
- 14–1 Reduced Visibility 327
- 14–2 Challenging Road Conditions 333
- 14–3 Hills and Mountain Roads 338

Chapter 15
Driving in Bad Weather 347
- 15–1 Rain 349
- 15–2 Snow and Ice 353
- 15–3 Other Severe Weather Conditions 359
- 15–4 Cold and Hot Weather 363

Chapter 16
Collisions 371
- 16–1 Collisions 373
- 16–2 What to Do at the Scene of a Crash 375
- 16–3 Preventing Collisions 383
- 16–4 Vehicle Restraint Systems 386

Chapter 17
Emergencies 395
- 17–1 Blowouts and Flat Tires 397
- 17–2 Mechanical Failures 399
- 17–3 Skids 406
- 17–4 Other Emergencies 409

UNIT 6
DRIVING RESPONSIBLY AT HOME AND AWAY FROM HOME 419

Chapter 18
Driving Under the Influence 421
- 18–1 Alcohol and Other Drugs 423
- 18–2 Preventing Drunk Driving 429
- 18–3 The Law and Driving Under the Influence 432

Chapter 19
Licensing and Vehicle Ownership 445
- 19–1 The Licensing Process 447
- 19–2 Registering and Insuring Your Vehicle 460
- 19–3 Your Driving Record 471
- 19–4 Buying and Selling a Vehicle 474

Chapter 20
Vehicle Maintenance, Equipment, and Security **485**

    20–1 Preventive Maintenance..487
    20–2 Selecting a Mechanic..495
    20–3 Vehicle Equipment and Accessories....................................497
    20–4 Vehicle Security 501

Chapter 21
Driving Away from Home **507**

    21–1 Road Trips..509
    21–2 Driving Unfamiliar Vehicles...513

Answers to "Guess the Vanity Plate,"
"Wild Wheels," and "Where Am I?" Questions...................................*523*
Glossary *524*
Index *536*
Figure Credits...*542*

Special Boxes

CELEBRITY SCENARIOS

Chapter	Title
1	James Dean
2	Bessie Smith
3	Jack Nicholson
5	Kelsey Grammer
6	Jackson Pollock
7	Rudy Tomjanovich
8	Trinidad Silva, Jr.
9	General George S. Patton, Jr.
10	Margaret Mitchell
11	Michael Hedges
12	Latrell Sprewell
13	Gary Busey
14	Sam Kinison
15	Derrick Thomas
16	Edward Givens
17	Alan Pakula
18	Howard E. Rollins, Jr.
19	Zsa Zsa Gabor

DEFENSIVE DRIVING

Chapter	Title
1	Reality Check: Driving with a Fake License
	Driving Tips: Behind-the-Wheel Jitters

Chapter	Title
2	Reality Check: Experience Counts
	Driving Tips: To Honk or Not to Honk?
	Driving Myths: Do Not Judge a Driver by His or Her Car
	Reality Check: Do Not Let the Pressure Get to You
3	Driving Tips: Preventing Fatigue
	Reality Check: "Flirting" with Danger
	Reality Check: Deadly Decibels
4	Reality Check: Know Your Nozzles
	Driving Tips: "Reading" by the Light of Your Turn Signals
	Driving Tips: Safe Signaling
	Reality Check: No Rain in the Forecast
	Driving Tips: "Preflight" Checks
5	Driving Myths: The Runaway Vehicle
	Driving Tips: Dry Steering
	Driving Tips: Prohibited Parking
	Driving Tips: Getting out of a Parked Car
6	Driving Tips: Do Not Drive in the Shadow of a Large Load
	Using SAFE: Curves
	Driving Myths: Speeding
	Reality Check: Radar Detectors
7	Reality Check: Sign Vandalism
	Driving Tips: "House" Signals
	Reality Check: Old Lane Markings
	Know Your Neighbor: HOV Lanes
8	Reality Check: The Right-Hand "Quick Starter"
	Driving Tips: Intersections
	Driving Myths: Timed Signals
	Using SAFE: Large Parking Lots
	Driving Tips: "Don't Be There, Be Square"
	Driving Tips: Right-of-Way
	Reality Check: Playing "Chicken" with a Train
	Driving Tips: Railroad Crossings
9	Driving Tips: Avoid the "Pickle"

Chapter	Title
	Using SAFE: "Stealing" Left Turns
	Driving Tips: Hesitant Left-Turners
	Reality Check: Do Not Cut Corners
	Driving Tips: Follow Your "Blocker"
10	Reality Check: Do Not Get "Booted"
	Reality Check: Squeaking By
	Reality Check: Lagging Left-Turn Arrows
	Reality Check: Do Not Let the "Windy City" Blow You Away!
	Driving Tips: The "Herd" Instinct
	Reality Check: The Constant Lane-Changer
	Reality Check: Detours
	Reality Check: Know Who *Not* to Pass
	Using SAFE: Passing
	Reality Check: Let the Fast Pass
11	Reality Check: Mailbox Bashing
	Reality Check: Do Not "Trash" Your Environment
	Driving Tips: Meeting a Line of Cars
	Driving Tips: Solo Signaling
	Using SAFE: Rural Roads
	Driving Myths: Do Not "Look for the Whites of Their Eyes"
12	Driving Tips: Facing Your Fear of the Freeway
	Know Your Neighbor: Freeway Potholes
	Using SAFE: Entering Freeways
	Reality Check: A Missed Exit Is Not the End of the World
	Driving Tips: Ferries
13	Reality Check: Help the Unseen Pedestrian
	Reality Check: Pedestrian "Don'ts"
	Driving Tips: Roads That Are Not "Pedestrian Friendly"
	Reality Check: Right-Turning Bicyclists
	Reality Check: Bicyclist "Don'ts"
	Know Your Neighbor: Motorcycle Helmets
	Using SAFE: Passing Big Rigs
	Reality Check: "Freeze" in Your Tracks!

Chapter	Title
14	Driving Tips: Dirty Windows and Lights
	Reality Check: The Blind Leading the Blind
	Reality Check: The Blinding Commute
	Driving Tips: Check Your Clearance
	Using SAFE: Construction Zones
15	Driving Tips: A Slick Combination
	Reality Check: Wet Leaves
	Driving Tips: Winter Driving
	Driving Tips: While You Were Sleeping
	Driving Tips: When "Hazard Lights" Are Hazardous
	Using SAFE: Driving in the Wind
	Driving Tips: "Locked" Out by the Cold
	Driving Tips: Summer Driving
16	Reality Check: "Set-up" Collisions
	Reality Check: Protect Yourself While Helping Others
	Using SAFE: Preventing Collisions
	Driving Myths: Safety Belts
	Driving Myths: Air Bags
	Reality Check: Child Safety Seats
17	Driving Tips: Freeway Breakdowns
	Reality Check: "Stuck in the Middle with You"
	Driving Tips: Emergency Traffic Patrols
	Driving Tips: When an Emergency Becomes Another Emergency
	Driving Tips: Unwelcome Passengers
18	Reality Check: Do Not Test "Fresh" Breath
	Driving Tips: If You Are Pulled Over by the Police
	Reality Check: Saturation Patrols and Roadside Safety Checkpoints
19	Driving Tips: Driving with an Instruction Permit
	Driving Tips: On Your Own
	Reality Check: License Refusal
	Reality Check: License Violations
	Driving Tips: Custom Plates
	Reality Check: Keep Your License Plates Visible
	Reality Check: Out of Sight, But Not Out of Mind!
	Driving Tips: Avoiding "Lemons"

Chapter	Title
20	Driving Tips: Tire Inflation in Cold Weather
	Driving Tips: Improving Fuel Efficiency
21	Driving Tips: Maps Are Not Always Reliable
	Driving Tips: Rental Car Insurance
	Driving Tips: Towing at Night

AUTO ACCESSORIES

Chapter	Title
5	Curb Feelers
8	"Round-the-Corner" Mirrors
9	Steering Wheel Grip
10	Fuel Additives
11	Mud Guards
14	Glare Guards
15	Weather Radio
	Heat Shields
17	Jump-Start Without Jumping Back!
	Preparing for Emergencies
18	Portable Breathalyzer
20	Tool Sets

Preface

The object of *License to Drive in Illinois* is to present you, the beginning driver, with a practical and realistic guide to the fundamentals of driving and the rules of the road. Unlike other textbooks, this book speaks *to* you, not *at* you. It focuses on situations that you are likely to encounter in the real world and offers useful suggestions on how to avoid common driving hazards. *License to Drive in Illinois* takes you step by step from the basics of vehicle control to the specific dangers of different driving environments to the responsibilities of vehicle ownership. This book also highlights the motor-vehicle regulations and potential penalties that all drivers in Illinois face when they get behind the wheel.

You are no doubt anxious to start driving as soon as possible. You have waited a long time for the opportunity to go where you want to, when you want to. In fact, if you are like many students, your only goal in taking a driver education course is to pass the tests required to obtain a driver's license. However, you must learn how to be a safe driver first. Safety, in all forms—from defensive driving to auto accessories—is thoroughly discussed in this book to provide you with the skills and knowledge you need when on the road. This book teaches you not only what you can and cannot do on the road, but why. By thinking creatively about the driving experience, you will reduce the risks of being involved in a collision.

The guiding philosophy of Alliance for Safe Driving, and of *License to Drive in Illinois*, is that by driving defensively you can avoid most, if not all, dangerous situations on the roadway. Rather than *react* to an impending collision or road hazard, you *actively* take steps to keep out of harm's way. This book presents a simple defensive-driving technique called SAFE that is easy to remember and apply. By constantly *scanning* ahead, *assessing* road and weather conditions, *finding* an "out" or escape route, and *executing* the best option available, you can become a safer driver. By becoming a safer driver, you will feel more comfortable and confident behind the wheel and will be able to enjoy more of the wonderful rewards that driving a motor vehicle can bring to your life.

License to Drive in Illinois includes several features to make the subject informative, interesting, and fun. Throughout the margins of the text are Factoids that present surprising statistics and facts. Highway Heroes portray law-enforcement officers who gave their lives to keep Illinois roadways safe. Yellow defensive-driving boxes provide helpful driving tips, debunk common myths, describe real-life situations that can get you in trouble, and the Know Your Neighbor feature highlights important differences in motor-vehicle laws and features of the driving environment in neighboring states. Most chapters include a red box on a collision or driving-related incident involving a celebrity, demonstrating that famous actors, sports figures, and artists face the same dangers that you do. Various auto accessories are keyed to the text in blue boxes. Because driving is not *all* work, the book includes amusing Bumper Sticker Sightings in the margins. If you cannot Guess the Vanity Plate, identify the Wild Wheels, or figure out Where Am I? in each chapter, you can find the answers after Chapter 21.

Metric equivalents are provided in parentheses after all units of measure. These are rounded off to avoid awkward numbers, fractions, or decimals. For example, the actual metric equivalent of 30 miles per hour is 48.28041 kilometers per hour, but in the text, it is presented as "(50 km/h)."

To focus your reading and prompt you to think about the material, *License to Drive in Illinois* offers a wide array of study aids. Chapter Objectives and Key Terms are listed at the front of each chapter. The Key Terms also appear in color in the text and are defined in the glossary at the back of this book. Chapters 2 through 17 include special Who's at Fault? boxes that challenge you to analyze an actual collision taken from real police records and determine who is responsible. Beginning with Chapter 2, each chapter has a comprehensive Self-Test, including two Critical Thinking questions that ask you what to do in a complex driving situation. Finally, two Projects that can be done individually or as classroom activities are provided to translate what you read in the chapter to the real world. All of these activities are designed to start you thinking about and preparing for your time on the road, which we hope will be rewarding.

Alliance for Safe Driving

Acknowledgments

The authors gratefully acknowledge the following individuals for their assistance in the development of this book:

Ricardo Cerna
Executive Secretary
Association of Driver Educators
　for the Disabled
Edgerton, Wisconsin

Mark McDonald
Public Information Officer
Illinois State Police
Springfield, Illinois

Mark A. Patrick
Curator, National Automotive
　History Collection
Detroit Public Library
Detroit, Michigan

Sandy Saunders
Museum Services Manager
National Automotive Museum
The Harrah Collection
Reno, Nevada

Dr. Donald L. Smith
Professor Emeritus, Highway
　Traffic Safety Program
Michigan State University
Lansing, Michigan

Alliance for Safe Driving and Delmar Publishers wish to express their appreciation to a dedicated group of professionals from Illinois who reviewed and provided commentary at various stages. Their insights, suggestions, and attention to detail were very important in guiding the development of this textbook.

Robert D. Carpenter, Jr.
Barrington High School
Barrington, Illinois

Norbert J. Chesta
Addison Trail High School
Addison, Illinois

Reg Green
Leyden High Schools
Franklin Park, Illinois

Mark Jadzak
Glenbard West High School
Glen Ellyn, Illinois

Kevin Wright
Glenbrook North High School
Northbrook, Illinois

Acknowledgments ◆ xxiii

The authors and Delmar Publishers would also like to acknowledge the following individuals for their contributions in the development of this textbook.

Lou Autry
Senior Consultant
Region 10, Education Service Center
Richardson, Texas

Terry Barnett
Driver Education Instructor
McClintock High School
Tempe, Arizona

James A. Brooks
Supervisor, Driver and Safety Education
Mississippi Department of Education
Jackson, Mississippi

Mackey D. Ervin
Coordinator, Driver Education
Region IV Education Service Center
Houston, Texas

Derek J. Ewing
Instructor/Manager
E&E Defensive Driving Associates
Loudonville, New York

Dr. Frank J. Gruber, IV
Associate Professor, Technology and Safety Studies
Northern Illinois University
DeKalb, Illinois

Alice Ann Klakos
Principal, Driver Education Instructor
Oak Glen High School
New Cumberland, West Virginia

Robert E. Lambert
Driver Education Instructor
Cabell Midland High School
Ona, West Virginia

Richard A. Mossey
Co-Owner
Defensive Driving School, Inc.
Seattle, Washington

John Papa
President, Illinois High School and College Driver Education Association
Lake Park High School
Roselle, Illinois

Sharon Postigo
President
D&D Driving School, Inc.
Kettering, Ohio

John R. Sawyer
Coordinator, Driver Education Services
Wellsville Local Schools
Wellsville, Ohio

Craig Westfall
Driver Education Instructor
Santa Rita High School
Tucson, Arizona

How to Use *License to Drive*

License to Drive is designed to help you learn to drive defensively, responsibly, and safely. The text has many unique features that also make learning how to drive interesting and fun.

1 Unit Openers

Each unit opener includes a photograph relating to the unit topic and a list of the chapters within that unit.

2 Chapter Openers

Each chapter opener includes a photograph for discussion that captures the theme of the chapter, a brief introduction to the chapter content, a list of Chapter Objectives tied to each section within the chapter, and a list of Key Terms in the order in which they appear in the chapter.

3 Chapter Objectives

Chapter Objectives identify key information to be gained from the chapters. Use these objectives, together with the Your Turn questions, to reinforce your understanding of the chapter's content.

4 Key Terms

Important terms to know are listed at the beginning of each chapter and appear in red the first time they are used in the text.

5 Color Photographs and Illustrations

Throughout the text, there are many visual images and detailed captions to aid in the explanation and understanding of the subject matter. Color photographs provide realistic examples of the content described. Full-color illustrations depict "real-life" driving scenarios and driving instructions, and reinforce chapter material.

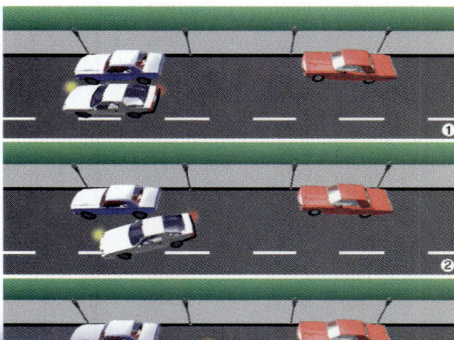

Figure 9–1 At some intersections, you may turn only when you have a green arrow.

9–1 TURNING BASICS

A protected turn is one made from a turn lane posted with signs, road marked with arrows, and accompanied by a traffic signal arrow. A green arrow or special green light for your lane allows you to turn while traffic from the oncoming lane(s) is halted with a red light. If the signal allows you to make a turn only when you have a green arrow (the green arrow turns to a red arrow or red light), it is a fully protected turn.

A semiprotected turn is one made from a turn lane but *not* accompanied by a special traffic signal light that directs your turn with a green arrow or a special green light. You are allowed to make such a turn either while you have a regular solid green light or when there is no light and a sufficient gap in traffic occurs.

Figure 9–2 When making a semiprotected turn, you may turn on a solid green light once a sufficient gap in traffic exists.

Figure 9–22 Making a three-point turn

xxvi ◆ *How to Use* License to Drive

Special Features

License to Drive includes many text boxes and marginal text features that provide advice, facts, and entertaining samples related to the driving experience.

- Celebrity scenarios are presented in red-shaded boxes throughout the text. The celebrity boxes convey the details of a collision or driving-related incident involving a famous artist, athlete, or musician. Highway Heroes are red-colored marginal items that portray police officers and state troopers who died in the line of duty in roadway collisions.
- Fun items are presented in green-shaded boxes and marginal text. They include Bumper Sticker Sightings, Guess the Vanity Plate, Wild Wheels, and Where Am I?

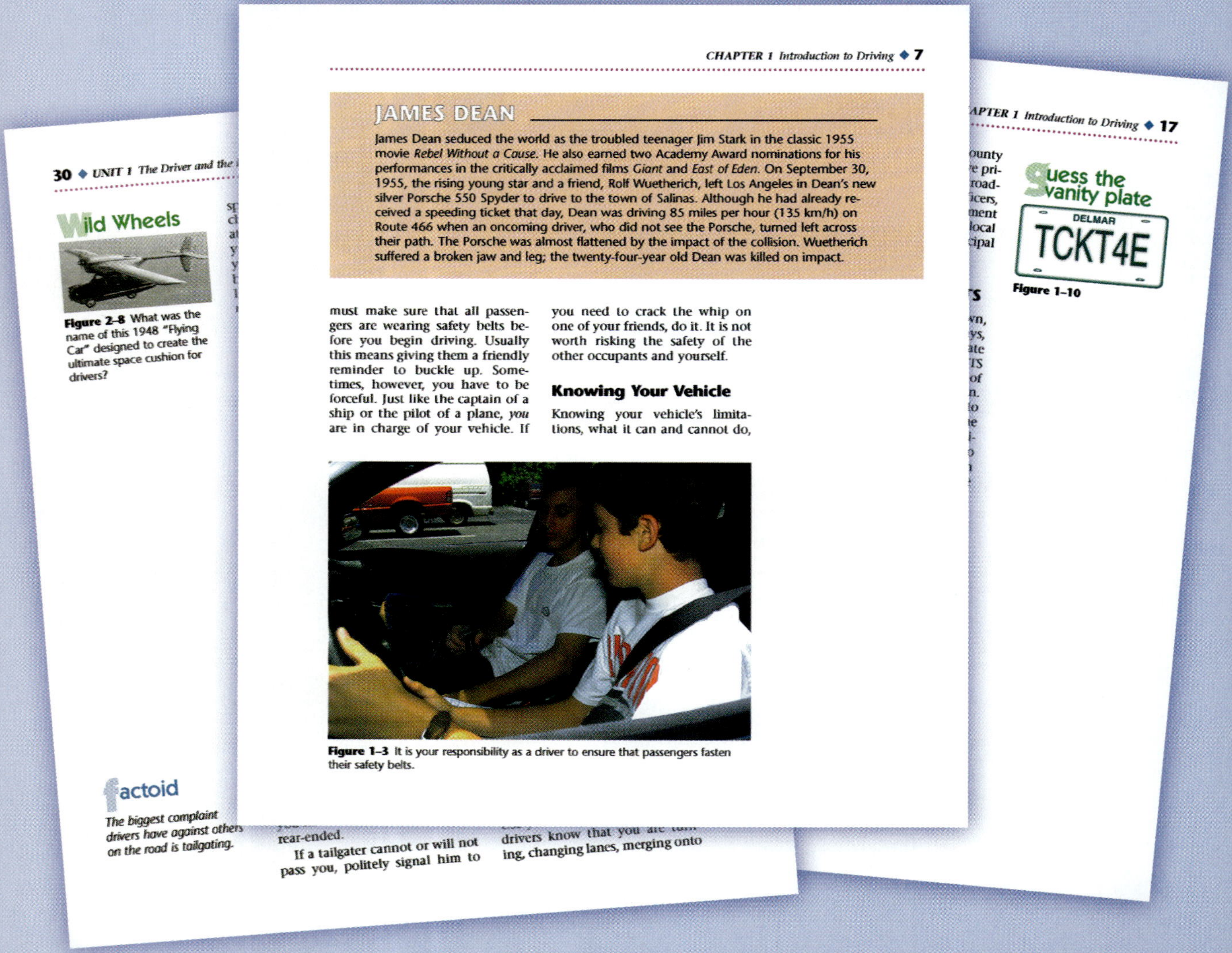

How to Use License to Drive ◆ **xxvii**

- Defensive-driving features are presented in three types of yellow-shaded boxes and marginal text: Driving Tips, Reality Check, Driving Myths, and Know Your Neighbor.
- Factual information appears in blue-shaded boxes and marginal text. Factoids present interesting data that reinforce points made in the text. Auto Accessories provide examples of optional equipment available for vehicles.

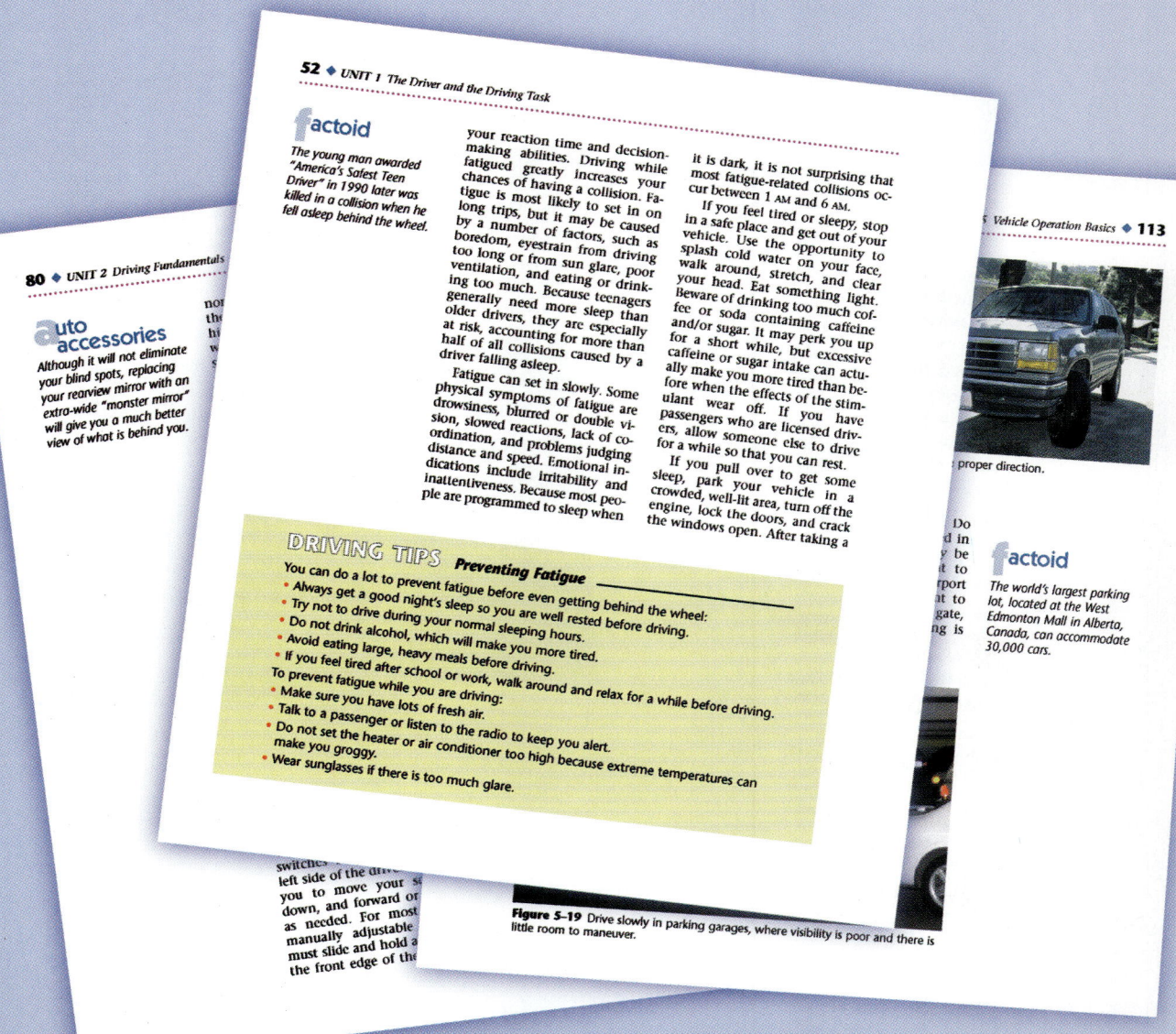

 **Who's at Fault?**
Chapters 2 through 17 each include an account of two collisions taken from actual police records. Based on your knowledge of the chapter content, you determine who is at fault.

 **Your Turn**
Use these questions, which mirror the Chapter Objectives, to test your understanding of the chapter content.

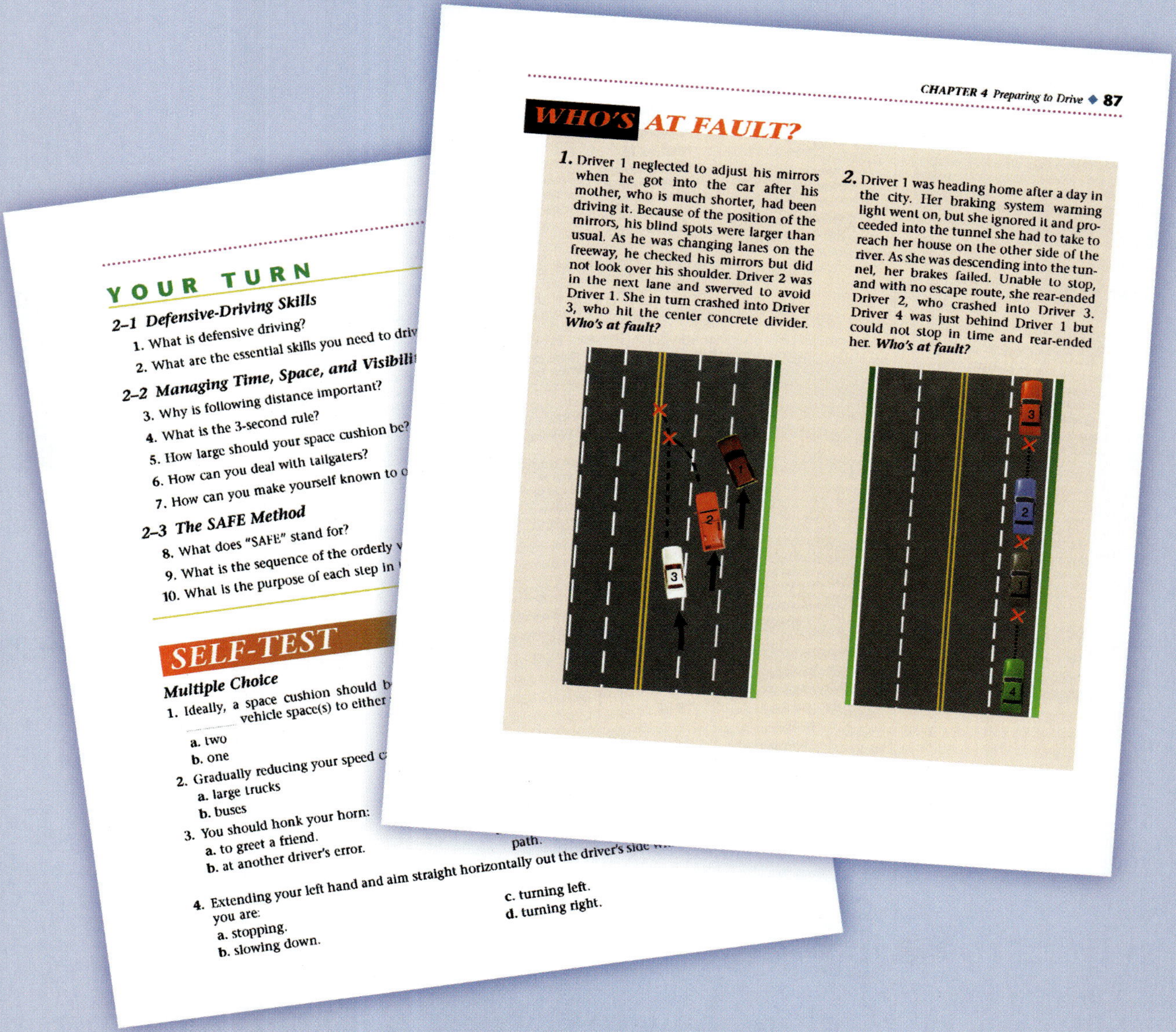

9 Self-Test

Every chapter, with the exception of Chapter 1, includes a test with five components (Multiple Choice, Sentence Completion, Matching, Short Answer, and Critical Thinking). Complete the Self-Test to be sure that you understand the chapter material.

10 Projects

Each chapter following Chapter 1 includes two Projects that you can complete outside of class individually or in groups. Also, ask the instructor about how to access the website **www.licensetodrive.com**, which offers you additional resources for learning.

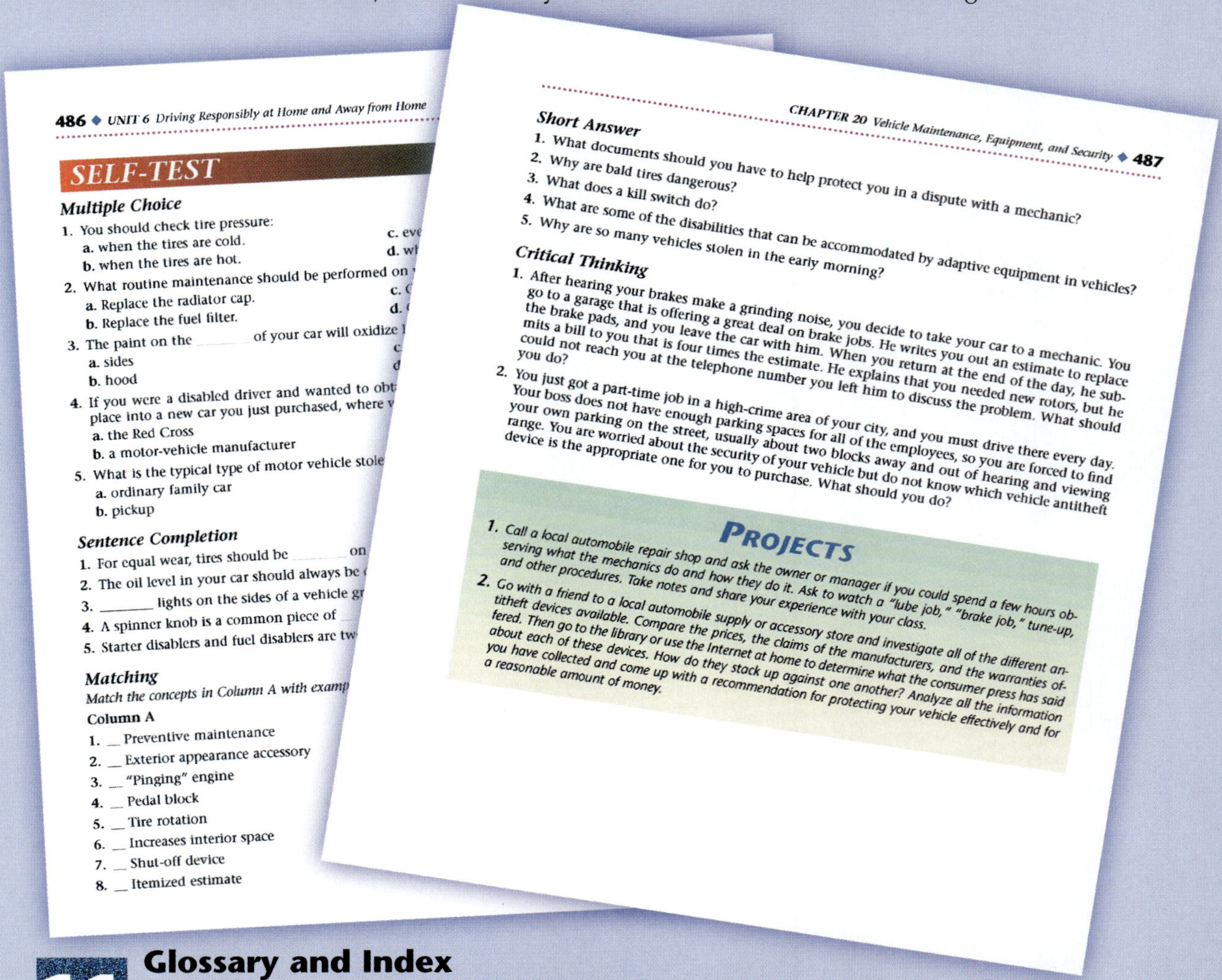
486 ◆ UNIT 6 *Driving Responsibly at Home and Away from Home*

SELF-TEST

Multiple Choice

1. You should check tire pressure:
   a. when the tires are cold. c. eve
   b. when the tires are hot. d. wh
2. What routine maintenance should be performed on
   a. Replace the radiator cap. c. C
   b. Replace the fuel filter. d.
3. The paint on the ______ of your car will oxidize
   a. sides
   b. hood
4. If you were a disabled driver and wanted to obt place into a new car you just purchased, where
   a. the Red Cross
   b. a motor-vehicle manufacturer
5. What is the typical type of motor vehicle stole
   a. ordinary family car
   b. pickup

Sentence Completion

1. For equal wear, tires should be ______ on
2. The oil level in your car should always be
3. ______ lights on the sides of a vehicle gr
4. A spinner knob is a common piece of
5. Starter disablers and fuel disablers are tw

Matching

Match the concepts in Column A with examp

Column A

1. __ Preventive maintenance
2. __ Exterior appearance accessory
3. __ "Pinging" engine
4. __ Pedal block
5. __ Tire rotation
6. __ Increases interior space
7. __ Shut-off device
8. __ Itemized estimate

CHAPTER 20 *Vehicle Maintenance, Equipment, and Security* ◆ 487

Short Answer

1. What documents should you have to help protect you in a dispute with a mechanic?
2. Why are bald tires dangerous?
3. What does a kill switch do?
4. What are some of the disabilities that can be accommodated by adaptive equipment in vehicles?
5. Why are so many vehicles stolen in the early morning?

Critical Thinking

1. After hearing your brakes make a grinding noise, you decide to take your car to a mechanic. You go to a garage that is offering a great deal on brake jobs. He writes you out an estimate to replace the brake pads, and you leave the car with him. When you return at the end of the day, he submits a bill to you that is four times the estimate. He explains that you needed new rotors, but he could not reach you at the telephone number you left him to discuss the problem. What should you do?
2. You just got a part-time job in a high-crime area of your city, and you must drive there every day. Your boss does not have enough parking spaces for all of the employees, so you are forced to find your own parking on the street, usually about two blocks away and out of hearing and viewing range. You are worried about the security of your vehicle but do not know which vehicle antitheft device is the appropriate one for you to purchase. What should you do?

PROJECTS

1. *Call a local automobile repair shop and ask the owner or manager if you could spend a few hours observing what the mechanics do and how they do it. Ask to watch a "lube job," "brake job," tune-up, and other procedures. Take notes and share your experience with your class.*
2. *Go with a friend to a local automobile supply or accessory store and investigate all of the different antitheft devices available. Compare the prices, the claims of the manufacturers, and the warranties offered. Then go to the library or use the Internet at home to determine what the consumer press has said about each of these devices. How do they stack up against one another? Analyze all the information you have collected and come up with a recommendation for protecting your vehicle effectively and for a reasonable amount of money.*

11 Glossary and Index

An extensive glossary of definitions for all of the Key Terms appears at the back of this book. Use the comprehensive index to easily locate subject areas throughout the text.

chapter 1

Introduction to Driving

You are about to learn how to do something that you have wanted to do for a long time. Perhaps you have an older brother or sister who is already driving or a friend who has already taken driver education and is operating a vehicle with a learner's permit. Whatever the case, now is your chance to finally get behind the wheel! The excitement that you feel now is very positive, and it will put you in the right frame of mind to learn all about driving and traffic safety.

Chapter Objectives

Upon completion of this chapter, you should be able to:

1–1 Driver Education
1. Understand the purpose of driver education.
2. Describe the basic components of a "driver ed" course.

1–2 Driving Is a Privilege
3. Understand why driving is a privilege, not a right.
4. Describe your basic responsibilities as a driver.
5. Understand why it is important for you to be courteous as a driver.
6. Understand the importance of being emotionally, physically, and mentally fit to drive.

1–3 The Highway Transportation System
7. Understand what the Highway Transportation System is.
8. List the three main classifications of roadways.
9. Identify who is responsible for enforcing traffic laws.

KEY TERMS

driver education

Highway Transportation System (HTS)

traffic laws

law enforcement

1-1 DRIVER EDUCATION

Driver education is designed to help unlicensed or inexperienced drivers become familiar with the basics of vehicle control and the rules of the road so that they can successfully pass the tests required to earn a driver's license. Even though you will be taught many skills and safe driving practices in your driver education course, you should remember that completing the course does not mean the end of your education.

A driver education course, or "driver ed," cannot teach you everything. The world of driving is constantly changing, and you must be prepared to handle new problems that may occur in the future. If you move from one location to another, for example, you will have to cope with driving situations you never experienced before. When crossing any state or provincial boundary, be aware that motor-vehicle laws, safety standards, registration rules, insurance requirements, and the costs of vehicle ownership change. With each new day of driving, you will gain more knowledge and experience that will make you a better and safer driver as time goes on.

Who Takes a Driver Education Course?

Some people cannot wait to begin driving, whatever their age. On the other hand, more people than you think never bother to get a license. The typical driver education student is a fifteen- or sixteen-year-old high school sophomore or junior. You, however, may fit into one of the following categories:

- A twenty-one-year-old recent graduate from college who never needed to drive, but who now has taken a job in a suburban area and needs a car to get there

Figure 1-1 "Driver ed" is just the start of your education in driving.

- A thirty-eight-year-old recent immigrant to North America who never drove before in your country of origin
- A seventy-six-year-old retiree who always depended on a spouse for transportation and so never got around to learning how to drive

Types of Driver Education Courses

Typically, you have three options if you want to enroll in a driver education course. You can take classes at a private or public high school, a community college, or a private driving school. Because of budget cuts, more and more high schools are contracting these courses out to specific private driving schools or suggesting that students pursue training at a private driving school of their choice. Depending on the area and the length of the program, the cost of a course at a private school can vary greatly.

Components of a Driver Education Course

The typical driver education course includes both in-class instruction and training behind the wheel. During this training, you will be accompanied at all times by the driving instructor. To supplement your course work, you should practice for many more hours with qualifed family members, friends, or another professional driving instructor. Practicing as much as you can before you take your driving test will increase your confidence. You also will not have to cope with the pressure of learning everything you will need to know in a short amount of time. Driving instructors generally believe that a minimum of 30 hours of training behind the wheel is necessary, and many recommend even more time.

1–2 DRIVING IS A PRIVILEGE

If you learn one thing from this book, always remember that *driving is a privilege, not a right.* There is no "natural" or constitutional right to operate a motor vehicle. Driving is a privilege granted to those who meet certain requirements and obligations. Society has determined that for practical reasons it should be up to the government to decide when and under what conditions a person can drive.

Those who wish to drive must demonstrate that they are worthy of a driver's license by passing tests and obeying the law. A license can be restricted, suspended, or even revoked (taken away) for committing various driving offenses. Even if your driver's license is necessary for you to get to work or school every day, this does not mean that the government *has* to give it to you.

Once you obtain your driver's license, you will be treated the same as any adult with a license who is also on the roadways. This includes drivers who have been driving for decades and are much more experienced than you are. Driving

---

bumper sticker sightings

EVACUATE THE ROAD!!!
STUDENT DRIVING!!

is an opportunity that society gives you to be treated like an adult, so take your obligations seriously.

Driving Is a Responsibility

Once you are given a driver's license, society expects you to give back something in return for your new freedom. Take responsibility for your driving. Your main obligations while driving a motor vehicle are to obey the law, to fulfill the financial obligations associated with a crash you may cause, to show courtesy to other roadway users, to protect your passengers, to know and properly maintain your vehicle, and to drive only when you are physically and mentally able.

Your Legal Duties

You have a responsibility to obey the law when you drive. This means that you must not break any of the laws in the vehicle code or any municipal laws or ordinances that may be in effect in the area in which you are driving. It also means trying to minimize the chances of colliding with another vehicle. If you think that this is a responsibility that does not need to be taken seriously, think again. You may be found liable or partly liable for causing a collision that was not strictly your fault but which you could have avoided.

Your Financial Obligations

When you drive a motor vehicle, you are interacting with countless other people, vehicles, and private property. If you commit an error while driving and injure someone else or damage property, you must pay for these injuries and damages if the collision is determined to be your fault. The primary reason for having automobile insurance is that if such an unfortunate event occurs, your insurance company will cover the costs. Keep in mind, however, that if you do not have enough insurance to cover the injuries or damage, you are not off the hook. The person or persons who are injured or whose property was

---

REALITY CHECK *Driving with a Fake License*

Just because you do not have a driver's license yet does not mean that you are not responsible for your actions if you drive illegally. Many teens obtain fake licenses, or borrow licenses from a relative or friend, just to have "a piece of paper" in case they get pulled over by the police. If you drive without having first received proper instruction and your license, however, you are endangering yourself and everyone else on the road. Driving is a privilege earned through education, practice, and the licensing process. If you drive illegally, you or your parents will be held accountable for whatever deaths, injuries, or property damage you cause.

Figure 1-2 Do not wait until it is too late to realize your responsibilities as a driver.

damaged can sue you in court for any damages over and above your insurance limits.

Courtesy to Others

If you were the only person on the roadway, driving would be a lot easier than it is. Not only do *you* have to drive safely, but you also have to watch for other drivers and pedestrians. By watching how other drivers or pedestrians react to different situations, you will soon discover that you can make fairly accurate predictions about similar situations you may encounter in the future. By driving predictably yourself, other drivers can respond and react to what you do easily and without creating unnecessary alarm, danger, and tension.

Every driver has a responsibility to be courteous to other roadway users. When drivers who interact with one other are courteous, traffic flows more smoothly, no one is injured, no private property is damaged, and everyone is able to get to their destination with a minimum of hassle. This does not mean that you have to be everyone's best friend. It does mean, however, that when you are in a motor vehicle, courtesy is more likely to get you to your destination in one piece than rudeness or aggression.

Protecting Your Passengers

The responsibility for the passengers in your vehicle is yours, the driver's. Most importantly, you

JAMES DEAN

James Dean seduced the world as the troubled teenager Jim Stark in the classic 1955 movie *Rebel Without a Cause*. He also earned two Academy Award nominations for his performances in the critically acclaimed films *Giant* and *East of Eden*. On September 30, 1955, the rising young star and a friend, Rolf Wuetherich, left Los Angeles in Dean's new silver Porsche 550 Spyder to drive to the town of Salinas. Although he had already received a speeding ticket that day, Dean was driving 85 miles per hour (135 km/h) on Route 466 when an oncoming driver, who did not see the Porsche, turned left across their path. The Porsche was almost flattened by the impact of the collision. Wuetherich suffered a broken jaw and leg; the twenty-four-year old Dean was killed on impact.

must make sure that all passengers are wearing safety belts before you begin driving. Usually this means giving them a friendly reminder to buckle up. Sometimes, however, you have to be forceful. Just like the captain of a ship or the pilot of a plane, *you* are in charge of your vehicle. If you need to crack the whip on one of your friends, do it. It is not worth risking the safety of the other occupants and yourself.

Knowing Your Vehicle

Knowing your vehicle's limitations, what it can and cannot do,

Figure 1–3 It is your responsibility as a driver to ensure that passengers fasten their safety belts.

is crucial to safe driving. A car with a small engine will have some problems keeping up with traffic in the fast lane on a steep hill. A car with low ground clearance will have a problem traversing a rutted dirt road. A car with high ground clearance will have problems negotiating sharp turns at the same speed as a normal passenger sedan. The better you know your vehicle, the more control you will have in an emergency situation and the easier it will be to park or execute potentially dangerous maneuvers such as passing.

For your own safety and comfort, as well as for the safety and comfort of other drivers on the roadways, you have a responsibility to perform regular maintenance on your vehicle. Some of this maintenance, such as adding oil or coolant, inflating your tires, or changing a windshield wiper blade, is basic and can be performed by you. Other tasks are more complicated and should be performed by a certified mechanic. Parts wear down, break, and lose lubrication over time. Day-to-day maintenance and periodic servicing will keep your repair costs to a minimum and reduce the risk of a breakdown.

Because your vehicle is a machine that releases emissions from the burning of fuel, it naturally has effects on the environment, some of which are quite harmful. You have a responsibility as a member of society to understand what kind of fuel emissions your car discharges and what you can do to minimize pollutants. Usually, this means ensuring that any emissions control device on the vehicle, such as a catalytic converter, is performing properly. It also means using public transportation (such as buses or light rail), carpooling, and taking every opportunity to walk or ride a bicycle for short trips.

Being Ready to Drive

Driving with the right attitude—a good attitude—is an important responsibility. Attitude directly affects your driving, which involves constantly making decisions, many of them with life-or-death implications. To make good decisions, you need to think clearly and not be burdened by your emotions. All of us have a "bad day" once in awhile, but if you have just broken up with someone, failed a test, or lost your job, you must either put these problems behind you or avoid driving. Having the right attitude means not being overconfident and cocky, which can also lead to poor decision making.

You are also responsible for being in good physical shape when you drive. This does not mean that you have to have the physique of a body builder or the stamina of a marathon runner. It does mean that you need to have the physical ability to perform all of the tasks associated with driving, such as steering, braking, and shifting gears. It also means that your vision, hearing, and hand-eye coordination (with or without the use of special equipment) must be adequate. Finally, it means that you

> **DRIVING TIPS** **Behind-the-Wheel Jitters**
>
> It is natural to be nervous the first time you are guiding a huge piece of machinery that weighs a couple of thousand pounds down streets, around corners, and into parking spaces. Driving it at high speeds with other, more experienced drivers flying past you or honking at you can literally cause you to shake. Remember that a healthy fear of the dangers of driving is a good thing, but if you feel overwhelmed by the requirements of driving your vehicle, pull over in a safe area so that you can calm down. Take a few deep breaths, replay in your mind the instructions for the particular driving task you are practicing, and when you feel ready, re-enter traffic.

should avoid driving when you are sick, under the influence of alcohol or other drugs, have an injury, or are taking medications that interfere with driving skills.

Driving is a mental, as well as an emotional and physical, exercise. Because all the vehicle's systems are controlled by the driver, your brain must constantly assess what systems are in use, how they are being used, and to what extent their use needs to be modified. Your mind's ability to correctly judge space and time is critical to your ability to negotiate a road and keep clear of other vehicles and obstacles.

Your capacity to memorize things is another aspect of being mentally fit. While you drive, you need to remember how your car functions, what the traffic laws are, and what you have just scanned ahead, to the sides, and to the rear of your vehicle. If an emergency develops or if you need to take immediate evasive action, your memory will help you out. For example, by remembering that a shoulder you scanned seconds earlier is clear of traffic, you know that you have an escape route should an obstacle suddenly appear in your path.

1–3 THE HIGHWAY TRANSPORTATION SYSTEM

The **Highway Transportation System (HTS)** is part of our interconnected North American transportation system, which includes rail, air, sea, underground, and pedestrian traffic, as well as motor vehicle transportation. It covers Canada, the United States, and parts of the Caribbean and Pacific Islands. By far, the most important component of the system, as it impacts our daily lives, is motor vehicles. As individuals we are likely to spend more of our time in a motor vehicle than in any other form of transportation. The purpose of the HTS is to move people and goods in a safe, efficient, and timely manner across state, provincial, and international boundaries with a minimum of bother. This facilitation of

More than 80% of vacation travelers use a personal vehicle for their trip.

Figure 1-4 The Highway Transportation System includes roadways, vehicles, and the people who use them.

travel and movement contributes significantly to the dynamic economy of North America.

As the population increases and society becomes more mobile, the HTS is getting more crowded and indispensable. Think for a moment where we would be without it. How would we get to a distant relative's house? How would food get to your local supermarket? How would an injured person get to a hospital? With very few exceptions, almost everyone depends on motor vehicles for transportation. Places where the motor vehicle is secondary in importance to subways, trains, ferries, and even airplanes are rare. Even in urban commuting situations in which people depend on public transportation to go back and forth to work, most either maintain or have access to a motor vehicle for their noncommuting needs.

The motor vehicle has become such an important aspect of our lives over the past century that the way our cities, towns, shopping malls, and other public places look has a lot to do with the automobile. The suburb, is largely a creation of the automobile, because without it, commuting back and forth to jobs in urban centers would have been impossible.

Of course, the HTS would not be what it is were it not for the people who make it function day in and day out. Many people depend on the HTS for their livelihood, and many jobs would be impossible without its existence. Truck drivers, car salespeople, au-

factoid

Only about 60% of the nearly 4 million miles of roadway in the United States are paved.

toworkers, law-enforcement officials, and auto mechanics are just some of the people who rely on the HTS to make a living.

How You Relate to the HTS

The HTS is a large grid or network much like the communication systems that connect people's telephones and computers to one another. In both cases, the overall system is made up of many smaller grids or networks. For example, each geographic region has its own telephone system and "area" code, but they are all linked together so that people in different regions can communicate. Similarly, your neighborhood is its own small grid connected to the larger HTS grid.

You have walked the sidewalks and streets in your neighborhood countless times, perhaps on a newspaper delivery route. Or perhaps you have ridden your bicycle or skateboard down to the local shopping mall. Undoubtedly, you have been a passenger in an adult's car while being driven out of town on a freeway.

In each case, you were actively playing a part in the HTS. When you delivered newspapers in the morning, you met the delivery truck that distributed a large bundle of papers to your driveway from the printing press at the other side of town. You then used the streets designed and built by road construction companies to deliver the newspapers on your bicycle. When you rode your bicycle to the shopping mall, you had to leave the neighborhood and ride down a busy main street to get to the mall, interacting with cars, motorcycles, buses, and trucks on your way. When you were a passenger in an adult's car, you left the neighborhood and the main street behind when you entered the freeway on-ramp and joined thousands of other vehicles on the high-speed thoroughfare that connects your local main street to the rest of North America.

Many people use several different parts of the HTS daily. Consider the person who drives his or her car through residential streets to the vanpool parking lot, commutes into the city on the freeway via vanpool to a central drop-off point, takes a bus from the drop-off point to the block where he or she works, walks the rest of the way down the sidewalk to the office building, and then takes a bike ride during the lunch hour for physical exercise. This person has used five different types of transportation; has covered three types of roadway; has been a driver, passenger, pedestrian, and rider; and has interacted with countless other HTS users—all in one day!

Each time that you use the HTS, you will inevitably face conflict of some kind. Not to worry, this conflict is almost always resolved, thanks to the way the HTS is maintained and ordered. Conflict is created between two or more users of the HTS when their goals or desires are different from one another. For example, when you are driving a car and you desire to

factoid

The average distance traveled by a car in the United States in one year equals a journey almost halfway around the Earth.

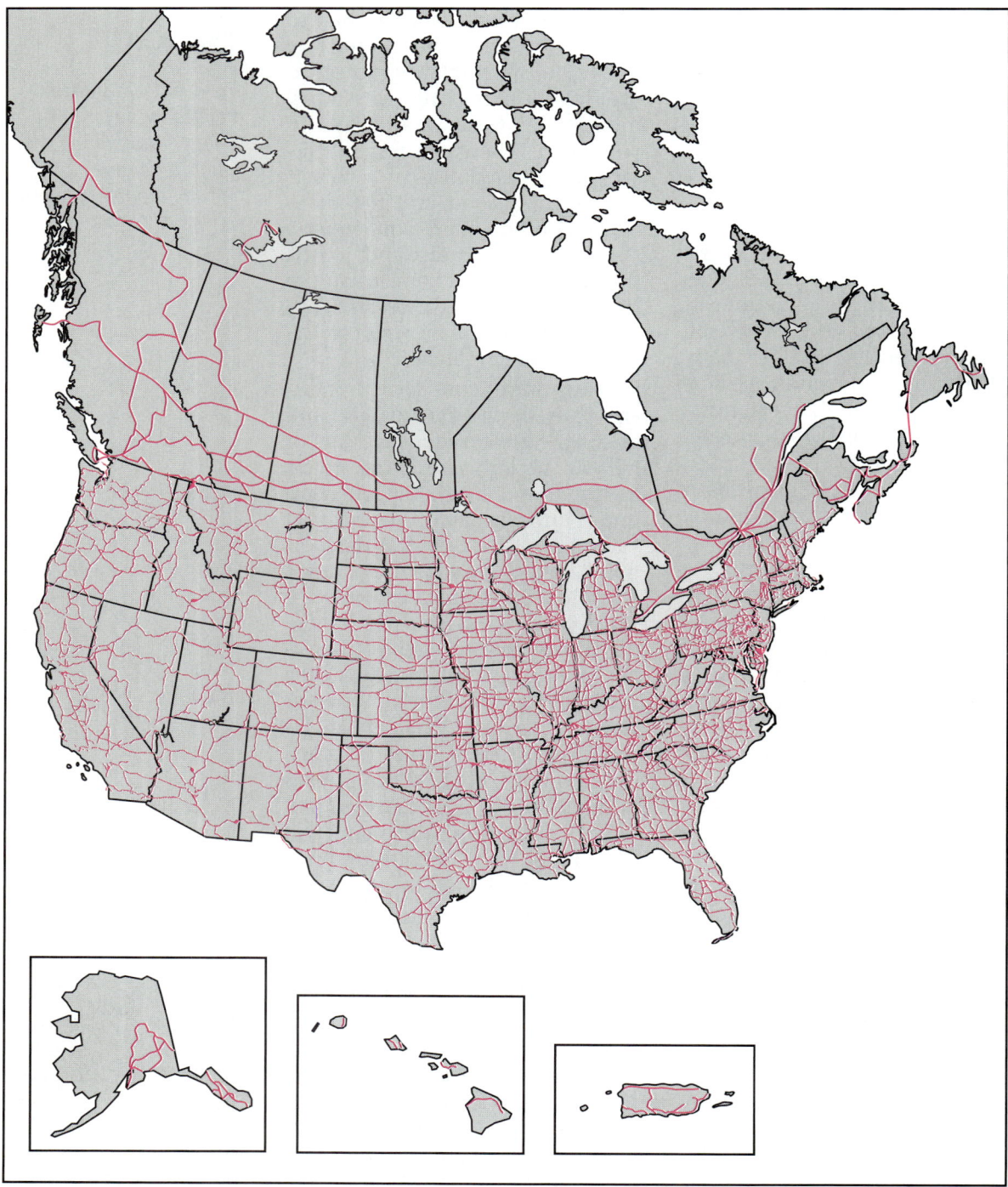

Figure 1–5 The Highway Transportation System links all of North America.

proceed forward on a straight road, the fact that a pedestrian has begun to cross the street in a crosswalk in front of you leads to conflict. You want to continue straight on your way. The pedestrian wants to cross the street. Your goals and desires are in conflict. The HTS has rules about who has priority to use the road at any one given point in time and place, and these rules tell you that you need to stop for the pedestrian. The conflict is resolved.

Roadways

The three main components of the HTS are roadways, vehicles, and people. There are almost as many types of roadway as you can imagine. Whether it is an eight-lane freeway or a small dirt access road, each has its own purpose and type of traffic. Roadways can be classified by condition, such as dirt roads, gravel roads, paved roads, and roads under construction. They can also be classified by function, such as freeways, undivided highways, and city streets. Finally, roadways can be classified by whether they are free or whether you must pay a usage fee, or toll, each time you use that road.

Vehicles

The HTS also includes many types of vehicles, ranging from lightweight children's bicycles to large tractor-trailer rigs. They share the HTS every day, usually using different parts of the system at the same time, but not always. For example, when a tractor-trailer truck moves a suburban family from one house to another, it must begin and end its journey in the same residential neighborhood that a child uses to ride his or her bike. When operators of vehicles in the HTS do not share space intelligently, the sometimes massive disproportion in weight between the vehicles can and does cause serious injury and death.

Although trucks, buses, and motorcycles all transport drivers and passengers, the most commonly used vehicles for passenger travel are cars, pickup trucks, sport utility vehicles, and vans. Because of the variety of vehicles using the roadway, it is always advisable to be aware of which vehicles are around you at any given time. A car will take longer to stop than a bicycle. A truck will need more room to turn than a car. A sports car can accelerate faster than a sedan. A motorcycle behind you is harder to spot than a bus. Because you must share the roadway with so many kinds of vehicles, you must learn their different limitations, characteristics, and capabilities to become a safe driver.

People

All types of people use the HTS. They may be young or old, tall or short, mature or immature. They may have quick reflexes or be slow to react. They may have excellent or poor vision. They may be sober or under the influence of alcohol or other drugs. They may be healthy or sick, injured, or disabled. They may be seasoned truckers or bus drivers who oper-

factoid

Nearly 220 million vehicles are registered in North America.

Figure 1–6 How long has this vehicle been roaming the HTS?

Nearly 90% of the driving-age population in the United States are licensed drivers.

Figure 1–7 Many types of people and vehicles share the HTS.

ate a vehicle for a living, or they may use the car once a month to make a shopping trip. The point is that you cannot rely on sharing the road with others who have your own strengths and weaknesses. Although you can easily learn to distinguish between different types of vehicles, it is much more difficult to identify different types of drivers. To be a safe driver, you must be a defensive driver.

Order in the HTS

Order in the HTS is maintained by a complex system of laws, rules, and regulations. These are necessary to help drivers predict what others will do, help preserve a smooth traffic flow, and act as a basis for the police and courts to determine who is and who is not contributing to order in the HTS. As a member of society, you are expected to know the motor-vehicle laws. *Ignorance of the law is no excuse.*

The federal governments in both Canada and the United States have established laws that create standards and structures that govern the roadways and those who use them. However, these federal governments by and large let the states and provinces determine their own laws and procedures within these structures. Apart from a few federal laws in Canada regarding dangerous driving, driving while under the influence of alcohol or other drugs, and vehicular negligence, almost all other laws regarding motor vehicles and drivers in both countries are state and provincial laws.

The state or provincial vehicle code is the single most important source of **traffic laws** within a jurisdiction. It covers the licensing of drivers, registering and titling of vehicles, financial responsibility laws, minimum safety equipment of vehicles, the various traffic and vehicle laws, and infractions and penalties. Local and municipal governments pass some traffic laws, also called bylaws in Canada, including parking rules, certain speed limits, and prohibited turns, but their influence is minimal. State and provincial governments are working more closely together in several areas with the goal of increasing safety in the HTS. They are coordinating their efforts with regard to drivers who have outstanding moving violations, drivers with suspended or revoked licenses, and drivers who have failed to properly register their vehicles.

Designing, Building, and Maintaining the HTS

All levels of government participate in the design and building of the roadways of the HTS. The federal government plays a large role in financing roadway construction and maintenance in addition to building and maintaining certain exclusively federal roads (such as those in national parks). State and provincial governments build and maintain their own roadway systems, and counties and municipalities build and maintain local roads.

The main elements of building a roadway are design, engineering, and construction, whether the roadway is new or an old or unsafe roadway that is being

Figure 1–8 Roadway construction and maintenance are funded by all levels of government.

improved. Modern technologies and engineering studies are used to make roads safer than they were in the past. Some improvements that have been applied to freeways and higher-speed highways in recent years are banked curves, breakaway signs and poles, carpool lanes, and widened traffic lanes. On urban streets, special turn lanes have been added at intersections to separate turning traffic from through traffic, while signs prohibiting parking in certain places or turning onto side streets help make traffic flow more smoothly. On certain residential and rural streets, STOP signs and traffic signals have been added to intersections that did not have them before, and speed-reduction devices have been installed in residential areas to improve safety.

Federal, state/provincial, and local maintenance crews maintain the roadways by keeping them clean; setting up warning signs, signals, and message boards; removing snow and ice; repairing potholes, broken guardrails, and signs; cleaning up after a collision; and removing obstacles and obstructions such as auto parts that have fallen from vehicles.

Enforcement of Traffic Laws

The traffic laws that govern conduct of users of the HTS are enforced by a wide range of **law-enforcement** agencies, depending on the road and location. The role of the police who patrol the HTS is to apply the laws in a uniform manner, resolve conflicts, act as a deterrent to potential lawbreakers, and catch offenders who make driving less safe for the rest of the users. This gives every driver a sense of security and stability, and it allows a system that functions smoothly. If we could not predict what we can and cannot do legally when we drive, the result would be total confusion on the roadways.

We are all familiar with law-enforcement officials from their role as "traffic cops" who issue moving violations. Sometimes a

factoid
William Phelps Engo, known as the "Father of Traffic Safety" for originating modern traffic regulations, never drove a car.

Figure 1–9 Without law-enforcement agencies, chaos would rule on the roadways.

peace officer stops a driver and issues a ticket after witnessing the violation. In other cases, the offense was recorded by radar, or some other device used to detect drivers exceeding the speed limit or breaking other traffic laws. Law enforcement also does its part to remove problems from the HTS, such as drivers under the influence of alcohol or other drugs and those operating substandard vehicles that do not comply with the minimum safety standards. The police also help drivers who are stranded on the side of the road because of a mechanical breakdown or other emergency, and they assist at collision scenes.

Because there are so many different law-enforcement agencies operating at any one time in the HTS, order is maintained by a system of "jurisdiction" in which certain agencies have primary enforcement responsibility over a given geographic area. The state and provincial police and highway patrols normally have primary jurisdiction on local, state, and interstate roadways within their state or province.

In the United States, park police and other federal law-enforcement agencies have primary jurisdiction on federally owned roadways. In Canada, federal police such as the Royal Canadian Mounted Police have authority not only on federal roadways but also on most roadways in provinces without their own police forces. Local and regional police forces such as county sheriff departments often have primary jurisdiction on local roadways. Parking enforcement officers, also known as bylaw enforcement officers in Canada, are usually local police who enforce municipal parking regulations.

Breakdowns in the HTS

Sometimes the HTS breaks down, resulting in traffic jams, delays, and road closures. Drivers create most of the problems in the HTS when they block traffic because of a mechanical failure or collision. Those who fail to show courtesy to other drivers or who violate the law often end up causing collisions. Drivers who never bother to perform routine maintenance on their vehicles will inevitably have breakdowns.

Occasionally, the HTS breaks down because of failures in the system itself. When urban planners fail to construct enough lanes on a freeway or offer alternative forms of public transportation, the result can be rush-hour gridlock. Sometimes roads must be closed for construction, repair, or snow or ice removal. In general, the United States and Canada have one of the best and most efficient roadway systems in the world. The next time you complain about a detour or traffic jam that causes you to be late to your destination, remember that drivers in other parts of the world would gladly trade places with you.

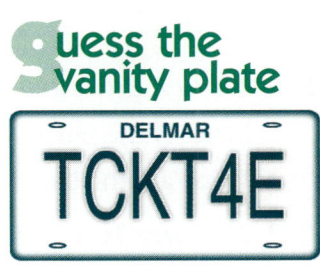

Figure 1–10

YOUR TURN

1–1 Driver Education

1. What is the purpose of driver education?
2. What are the basic components of a "driver ed" course?

1–2 Driving Is a Privilege

3. Why is driving a privilege, not a right?
4. What are your basic responsibilities as a driver?
5. Why is it important for you to be courteous as a driver?
6. What is the importance of being emotionally, physically, and mentally fit to drive?

1–3 The Highway Transportation System

7. What is the Highway Transportation System?
8. What are the three main classifications of roadways?
9. Who is responsible for enforcing traffic laws?

The Driver and the Driving Task

unit **1**

Chapter 2   Defensive Driving
Chapter 3   The Condition of the Driver

Chapter 2

Defensive Driving

Defensive driving is a method of driving that emphasizes anticipating and avoiding danger on the roadway. By staying alert and being prepared for the worst, you will be able to manage almost any hazardous situation with confidence and control. Defensive drivers actively observe, analyze, and plan. They evaluate road and traffic conditions, assess what other drivers and pedestrians will do, weigh their options, and decide on the best solution *before* they act. They do not blindly react to whatever happens around them.

CHAPTER OBJECTIVES

Upon completion of this chapter, you should be able to:

2–1 Defensive-Driving Skills
1. Understand what defensive driving is.
2. Describe the essential skills you need for defensive driving.

2–2 Managing Time, Space, and Visibility
3. Understand the importance of following distance and space cushioning.
4. Describe the 3-second rule.
5. Describe how large your space cushion should be.
6. Describe ways to deal with tailgaters.
7. Describe ways to make yourself known to others on the roadway.

2–3 The SAFE Method
8. Explain what "SAFE" stands for.
9. List the sequence of the orderly visual search pattern.
10. Describe the purpose of each step in the SAFE process.

KEY TERMS

defensive driving	tailgate	orderly visual search (OVS)
following distance	hand signals	assess
3-second rule	SAFE	find
space cushion	scan	execute

2–1 DEFENSIVE-DRIVING SKILLS

Defensive driving does not come naturally. You must train yourself to stay alert while monitoring the driving scene. Because of the large amount of information you must process, it is easy to become distracted. Once you lose your concentration, you become at risk for a collision. A driver who is tired, lazy, or unfocused is a dangerous driver. Therefore, it is important to develop sound habits early on so that defensive driving becomes a routine rather than a chore.

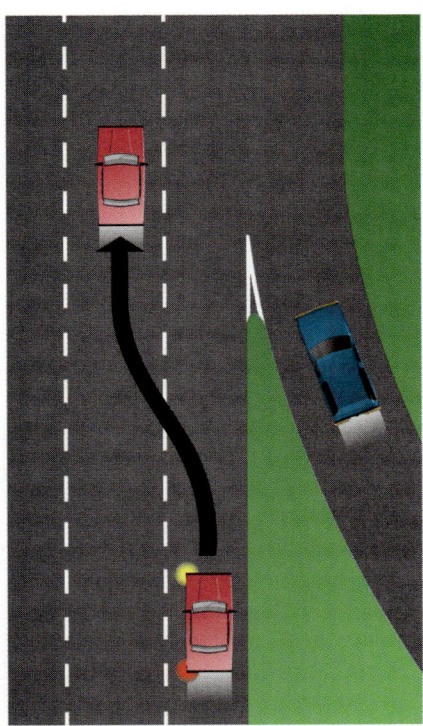

Figure 2–1 A defensive driver *actively* avoids potentially dangerous situations.

Knowledge

Defensive driving relies on several skills. To be able to safely maneuver your car or truck in various situations, you must have a thorough working knowledge of all its operating devices and controls. If you are thinking about how to steer or shift gears while driving, you are not watching the road.

Figure 2–2 To obey traffic laws and controls, you must understand them.

Knowledge of traffic laws, signs, signals, and roadway markings is also essential to defensive driving. To obey traffic laws and controls, you must first understand them. Although you cannot be expected to memorize the vehicle code, you have a responsibility to become as informed as possible. The more you know about the rules of the road, the fewer tickets you will receive and the less chance you will have of getting involved in a collision. You will also experience less stress behind the wheel.

Preparation

Your safety depends on your being prepared to drive. You must be not only physically and mentally ready, but also able to respond when things do not go as planned. What if your car breaks down on the road? What if you are on an unfamiliar freeway and miss your exit? What if the weather changes for the worse? Asking "what if?" questions prepares you for possible emergencies and helps you avoid them entirely.

For example, by asking "What if my car breaks down?" you will remind yourself that your vehicle has needs beyond gas. When was the last time you checked your oil or antifreeze? Have you had your brake pads checked recently? You will also be forced to rethink your driving schedule and route. Is today the best day to go on a long drive? Are there service stations along the road you are taking? If your car's timing belt blows out far from a service station on Sunday night, you will have no one to blame but yourself because you failed to plan properly.

Awareness

Awareness is essential to defensive driving. Being aware of what is happening around you allows you to spot potentially dangerous situations early. If you see a line of parked cars on the road, for example, you should be ready for surprise door openings. Brake lights on cars ahead tell you to prepare for sudden stops or changes in speed. If you encounter a road construction detour, you should expect confused or frustrated drivers to dart in and out of lanes. By remaining aware of the driving environment, you can detect even the smallest indications of trouble before it occurs.

Anticipation

No matter where you are driving or what the road conditions are, you should always "expect the unexpected." The ability to anticipate problems before they happen is fundamental to developing an overall defensive-driver attitude. Make a habit of anticipating what drivers and pedestrians around you are about to do and how your car will respond in both normal and emergency situations. No matter what driving decision you make, always have in mind an alternative you can implement if you get into trouble.

Good Judgment

Good judgment in driving situations involves choosing the safest and most effective option available to you. For example, you must decide when to start applying your brakes, how early you should signal, whether you have enough room to pass, and how close you should follow another vehicle. You must consider a number of factors when determining how to avoid potential hazards:

- *The type of road you are on:* Is it a smooth, straight, and well-marked city street with

Figure 2–3 Driving on rain-slick or icy roads calls for good judgment.

clearly painted lines and helpful signs or a poorly lit, narrow country road with soft shoulders and potholes?

- *The weather:* Is it a warm day with a clear, blue sky and good visibility or is rain coming down in sheets too thick for your windshield wipers to handle? Is fog obscuring the road ahead? Is wind whipping your vehicle dangerously close to the guardrail? Are freezing temperatures transforming gentle bends in the road into icy, hairpin turns?
- *Visibility:* Maybe you have 20/20 vision, but what about the driver about to make the left turn in front of you? Is that driver someone who could not see the "E" at the top of the vision test chart even with glasses—which, by the way, the driver is not wearing right now! Can that driver on your right who is about to move into your lane see your car through the mud-caked windows?
- *Vehicle condition:* Are you driving a brand-new car outfitted with the latest antilock brakes or an old, hand-me-down "junkmobile" that needs new tires and a brake job? What about the cars around you?
- *Traffic conditions:* Are you on a six-lane highway with few cars in sight or a four-lane freeway with bumper-to-bumper, rush-hour traffic?
- *Other drivers:* Are other drivers around you courteous and law-abiding or are they drunk, careless, rude, or angry at the world and ready to prove a point at your expense?

Practicing Defensive Driving

You have likely heard the expression, "practice makes perfect." Although practicing driving will not necessarily make you a perfect driver, it will definitely make you a safer driver. The more you practice, the faster you will gain confidence behind the wheel. The best way

REALITY CHECK — *Experience Counts*

Compared with the average driver, young drivers have superior physical skills. They are generally in excellent health and have both good eyesight and fast reflexes. However, teenagers are involved in more motor vehicle collisions than any other age group. In fact, traffic-related injuries are the leading cause of death among people between the ages of fifteen and nineteen. Why? Teenagers have less experience, maturity, and motor-vehicle training than most adults. Some feel uncomfortable behind the wheel and drive in constant fear of getting into a collision or being pulled over by the police. Teenagers also have a more difficult time analyzing different variables at once and can easily become distracted. They are more likely to drink, use drugs, and engage in reckless behavior because of peer pressure. For all of these reasons, it is especially important for teens to drive defensively.

---

to develop your driving skills is to practice them in a low-density, low-stress environment like an empty parking lot or on quiet suburban streets. Once you master the basics, you will have the self-confidence to move to more challenging environments where your alertness and concentration are even more critical.

Defensive driving means developing the ability to perform evasive driving maneuvers. To safely execute such a maneuver, you must be comfortable with using all of your car's controls. You have to be "in synch" with your vehicle's limits and capabilities. In most cases, you can resolve a problem with basic actions like braking, stopping, turning, flashing your headlights, or tapping your horn. In more complex situations, however, you may have to use a combination of well-timed and smoothly coordinated maneuvers to move your vehicle out of harm's way or to minimize an impending impact. The only way to be good at executing such maneuvers is to practice, practice, practice!

2–2 MANAGING TIME, SPACE, AND VISIBILITY

Defensive driving involves managing time, space, and visibility as much as possible. You can never completely control these factors in a particular driving situation, but the better you manage them, the more flexibility you will have in choosing a course of action.

Following Distance

One of the greatest errors committed by drivers is following the vehicle ahead too closely. By increasing your *following distance,* the distance between your vehicle and the vehicle directly ahead of you, you can significantly reduce the chance of becoming involved in a collision. A long following distance allows you to scan farther ahead, makes it easier for drivers ahead to see you in their mirrors, gives you more time to react if the vehicle in front of you suddenly comes to a stop, and provides you with an escape path if another vehicle is about to rear-end your own car. Because rear-end collisions are one of the most common types of motor-vehicle collisions, increasing your following distance is critical to defensive driving.

In ideal low-speed driving conditions, with little traffic congestion and good visibility, you should maintain a minimum following distance of 3 seconds behind another vehicle. To check your following distance, you can use the **3-second rule.** Pick a fixed object ahead, such as a tree, sign, or telephone pole. Count the number of seconds that pass between the time that the car ahead of you passes this object and the time that you pass it. Use *full* seconds: "one thousand and one . . . one thousand and two . . . one thousand and three." If you reach the landmark before you count to three, you are driving too closely to the vehicle ahead of you.

Figure 2–4 Maintaining a safe following distance increases your options in an emergency.

BESSIE SMITH

Born in Chattanooga, Tennessee, in 1894, Bessie Smith reached the peak of her fame as a blues singer in the 1920s. Her style influenced other artists such as Billie Holiday, LaVern Baker, and Janis Joplin. After the huge success of her first record, "Downhearted Blues," she was dubbed "The Queen of the Blues." On September 26, 1937, Smith was badly injured when her car, driven by her friend Richard Morgan, crashed into the back of a truck on Route 61 near Clarksdale, Mississippi. Smith suffered a shattered left elbow, crushed ribs, and internal injuries. While they were waiting for an ambulance, another driver, not seeing the collision scene, crashed into the rear of Smith's car and injured her further. Despite the best efforts of a doctor at the scene and at a nearby hospital, Smith died from shock several hours later. By increasing your following distance from vehicles ahead of you and giving yourself an escape route, you can avoid rear-end collisions like the one that killed Bessie Smith. By scanning ahead down the roadway, you will be able to spot a collision scene or other hazard far enough in advance to allow you to stop or drive around it safely.

bumper sticker sightings

IF YOU CAN READ THIS, YOU'RE TOO CLOSE

Because at higher speeds your vehicle travels farther in the same amount of time and it takes longer to stop, you should increase your following distance to 4 or 5 seconds the faster you travel. In certain low-speed driving situations, you should also increase your following distance to 4, 5, or even 6 seconds:

- If you are a new driver
- When driving in severe weather
- If traction is poor
- When driving at night or anytime visibility is reduced
- If your view ahead is blocked by a large vehicle such as a truck or bus
- When following a motorcycle
- When following an obviously unsafe vehicle
- When following vehicles with license plates from another jurisdiction
- When traveling on unfamiliar roadways
- When a driver ahead of you is driving erratically or unsafely
- When a driver behind you is following too closely
- When pulling a trailer or a heavy load
- When driving downhill
- When you are stopped in traffic going uphill
- When you sense trouble ahead
- If you feel sick or tired

A second way to gauge your proper following distance is to maintain a space between you and the vehicle in front of you equal to

Figure 2–5 Use full seconds when testing your following distance to vehicles ahead of you.

Figure 2–6 Space cushioning increases visibility and leaves you potential escape routes.

one car length for every 10 miles per hour (15 km/h) that you are traveling. This method works well for all speeds that you are driving. For example, if you are driving 30 miles per hour (50 km/h), you should be at least three car lengths behind the vehicle in front of you. If you are on a freeway traveling 55 miles per hour (90 km/h), you should be five and a half car lengths back.

Space Cushioning

As you drive, try to surround all sides of your vehicle with a **space cushion,** an empty space between you and the cars and other objects on the roadway around you. This unoccupied space helps you in two important ways. First, it allows you to have a better view of your driving environment so that you can recognize potential problems early. Second, space cushioning provides escape routes that you can use to avoid a potential hazard.

Ideally, a space cushion should be at least one vehicle space to either side of you and at least three to seven vehicle lengths ahead of you and behind you. Often, as the driver, you must *create* a space cushion around your vehicle by increasing and decreasing your speed or switching lanes to

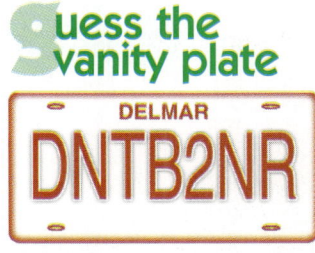

Figure 2–7

Wild Wheels

Figure 2–8 What was the name of this 1948 "Flying Car" designed to create the ultimate space cushion for drivers?

factoid

The biggest complaint drivers have against others on the road is tailgating.

change your vehicle's position relative to the cars traveling around you. This requires communicating your intentions to other drivers by using your turn signals, headlights, brake lights, horn, body movements—whatever it takes.

In heavy traffic, it can be nearly impossible to maintain a large space cushion. However, this does not mean you are completely helpless. In these situations, you can still increase your following distance. At the very least, you will be able to maintain a cushion in front of you, even if you are hemmed in from the rear and sides by other drivers.

Tailgating

Drivers who follow other cars too closely pose one of the most common and dangerous challenges to maintaining a proper space cushion around your car. Fortunately, there are ways to deal with those who **tailgate.** Increase your own following distance from the vehicle in front of you to 6 seconds or more. If the driver ahead has to stop quickly, you will have more time to gradually apply your own brakes and reduce the chance of the tailgater hitting you. Move to the right side of your lane to give the tailgater a better view of the road ahead so that if he wants to pass you he can do so safely. Plan an escape route in the next lane or on the road shoulder in case you have to swerve to avoid being rear-ended.

If a tailgater cannot or will not pass you, politely signal him to "back off" by tapping gently on your brakes to flash your brake lights. Do *not* slam on your brakes—this can easily cause a collision or provoke an angry response from the driver behind you. If tapping your brakes is ineffective, try very gradually reducing your speed to encourage the tailgater to move into another lane. If this still does not work, change lanes yourself at the first opportunity. Finally, if you think that the tailgater is a real threat to you or other drivers on the road, pull over as soon as it is safe to do so to let the driver get by you.

Making Yourself Known

Always remember that you must be visible to other drivers. When positioning yourself on the road, consider *their* field of vision. It is just as important that you can be seen at all times as it is for you to see what is happening around you. Many collisions can be avoided simply by making your presence known to other drivers. When your view is blocked and you are not sure if others are aware of you, or when you see another vehicle but are not sure if the other driver sees you, use all available means to *make yourself known!* Other drivers will be able to respond appropriately to your actions if they know exactly what you are doing.

Communicating Your Intentions

Use your turn signals to let other drivers know that you are turning, changing lanes, merging onto or exiting a roadway, parking, or pulling into or out of a driveway. Signal far enough ahead of time that drivers near you can respond to your change of direction and speed. Signaling just as you turn is too late. On the other hand, if you signal too early the driver behind you may misinterpret your intentions. For example, suppose you activate your left-turn signal two blocks ahead of the cross street you plan to turn onto. The driver behind you will expect you to turn at the first intersection and may not slow down when you actually turn because he or she no longer trusts the meaning of your signal.

Use your brake lights to signal your intention to stop. Drivers behind you may not be paying close attention to the road ahead or cannot see what you can. Do not simply hit your brakes when you have to stop. Instead, tap your brake pedal to warn other drivers that you are slowing down.

Hand signals, sometimes called arm signals, are another way to inform other drivers and pedestrians that you intend to turn, abruptly slow down, or stop. You might use hand signals in addition to turn indicators and brakes when bright sunlight decreases the visibility of your

DRIVING TIPS To Honk or Not to Honk?

Your horn is a valuable safety feature that is both overlooked and overused as a communication tool.

- Honk when approaching blind or narrow curves or when exiting a narrow alley to alert approaching drivers of your presence.
- Use a light tap on your horn to get the attention of another driver or a pedestrian who cannot see you.
- Use your horn, loudly if necessary, to avoid a collision. Use a sharp blast when a pedestrian or bicyclist is about to walk or ride into the street in front of you, or when another vehicle is in danger of hitting you.
- Use your horn when you lose control of your vehicle and are moving toward someone or another vehicle. This may give the other person enough warning to get out of the way.
- Do *not* use your horn to encourage someone to drive faster.
- Do *not* honk if you can avoid entering into a blind spot by speeding up or slowing down.
- Do *not* use your horn to greet friends.
- Do *not* use your horn if a blind pedestrian enters the road, even against traffic signals. Stop and allow the person to cross.
- Do *not* honk if you can make eye contact with the other driver or communicate with a hand gesture.

Figure 2–9 Use hand signals to help communicate your intention to turn, stop, or slow down.

rear lights or if you think the driver behind you is not paying enough attention to the road. Hand signals should not be used instead of your vehicle's mechanical indicators, but as an extra warning or in case your turn signals or brake lights malfunction.

When using hand signals, make sure that you begin to signal at least 100 feet (30 m) before turning, stopping, or suddenly decreasing in speed. To signal a left turn, extend your left hand and arm straight horizontally out the window of the vehicle. To signal a right turn, extend your left arm out the window and bend it at the elbow so that your left hand points skyward at a 90-degree angle. To signal a stop or abrupt slowing, extend your left hand and arm out the window and bend it at the elbow so that your left hand points downward at a right angle.

The position of your vehicle also communicates your intentions to other drivers. Moving to the right side of the lane, for example, may indicate that you are preparing to turn right or park by the curb. Moving to the left of the lane may signal your desire to pass a driver ahead of you.

Make eye contact with other drivers or pedestrians when possible. Although you should never depend on eye contact alone to communicate your intentions, it may help reduce the risk of conflict. You can use body movements such as waving to tell a pedestrian to cross. You can also motion clearly with your arm or hand to other drivers, indicating to them to proceed before you or to wait.

Increasing Your Visibility

Your headlights improve your view of the road, but they also help others see you. You should always use your headlights at dawn, dusk, and in bad weather such as rain, snow, or fog, as well as at night. As an added measure

Figure 2–10 Where you position your vehicle tells other drivers what you intend to do.

of safety, use your headlights during regular daytime hours. "Flash" your high beams when approaching a blind or narrow curve at night to alert drivers approaching from the opposite direction who might turn wide into your lane of travel. As a gesture of courtesy, you might also briefly flick on your "brights" at oncoming drivers who have forgotten to turn on their headlights at night or when visibility is poor.

Use your hazard lights to tell other drivers that your vehicle cannot move or is moving very slowly. This will encourage them to give you extra space on the road and be patient while you get your vehicle to a safe stopping point. Once you are stopped, your emergency flashers will help other drivers see you on the side of the road, especially at night. They will also make it easier for the police, emergency vehicles, and tow trucks to find you if you need help.

2–3 THE "SAFE" METHOD

SAFE is an innovative defensive-driving strategy that you can use to evade potential danger on the roadway. By helping you to manage time, space, and visibility in a manner that is simple and easy to remember, it prevents conflicts and makes for safer, less stressful driving. The SAFE method builds on the defensive-driving skills described at the beginning of this chapter. The more knowledge, preparation, awareness, anticipation, and good judgment you have, the more effective it will be.

SAFE stands for *scan, assess, find,* and *execute*. Following this four-step sequence gives you an organized way to gather, interpret, and act on information about the driving environment. When driving you should constantly scan for clues, assess what others are likely to do and what your options are, find a solution or "out," and execute any necessary driving maneuvers successfully. In some situations, you might repeat this process dozens of times. With practice, the SAFE system will help you avoid risky situations and escape from sudden and unexpected dangers.

Scan

The first principle of SAFE is that you **scan** ahead to gather as much information as possible about the complete driving scene. Your eyes are your best tool to identify traffic situations and road conditions. You want to be aware of other vehicles, pedestrians, bicycles, and possible hazards around you as soon as possible. By scanning ahead, you give yourself time to slow down gradually and to change lanes smoothly while avoiding unnecessary braking. When you eliminate the need to stop or turn suddenly, you are less likely to be involved in a collision.

Look far ahead down the road to spot potential problems. This allows you to analyze traffic situa-

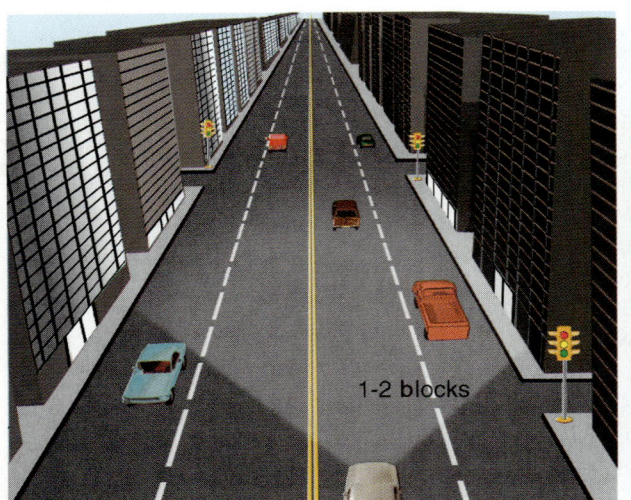 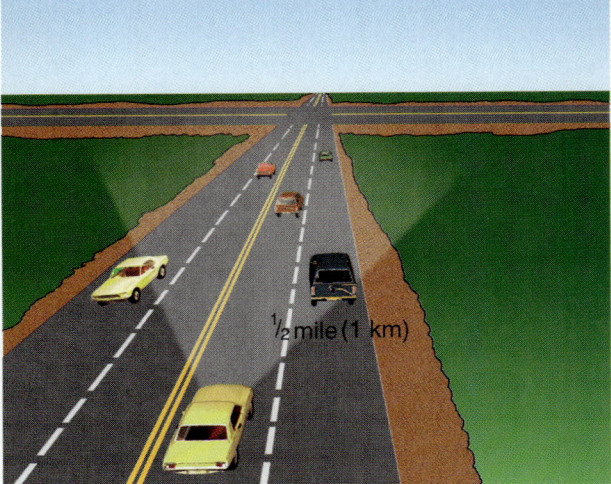

Figure 2–11 Your speed determines how far ahead you should look.

tions and road conditions and to predict what might happen long before a driving conflict arises. As a general rule, you should look between 20 and 30 seconds ahead of you. In urban driving, in which you are typically moving from 25 to 30 miles per hour (40 to 50 km/h), this is equivalent to 1½ to 2 average city blocks. On highways and freeways, when you are moving up to 50 miles per hour (80 km/h) or more, you should look between ⅓ and ½ mile (500 m to 1 km) down the road.

Identifying traffic controls is one of the essential purposes of scanning. Traffic signals and signs may be located overhead, in the center of the road, or off to the side. Look for anything unusual such as temporary detour signs, orange cones, or flashing lights that can alert you to an upcoming change in traffic flow. It is espe- cially important to identify and interpret traffic controls that are hard to read or see because they are located in a shady area, are covered by natural growth, or are reflecting bright sunlight.

As you scan with your eyes, make sure that you get the "big picture" of what is ahead of you.

Figure 2–12 Not all traffic controls are easy to identify.

Search the whole scene, not just part of it. Try to gain a sense of what vehicles and pedestrians in *all* directions are doing. Never gaze straight ahead or "fixate" on an object for too long. When scanning the areas near your vehicle, look to the sides of the driving lanes for hazards such as pedestrians stepping into the street or cars pulling away from the curb. Alternate glances ahead and to the sides with checks of your vehicle's mirrors and dashboard instrument panel. If you are constantly scanning, you should be able to pick up the movement of hard-to-see vehicles such as bicycles and motorcycles. Once you have identified someone outside your normal field of vision, keep track of him or her until the person turns off your path of travel. Occasionally scan at ground level to check the orientation of other vehicles' tires to determine which way a driver is going to turn.

One way to make sure that you scan in all directions is to make an **orderly visual search (OVS)** of the scene around you. In the late 1950s, Howard L. Smith of the Ford Motor Company developed a systematic technique that requires using selective glances in a constantly repeating pattern to moni-

factoid

It takes only one-tenth of a second for a collision to occur, half the time it takes just to blink your eyes.

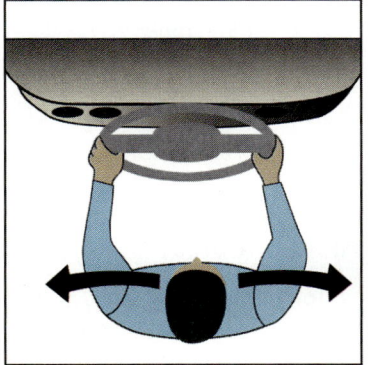

Figure 2–13 The OVS pattern is an effective scanning technique.

tor all the different areas around your vehicle:

1. Look at the road near you.
2. Look ahead in the distance.
3. Alternate glances at each of the following with looking ahead down the road:
   - To your right
   - To your left
   - At your rearview mirror
   - At the driver sideview mirror
   - At the passenger sideview mirror
   - At the instrument panel

By repeating this sequence as you drive and by remaining alert, you will spot most changing road conditions and potential dangers quickly. You can then adjust your speed and position in plenty of time. Whether you use the OVS pattern or create your own system, remember to *always keep your eyes moving*.

Assess

The next step in the SAFE process is to **assess** potential threats on the roadway. The ability to predict problems before they happen is fundamental to defensive driving. Once you are able to consistently and accurately anticipate what others are going to do in a dangerous situation, the options available to you, the probable consequences of your actions, and how your vehicle will respond, you can make informed decisions to prevent a collision.

The knowledge needed to make good predictions partly comes from the study of traffic laws, the experiences of friends and family, and common sense. There is no substitute, however, for personal experience. The more time you spend behind the wheel, the better you will be at predicting which hazard is the most critical

Figure 2–14 The condition of other vehicles on the road, as well as your own, may affect what you decide to do in a hazardous situation.

DRIVING MYTHS Do Not Judge a Driver by His or Her Car

Just because someone drives a vehicle with a high safety rating does not mean he or she is a safe driver. Always watch others on the road equally closely, no matter what they are driving.

Figure 2–15 Clues can help you predict whether this driver is entering or pulling out of the parking space.

to avoid. As time goes by, your ability to forecast outcomes will improve with the level of your experience.

Threats come in many forms. Other vehicles may suddenly enter your lane of travel and cause you to maneuver around them or stop suddenly. An oncoming car may unexpectedly turn left in front of you. A pedestrian may ignore a flashing DON'T WALK sign and step off the curb right in front of you. By looking for certain clues, you can anticipate the actions of others on the roadway. For example, if you identify a car at the curbside with its left-turn signal on, the wheels turned toward the street, and smoke coming out of the exhaust pipe, the driver may be preparing to pull away from the curb into your path of travel. A driver on a cross street who is looking at *your* signal may try to "get a jump" on the light and enter the intersection too early. Be on the lookout for drivers who speed, change lanes often to get ahead of normal traffic flow, or drive slowly because they are lost and confused. Assess whether drivers using cellular phones, drinking coffee, or doing something else distracting are possible threats.

The dangers you face largely depend on your driving environment. On city streets, for example, pedestrians and bicyclists, vehicles pulling in and out of parking spaces, and double-parked cars or delivery vehicles that may block your lane of travel are common hazards. In residential or suburban neighborhoods, you must watch for children playing, pets, and cars backing out of driveways. On rural roads, dangers include hidden intersections, downed tree limbs, and wild or domestic animals. On freeways, you have to keep track of multiple vehicles in several lanes, as well as those entering and exiting the roadway, to maintain a safe space cushion.

As you travel, roadway conditions constantly change. You must determine which of these changing conditions might affect your position on the road or pose a danger. Look for information that will help you identify upcoming intersections or changes in the width of your lane. If you are approaching a construction zone or a bridge on a rural road, for example, you should expect less space for your vehicle to pass oncoming traffic.

When driving in heavy rain, at night or in conditions of reduced visibility, or on very hilly or curving roads that restrict your ability to scan far ahead, you must assess possible hazards as early as possible. Sometimes the road surface itself is a problem. A street that is slick because of the weather or is poorly maintained will affect how your vehicle handles. If you skid on a slippery road or hit a large pothole, you can lose control of your vehicle and cause a collision. By evaluating how standing water, patches of snow or ice, leaves, potholes, debris, and warning signs and signals affect *other* drivers, you can adjust your driving in time to avoid a problem.

Find

Once you have identified a potential danger and assessed your options, you must **find** an "out." An out is an escape route that you have identified as the best means of avoiding a conflict on the road. Always position your vehicle so that there is a margin of space around it to provide a cushion between you and other vehicles. Constantly adjust your position in changing traffic conditions to keep that space cushion around your car. This will give you the extra time needed to stop suddenly or to move to the side to avoid a hazard.

Do not assume that others on the road will always take the correct evasive action in a driving conflict. Consider all the possible paths that other drivers may take

Figure 2–16 Change your position to leave yourself as many "outs" as possible.

REALITY CHECK   Do Not Let the Pressure Get to You

As a new driver, you may drive more slowly or need extra time to decide when it is safe to make a difficult driving maneuver, such as taking a left turn in front of oncoming traffic. If other drivers honk their horns at you or express impatience in other ways, do not let them pressure you into taking a risk. You should execute a difficult maneuver only when *you* know it is safe to do so.

Figure 2–17 Many roadways offer limited room to find an out.

and think of how you might respond to each one. Look for escape routes in each of the possible outcomes. Try to predict when and where another vehicle will intrude on your space cushion, cross your path, or make contact with your vehicle. Not every emergency has a perfect out. Your only options may be a rear-end collision or being sideswiped by a car in an adjoining lane. If you see no clear way of avoiding a crash, do not panic and throw in the towel. Accept the situation and find the safest alternative.

Execute

The final step in the SAFE method is to **execute,** or carry out, your decision to avoid an upcoming conflict. You always have at least two options if you encounter danger on the road. You can change your speed, and you can change your direction. In most cases, if you change your speed you will choose to slow down or stop. If you have maintained a safe following distance, you should have plenty of room to slow down before hitting the vehicle ahead of you. If another car is about to rear-end you or hit you from the side, however, it may be to your advantage to speed up if the road ahead of you is clear of other vehicles and pedestrians.

In some situations, you may decide to change direction if you cannot stop in time and there is an escape path to either side of your vehicle. By swerving or making a sharp right or left turn, you may be able to avoid a hazard with less risk to yourself or others.

WHO'S AT FAULT?

1. Driver 1 did not notice the merging traffic warning sign and continued driving in the right lane. At the last moment, he was forced to swerve into the middle lane. This in turn caused Driver 2 to swerve into the left lane, and he hit Driver 3. **Who's at fault?**

2. Driver 1 was approaching the end of an on-ramp to a freeway. Driver 2 was coming down the ramp right behind him. Driver 3 was approaching in the right lane of the freeway. Just as Driver 1 was about to merge with freeway traffic, she noticed that Driver 3 was too close, so she stopped to let Driver 3 pass. Driver 2 thought that Driver 1 was going to enter the freeway without stopping. He turned his head to check traffic on the freeway. When he turned back and saw that Driver 1 had stopped, he slammed on his brakes but could not stop in time and rear-ended Driver 1. **Who's at fault?**

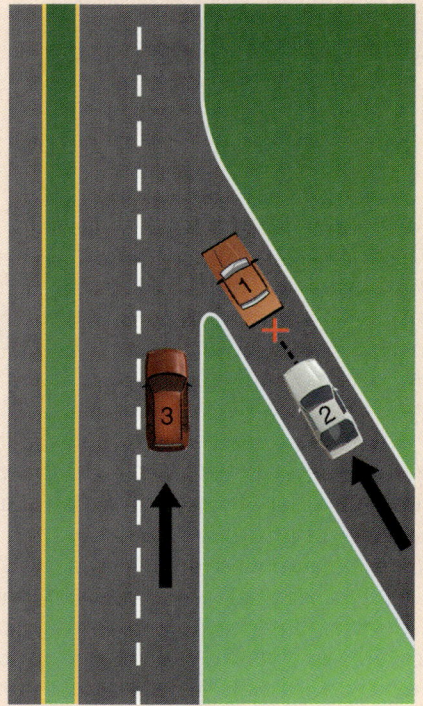

YOUR TURN

2–1 Defensive-Driving Skills

1. What is defensive driving?
2. What are the essential skills you need to drive defensively?

2–2 Managing Time, Space, and Visibility

3. Why is following distance important?
4. What is the 3-second rule?
5. How large should your space cushion be?
6. How can you deal with tailgaters?
7. How can you make yourself known to others on the roadway?

2–3 The SAFE Method

8. What does "SAFE" stand for?
9. What is the sequence of the orderly visual search pattern?
10. What is the purpose of each step in the SAFE process?

SELF-TEST

Multiple Choice

1. Ideally, a space cushion should be at least three to seven vehicle lengths ahead of you and _____ vehicle space(s) to either side.
   - a. two
   - b. one
   - c. zero
   - d. three

2. Gradually reducing your speed can encourage _____ to move into another lane.
   - a. large trucks
   - b. buses
   - c. tailgaters
   - d. police cars

3. You should honk your horn:
   - a. to greet a friend.
   - b. at another driver's error.
   - c. to avoid a collision.
   - d. to warn a blind pedestrian who is in your path.

4. Extending your left hand and aiming straight horizontally out the driver's side window indicates that you are:
   - a. stopping.
   - b. slowing down.
   - c. turning left.
   - d. turning right.

5. If you are traveling on a highway at 50 miles per hour (80 km/h), allow a following distance of at least _____ car lengths.
   a. three
   b. four and a half
   c. five
   d. seven

Sentence Completion

1. You will spot most changing road conditions and potential dangers quickly by constantly repeating an _____ sequence.
2. Increasing your _____ significantly reduces the chances of rear-ending the vehicle in front of you.
3. A _____ allows a better view of your surroundings so that you can predict problems early and provides you with an escape route.
4. An _____ is an escape path that you have identified as the best means of avoiding a conflict on the road.
5. SAFE stands for _____, _____, _____, and _____.

Matching

Match the concepts in Column A with the examples of the concepts in Column B.

Column A
1. __ Assess
2. __ Keep your eyes moving
3. __ Find
4. __ Communicate your intentions
5. __ Increase your following distance
6. __ Scan ahead
7. __ Create a space cushion
8. __ Make yourself known

Column B
a. Escape path
b. Orderly visual search
c. Look two city blocks ahead
d. When being followed by a tailgater
e. Flash high beams on blind curves at night
f. Double-parked car
g. Decrease your speed
h. Tap brake lights before stopping

Short Answer

1. What two choices do you always have when you encounter a road hazard?
2. How does asking "what if?" questions help you?
3. Describe a situation in which your vehicle position communicates your intention to another driver.
4. What does scanning ahead allow you to do?
5. What is meant by the phrase, "Not every emergency has a perfect out"?

Critical Thinking

1. You are driving on a two-lane road and want to make a left turn into a local shopping center. You notice, however, that there is a large truck blocking the driveway as it tries to exit. Just as the truck pulls out to make a left turn, you observe the car behind you trying to pass you on the right. What should you do?

2. You are in the right lane of a four-lane highway driving the speed limit. As you approach a sharp curve, you observe a speed reduction warning sign and slow down. Rounding the curve, you notice that traffic is backed up nearly to the end of the curve because of a collision farther down the road. The driver in a car behind you appears to have ignored the speed reduction sign and is coming on fast without being aware of the danger ahead. What should you do?

Projects

1. As a passenger in a motor vehicle, watch how other drivers communicate with one another. Based on what you see, predict what different drivers will do. Did you predict correctly? Note how many vehicles turn, pull out of parking spots, change lanes, and execute other maneuvers without signaling properly. Were you able to correctly predict their intentions using such clues such as wheel direction, lane position, and so on?

2. As a passenger, observe how the following distance and space cushion around the vehicle you are in changes. As you move into various situations, determine what your best escape route would be if you were the driver.

chapter 3

The Condition of the Driver

Your ability to drive safely is determined by both your physical condition and your emotional well-being. The quality of your eyesight and hearing, whether you have an illness or disability, and even your size and height are factors that you must consider when you get behind the wheel. Your mental state is equally important. It takes a good attitude to be courteous to other drivers and to drive defensively. A negative attitude, overconfidence or underconfidence, and any other feelings that affect your driving do not belong in the car. Anger, stress, anxiety, and other emotions can and do lead to crashes.

CHAPTER OBJECTIVES

Upon completion of this chapter, you should be able to:

3–1 The Physical Condition of the Driver

1. Describe permanent physical conditions that can limit your ability to drive safely.
2. Describe the four main components of vision.
3. Describe temporary physical conditions that can limit your ability to drive safely.
4. Understand how you can prevent carbon monoxide poisoning.

3–2 The Mental Condition of the Driver

5. Understand how your mental condition impacts your ability to drive safely.
6. Understand how to avoid becoming a victim of road rage.
7. Describe activities you should avoid while driving.

KEY TERMS

visual acuity	depth perception	carbon monoxide
field of vision	color vision	road rage
central vision	color blindness	rubbernecking
peripheral vision	fatigue	

3–1 THE PHYSICAL CONDITION OF THE DRIVER

Your individual makeup, the natural process of aging, and chronic illness or disability can all impose physical limitations on your driving skills. Although you can correct for poor vision or compensate for deficiencies in hearing, you must always take such permanent conditions into consideration when you make choices behind the wheel. Other physical conditions such as fatigue or injury may only temporarily affect your ability to drive safely, but they are equally deadly if ignored.

Vision

Because so many of your decisions as a driver are based on what you see and when you see it, good vision is one of the most important physical qualities necessary for driving. You need to see far enough ahead to clearly read traffic signals and signs and to allow you to make adjustments in speed and direction to avoid potential hazards. If you have vision problems, you should get your eyes checked at least every two years to make sure that you have the best prescription to provide clear vision. *If you need corrective lenses to pass the driver's license eye exam, you must wear them while driving.*

Visual Acuity

You must meet a minimum vision standard to obtain a driver's license. **Visual acuity** is how sharply you can see an object at a specified distance. An eye exam will determine your visual acuity and whether you need corrective glasses or contact lenses to drive.

Field of Vision

Your **field of vision** includes the area you can see directly in front of you, to the sides, and when looking straight ahead. The area directly in front of you where everything is clear is your **central vision**. This area is only about 3 degrees wide. To give you an idea of how narrow that is, sit in a chair and fix your eyes on a point straight ahead of you. Take a pencil and hold it at arm's length directly in front of your eyes. The pencil will look very clear, and you should be able to read the brand name and number of the pencil. Keeping your eyes fixed straight ahead, begin to move the pencil slightly to the right or left. Almost immediately you can no longer see details of the pencil—just its general shape, color, and location. As you move the pencil farther to the sides, it will become less visible. When your arm is even with your shoulder you will barely be able to see it at all, even if you wiggle it back and forth.

The unfocused areas to the sides of your central vision form your **peripheral vision,** which extends to 180 degrees. Although objects to the sides of your peripheral vision are not focused, you will still be able to detect them. The faster you drive, the harder it is to see to the sides. At

driving tips

If you wear glasses or contact lenses, keep a spare pair of glasses in your vehicle in case your regular glasses are lost or broken, or if you need to remove your contact lenses.

48 ◆ UNIT 1 The Driver and the Driving Task

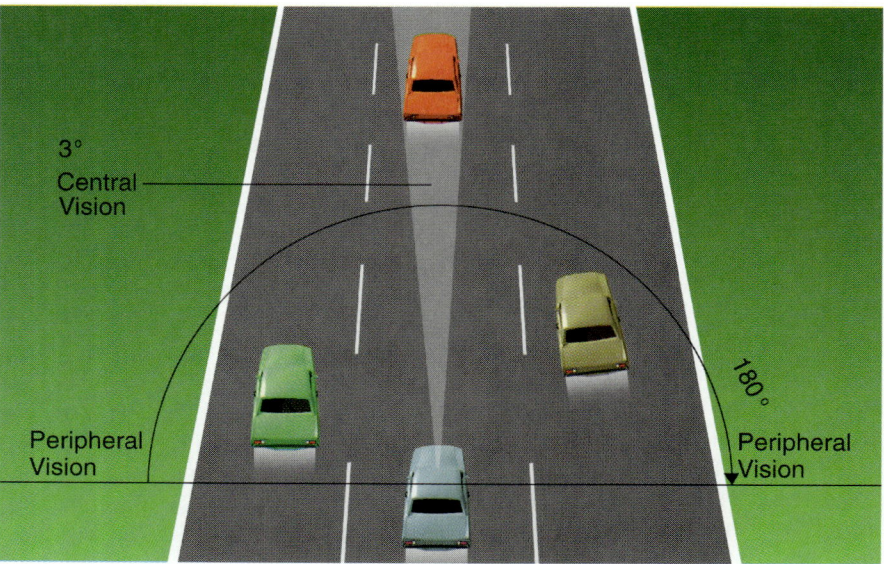

Figure 3–1 Your central vision is extremely narrow.

30 miles per hour (50 km/h), peripheral vision may be reduced by 25%, whereas at 60 miles per hour (100 km/h) it can be reduced by 90%. Because your field of vision narrows while driving, it is essential to move your head *constantly* to see what is happening around you.

Depth Perception

Depth perception is the ability to judge the distance between two objects. We can perceive depth well because we have binocular vision that combines the two separate images from our two eyes into a three-dimensional image. When driving, you use depth perception to evaluate the distance between you and other objects. It is important to keep in mind that depth is more difficult to judge when you are moving than when you are stationary.

Depth perception is also used to judge your following distance behind other vehicles and to evaluate your stopping distance. Those with impaired depth perception should increase their following distance behind other vehicles and drive more slowly than those with normal depth perception.

Color Vision

Drivers use **color vision,** or the ability to see color, to help determine the meaning of various signs and traffic signals. The colors used most often in signs and signals are red, green, and yellow. **Color blindness** is an inability to differentiate between certain colors such as red and green, the most common form of color blindness. If you are color blind, you can still drive, but you must interpret signs and signals by memorizing their shape and meaning. For example, a color-blind person can identify a red traffic light by knowing that it is either at the top of a vertical signal or at the far left of a horizontal signal.

Age

Age is another physical condition that can have an impact on a person's driving abilities. Older drivers generally are better at detecting and avoiding potential hazards than younger drivers. However, they also tend to have decreased reaction time, weaker eyesight, and poorer hearing. To compensate for such limitations, older drivers may reduce their speed and try to avoid situations that require quicker reflexes, such as driving on heavily congested streets or in the fast lane of a freeway.

Coordination

Your coordination is determined by how quickly and efficiently your nervous and muscular systems work together. When driving you must constantly interpret what you see and make adjustments. Coordination affects how quickly and efficiently you synchronize your hands and feet for various tasks such as steering, manually shifting gears, and pressing the accelerator, brake pedal, and clutch with your foot.

Some people are naturally more coordinated than others, but even those with average coordination can develop good driving skills. If you are not very coordinated, you can compensate by staying more focused and attentive. Increase your space cushioning and following distance if necessary. With experience, your ability to smoothly work your vehicle's controls in response to hazardous situations will improve.

Hearing

Hearing greatly increases your awareness while driving. Your ears detect sounds of potential hazards outside the car such as horns, sirens, train whistles, nearby vehicles you cannot see, trucks, pedestrians, and bicycles. Hearing tells you when to shift gears on a manual transmission. Sounds may warn you of an impending engine breakdown or a flat tire on your own or another driver's vehicle. Screeching brakes can warn you that someone is about to rear-end you. The information you obtain by hearing can be crucial in some of the split-second decisions you have to make while driving. Drivers who are hearing impaired can still drive safely with the help of hearing aids.

Size and Height

Size and height may limit a driver's ability to comfortably or safely operate a motor vehicle. A tall or large person may feel too restricted in a subcompact. A short driver may be unable to see the road well from the seat of a very large sedan. When you get behind the wheel, take a minute to adjust your seat and mirrors so that you can easily reach the controls while maintaining good visibility of the road. You might need to obtain special equipment

Deaf drivers are among the safest on the road because they develop better visual skills to compensate for their loss of hearing.

Figure 3–2 Drivers with hearing problems must compensate for their impairment by scanning ahead and using their mirrors more often.

for your car to enable you to see clearly over the dashboard and out the windows, or to allow you to reach all the foot pedals without stretching.

Chronic Illness and Disability

Many drivers with chronic illnesses or permanent disabilities are able to drive because of medical and technological advances. Those with an illness such as diabetes, asthma, epilepsy, and heart disease can regulate their symptoms with medication. Drivers have a responsibility to carefully monitor their own medical condition and refrain from driving if they experience side effects, such as dizziness or fatigue, that adversely affect their ability to drive.

People who have permanent physical disabilities that limit their use of arms, legs, or hands can drive vehicles adapted with special equipment such as hand-operated foot pedals and steering wheel "spinners."

Fatigue

Not all physical limitations are the result of permanent conditions. Fatigue severely affects

CHAPTER 3 *The Condition of the Driver* ◆ **51**

Figure 3–3 Do not drive a vehicle if you cannot comfortably see out the windows or reach the pedals.

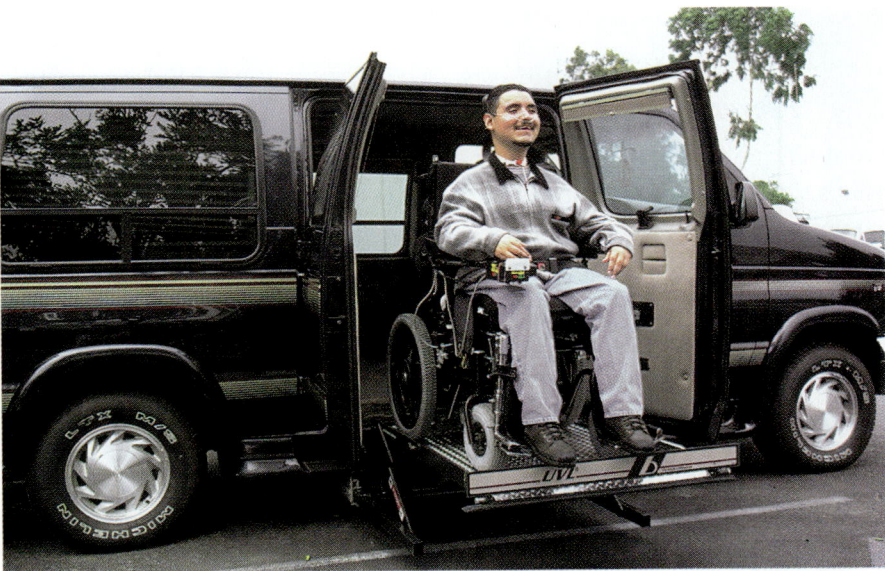

Figure 3–4 Specially modified vans allow wheelchair-bound people to drive.

factoid

The young man awarded "America's Safest Teen Driver" in 1990 later was killed in a collision when he fell asleep behind the wheel.

your reaction time and decision-making abilities. Driving while fatigued greatly increases your chances of having a collision. Fatigue is most likely to set in on long trips, but it may be caused by a number of factors, such as boredom, eyestrain from driving too long or from sun glare, poor ventilation, and eating or drinking too much. Because teenagers generally need more sleep than older drivers, they are especially at risk, accounting for more than half of all collisions caused by a driver falling asleep.

Fatigue can set in slowly. Some physical symptoms of fatigue are drowsiness, blurred or double vision, slowed reactions, lack of coordination, and problems judging distance and speed. Emotional indications include irritability and inattentiveness. Because most people are programmed to sleep when it is dark, it is not surprising that most fatigue-related collisions occur between 1 AM and 6 AM.

If you feel tired or sleepy, stop in a safe place and get out of your vehicle. Use the opportunity to splash cold water on your face, walk around, stretch, and clear your head. Eat something light. Beware of drinking too much coffee or soda containing caffeine and/or sugar. It may perk you up for a short while, but excessive caffeine or sugar intake can actually make you more tired than before when the effects of the stimulant wear off. If you have passengers who are licensed drivers, allow someone else to drive for a while so that you can rest.

If you pull over to get some sleep, park your vehicle in a crowded, well-lit area, turn off the engine, lock the doors, and crack the windows open. After taking a

DRIVING TIPS *Preventing Fatigue*

You can do a lot to prevent fatigue before even getting behind the wheel:
- Always get a good night's sleep so you are well rested before driving.
- Try not to drive during your normal sleeping hours.
- Do not drink alcohol, which will make you more tired.
- Avoid eating large, heavy meals before driving.
- If you feel tired after school or work, walk around and relax for a while before driving.

To prevent fatigue while you are driving:
- Make sure you have lots of fresh air.
- Talk to a passenger or listen to the radio to keep you alert.
- Do not set the heater or air conditioner too high because extreme temperatures can make you groggy.
- Wear sunglasses if there is too much glare.

nap, make sure you are fully awake before getting back on the road.

Illness

Avoid driving when you are sick. Cold, allergy, and flu symptoms can slow your reactions and make you uncomfortable, drowsy, and inattentive. Serious symptoms such as dizziness, sneezing, or constant coughing may even cause you to lose control of your vehicle. If it is absolutely necessary for you to drive, concentrate fully on your driving, keep your speed lower than normal, and stay clear of heavy traffic and high-speed roadways.

Injury

If you are injured, you may or may not be able to drive, depending on the location and extent of your injury. A broken hand, a severe sprain to the right foot, or an eye injury, for example, could all make it difficult, if not impossible, to drive. Before you drive, evaluate how much mobility or strength you may have lost, whether your senses are impaired, and how much pain and discomfort you are feeling because of the injury.

You must also consider what kind of vehicle you are driving. If you have a moderately sprained left ankle, you should have little trouble driving a car with an automatic transmission. However, with the same injury you might not be able to safely operate the clutch on a car with manual transmission. If an injury seriously affects your abilities, do not risk a collision. Ask someone else to drive you, take a bus, or call a taxi.

Carbon Monoxide Poisoning

Exhaust fumes from all gasoline engines contain **carbon monoxide,** a colorless, odorless, tasteless gas that under certain circumstances can seep undetected into the passenger compartment of a vehicle. If the levels become high enough, you can develop carbon monoxide poisoning and not know until it is too late. Carbon monoxide poisoning can have a number of causes:

- A damaged exhaust system
- Driving in an area with insufficient ventilation such as enclosed parking garages and tunnels
- Starting a vehicle in a garage with the garage door closed
- Driving a sport utility vehicle, van, hatchback, or station wagon in heavy traffic with only the back window open
- Smoking with the windows closed

Some recognizable early warning signs of carbon monoxide poisoning are headaches, nausea, drowsiness, confusion, and/or loss of strength. Ignoring these symptoms and neglecting to do something about them can ultimately lead to unconsciousness and even death.

factoid

At least 100,000 traffic collisions in the United States annually involve driver fatigue.

Figure 3–5 Do not try to drive if an injury prevents you from safely operating a motor vehicle.

Wild Wheels

Figure 3–6 What is the name of this car, which pushed the physical and mental limits of its American driving team in the 1908 Round-the-World Race, logging 12,427 miles (20,000 km) in 170 days?

Always try to keep your car windows open, at least slightly, when the engine is running. If your heater or air conditioner is on, keep your distance from vehicles in front of you while you are in heavy traffic to prevent their exhaust from entering your ventilation system. Remember to turn off the engine when you are motionless in an enclosed area. It is a good idea to have your exhaust system checked regularly for leaks, especially if at any time you feel any of the symptoms described above. When you suspect high levels of carbon monoxide, get out of any confined space as soon as possible and ventilate your car with fresh air.

3–2 THE MENTAL CONDITION OF THE DRIVER

Your emotional state is a very important factor in your ability to drive well. When you are angry, excited, or depressed, you cannot allow your emotions to distract you. Giving in to powerful emotions while driving diminishes your coordination, concentration, and reaction time, all of which are fundamental driving tools. If you are upset by a personal issue or something that happens on the road, remind yourself to focus on the driving task. Wait to deal with the problem affecting you once you have arrived safely at your

Figure 3–7 Strong emotions and driving do not mix.

destination. Do not abandon good judgment and take risks that you would normally avoid.

Anger

One of the most powerful emotions that can affect driving is anger. You may become angry at another driver who performs unexpected, illegal, or dangerous maneuvers. Your temper might flare at drivers who follow your vehicle too closely, drive too slowly ahead of you, fail to go forward as fast as you think they should at a green light, or "cut you off" by suddenly turning into your lane without signaling. If you are under stress, tired, or irritated in any way, threats to your safety and peace of mind may cause you to respond more angrily than you would otherwise. If you are in the wrong frame of mind, just getting stuck in traffic can cause you to blow your stack.

Dealing with Anger

Do not be tempted to take out your aggression on the road. If you are angry, cool down before you get in your car. Compose yourself by talking to friends or taking a walk and thinking things out. When you are driving, *expect* others to exhibit poor judgment. If another driver makes an error, swears at you, or just gets under your skin, remain calm and patient. By simply increasing your space cushioning and following distance, you will go a long way toward avoiding situations that can lead to anger. If you are tempted to act on your anger by driving recklessly yourself, take a deep breath and ask yourself, "Is it worth it?" The answer is no!

Road Rage

The phrase road rage is used to describe the increasingly common phenomenon of aggressive driving. Examples of road rage include deliberate tailgating, yelling at other drivers, using obscene gestures, purposely blocking other drivers' paths, and in extreme cases, assaulting others. Road rage is often set off by a minor event, such as an argument over a parking space, that acts as a "final straw" for a motorist already under stress. It is often aggravated by hot temperatures and overcrowding on the roadways. Although most "road ragers" are young males, usually with histories of emotional problems and drug or alcohol use, anyone can

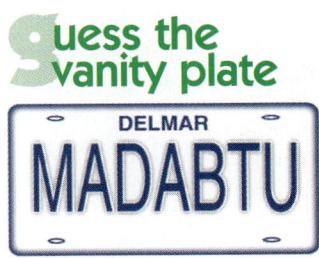

Figure 3–8

factoid

An average of 1,500 people are injured or killed annually in the United States because of aggressive driving.

Figure 3–9 Do not risk a confrontation with other drivers.

lose their cool under the right circumstances.

There are things you can do to avoid becoming a victim of road rage. Avoid the urge to honk your horn or flash your "brights" at other drivers out of anger or frustration. Do not tailgate, block the

JACK NICHOLSON

A two-time Oscar winner for Best Actor, Jack Nicholson has starred in a number of popular films over the course of his career. His uncanny ability to portray emotionally disturbed characters has made him an icon of American cinema, perhaps best symbolized by his role as the murderous hotel caretaker Jack Torrance in *The Shining*. On February 8, 1994, Nicholson gave another chilling performance, in real life this time, in the role of a driver suffering from road rage. Nicholson stepped out of his car at a red light in North Hollywood and, with a golf club, repeatedly struck the windshield of the car behind him, the driver of which he believed had cut him off. After the attack, Nicholson returned to his car and drove away. The victim of the attack was injured by flying glass from the windshield and feared for his life. Nicholson, who was easily identified by witnesses at the scene, was charged with misdemeanor vandalism and assault. Never drive if you are angry or upset in any way. As a defensive driver, you should expect other drivers to make mistakes or break road rules, and you should be as prepared as possible to react to these errors to protect your own safety. However, no matter what other drivers have done, you do not have the right to verbally or physically assault them or their property.

passing lane or merging lane, blare your music out of open windows, or switch lanes without first signaling. When parking, make sure that you do not take up more than one spot and that you do not hit the car parked next to you when opening your door. Do not use bumper stickers or vanity plates that may offend others. Always give other drivers the benefit of the doubt if a conflict develops.

If you find yourself the "target" of an extremely aggressive driver, do not allow the incident to escalate. Avoid making eye contact with angry drivers, gesturing, or participating in challenges of any kind. If you recognize that the other driver is getting violent, either seek help or protect yourself as best you can. Drive to a police station or an area crowded with people. If you cannot drive your car to safety, make sure your doors are locked and your windows are rolled up.

Stress

If you have a busy schedule with a lot to accomplish in a limited amount of time, you are acquainted with stress. Extreme stress from being too busy, not getting enough sleep, or personal problems at home, work, or school can seriously impact your physical and emotional state. Stress can cause adrenaline "rushes," muscle tension, increased breathing and heart rates, sweaty palms, headaches, and extreme fatigue. All of these can make it difficult for you to perform the physical action of driving. You should avoid getting behind the wheel when you are under enormous stress. If you

Figure 3–10 To reduce stress when traveling to an unfamiliar place, allow extra time to reach your destination.

have to drive, keep your thoughts on your driving and not what is causing you stress. Increase your space cushioning and decrease your speed.

Because certain driving experiences themselves can provoke stressful reactions, try to plan ahead to reduce their stress factor. For example, allow extra time during rush hour and bad weather, map out your route before traveling to unfamiliar areas, and call ahead if you are going to be late for an appointment so you will not be rushing to get there.

Anxiety, Excitement, and Depression

A major source of anxiety is driving in unfamiliar surroundings. If you get lost you may panic or get confused, causing you to overlook critical information or potentially threatening situations. Once panic sets in, it is easy to become indecisive or miss the signals of other vehicles. If you are too preoccupied looking for a street address, for example, you might rear-end the car in front of you because you did not notice its brake lights. Being overly excited can also affect driving by decreasing your attentiveness and increasing your willingness to take risks. Depression, or feelings of hopelessness and despair, can likewise diminish your concentration and coordination.

If you are coping with anxiety, excitement, or depression, try to have another licensed driver drive for you. If you must drive, make an extra effort to concentrate on the road, not on your emotions. Stay away from heavy traffic. If you are overwhelmed, pull over and try to relax until you feel able to concentrate again.

Distractions

There is no substitute for a mentally alert driver who is aware of the surroundings and can react quickly and clearly to every driving situation encountered. Anything that takes your attention away from driving can cause you to make poor decisions and impair your reactions at a crucial moment. Trying to simultaneously drive and perform unrelated tasks reduces your concentration and may cause you to take your eyes off the road. Activities you should avoid while driving include eating, retrieving and replacing drinks in cup holders, searching for objects or pocket change for tolls, reading maps or directions, looking at yourself in the rearview mirror, putting on makeup, shaving, or brushing your teeth.

If you have a passenger, ask him or her to help you with any

Figure 3–11 When you are behind the wheel, avoid distracting tasks unrelated to driving.

REALITY CHECK "Flirting" with Danger

As a passenger, you can indulge yourself by smiling at every good-looking person you see out of the car windows. However, as a driver, every second you spend making eyes at your new love is a second when you might miss some crucial information about what is happening around you. Keep your eyes on the road, and do not flirt with danger.

---

necessary tasks such as pulling out money for tolls or reading maps. Allow time to eat your meals and get ready before you get in your car, or plan to do other tasks once you arrive at your destination. If something draws your attention either inside or outside the vehicle, quickly assess whether it is a hazard and return your attention to the road.

Car Stereos

Music or talk radio can help you pass the time and stay relaxed and alert on long trips or in slow traffic. In addition to being good entertainment, your car stereo is a great means to get information about local traffic conditions and weather reports. However, under certain conditions, your stereo can become a distraction. If you play music too loudly, you may not hear approaching emergency vehicles or warning sounds such as horns or screeching brakes. You may also irritate your neighboring drivers to the point of anger. Fumbling with tapes or CDs or looking down at the radio and not at the road can cause you to crash or veer out of your lane even at slow speeds.

You can listen to music while driving, but make sure it is not so loud that you cannot hear the sounds of your car or what is happening around you. If your radio has preset buttons, set them on your favorite stations before you start driving. Change CDs or tapes before you drive or only when safely stopped, and do not try to

REALITY CHECK Deadly Decibels

Many drivers insist on playing their stereos at maximum volume. Their cars are like speakers on wheels, blaring music so loudly that it can be heard several blocks away. These drivers can go right through an intersection without having any idea that an emergency vehicle is approaching along a cross street. When they are at a stop light, everyone else has to hear their music whether they like it or not. This kind of behavior is both dangerous and irritating to other drivers. It is also illegal to exceed specified noise limits in many communities.

pull them out of hard-to-reach places like the back seat! It is not worth the risk. Listen to what is already on the stereo, or turn it off and focus on your driving. Never wear headphones while driving.

Cellular Phones

Speaking on a cellular mobile phone, or "cell phone," while driving can increase the chances of a collision by as much as four times. Talking on a cellular phone is distracting in general because your attention is divided between two separate tasks. You might have to take your eyes off the road when dialing, for example, which can cause you to weave or erratically vary your speed. Handheld cellular phones are particularly dangerous because you will have only one hand on the steering wheel while calling and talking. It is always best to make calls only when stopped and parked. If you must talk while driving, it is safest to use a speaker phone system, which allows your hands to be free to drive.

Figure 3–12 You should use a car phone only when you are stopped or parked.

Passengers and Kids

Passengers can be a great support, especially on long trips, by helping with tasks such as tuning the radio, giving you directions, or keeping you alert. Passengers, however, can also be distracting, especially "back seat drivers" who try to influence your driving decisions. Even passengers who do

bumper sticker sightings

HANG UP AND DRIVE!

Figure 3–13 Out-of-control children can be a major distraction to driving.

not address you directly can distract you by talking too loudly to one another or by "roughhousing" with one another, hanging out of windows, or blocking your rear visibility. Small children can also get noisy and excitable, making it hard to concentrate on the road.

When you are behind the wheel, *you* must take full responsibility for driving safely. Even though it is natural and polite to look at people when talking to them, you should avoid taking your eyes off the road when conversing with passengers. Calmly ignore back seat drivers. If passengers are getting out of control, pull over to the side of the road and ask them to settle down or get out. To prevent arguments with children while driving, establish a firm set of "car rules" and try to consistently follow them.

Smoking

Smoking is one of the most distracting things you can do while driving. Because you have less use of your hands, smoking decreases your ability to coordinate controls. Lighting cigarettes or cigars causes you to look away from the road while handling an open flame or a hot cigarette lighter. There is also the constant danger of dropping a lit cigarette or match in your lap or somewhere else in the car. Cigarettes thrown out of a front window can make their way into the back of the car through an open rear window. Even worse, a discarded cigarette can cause a fire if it ignites brush, leaves, dry wood, or other flammable material on the roadside. Smoking in your vehicle also causes a film to build up on your windows that can reduce visibility.

Pets

Pets should always be secured in a carrier that is well made and the right size, either inside the car or in the open bed of a truck. Specially designed harnesses are available to secure medium- and large-sized dogs. Do not put an animal in the back of a pickup with an unventilated camper shell—you may be putting the animal at risk for carbon monoxide poisoning. If you do not have a carrier or harness, and it is urgent to transport your pet, have a passenger hold and watch the animal. Never let a pet sit on your lap or wander around your vehicle where it may interfere with your driving in some other way.

Rubbernecking

Do not slow down to observe a crash site. So-called **rubbernecking** is a dangerous habit that contributes to traffic congestion on both sides of a roadway and often leads to further collisions. If you pass the scene of a crash, keep your attention focused on driving.

Other Distractions

Driving in unfamiliar areas can be distracting because you are taking in a great deal of information for the first time, but you can be equally distracted in your own

Figure 3–14 Rubbernecking causes traffic congestion and can lead to other collisions.

neighborhood or town. For instance, if you run into a friend or neighbor on your way to school or when running an errand, you may become concerned with getting their attention and not be attentive to your driving. If you want to get a pedestrian's attention, pull over and park before you call out to him or her. If you are talking to friends in another vehicle, make sure that you both move into a parking spot or to the side of the road and that you are not blocking traffic.

Observing scenery rather than paying attention to the road can be extremely dangerous. Stay focused on driving and save any sightseeing for a rest stop or scenic overlook. If you can, switch drivers periodically so that one person can focus on driving while the others can enjoy the sights.

WHO'S AT FAULT?

1. Driver 1 was driving late at night in the left-hand lane of a straight, flat highway. He fell asleep and began to drift across the right-hand lane in front of Driver 2 and off the side of the road. Driver 2 thought that Driver 1 was pulling off the roadway. She slowed down but did not change lanes as she went past Driver 1. At that moment, Driver 1 woke up and veered suddenly to the left, back onto the roadway, colliding with Driver 2. **Who's at fault?**

2 Driver 1 thought that Driver 2 had cut him off. He became angry and decided to give Driver 2 a taste of his own medicine. He accelerated alongside Driver 2 and cut in front of him. Driver 2 slammed on his brakes but could not stop in time and crashed into Driver 1. **Who's at fault?**

YOUR TURN

3–1 The Physical Condition of the Driver

1. What permanent physical conditions can limit your ability to drive safely?
2. What are the four main components of vision?
3. What temporary physical conditions can limit your ability to drive safely?
4. How can you prevent carbon monoxide poisoning?

3–2 The Mental Condition of the Driver

5. How does your mental condition impact your ability to drive safely?
6. How can you avoid becoming a victim of road rage?
7. What activities should you avoid while driving?

SELF-TEST

Multiple Choice

1. Road rage is a common form of what type of driving?
   a. beneficial
   b. aggressive
   c. safe
   d. defensive
2. Your central vision is typically how wide?
   a. 45 degrees
   b. 90 degrees
   c. 10 degrees
   d. 3 degrees
3. Visual acuity is the ability to:
   a. drive only in daylight.
   b. distinguish colors of road signs.
   c. see objects far away.
   d. drive only at night.
4. If you are driving, the best way to deal with a back seat driver is to:
   a. pull off the road and let the person drive.
   b. ignore the person.
   c. ask the person to sit in the passenger's seat.
   d. stop the car and wait for the person to finish his or her thoughts.
5. Most fatigue-related collisions occur:
   a. between 8 PM and 1 AM.
   b. between 1 AM and 6 AM.
   c. during rush hour.
   d. during daylight savings time.

Sentence Completion

1. The ability to judge distance between two objects is called _____ .
2. Boredom, eye strain, and poor ventilation all contribute to _____ .
3. Your _____ is determined by the quickness and efficiency that your nervous and muscular systems work together.
4. If you are unable to distinguish between red and green, you will have to depend on the _____ and meaning of road signs to drive safely.
5. Smoking with the windows closed can cause _____ .

Matching

Match the concepts in Column A with examples of the concepts in Column B.

Column A
1. __ Road rage
2. __ Carbon monoxide
3. __ Fatigue
4. __ Preventing fatigue
5. __ Central vision
6. __ Distractions
7. __ Peripheral vision
8. __ Depth perception

Column B
a. Distance between two objects
b. Eating, shaving, reading
c. Extends 180 degrees
d. Odorless, colorless, tasteless
e. Talk to a passenger
f. Area directly ahead
g. Temporary impairment
h. Deliberate tailgating

Short Answer

1. What are some of the early signs of carbon monoxide poisoning?
2. What should you do if another driver becomes aggressive toward you?
3. How can car stereos be distracting to driving?
4. What can you do to prevent fatigue while driving?
5. How does driving affect peripheral vision and depth perception?

Critical Thinking

1. You are driving down a busy street when the door of a parked vehicle immediately ahead of you flies open. You swerve quickly to the left side of your lane to avoid hitting the door, but a driver just behind you in the left lane is angered that you nearly cut him off. He pulls up next to you honking his horn and making gestures at you, then cuts in front of you and slows down. What should you do?
2. At the last STOP sign, an elderly woman driving a white sedan just in front of you turned left. You made a left turn into the same side street behind her. It is dark, and you are in an unfamiliar neighborhood looking for a friend's house that you are pretty sure is coming up on your left. The woman is driving very slowly and you are late for dinner. You want to pass her, but you notice that her left-turn signal is still on. What should you do?

Projects

1. Your sight is your most valuable sense when driving, but other senses such as your hearing help you know what is happening around your vehicle. The next time that you are riding as a passenger in a vehicle, close your eyes and listen for sounds that clue you in to what other vehicles are doing on the road. Do you hear horns, an accelerating engine, emergency sirens, or the squeal of a truck's brakes as it comes to a stop beside you?

2. As a passenger, observe the drivers of other vehicles and take note of how many people are doing other distracting things while driving such as eating, drinking, looking at their passengers rather than the roadway, smoking a cigarette, and talking on the phone. Share your results with the rest of your class.

Driving Fundamentals
unit 2

Chapter 4   Preparing to Drive
Chapter 5   Vehicle Operation Basics
Chapter 6   The Speed Factor: Negotiating Curves and Braking

4 chapter

Preparing to Drive

The first step in learning how to drive defensively is to understand the various instruments, operating devices, and controls of your car. When you are behind the wheel, you have the ability to change everything from your speed to the angle of your mirrors to the temperature inside the passenger compartment. Over time, the use of this equipment will become second nature, but until then you must gradually familiarize yourself with each component.

CHAPTER OBJECTIVES

Upon completion of this chapter, you should be able to:

4–1 Vehicle Instrumentation
1. Identify the gauges and warning lights on a typical instrument panel.
2. Understand what might cause the temperature and oil-pressure warning lights to come on.
3. Understand what an alternator does.

4–2 Operating Devices
4. Describe the different types of exterior lights found on most motor vehicles.
5. Identify the three settings on a typical headlight switch.
6. Know how to properly adjust mirrors.
7. Know what blind spots are and why they exist.

4–3 Vehicle Controls
8. Understand why seat position is important to driving.
9. Identify the controls found on both automatic- and manual-transmission vehicles.

KEY TERMS

gauge	running lights	accelerator
warning lights	parking lights	power brakes
speedometer	turn-signal lights	brake lights
tachometer	hazard lights	parking brake
odometer	rearview mirror	gearshift
alternator	sideview mirrors	clutch
headlights	blind spots	cruise control
taillights	steering wheel	
back-up lights	ignition	

4–1 VEHICLE INSTRUMENTATION

Various gauges and warning lights are arranged on the dashboard within easy view of the driver's position behind the wheel to provide you with accurate and important information about the status of your vehicle. The exact layout varies by vehicle type and model, but these instruments serve the same functions in every car.

Figure 4–1 The instrument panel contains gauges and warning lights.

A **gauge** has a scale with an indicator needle or numerical marker that keeps track of a changing condition like fuel level or speed. **Warning lights,** which are usually red or yellow to attract your attention, indicate a more serious problem or safety concern. At night, the dashboard instruments are usually backlit to help you see them better. Some vehicles have a feature called a rheostat that allows you to regulate the intensity of backlighting in the instrument panel to the level most visible and comfortable for you.

The **speedometer** is a gauge that indicates how fast your vehicle is traveling, usually in both miles and kilometers per hour. A **tachometer** is a gauge that measures your engine in revolutions per minute (rpm). The higher the rpm, the faster your engine is "turning" and the harder it is working. Tachometers are useful to help you determine when to shift gears if you have a manual transmission. The red zone on the tachometer specifies an unsafe range of rpms for your vehicle's engine. If your gauge is in the red area, you should either slow down or shift to a higher gear.

The **odometer** is a meter that displays the total number of miles a vehicle has been driven since it was manufactured. Because this gauge is an important measure of the wear and tear on a vehicle, it is strictly illegal to alter it. The odometer is useful for calculating fuel consumption and trip mileage, and for keeping track of your vehicle's maintenance schedule. Most cars also have a separate "trip odometer" that can be set back to zero at any time with the press of a button.

The fuel gauge displays the amount of fuel in the fuel tank. The fuel gauge in some vehicles displays the fuel level whether the engine is on or off. In other cars, you must turn on the ignition to activate the gauge. Consult your owner's manual to find out how many gallons or liters your vehicle's tank holds. The fuel gauge usually has at least five levels of fuel indicated: empty, ¼ of a tank, ½ of a tank, ¾ of a tank, and full.

Running out of gas is a common mistake for new drivers. As a

factoid

The first car with computerized digital instrumentation appeared in 1976.

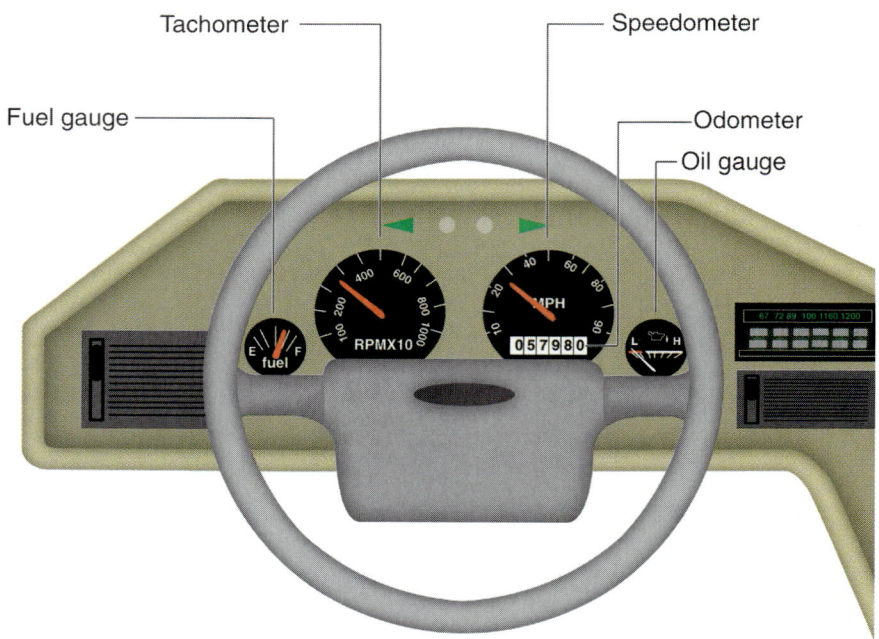

Figure 4–2 Gauges

rule of thumb, it is best to not let your tank get below one-quarter full. In cold weather, you should try to keep the tank at least half full. This will help prevent "fuel-line freeze" caused by moisture that condenses and freezes inside the tank and fuel line, forming ice particles that can block the fuel line.

The temperature warning light goes on if the engine temperature is too high or if coolant in the radiator is getting too hot. Other reasons this light may go on include a loss of coolant, a clogged radiator, a slipping or broken belt, a defective thermostat, the need for oil in the crankcase, or an unusual strain on the engine. On some vehicles, a gauge is used instead of a warning light. The gauge will read "hot" or the needle will be in the red zone if the

REALITY CHECK Know Your Nozzles

When you refuel, do not get distracted and put the wrong fuel in your car. If you put diesel fuel in an unleaded engine or unleaded fuel in a diesel engine, the engine will stop. Be especially careful not to put diesel fuel in an unleaded engine, which can cause serious damage.

CHAPTER 4 Preparing to Drive ◆ 73

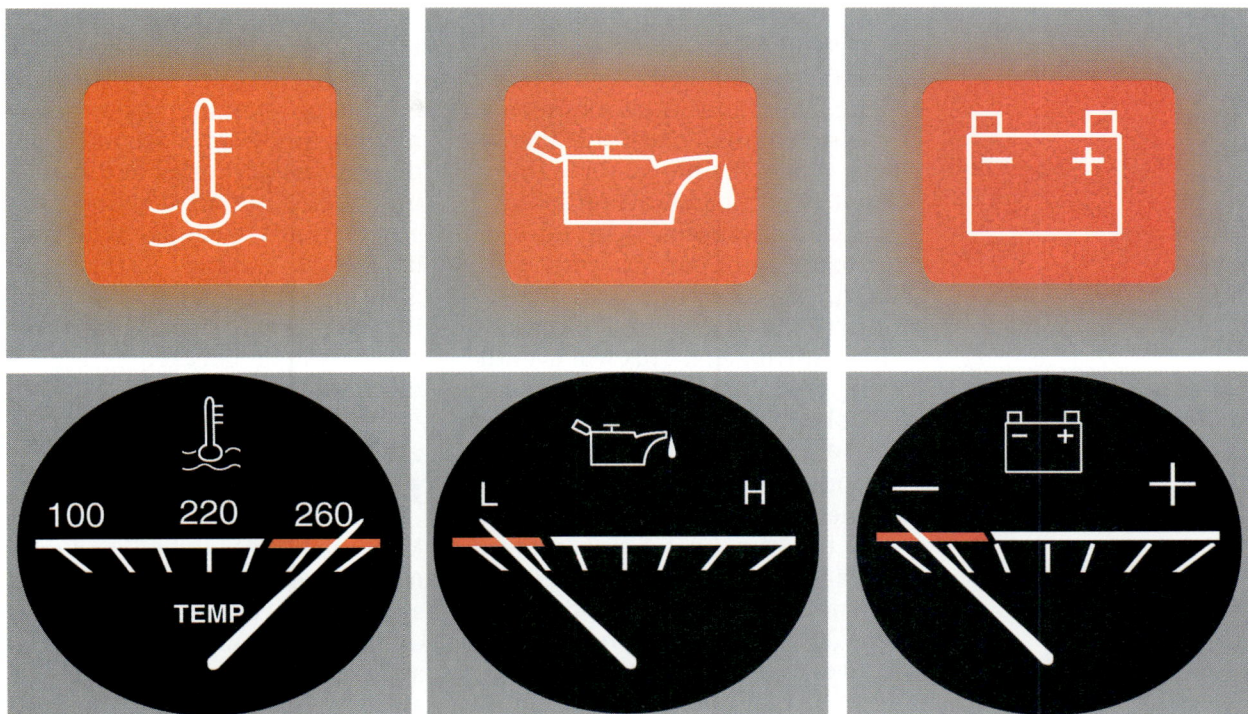

Figure 4–3 Engine temperature, oil pressure, and the car's electrical system may be monitored by warning lights and/or gauges, depending on your vehicle.

car is overheating. If you have a gauge indicator, learn where the needle normally points so that you can recognize when the engine is getting too hot.

The oil-pressure warning light or gauge warns you when the engine oil is not circulating at the right pressure. If the warning light goes on or the gauge reads "low," stop immediately. Because oil is needed to reduce friction between engine parts, losing oil pressure can seriously damage your engine. Oil pressure loss may be caused by a slow leak, a puncture of the oil pan, or natural burning as your car gets older. It is your responsibility to check your oil level regularly, using the dipstick located under the hood.

The oil-pressure light will go on briefly when the ignition is turned to the "on" position to show that it is working and will go off again once the car is started. If the light does not go on, or remains on, check for problems as soon as possible. The oil pressure should be zero when the engine is at rest, and it may remain low when the engine is idling.

In some vehicles, there is a red light labeled "alt" for "alternator" or "gen" for "generator" on the instrument panel. The **alternator** is a generator that produces electricity to power the car's electrical system. Everything that uses electricity in the car, including the ignition system, lights, and accessories such as the stereo, runs off of the alternator. When the alternator warning light is on, it indicates trouble in the car's electrical system. If the alternator is not putting out enough electricity to run the car, the engine must use stored electricity from the battery. This will eventually drain the battery, and once that happens the car can no longer run at all.

Other vehicles have a gauge rather than a warning light. Usually the gauge has a scale marked by a positive sign (+) at one end and a negative sign (−) at the other end. If the indicator on the gauge moves to the (−) side of the gauge, shut off all nonessential equipment such as the radio and go to the nearest service station to have your electrical system checked.

Many cars have a braking system warning light that serves two functions. First, it reminds you to release the parking brake before moving the vehicle. Second, it indicates that part or all of the braking system is not working properly or that your brake fluid may be low or leaking. If this light goes on while your foot is pressing the brake pedal, have the brake system checked out by a mechanic right away.

4–2 OPERATING DEVICES

Motor vehicles are equipped with many devices that must be operated by the driver. Some are required for the safety of the driver and passengers, whereas others are designed to increase comfort and convenience. Most of these devices are within easy reach of the driver's seat: on the steering wheel column, on the dashboard, on the console or "hump" between the front seats, on the driver's door, and near the floor to the driver's left.

Lights

Making yourself visible to others is an essential part of driving defensively. All vehicles are required to have two **headlights** giving off white light, one on each side of the front of the vehicle. Headlights must be lit at designated times or when your visibility is limited such as in rain or snow. Law enforcement takes driving without headlights very seriously,

Figure 4–4 If you see this light and your parking brake is not on, have your braking system checked immediately.

CHAPTER 4 Preparing to Drive ◆ 75

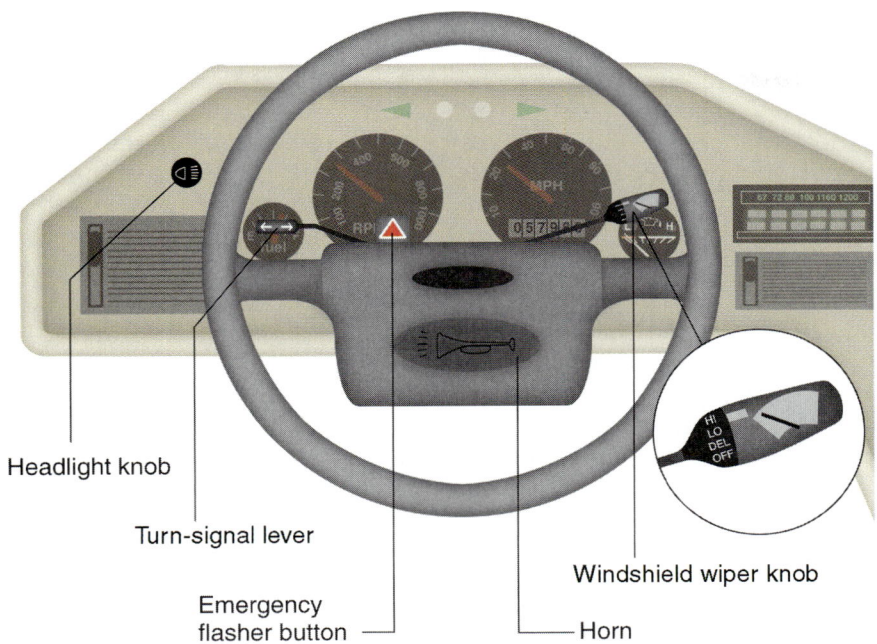

- Headlight knob
- Turn-signal lever
- Emergency flasher button
- Horn
- Windshield wiper knob

Figure 4–5 Operating devices

and you will be cited if your headlights are not functioning properly. All vehicles are also required to have two **taillights** giving off red light. Taillights come on whenever you turn on your headlights.

Vehicles may have no more than two white or amber **back-up lights.** These lights are activated when the transmission is set to REVERSE, indicating your intention to back up.

Daytime **running lights** have been installed on many vehicles manufactured after 1989. These lights are on the front of a vehicle and automatically illuminate upon ignition. They can be a dimmer setting of your headlights or a separate set of lights located next to your headlights. **Parking lights** are white or amber-colored on the front of the vehicle, and red on the rear of the vehicle. Parking lights must be used only for parking purposes. *Never use your parking lights as an alternative to headlights and taillights while driving your vehicle.*

Figure 4–6 You can be cited for driving with a burned-out headlight.

factoid

The first electric "winker" turn indicators were offered by Buick in 1939.

The headlight switch is usually a knob at the end of the turn-signal lever. It may also be located on the dashboard instrument panel to the left of the driver. When you turn your headlights on, your instrument panel and license plate should light up as well.

No matter where it is located on your vehicle, the headlight switch usually has three settings. The first turns on only the parking and side-marker lights, the second turns on your low-beam headlights, and the third turns on your high beams. If your headlight switch is on the steering column, you may also be able to activate your high beams by pulling the lever toward you. Pull it again to click the high beams off. To "flash" your high beams at another driver as a warning, pull the lever back slightly and release it. In some older cars, the high-beam switch may be a foot-activated button on the floor to the far left of the pedals.

When your high beams are on, a blue indicator light appears on the instrument panel of most vehicles. High beams generally allow you to see twice as far as low beams, but with less clarity immediately in front of your vehicle. Their use should be restricted to very dark, uncrowded, or unfamiliar roadways. Always "dim" your high beams back to the normal low-beam setting when you see an oncoming vehicle or are following another vehicle in traffic to prevent blinding the other driver.

Turn Signals

Your vehicle must have **turn-signal lights** on all four corners of the vehicle. They normally cast off white or amber light in front, and red or amber light in the rear. Turn signals usually appear as two green arrow-shaped lights on the instrument panel that flash to indicate the direction of your turn. These lights are synchronized with the signal lights on the exterior of your vehicle.

The turn-signal lever is almost always located on the left side of the steering column. You move the lever up to signal a right turn

---

DRIVING TIPS — "Reading" by the Light of Your Turn Signals

Whenever it is dark and you are searching for an address on a poorly lit residential street, take advantage of address numbers painted in white or fluorescent paint on the side of the curb. When you approach a residence to find out its number, pull out of traffic safely and approach the number on the curb. Park slightly behind it, and keep your turn signal on. The signal will cast a glow against the curb and illuminate the number so that you can read it.

CHAPTER 4 Preparing to Drive ◆ 77

Figure 4–7 Turn signals are located at the front and rear of the vehicle.

Hazard lights, also called emergency flashers, are used to indicate that you are experiencing an emergency situation or to warn other drivers that your vehicle is not functioning properly. The emergency-flasher switch or button activates all four exterior turn-signal lights at once and may be located on the steering column, on the dashboard, or even on the console. On most vehicles, the emergency flashers are indicated on the dashboard by both turn-signal lights flashing at the same time. Some cars have a hazard-light indicator independent of the dashboard turn-signal indicators.

Windshield Wipers

You must have windshield wipers to remove rain, snow, and other moisture from the windshield to see clearly while driving. Windshield wipers are activated by a lever on the right side of the steering column or a switch located either on the dashboard or on the turn-signal lever. They can usually be set at one of several variable speeds, according to the severity of the weather. Some cars have rear-window wipers that may be activated by another

and down to signal a left turn. When you push the lever up or down far enough, you will feel it click into place. Both the exterior and dashboard turn-signal lights are designed to stop automatically after a turn is made. However, sometimes if you are changing lanes, merging, or making a very wide turn, the sensor may not detect the "turn" and you must cancel the signal manually.

DRIVING TIPS Safe Signaling

For lane changes, lightly hold the turn-signal lever up or down with one finger until the lane change is complete, then release it back to the standard position. This will allow you to keep both hands on the wheel.

REALITY CHECK No Rain in the Forecast

On a clear, sunny day, just when you least expect it, a blinding spray of water smacks your windshield. This "rain" is not from the sky but rather from the vehicle ahead of you whose driver decided to wash the windshield while the car was moving. If this happens, do not panic. Slow down and clear your view with your windshield wipers. The next time *you* feel the urge to squirt your windshield while you are driving, think of the person behind you and wait until you are safely stopped at a light or a STOP sign.

setting on the windshield-wiper lever or by a separate switch on the dashboard instrument panel. The windshield-wiper switch or lever can also be used to dispense washer fluid for cleaning the windshield when necessary.

Horn

A working horn is required by law and can be used to warn others of your presence or an impending collision. The horn is usually located on the steering wheel, but on some vehicles it is found on the turn-signal lever. This allows you to honk without taking your hands completely off the wheel. Make sure that you know where the horn is located in your vehicle and how much pressure it takes to activate it.

Mirrors

To drive defensively, you must be aware of what is happening around you at all times. By revealing what you cannot see out of your front windows, your mirrors help you respond to traffic events quickly and safely. Most vehicles have two types of mirrors. The interior **rearview mirror** is a wide, rectangular mirror either suspended from the roof or attached to the windshield that allows you to see directly behind you. Exterior **sideview mirrors** mounted on the doors allow you to see along the sides of your vehicle and neighboring lanes of traffic.

Every time you get into your vehicle, you should make sure that all of your mirrors are properly adjusted. If someone else used your car, he or she may have shifted the mirror settings. You or a passenger may have knocked the rearview mirror out of alignment getting out. A person walking past your car may have knocked a sideview mirror out of place.

To adjust your rearview mirror, sit as you would to drive. Without moving your head, grasp the mirror by the frame and adjust its position so that you can see out of the rear window with the right edge of the mirror aligned with the right edge of the rear window. You should use only your eyes to properly align the rearview mirror. Adjust the sideview mirrors so

Wild Wheels

Figure 4–8 What is the name of this American vehicle, built in 1911, that was the first car to have a rearview mirror?

CHAPTER 4 Preparing to Drive ◆ **79**

Figure 4–9 Remember to adjust your sideview mirrors before driving.

that you can barely see the sides of your car from the driving position. The right sideview mirror often takes a little more effort because you must lean over to adjust it (unless you have electric controls). The important thing to remember is that vehicles passing you should be in your sideview mirrors before they leave your rearview mirrors and in your peripheral vision before they leave your sideview mirrors.

Most rearview mirrors have a day/night lever on the back that allows you to switch the angle of the mirror to reduce glare when driving at night or if the sun is directly behind you. Always make sure your mirror is set for the time of day you are driving to ensure maximum visibility.

Blind Spots

The areas *not* reflected in your mirrors are called **blind spots.** The size of a blind spot varies with the design of the vehicle and the physical characteristics of the driver. A driver who is short and drives a huge sedan, for example, will have larger blind spots than a tall driver in a compact. The two main blind spots on a car extend from just behind the driver's

Figure 4–10 Your rearview mirror can be adjusted for both daytime and nighttime driving.

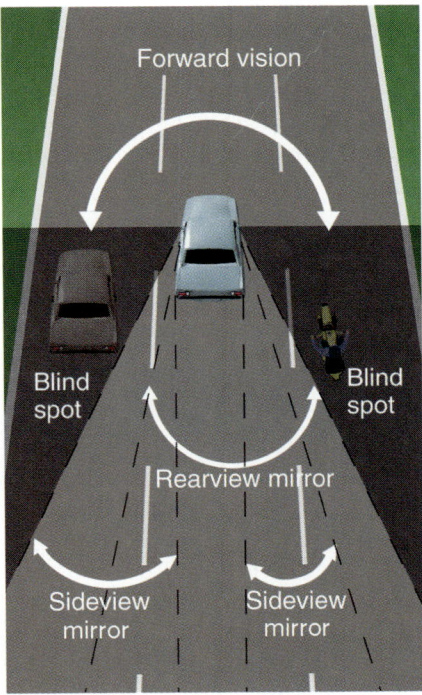

Figure 4–11 Your mirrors cannot reveal what is in your blind spots.

auto accessories

Although it will not eliminate your blind spots, replacing your rearview mirror with an extra-wide "monster mirror" will give you a much better view of what is behind you.

normal field of forward vision to the right and left sides of the vehicle. Proper mirror adjustment will help reduce the size of blind spots, but it will not completely eliminate them. *Never trust only your mirrors. Always turn your head and look before you change your position!*

As a defensive driver, you have the responsibility not only to drive outside of other vehicles' blind spots but also not to allow other drivers to stray into *your* blind spots. If you pay attention to the traffic patterns to your sides, you will know when a car approaching from the rear or slowing down in front of you is moving into your blind spot. When this happens, use gentle increases and decreases of speed by momentarily taking your foot off the gas pedal or by lightly applying some pressure to the brakes to position yourself so that these cars are no longer in your blind spots.

Other Operating Devices

Seat adjustment controls allow you to move your seat to the best and most comfortable driving position. For vehicles with electrically adjustable seats, a series of switches usually located on the left side of the driver's seat allows you to move your seat up or down, and forward or backward, as needed. For most cars with manually adjustable seats, you must slide and hold a lever under the front edge of the seat as you use your weight to push the seat forward and backward. Always make sure that your seat is securely locked into place before you start driving. On the door side of the front seats near the floor are levers that control the angle of the seats.

Door locks protect your car from theft or unauthorized entry, and they prevent doors from opening in a collision. To prevent drivers from accidentally locking themselves out of their cars, the driver's door on many vehicles can typically be locked only from the outside. Most door locks can be unlocked only with a key, but some vehicles have electronic locks. Door locks are operated from the inside of the vehicle either manually or electronically.

Windows are controlled from the inside of the vehicle either manually or electronically, depending on the vehicle. Manual window controls are cranks that you turn counterclockwise to "roll down," or open, and clockwise to "roll up," or close. On cars with electric, or "power," windows, the controls for each door are located on the door handles.

Sun visors are hinged panels located at the top of the windshield over the head of the driver and front-seat passenger. When the sun is directly in your eyes, you can lower the sun visor to help cut glare. Most sun visors can also be rotated to the side to shield you from sun glare entering through the side windows. Position the visor in such a way that it effectively blocks the sun without interfering with your view of the roadway.

The heater, air conditioner, and defroster controls are usually located in the center of the dashboard to be within easy reach of both the driver and front-seat passenger. The "defrost" setting is used in cold or rainy weather to remove condensation from the inside of the windshield, which can partially or completely block your view outside. Some vehicles also have a "defrost" setting or switch for the rear window.

Your vehicle also includes devices that provide access to the engine, trunk, and fuel tank. Check your owner's manual to see where they are located.

4–3 VEHICLE CONTROLS

The five basic controls found on all cars are the steering wheel, ignition, accelerator, brakes, and gearshift. Vehicles with manual transmissions also have a clutch. These controls are simple to use with experience, but it takes practice to learn to smoothly coordinate their movements. When driving a car for the first time, always take the time to get a "feel" for the controls. Each vehicle is different, and the controls of one car may be more or less sensitive than the same controls on another car.

When driving, sit in a comfortable, erect position squarely behind the steering wheel with your back firmly against the seat. Make sure that you are sitting high enough to see over the steering wheel and the hood. If you cannot clearly see the road, even at your seat's highest setting, get a firm wedge-shaped seat cushion to raise you higher. If necessary, adjust your head restraint to reach the middle of the back of your head.

If your seat is positioned properly, you should be able to easily reach all of the vehicle's controls. Your right heel should rest on the floor at the base of the accelerator. You should be able to pivot your right foot from the accelerator to the brake pedal without lifting your heel from the floor. Your left foot should rest on the floor to the left of the brake pedal if you have an automatic transmission, or at the base of the clutch if you have a manual transmission. Your legs should be slightly bent, not stretched out as far as they will go. If your legs are straight, you will not be able to push the brake pedal down all the way if necessary. If you cannot reach the pedals comfortably after your seat is adjusted to the right height, you may need to buy extensions for the pedals.

Steering Wheel

The **steering wheel** controls the front wheels of the vehicle, and turning the wheel determines the direction the vehicle will take. Some cars have an adjustable steering wheel, or "tilt-wheel," for better comfort and control. Most vehicles today have a power steering system that mechanically assists the driver so that turning the wheel requires a minimum of

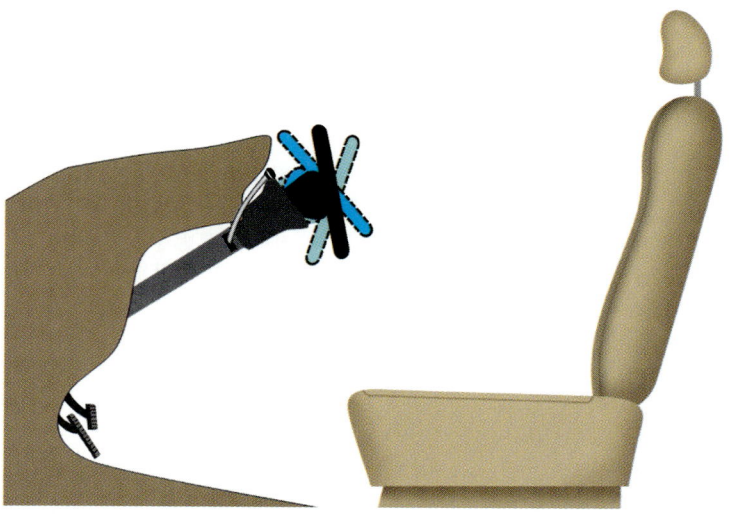

Figure 4–12 The angle of "tilt-wheel" steering wheels can be adjusted.

physical effort. If your vehicle has power steering and your steering wheel is extremely hard to turn, you should have the steering system checked by a mechanic.

Ignition

The ignition, where you insert your key to start the engine, is usually located on the right side of the steering column. The ignition has at least three positions. When the transmission is in "park" and the engine is off, the ignition is in the "lock," or "off," position. Insert your key and turn it clockwise one notch to the "on" position. The dashboard instruments should come on to show that they are working properly. Turn the key one more notch to the "start" position to start the engine. To turn the engine off, turn the key back to the original "lock"/"off" position. To remove the key in some vehicles, you must push the ignition in before turning it all the way off, or press a button located right next to the ignition. These safety catches prevent you from removing the key accidentally while driving.

You can lock your steering wheel when the ignition is off and the key is out by turning the wheel to one side until you feel it lock. To unlock the wheel, you usually insert the key and turn it to the "on" position and then turn the wheel to release it. Locking your steering wheel is a deterrent to theft and will keep your tires pointed in the right direction when parked on a hill.

Some vehicles have an additional "accessory" position on the ignition that activates the electrical system, allowing you to operate the radio and other electrical equipment without starting the engine. This position is usually one notch away from "off" in the clockwise direction. Be careful not to use the accessory setting for too

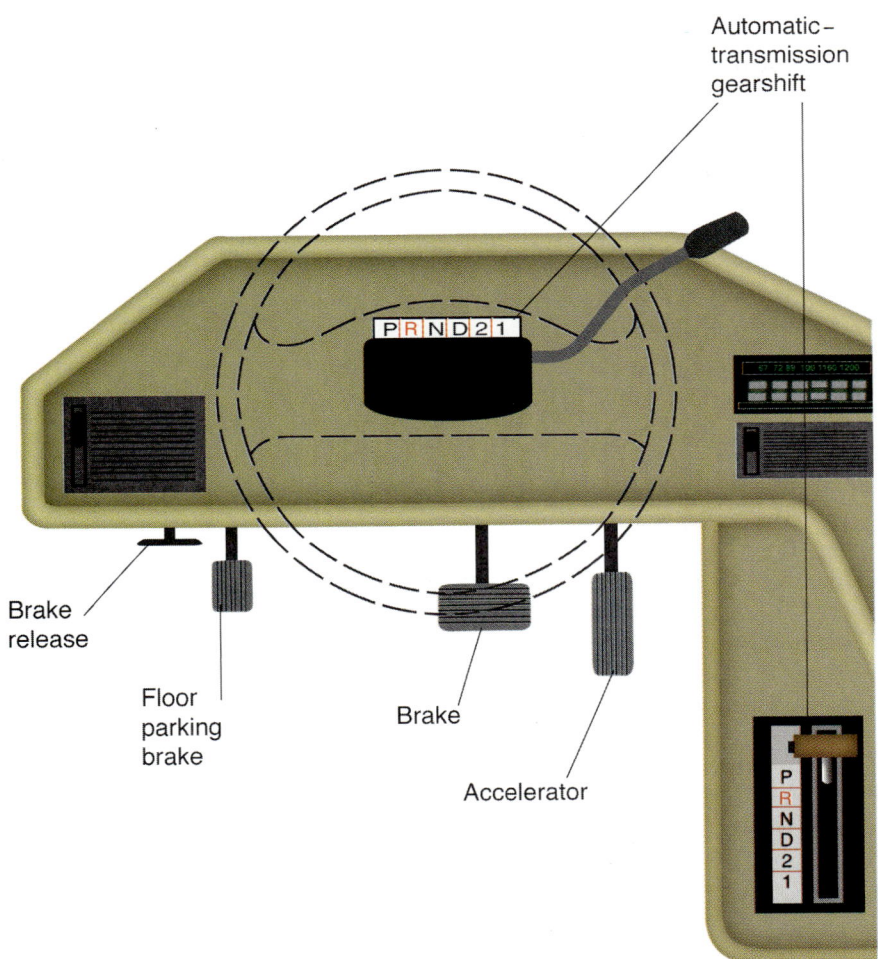

Figure 4–13 Typical vehicle with automatic transmission

long. Depending on which accessories you are running and the condition of your battery, you might drain the battery.

Accelerator

The **accelerator,** also called the gas pedal, is the far-right foot pedal. The accelerator controls the amount of fuel your car's carburetor or fuel injectors feed to the engine. You operate the accelerator with your right foot. Resting your heel on the floor, gradually depress the gas pedal to increase speed and let up on the gas pedal to let the car slow down on its own. Because vehicles accelerate differently, you must develop a sensitivity to your own accelerator.

actoid

The floor-mounted accelerator was developed in 1899.

Brakes

The brake pedal, which allows you to slow or stop your car, is the pedal located just to the left of the accelerator. Like the accelerator, you operate the brake pedal with your right foot. Keeping your heel on the floor, exert pressure on the pedal with the ball of your foot to activate your car's braking system.

Your vehicle may have standard brakes or power-assisted brakes. If you have standard brakes, the pedal will move a couple of inches (5 cm) until it meets resistance. **Power brakes** require less foot pressure to operate than standard brakes, but they do *not* shorten the distance needed to stop the car. When driving a car with power brakes for the first time, you should depress the pedal gradually and gently to get a feel for how much pressure is needed to bring the vehicle to a smooth stop. If you apply pressure too suddenly, you may stop so abruptly as to cause a driver behind you to rear-end your vehicle. If the power-brake system fails or your engine stalls, you will still be able to stop the car but you will have to put heavy pressure on the brake pedal.

Your brakes are also a very important communication device. When you step on the brake pedal, you activate your **brake lights.** In some vehicles, the taillights also function as brake lights by getting brighter when the brakes are applied. If you see that traffic ahead of you is slowing or stopping, you can warn other drivers behind you to prepare to slow or stop by tapping lightly on your brakes a few times.

A feature required on all passenger cars manufactured since 1985, and on all light-duty trucks, vans, and sport utility vehicles manufactured since 1993, is the center high-mounted rear brake light. This "extra" brake light is highly effective in reducing rear-end collisions because drivers are less likely to confuse it with the taillights of the vehicle. If you have a vehicle made before 1985, you should consider having a center high-mounted brake light installed.

The **parking brake** keeps your vehicle in place when parked. It can also be used to stop your car in emergency situations in which the normal brakes fail. In most vehicles, the parking brake is a large handle located on the console. To set the brake, pull the handle up until it feels tight and locked into place. To release the brake, push the button at the tip of the handle with your thumb and lower it down as far as it will go.

In some cars and trucks, you must activate the parking brake using a foot pedal. Usually this pedal is smaller than the other foot pedals and is located on the far-left side. Push the pedal down as far as you can to set the brake. This type of parking brake is usually released by pulling a lever or handle located on the driver's left side, just under the bottom edge of the instrument panel.

Figure 4–14

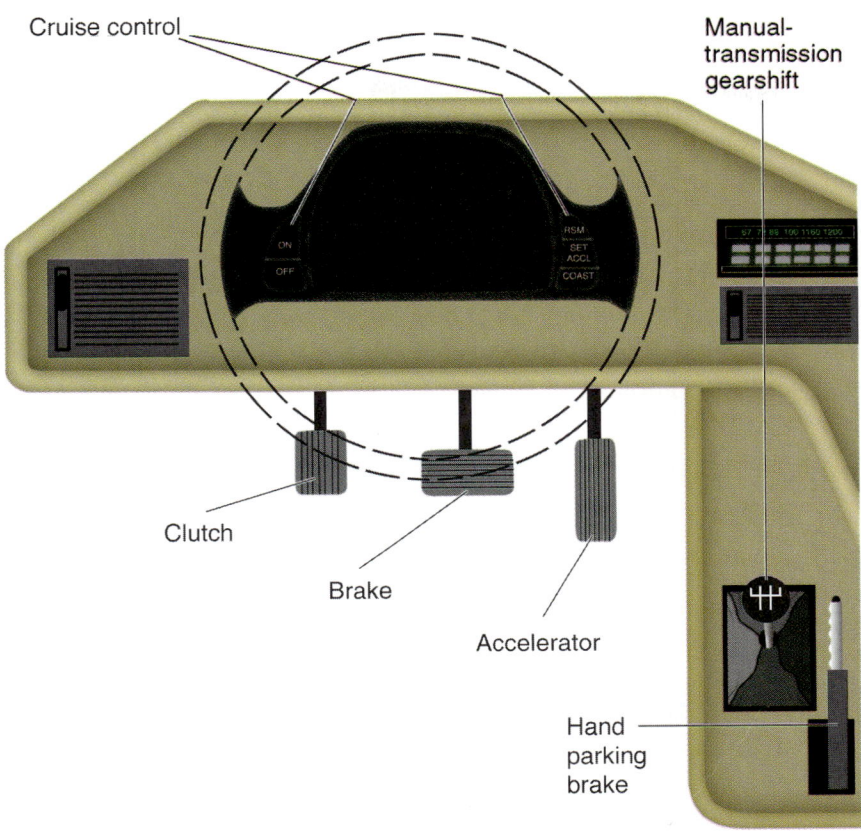

Figure 4–15 Typical vehicle with manual transmission

Driving with the parking brake engaged, even partially, can damage the brake lining, so always check to make sure that the parking brake is released before you start moving.

Gearshift

The **gearshift** is mounted either on the console or on the right side of the steering column, and it is attached to the transmission, allowing the driver to drive forward, reverse, or remain in neutral. In automatic-transmission vehicles, if the gearshift is on the console, you must move the handle backward and forward to switch from one gear to another. If the gearshift is on the steering column, you must pull the handle toward you and then slide it to engage the gear you want.

In vehicles with a manual transmission, the stickshift, or "stick," is a handle with a knob on top that rests on the console. To switch to different gears, you must learn to coordinate moving

> ### DRIVING TIPS — "Preflight" Checks
>
> Aircraft pilots conduct a preflight check before takeoff, and so should you when you get behind the wheel of an automobile. Driver and passenger safety depends on what you do before driving, including making proper adjustments to your seat and mirrors, using safety belts, checking outside your vehicle for obstacles, and securing items in and on the vehicle. It is important to develop a sequence of predriving checks to run through your mind each time you drive.

the gearshift handle with operating the clutch pedal.

Clutch

The **clutch** is a pedal found only on cars with a manual transmission and is located to the left of the brake pedal. The clutch must be pressed each time you shift gears. To operate the clutch, rest the heel of your left foot on the floor of the car and use the ball of the foot to depress and release the pedal each time you change gears.

Cruise Control

Cruise control is a device that allows you to regulate your vehicle's speed for highway or freeway driving without using the accelerator. It is usually activated by a button on the end of the turn-signal lever or on the steering wheel. To set most cruise controls, accelerate to the speed desired and then allow the speed of the vehicle to decrease by 2 to 3 miles per hour (3 to 5 km/h). Push the "set cruise control" button. Usually a light on your instrument panel will indicate that cruise control is on. Remove your foot from the accelerator. Your speed will stay constant. To cancel cruise control at any time, turn the control switch off or lightly tap the brake pedal.

Exercise caution when using the cruise control feature because it reduces the amount of control you have over your speed. You will not be able to quickly slow down by easing up on the accelerator, or you may skid if you hit the brake to disengage the control while on a slick surface. You also risk developing a false sense of security and becoming less alert. If you get into a situation in which you have to brake quickly, you may not be able to react fast enough because you are too relaxed or your foot is too far from the brake pedal. Even when cruise control is on, keep your right foot on the floor at the base of the accelerator, ready to either engage the accelerator or pivot to the brake pedal instantly.

WHO'S AT FAULT?

1. Driver 1 neglected to adjust his mirrors when he got into the car after his mother, who is much shorter, had been driving it. Because of the position of the mirrors, his blind spots were larger than usual. As he was changing lanes on the freeway, he checked his mirrors but did not look over his shoulder. Driver 2 was in the next lane and swerved to avoid Driver 1. She in turn crashed into Driver 3, who hit the center concrete divider. *Who's at fault?*

2. Driver 1 was heading home after a day in the city. Her braking system warning light went on, but she ignored it and proceeded into the tunnel she had to take to reach her house on the other side of the river. As she was descending into the tunnel, her brakes failed. Unable to stop, and with no escape route, she rear-ended Driver 2, who crashed into Driver 3. Driver 4 was just behind Driver 1 but could not stop in time and rear-ended her. *Who's at fault?*

YOUR TURN

4–1 Vehicle Instrumentation
1. What gauges and warning lights are found on a typical instrument panel?
2. What might cause your temperature and oil-pressure warning lights to come on?
3. What does an alternator do?

4–2 Operating Devices
4. What are the different types of exterior lights found on most motor vehicles?
5. What are the three settings on a typical headlight switch?
6. What is the proper way to adjust mirrors?
7. What are blind spots and why do they exist?

4–3 Vehicle Controls
8. How is seat position important to driving?
9. What controls are found on all automatic-transmission vehicles? manual-transmission vehicles?

SELF-TEST

Multiple Choice

1. A tachometer measures:
    a. the total number of miles/kilometers on your car.
    b. oil pressure.
    c. gasoline level.
    d. engine revolutions per minute.

2. What is the main purpose of the alternator?
    a. to regulate the speed of the car
    b. to keep the engine cool
    c. to maintain constant oil pressure
    d. to power the car's electrical system

3. Back-up lights come on when:
    a. the transmission is set to REVERSE.
    b. you are parking.
    c. you activate your high beams.
    d. you activate your emergency flashers.

4. Blind spots are those areas not reflected in your:
    a. sideview mirrors.
    b. rearview mirror.
    c. rearview and sideview mirrors.
    d. sun visor vanity mirror.

5. Power brakes:
   a. require less foot pressure to operate than standard brakes.
   b. increase the distance needed to stop the car.
   c. keep your vehicle in place when parked.
   d. have been required on all passenger cars manufactured since 1985.

Sentence Completion
1. The _____ shows the total number of miles/kilometers that a car has been driven since it was new.
2. _____ are on the front of a vehicle and automatically illuminate upon ignition.
3. The _____ controls the amount of fuel that the carburetor or fuel injectors supply to the engine.
4. The _____ setting is used in cold or rainy weather to remove condensation from the inside of the windshield.
5. The _____ is effective at reducing rear-end collisions because they are less likely to be confused with the taillights of a vehicle.

Matching
Match the concepts in Column A with examples of the concepts in Column B.

Column A
1. __ Kilometers per hour
2. __ Reduce blind spots
3. __ Ignition
4. __ Alternator
5. __ Maintain a constant speed
6. __ Reduce glare
7. __ High beams
8. __ Communication device

Column B
a. Speedometer
b. Adjust mirrors
c. "Discharge"
d. "Lock," "on," "start"
e. Sun visor
f. Brake lights
g. "Set cruise control"
h. Blue indicator light

Short Answer
1. Why is it illegal to alter the odometer on a vehicle?
2. How do you "flash" your high beams?
3. What affects the size of blind spots?
4. Why is it a good idea to lock your doors while driving?
5. What are the chief advantages of power brakes over standard brakes?

Critical Thinking
1. You start your car and begin to pull out of a parking space. After leaving the space, you notice that the brake light on your instrument panel is still on. What should you do?
2. Come up with your own "preflight" check before driving a car. Briefly describe the importance of each step.

PROJECTS

1. Make a diagram of the instrument panel of your car to help you remember the location of all the gauges and warning lights. Study it for a day or two and then put it away. Take a blank piece of paper and draw the panel again from memory. How did you do?

2. Sit in the driver's seat of your vehicle while parked in your driveway or an empty parking lot. Adjust your mirrors so that you can see the maximum area behind you and to the sides. Ask a friend to move around the vehicle one step at a time and, with a piece of chalk, mark your blind spots. Switch places and repeat the same exercise. How do your blind spots compare with your friend's?

Vehicle Operation Basics

There is more to driving a car than getting behind the wheel and hitting the gas. If you have been around cars your entire life, you might think that steering, changing gears, parking, and other basic operations will come quickly to you. In fact, many adult drivers do not know how to engage a manual transmission, and even experienced drivers can have trouble parking. Before you are ready to handle the more difficult challenges that face today's drivers, you must master the fundamental skills of operating a vehicle.

Chapter Objectives

Upon completion of this chapter, you should be able to:

5–1 Starting the Engine and Engaging the Transmission
1. Identify the different gears on an automatic and manual transmission.
2. Describe how to engage an automatic and manual transmission.
3. Describe how to shift gears on a manual transmission.
4. Understand what precautions you should take when putting your vehicle in motion on a hill.

5–2 Steering
5. Understand how steering is different from turning.
6. Describe the two basic steering methods.

5–3 Backing up
7. Describe the proper procedure for backing up.
8. Understand what precautions you should take when backing up to the left or right.

5–4 Parking
9. Identify the four main types of parking.
10. Describe the proper procedure for entering and exiting a parallel parking space.
11. Understand how you should turn your wheels when parking on hills.

KEY TERMS

choke	tracking	curb parking
flooded engine	hand-over-hand method	parallel parking
friction point	push-pull method	angled parking
downshift	double parking	perpendicular parking
steer		

5–1 STARTING THE ENGINE AND ENGAGING THE TRANSMISSION

The steps you take to start the engine depend on whether your vehicle has an automatic or manual transmission. Although manual transmissions require more effort and skill from a driver, they give you more choice in gear selection, require less maintenance, result in better gas mileage, and put less wear and tear on the brakes.

Take time to familiarize yourself with the gear configuration of your vehicle. The main difference between automatic and manual transmissions is the number of driving gears. Lower gears deliver more power, and higher gears allow you to maintain higher speeds using less power on level roads. Keep in mind that all engines are different, and your transmission may have different optimal speeds for each gear.

Starting the Engine

To start the engine, first confirm that the parking brake is set. If you have an automatic transmission, shift to PARK. If you have a manual transmission, fully depress the clutch with your left foot while pressing the brake with your right foot, and shift into NEUTRAL. On some vehicles, you do not have to depress the clutch while starting the engine.

If the engine is cold and you do not have a fuel-injected car, press and release the accelerator once to set the automatic **choke,** a device that controls the amount of air that enters the carburetor. If your vehicle has fuel injection, you do not need to press the accelerator.

Turn the key to start the engine, and release it as soon as the

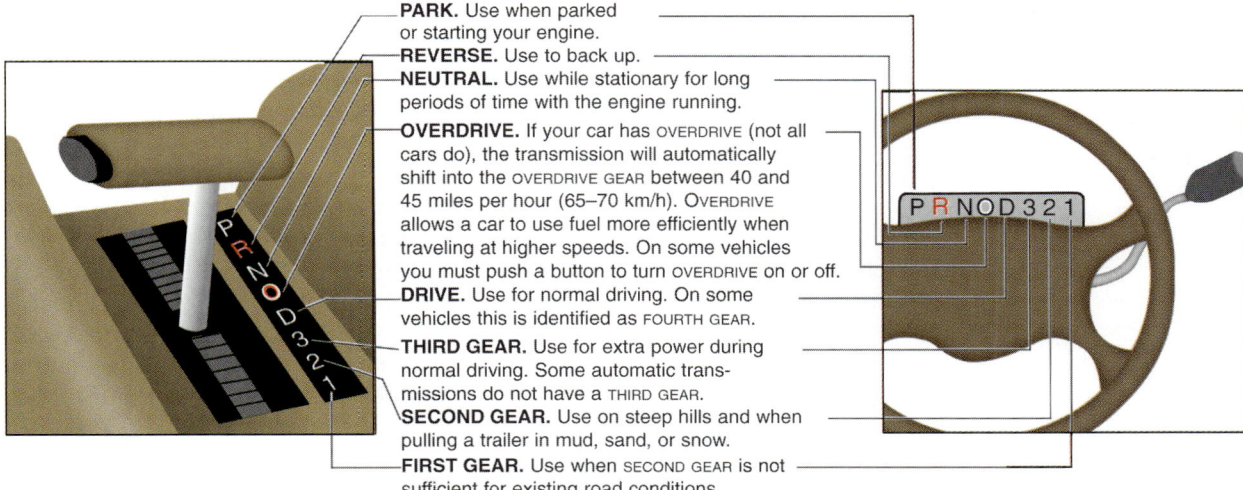

PARK. Use when parked or starting your engine.
REVERSE. Use to back up.
NEUTRAL. Use while stationary for long periods of time with the engine running.
OVERDRIVE. If your car has OVERDRIVE (not all cars do), the transmission will automatically shift into the OVERDRIVE GEAR between 40 and 45 miles per hour (65–70 km/h). OVERDRIVE allows a car to use fuel more efficiently when traveling at higher speeds. On some vehicles you must push a button to turn OVERDRIVE on or off.
DRIVE. Use for normal driving. On some vehicles this is identified as FOURTH GEAR.
THIRD GEAR. Use for extra power during normal driving. Some automatic transmissions do not have a THIRD GEAR.
SECOND GEAR. Use on steep hills and when pulling a trailer in mud, sand, or snow.
FIRST GEAR. Use when SECOND GEAR is not sufficient for existing road conditions.

Figure 5–1 Automatic transmission

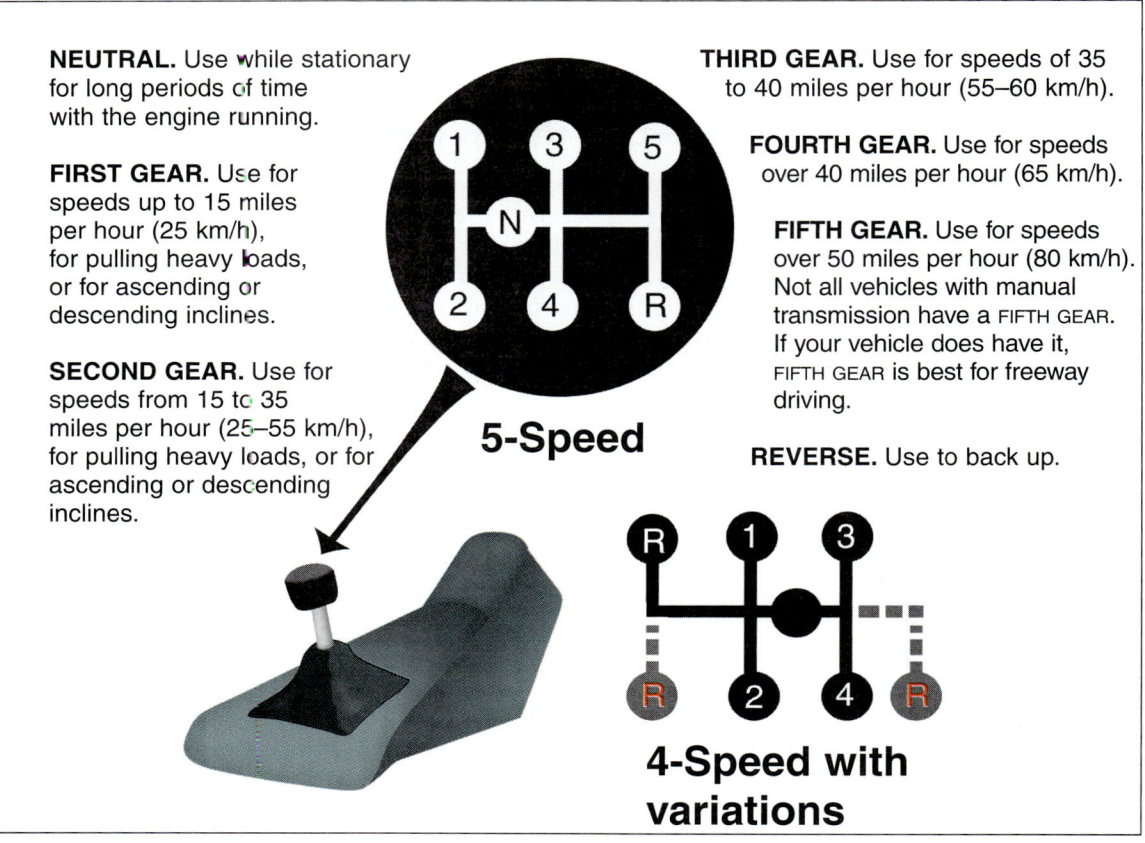

Figure 5–2 Manual transmission

engine starts. If you continue to turn the key after the engine has started, you may damage the starter. In some vehicles, you must first push the key and then turn it to start the engine. If you are driving an older car with a manual choke, pull the choke out before turning the key in the ignition and push it in after the engine has started. When the car is idling smoothly, release the choke. If you have a manual transmission, release the clutch after the engine has started.

If you turn the ignition key and can hear the starter clicking but the car does not start, the engine may be **flooded.** Flooding occurs when there is too much gasoline in the engine, which can be caused by depressing the accelerator too many times before starting the car. If you think your engine is flooded, press down continuously on the accelerator as you turn the key in the ignition for about 10 seconds, making sure to let off the accelerator if the car begins to start. If this does not work, wait several minutes to allow the excess fuel to evaporate and then try starting the engine again.

Putting the Vehicle in Motion

Once you start your vehicle and the engine is running, you must engage the transmission to put the car in motion. The procedure you follow depends on what type of transmission you have.

Automatic Transmission

If you are a new driver, many instructors recommend that you first drive a vehicle with an automatic transmission. Automatic transmissions are designed to be "user friendly." Because they have no clutch, less coordination of hands and feet is required to operate the vehicle. In addition, gears change automatically during normal driving, so you can concentrate more on the driving task. To engage an automatic transmission, first press firmly on the brake pedal. Shift to either DRIVE or REVERSE, depending on which way you want to go. Release the parking brake. Check for traffic in your mirrors, and look over your shoulder to make sure that your blind spots are clear. If you are driving forward, use your turn indicator to signal to others which direction you want to move. When your intended path is clear, remove your foot from the brake and apply gradual pressure to the gas pedal, accelerating smoothly into traffic.

Manual Transmission

Manual transmissions are more complicated to operate than automatic transmissions. You must coordinate the clutch, gearshift, and accelerator to engage the gears. Depending on driving conditions, you may have to change gears often. In heavy traffic, on city streets, and on windy roads, for example, you have to shift gears constantly. Although anybody can learn to use a stickshift with practice, it takes time to do it so that you can shift gears in one fluid motion without having to think about what you are doing.

To engage a manual transmission, follow these steps (see Figure 5–3):

1. While pressing on the brake pedal, fully depress the clutch.
2. Shift into FIRST gear or REVERSE, depending on which way you want to go. To shift into

The first production car with an automatic transmission, called Dynaflow, was offered by Buick in 1948.

DRIVING MYTHS **The Runaway Vehicle**

Some people believe that if you have an automatic transmission and you take your foot off the brake while the vehicle is idling in DRIVE, it will begin to move and continue to increase in speed automatically on its own. In fact, if you are on level ground and you take your foot off the brake while the automatic transmission is in DRIVE, the car may begin to creep forward. The speed at which it will move depends on the setting for the engine idle and varies among vehicles, but it will not *increase* until you actually press the accelerator.

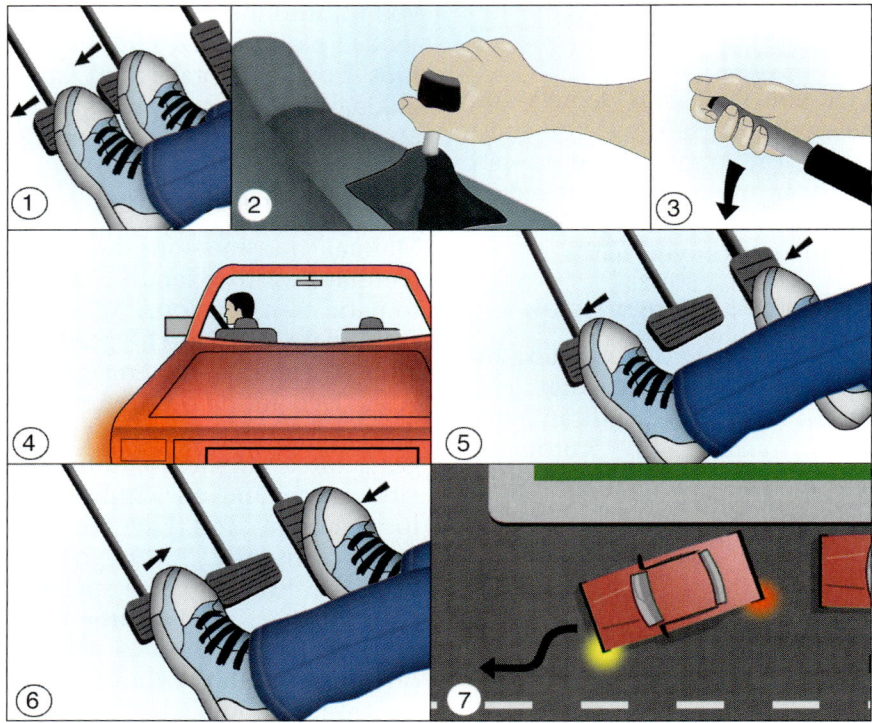

Figure 5–3 Engaging a manual transmission

REVERSE, you may have to use extra effort to engage the gear. Most manual transmissions have a safeguard to prevent accidental shifting into REVERSE from a driving gear, and to engage REVERSE you may have to push the gearshift down first.
3. Release the parking brake.
4. Check for traffic in your mirrors and by looking over your shoulder. Make sure that your blind spots are clear. If driving forward, use your turn indicator to signal to others in which direction you intend to move. Get ready to accelerate.
5. Lightly press the accelerator and hold.
6. With your eyes on the road, slowly release the clutch until it reaches the **friction point,** the point at which the engine begins to engage the transmission. At first you may find this difficult, but in time you will recognize the friction point through the "feel" of the clutch pedal and the sound of the motor.
7. Once you have reached the friction point, continue to slowly release the clutch while gently accelerating. Accelerating too quickly might cause

the wheels to spin, and cause you to lose control if the car moves forward too quickly.

If the car jerks when you try to engage the transmission, try releasing the clutch more slowly or putting less pressure on the accelerator. However, the engine will stall if you do not give it enough gas. A stall may also occur if you do not fully depress the clutch. Remember that every clutch is different. If you learn how to drive a manual transmission on one car, it may take time to get used to the clutch on another car.

Shifting to Higher Gears. As you accelerate, you must engage higher gears to efficiently transfer power from the engine to the transmission. Each gear has a maximum speed, which varies by vehicle, and to reach a higher speed you must switch to a higher gear. If you are driving in a gear that is too low, your engine will "rev" too high—that is, it will operate at too many revolutions per minute (rpm).

If you are driving a vehicle with an automatic transmission, the gears will shift automatically for you. Although you may want to manually shift to a higher gear—for example, from "1" to "2" to "D"—under certain conditions, you normally do not have to "shift" to a higher gear if you stay in "D."

To shift to a higher gear in a vehicle with a manual transmission, release the accelerator and fully depress the clutch. Briefly pausing in NEUTRAL, shift to the next gear.

You should always shift up one gear at a time in sequence—from 1 to 2, from 3 to 4, and so on. Gradually release the clutch while slowly accelerating. Avoid "riding the clutch," or driving with your foot continuously resting on the clutch. Once you have released the clutch, lift your left foot completely off the pedal.

In addition to the feel of the clutch and the sound of the engine, you can refer to your speedometer and the tachometer to help you decide when to change gears. Just as it takes experience to recognize the friction point when first engaging the transmission, it will take time to learn when to switch gears.

Downshifting. To downshift means to shift from higher to lower gears. There are a number of reasons for downshifting: to slow down or prepare to stop, to gain more power from the engine, or to gain more control—for example, on curves and downward slopes.

For normal driving, you do not need to downshift an automatic transmission. To get more power—for example, during passing—you need only press the accelerator further. You will feel and hear the gears "kick in." If conditions require you to manually downshift, be careful to do so at low speeds. Slow down by either easing off the accelerator or braking to around 30 to 40 miles per hour (40 to 65 km/h) if you are switching to SECOND gear, and even slower if you are switching to FIRST gear.

If you are driving a vehicle with a manual transmission, downshifting is an integral part of driving at all speeds. When you are traveling at higher speeds in higher gears, for example, downshifting can help you slow down without using your brakes, which will extend the life of your brake pads. To downshift, release the accelerator, fully depress the clutch, shift to the best gear for the speed at which you are traveling, slowly release the clutch pedal to the friction point, and press the accelerator as necessary.

You do not always have to shift to the next lowest gear when downshifting. For example, you can downshift from FIFTH to THIRD gear, or from FOURTH to SECOND gear, as long as the gear you shift into is consistent with your speed. The most difficult downshift to make is from SECOND to FIRST gear because it requires the vehicle to be slowing down rapidly.

Putting Your Vehicle in Motion on a Hill

If you have an automatic transmission, starting a vehicle and putting it in motion on a hill is virtually the same as it is on a level surface. Remember to keep your foot firmly on the brake as you shift into DRIVE and release the parking brake with your hand. Once the vehicle is in DRIVE, you can safely move your right foot to the accelerator without danger of the car rolling backwards.

Starting a vehicle with a manual transmission and putting it in

Figure 5–4 When starting your vehicle on a hill, make sure the parking brake is on first.

motion on a hill can be tricky if you are facing uphill. If there is another vehicle parked behind you, be aware of how much space is behind you before you get inside your car and start the engine. Because you need your right foot for the accelerator, you will not always be able to use the brake while starting on a hill, so *make absolutely sure the parking brake is securely set.*

Fully depress the clutch while firmly pressing the brake pedal, and shift into NEUTRAL. After starting the engine, press the clutch and shift into FIRST gear. Slowly begin to release the clutch until you reach the friction point, and then fully release the parking brake and apply pressure to the accelerator. Accelerate as evenly as possible to minimize the distance you might roll back downhill and to move forward steadily without jerking or stalling.

5–2 STEERING

Once you have put your vehicle in motion, you must control your direction, as well as your speed.

> **DRIVING TIPS** **Dry Steering**
>
> You should turn the steering wheel only while the car is in motion. "Dry steering," or turning the steering wheel when the car is stopped, can prematurely wear out your tires.

To direct, or **steer,** your automobile, you must use the steering wheel. Steering is not exactly the same thing as turning, which means to change directions. However, you still do "steer" through a turn. Even if you are driving on a straight road and your tires are perfectly aligned, your vehicle may veer in one direction or another because of engine torque, uneven pavement, wind, and other factors. Therefore, you should always steer with both hands on the wheel.

Figure 5–5 Driving with one hand is dangerous because you have less control if you have to react quickly to a hazardous situation.

Hand Position

The safest and most practical position for your hands on the steering wheel is at opposite sides of the wheel. Generally, the left hand should be placed at the 9 o'clock position on the wheel, and the right hand should be placed at 3 o'clock. Make sure you are sitting at a comfortable distance from the wheel and that your arms are relaxed and free to move. This will also allow you to turn at higher speeds without taking your hands off the wheel. Grip the wheel with your fingers rather than your palms, and keep your thumbs pointing up along the face of the steering wheel.

Tracking

Steering requires constant adjustments of the steering wheel to maintain a smooth and steady course. **Tracking** is a method of steering that allows you to keep your vehicle on the intended path of travel whether you are traveling on straight or curving roads. To track, look toward the center of the lane ahead of you, making only the slightest of movements with the steering wheel to adjust your course. When driving in a straight line, it is important to steer toward the center of your path of travel. If you do this, you should have to make only minor adjustments to keep your vehicle in your lane

Figure 5–6 Keep your "sight" focused on the center of the road to maintain a smooth and steady course.

and on the correct path. Looking continuously to either side of the road, rather than toward the center, while driving straight can cause you to drift outside of your lane.

Steering Methods for Turns

There are two basic ways to steer during a turn. Each method has its own advantages, and you should experiment with both to deter-

KELSEY GRAMMER

Actor Kelsey Grammer first gained recognition as the neurotic psychologist Frasier Crane on the hit TV series *Cheers*. In 1993, he first appeared in his own television comedy, *Frasier*, based on the same character. *Frasier* has earned dozens of awards, including multiple consecutive Emmys for Outstanding Comedy. On the night of September 21, 1996, Grammer was driving on Cornell Road in the Malibu Lake area near Los Angeles when he allowed his Dodge Viper to drift to the right shoulder of the roadway. Trying to correct his steering, he swerved across the road, hit a dirt embankment, and flipped his sports car. Grammer was lucky to suffer only minor abrasions, but his Viper had major damage. Even experienced drivers can make a costly mistake if they do not remain attentive to the road and smoothly adjust their steering in response to changing road conditions.

CHAPTER 5 Vehicle Operation Basics ◆ 101

mine which is best for you in different situations.

Hand-over-Hand Method

The **hand-over-hand method** requires you to have only one hand on the wheel at times, so you must always be attentive to keeping control of the car. This method is generally more comfortable, especially on wide, slow turns, and allows you to keep your shoulders straight. To steer hand-over-hand, you use one hand to push the steering wheel around and down as the other hand crosses over to pull the wheel even further down. For example, to execute a left turn you would do the following (see Figure 5–7):

1. With your hands positioned at 9 o'clock and 3 o'clock, begin turning the wheel toward the left with both hands.
2. With your right hand, continue pushing the wheel past the top and then down to the left. Release your left hand.
3. Cross your left hand over your right arm, grabbing the far side of the wheel.
4. Once you are sure you have a firm grasp of the far side of the wheel with the left hand, release your right hand and bring it back to the right side of the wheel while at the same time pulling the wheel to the left past the top and then down to the left.
5. Return to the 9 o'clock and 3 o'clock position or repeat

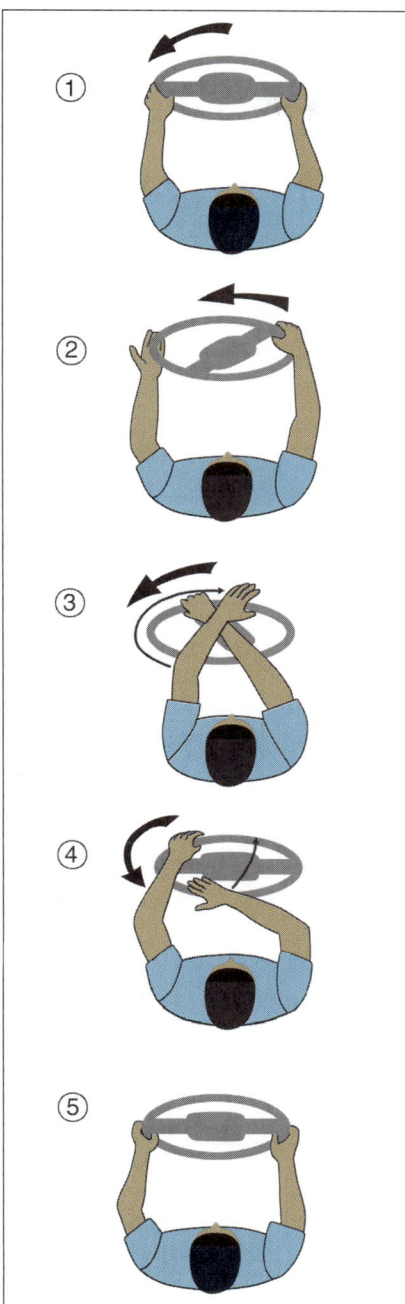

Figure 5–7 Hand-over-hand method

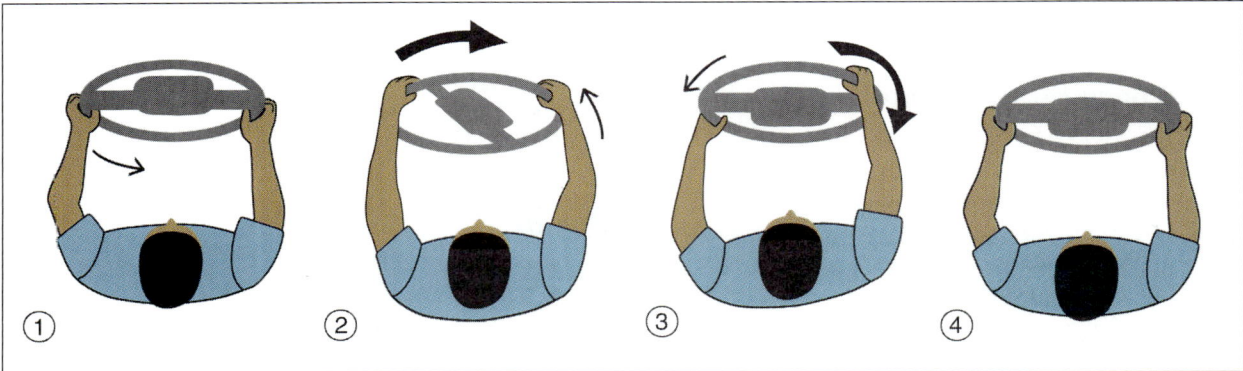

Figure 5–8 Push-pull method

steps 1 to 4 as necessary until the turn is complete.

Push-Pull Method

The advantage of the **push-pull method** of turning is that it allows you to make sharp turns quickly while keeping both hands on the wheel. To use this method to make a sharp right turn, you would do the following (see Figure 5–8):

1. Slide your left hand to the 7 o'clock position and firmly grip the wheel.
2. With your left hand, push the wheel up to the 11 o'clock position, and slide your right hand along the wheel back to 1 o'clock.
3. Pull the wheel down with your right hand. As you do so, allow your left hand to slide over the wheel back to 9 o'clock.
4. From the 9 o'clock and 3 o'clock position, repeat steps 1 to 3 as necessary until the turn is complete. On most turns, you will only need to complete the push-pull cycle once to make the turn.

Straightening Out

After any turn, make sure that you straighten out the wheel yourself by reversing either the hand-over-hand method or the push-pull method. Letting the wheel straighten itself out by completely letting go of the wheel or letting it "slip through your fingers" as you accelerate could cause you to lose control of the car. When you are back on course, remember to adjust your hands back to their usual position on the steering wheel.

5–3 BACKING UP

When backing up, or driving in REVERSE, it is important to realize that your visibility through the back window is limited. Your vision might also be restricted by head restraints, passengers, and

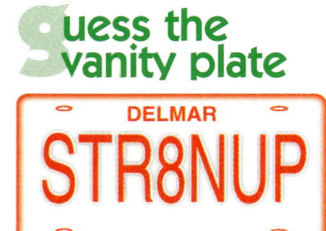

Figure 5–9

child seats. Be careful to keep control of the steering wheel and back up slowly, while constantly being aware of what is around you. You must be prepared at all times to brake if someone or something enters your path. Never back up out of a driveway or on a street unless you are sure that the road you are backing into is clear of traffic.

Before entering your vehicle, inspect the surrounding area for pedestrians, signposts, trash cans, parked cars, gardening tools, and other potential obstacles. Make sure nothing is behind your vehicle that you cannot see through your rear window. Follow these steps to back up straight (see Figure 5–11):

1. Once inside your vehicle, turn your head and look behind you to make sure that your path of travel is clear. Also, look to each side for anything that may enter your intended path.

2. Depress the brake and shift into REVERSE. If you have a manual transmission, remember to depress the clutch before shifting into REVERSE.

3. Turn your body to the right, and with your right arm over the back of the passenger seat or head restraint, look through the back window, and identify where you intend to move. *Never use only the rearview mirror while backing up.* Place your left hand at the 12 o'clock position of the steering wheel.

4. Back up slowly. If you have an automatic transmission, you can usually idle backwards without using the accelerator. Ease up on the brake until the car slowly begins to move, keeping your foot over the brake pedal in case you need to stop to avoid hitting someone or something. If you have a manual transmission, release the brake and ease up on the clutch until you reach the friction point. Maintain the clutch at the friction point to slowly back up, and keep your foot over the brake if you can back up without using the accelerator. Otherwise, lightly press the accelerator and carefully release the clutch a little at a time. Fully releasing the clutch all at once can cause the car to jerk backwards or stall.

5. Turn away from anything you get too close to by making slight steering adjustments. Continue looking out the back window, while periodically checking in all other directions.

6. Once you have finished backing up, come to a complete stop by braking smoothly. If you have a manual transmission, remember to fully depress the clutch first.

When backing up, keep in mind that the turning radius of the front of the vehicle is wider than that of the rear. As a result, the front wheels swing out farther

driving tips

Always keep a window open while backing up to hear any warning yells from other drivers or pedestrians.

Wild Wheels

Figure 5–10 What was the name of this 1903 car that used a handle instead of a steering wheel to turn and that could be operated from either the left or right seat?

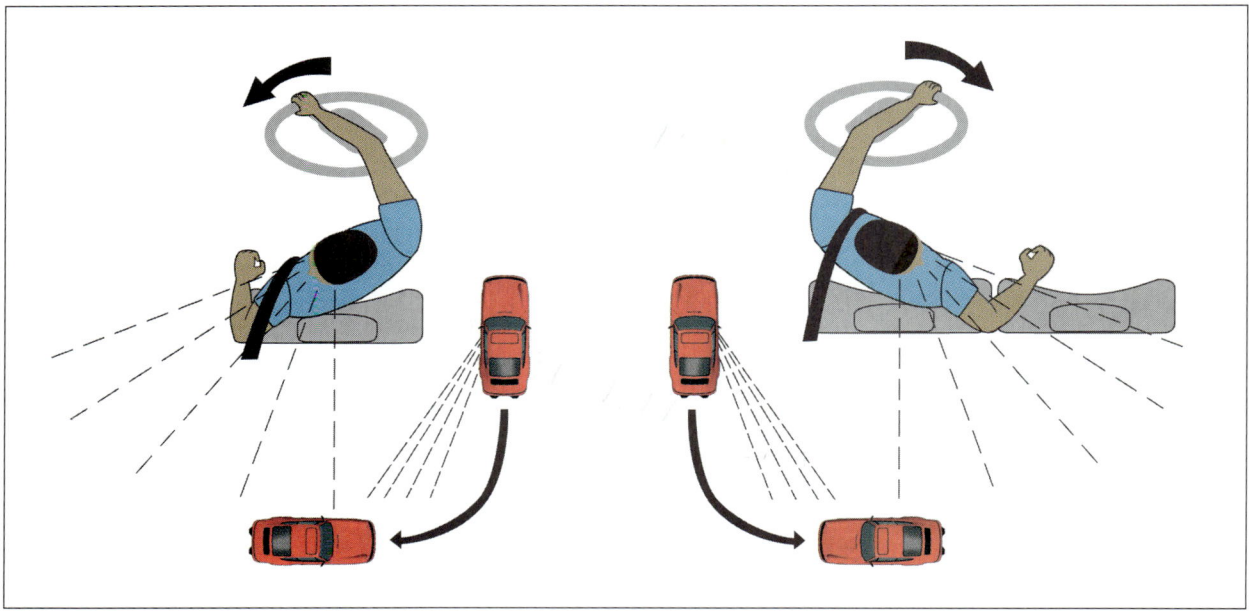

Figure 5–11 When backing up, always turn and look over your shoulder to make sure that your intended path is clear.

than the rear wheels do. When backing up to the left, you must leave extra space for the car to swing right. When backing up to the right, you must leave extra space for your car to swing left. By anticipating the room you will need to back up, you will avoid striking pedestrians, bicyclists, other vehicles, and stationary objects.

When backing up to the left or right, turn your wheel in the direction you are backing. In other words, if you are backing up to the left, turn the wheel to the left. If you are backing up to the right, turn your wheel to the right. Until you gain more experience, avoid backing up around corners or sharp turns. Try to approach the space you are moving into head-on, even if this means driving around the block to position yourself properly.

5–4 PARKING

Parking is a difficult skill to master. You need to be able to steer accurately into spaces that are often tight and narrow. You must be aware of the space around your car at all times so that you can judge the distance from curbs you cannot see up close, other parked cars, and passing traffic. You must be on the lookout for pedestrians and other vehicles that may get in your path. No matter how experienced a driver you are, you should always enter and exit parking spaces slowly. Do not be pressured into pulling out too quickly or racing to a space to beat someone else to it.

CHAPTER 5 ◆ Vehicle Operation Basics ◆ **105**

Figure 5–12 It takes practice to determine how much space you need to park your vehicle.

Before you park, ask yourself if there are any parking restrictions. Are there curb or roadway markings, prohibited parking signs, or posted time limits? Make sure that there is enough space on all sides of your car to enter and exit the space and that you have enough time to park in the space given the speed of traffic.

When parking, always remember to set the parking brake when

DRIVING TIPS *Prohibited Parking*

Parking prohibitions are different in every jurisdiction, but in general, you should never park in the following places:
- On the roadway side of another parked or stopped vehicle—this is called **double parking**
- Too far from the curb—as a rule of thumb, more than 12 inches (30 cm)
- Within or near an intersection, railway crossing, or crosswalk
- In any designated fire lane or near a fire hydrant
- In front of driveways
- On sidewalks or bicycle paths
- In tunnels or on bridges
- In any zone or space where parking is specifically prohibited
- In bus zones or loading zones
- In a lane or any other paved portion of a highway

You should also never park in such a way that you are creating a hazard for other motorists.

you park and release it before you exit a parking space. Put the gear in PARK if you have an automatic transmission, or in FIRST or REVERSE if you have a stickshift. Park so that other drivers can both see and avoid hitting your vehicle. Check that your headlights, radio, air conditioner, heater, and any other devices are turned off. Finally, remember to lock your doors and take the keys with you.

Entering and Exiting Parking Spaces

As you approach a parking space, always check traffic ahead of you and then behind you by using your rearview and sideview mirrors to make sure that your path is clear. Signal your intention to park using your turn signal. Tap your brakes to warn drivers behind you that you will be slowing down. Watch for vehicles behind you as you gradually reduce your speed, and keep an eye out for parked vehicles that may be exiting nearby spaces. Before turning out of traffic, check your mirrors again and look over your shoulder to make sure that your blind spots are clear. After parking, remember to cancel your turn signal.

When exiting a parking space, first check for traffic in your rearview and sideview mirrors. Turn your head to make sure that your blind spots are clear. Warn other drivers that you are preparing to exit the parking space by signaling with your turn indicator. When it is clear and safe to move, slowly pull away from the curb and into the nearest lane on the street. Gradually accelerate to a normal driving speed and cancel your turn signal.

Curb Parking

There are four basic types of parking. The easiest is **curb parking,** which involves parking alongside a curb where there are no other surrounding vehicles. You have more room to maneuver, and unless another driver parks ahead of you or you are parked at the end of a curb, you do not have to back up to exit the space. Park as close to the curb as possible so that other vehicles will not hit you.

Parallel Parking

Parallel parking involves parking alongside a curb between two other parked vehicles and can be extremely difficult even for seasoned drivers. You must back up into traffic, which can be danger-

---

AUTO ACCESSORIES *Curb Feelers*

Curb feelers are flexible metal rods that attach easily to your vehicle near the tires and act as "whiskers" to let you know when you are close to the curb. This product is especially helpful if you are driving a station wagon, pickup, camper, van, or other large vehicle in which it is difficult to see the curb or judge your distance from it.

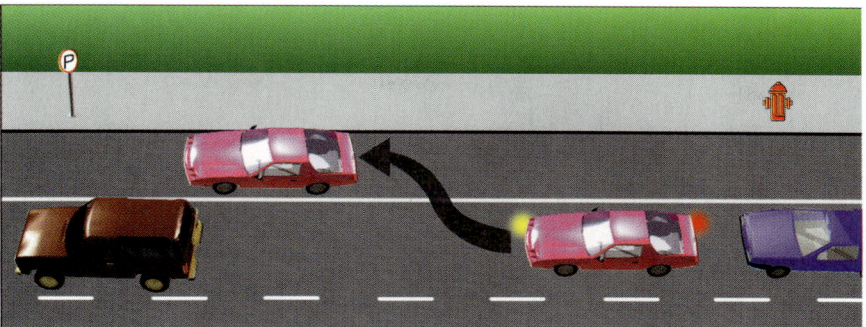
Figure 5–13 If you can, park on a curb without surrounding vehicles.

ous and frustrating if there is congestion. It takes time and practice to develop the ability to judge space and distance accurately enough to be able to parallel park correctly. Good steering and speed control skills are also essential.

It is your responsibility to avoid collisions while entering and exiting a parallel parking space. To reduce the risk of a collision, choose a parallel parking space that is at least 6 feet (20 m) longer than your vehicle to allow enough room for you to maneuver. Be aware of passing traffic and of how far the front of your car is projecting out into the nearest lane.

Follow these steps to parallel park in a space on the right side of the street (see Figure 5–14):

1. Pull up alongside the vehicle in front of the empty space so that the right side of your car is about 3 feet (1 m) away from the left side of the parked car. Stop when your rear bumper is even with that of the parked car.
2. As you look over your right shoulder, slowly back into the space while sharply turning the steering wheel to the right. You should aim the rear of your car for the rear right corner of the parking space. *As you back up, always be prepared to stop suddenly for passing vehicles, bicyclists, and pedestrians.*
3. When your steering wheel is even with the front vehicle's rear bumper, straighten out. Keep looking behind you, but make repeated glances ahead to make sure that your front bumper clears the vehicle in front of you.
4. Make a sharp left turn when the front end of your bumper is even with the rear bumper of the front vehicle.
5. Slowly back up until your vehicle is parallel to the curb. Straighten the steering wheel. Be sure to stop before your rear bumper hits the vehicle behind you. Drive forward slowly to center your car within the parking space.

If you misjudge the distance between your car and the front car or if you make your turns too early or too late, you may not be able to enter the parking space all

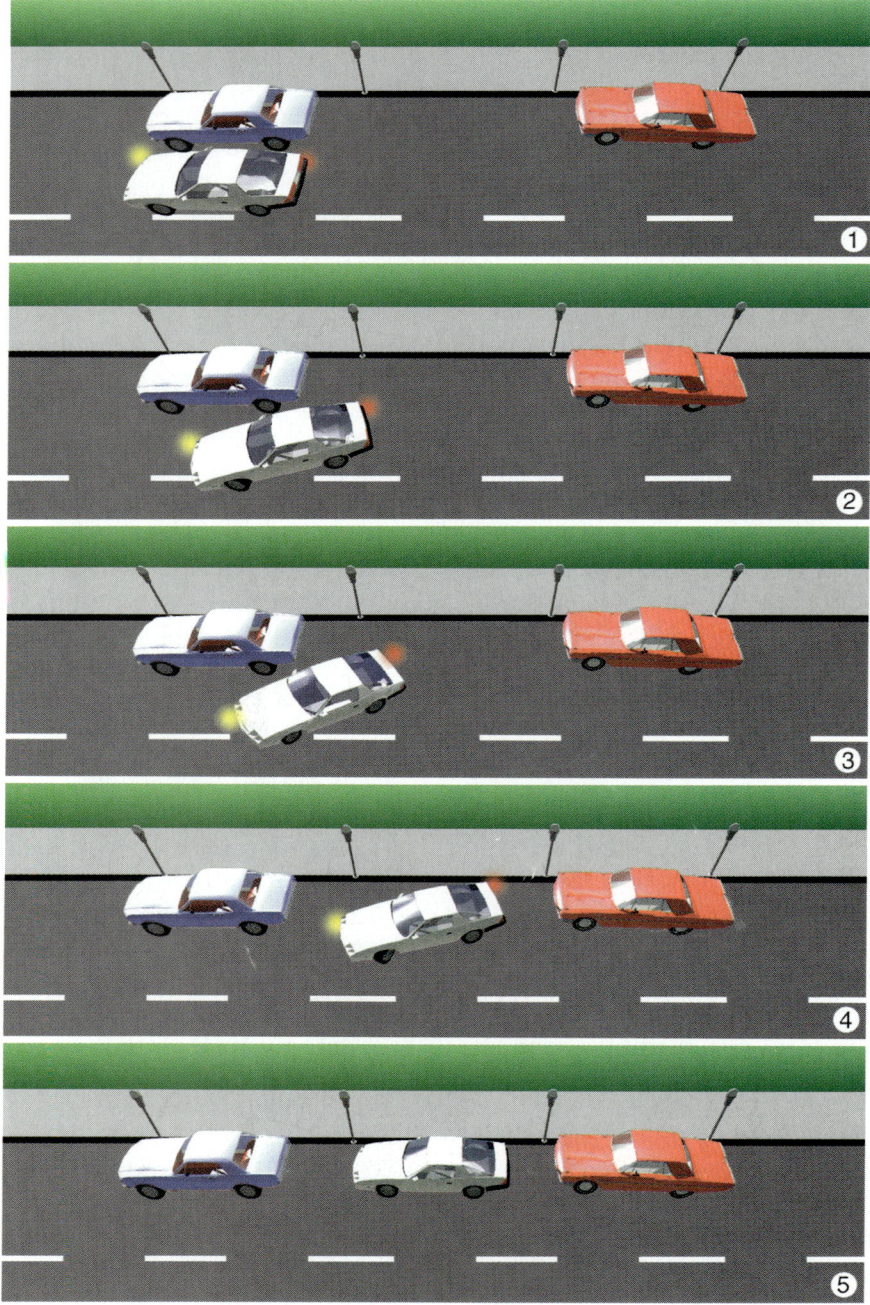

Figure 5–14 Entering a parallel parking space

the way. If this happens, pull out of the spot completely and start over again. If you try to make adjustments when you are partially or incorrectly positioned in a space, you risk hitting the other parked vehicles or confusing approaching traffic and causing a collision.

Follow these steps to exit a parallel parking space on the right side of the street (see Figure 5–15):

1. Turn your body to the right, and with your right arm

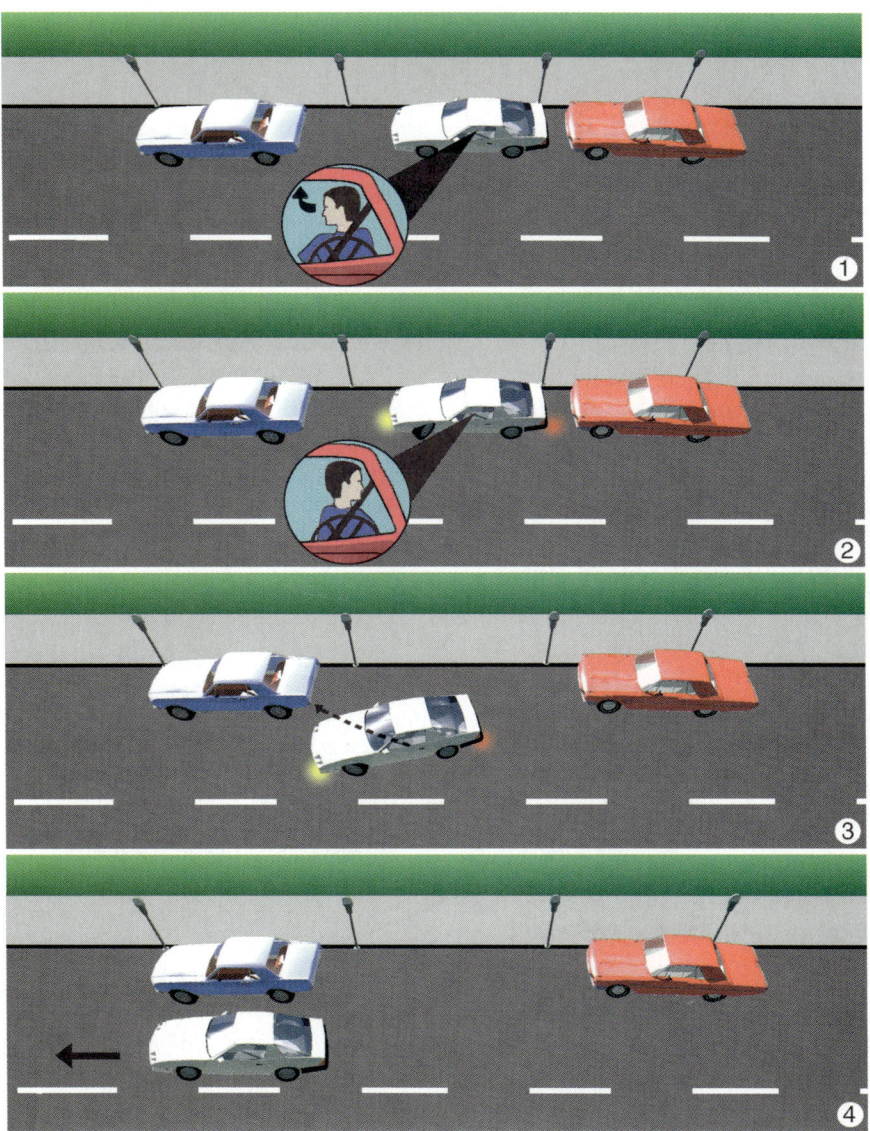

Figure 5–15 Leaving a parallel parking space

DRIVING TIPS Getting Out of a Parked Car

When curb or parallel parking, *always look before you open the door!* Watch out for approaching vehicles, pedestrians, and bicyclists. Do not assume that it is up to other drivers to make room for you. You also risk getting hit by passing cars if you stand in an open doorway next to traffic to get a jacket, backpack, or anything else out of your vehicle. Instead, retrieve your belongings from the sidewalk side of your car.

over the back of the passenger seat or headrest, look through the back window and make sure there is nothing between your vehicle and the vehicle behind you. Slowly back up straight until your rear bumper is close to the other car.

2. Sharply turn the steering wheel to the left as you stop. Activate your turn signal.
3. When it is clear, slowly move forward while repeatedly checking your right front bumper to make sure you are not in danger of hitting the vehicle in front of you.
4. When your car is halfway out of the space, slowly turn right to center your car in the nearest lane of the street. Cancel your turn signal.

Angled and Perpendicular Parking

Angled parking involves parking in a space that is angled to the curb, usually at about 30 degrees. **Perpendicular parking** involves parking at a 90-degree angle from the curb, in which the front or rear of your vehicle is aligned with the curb. Angled and perpendicular parking spaces are most often found in parking lots and garages and on some wide streets. Because spaces can be narrow, you will need to keep precise control of your steering and an awareness of the space around your vehicle. This is especially true if your vehicle is large and has a wide turning radius.

To avoid hitting other parked vehicles, you should choose an empty parking space with as much room as possible between the neighboring vehicles. You should also be aware of how the cars on either side of an empty space are parked. Are they parked to one side or the other of their own space? Are they parked at a crooked angle within their own space? Both of these factors will affect how you will have to maneuver your vehicle.

If it is an angled space, leave about 5 feet (1½ m) between your vehicle and the parked car to your right to allow sufficient steering space and visibility. When you can see the left side of the vehicle parked on the right side of the space you want to pull into, make a hard right. If it is a perpendicular space, approach the empty space as far to the left in your lane as possible and slowly brake.

CHAPTER 5 Vehicle Operation Basics ◆ **111**

Figure 5–16 Why should you not park in this space?

When your front bumper passes the back of the vehicle just before the open space, make a hard right.

As you repeatedly check your clearance on the left and right side with quick glances, slowly pull into the space. Make sure your right rear bumper does not hit the vehicle to your right, especially in a perpendicular parking space. Straighten your wheel once you are in the center of the space, and brake just short of the curb or painted line.

When exiting an angled or perpendicular parking space, position yourself so that you can look back over your right shoulder while constantly checking to the sides for pedestrians or oncoming traffic. When the area behind you is clear and it is safe to do so, slowly back up straight. Other vehicles may restrict your visibility as you back up, so be ready to brake for passing cars. *Always give the right-of-way to oncoming traffic.* Turn the steering wheel to the right when the front of your vehicle begins to pass the rear bumper of the vehicle parked to your left. As you back into the nearest lane of traffic, straighten out your wheels and come to a complete stop before accelerating into traffic.

Parking on Hills

If you park on a hill, you have to take special precautions to prevent your vehicle from rolling away. Whether you park uphill or downhill, make extra sure that the vehicle is in FIRST gear or REVERSE for manual transmissions, or in PARK for automatic transmissions, and that the parking brake is engaged.

When parking with your vehicle facing downhill, turn the front wheels *toward the curb* or sidewalk. If the parking brake fails, the curb will prevent the car from rolling downhill. If there is no curb, turn the front wheels in the same direction—away from

factoid

The world's first parking meter was introduced in Oklahoma City in 1935.

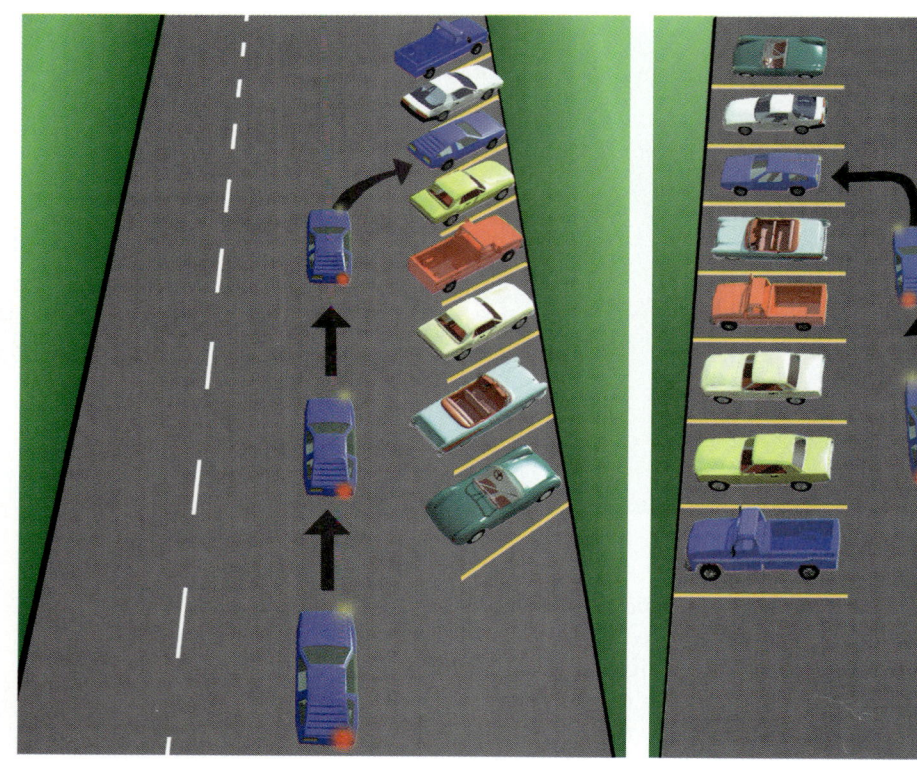

Figure 5–17 You need to steer precisely when navigating into narrow angled and perpendicular spaces.

the road—to prevent the vehicle from coasting into the street if your parking brake fails.

When parking with your vehicle facing uphill, turn the front wheels *away from the curb* or sidewalk. If the parking brake fails, the car will roll backwards into the curb and not into the street. If there is no curb, turn the front wheels the other away—away from the road—to prevent the vehicle from coasting backward into the street if your parking brake fails.

Parking Lots and Garages

Parking lots and garages usually have restricted maneuvering space. Visibility can also be limited, especially in dark parking garages. Drivers preparing to back out of parking spaces may not see you approaching. Other cars may surprise you coming around blind corners. A space that appears to be empty may have a small car, motorcycle, or shopping cart in it that you do not discover until you have already started to pull in. Always drive slowly and watch constantly for vehicles backing out of spaces and pedestrians walking to and from their vehicles, especially children. Observe all traffic arrows and signs, and stay to the right. Do not be tempted to cut across a row of empty spaces to get to a space or get out of a lot.

CHAPTER 5 Vehicle Operation Basics ◆ 113

Figure 5–18 When parking on a hill, make sure your wheels are turned in the proper direction.

Airport Parking

Parking can be an inconvenience at major airports. At terminals there are loading zones where you can pull over for a brief period to load and unload passengers and their luggage. Because of problems with traffic congestion in many airports, loading zone rules are usually strictly enforced. Do not leave your car unattended in these areas or your car may be quickly towed. If you want to leave your vehicle at the airport during a trip or if you want to greet guests at the airplane gate, short- and long-term parking is usually available nearby.

The world's largest parking lot, located at the West Edmonton Mall in Alberta, Canada, can accommodate 30,000 cars.

Figure 5–19 Drive slowly in parking garages, where visibility is poor and there is little room to maneuver.

WHO'S AT FAULT?

1. Driver 1 pulled away from a parking spot at the curb just as Driver 2 approached at a legal speed. To avoid hitting Driver 1, Driver 2 swerved across the center line and hit Driver 3 head-on. Driver 3 had just turned right at the intersection. ***Who's at fault?***

2. Driver 1 was driving in the center lane of a three-lane road. She was watching two drivers in the far-right lane who were arguing over a fender bender. Without realizing it, Driver 1 drifted left into the far-left lane, causing Driver 2, who was traveling slightly faster, to abruptly hit his brakes. Driver 2 was not able to stop in time to avoid clipping the side of Driver 1's car. ***Who's at fault?***

YOUR TURN

5–1 Starting the Engine and Engaging the Transmission
1. What are the gears on an automatic and manual transmission?
2. How do you engage an automatic and manual transmission?
3. How do you shift gears on a manual transmission?
4. What precautions should you take when putting your vehicle in motion on a hill?

5–2 Steering
5. How is steering different from turning?
6. What are the two basic steering methods?

5–3 Backing up
7. What is the proper procedure for backing up?
8. What precautions should you take when backing up to the left or right?

5–4 Parking
9. What are the four main types of parking?
10. What is the proper procedure for entering and exiting a parallel parking space?
11. How should you turn your wheels when parking on hills?

SELF-TEST

Multiple Choice

1. Engine flooding is an indication that there is:
   a. not enough oil in the car.
   b. rain on the pavement.
   c. too much air in the fuel injector.
   d. too much gasoline in the engine.

2. THIRD gear on a manual transmission is best for:
   a. freeway and highway driving.
   b. backing up.
   c. city driving.
   d. for speeds from 35 to 40 miles per hour (55 to 65 km/h).

3. The push-pull method of driving:
   a. requires you to have only one hand on the wheel at times.
   b. is best on wide, slow turns.
   c. allows you to make sharp turns quickly while keeping both hands on the wheel.
   d. requires you to start with your hands at 12 o'clock and 6 o'clock.

4. Tracking refers to:
   a. monitoring how far you are traveling.
   b. a method of steering.
   c. a method of backing around corners.
   d. changing gears on an automatic transmission.
5. When parking your vehicle facing downhill, keep the front wheels:
   a. turned toward the curb.
   b. straight ahead.
   c. turned away from the curb.
   d. in the original position that you parked the car.

Sentence Completion

1. The _____ is the point at which the engine begins to engage the transmission.
2. You do not always have to shift to the next lower gear when _____.
3. The _____ controls the amount of air entering the carburetor.
4. When backing up to the left, turn your wheel to the _____.
5. _____ parking involves parking between two other parked vehicles.

Matching

Match the concepts in Column A with examples of the concepts in Column B.

Column A
1. __ Manual transmission
2. __ Parking on a hill
3. __ Downshifting
4. __ Dry steering
5. __ Hand-over-hand method
6. __ Double parking
7. __ Push-pull method
8. __ Perpendicular parking

Column B
a. Parking in the roadway
b. Grab the far side of the wheel
c. Slide your hand along the wheel
d. Turning wheels toward curb
e. Parking lot
f. Requires a stick shift
g. Shifting from "4" to "1"
h. Turning wheels when stopped

Short Answer

1. If you are driving a car with a manual transmission, how can you prevent rolling backwards when leaving an uphill parking space?
2. What should you always do before backing up?
3. What are the advantages and disadvantages of a manual transmission?
4. What makes parallel parking so challenging?
5. What is the proper way to straighten out the wheel?

Critical Thinking

1. You are preparing to back out of a perpendicular parking space in a busy lot. The driver directly behind you, across the aisle, has just entered a parking spot also, but you are not sure if he has seen you. Several drivers are lined up to your right anxiously waiting for an empty spot. What should you do?
2. You have just climbed a very steep hill in a car with a manual transmission and are stopped at a red light at the top. A car pulls up right behind your bumper, and the driver starts "revving" the engine in anticipation of the light turning green. He is so close that you are afraid you will roll back into his car no matter how quickly you engage the transmission. What should you do?

Projects

1. *Have two friends park their cars along a curb, leaving a parallel parking space. Measure the distance between the vehicles. Practice entering and exiting the space several times, having your friends move their cars slightly closer together each time to reduce the available space. Drive very slowly when parking and pulling out, and make sure that they stand nearby to caution you if you get too close to either vehicle. Determine the smallest space that you can fit in, and share the results with your class.*
2. *One of the most challenging places to engage gears in a vehicle with a manual transmission is on a hill. If you cannot engage your transmission smoothly, your car will stall or you will roll back too far, possibly hitting the vehicle behind you. Practice this skill on an incline of a quiet street by having someone stand behind your vehicle on the sidewalk to determine how far back you roll each time.*

chapter 6

The Speed Factor: Negotiating Curves and Braking

Although you do not need to be a scientist to drive a car, the more you understand about the basic physical laws acting on you and your vehicle the better you will be able to maintain control under different circumstances and avoid becoming involved in a collision. How long will it take you to stop on a wet road? How much traction will you have going around a sharp curve? What will happen if you accelerate too rapidly? Knowing the answers to these kinds of questions will make you a safer and more skillful driver.

Chapter Objectives

Upon completion of this chapter, you should be able to:

6-1 Physical Laws Affecting Driving
1. Understand the importance of gravity as it relates to driving.
2. Describe the factors that affect vehicle traction.
3. Understand the importance of inertia as it relates to driving.
4. Understand why it is better to collide with a soft object than a hard one.

6-2 Negotiating Curves
5. Understand what centrifugal force is.
6. Understand the importance of center of gravity and loading in negotiating curves.
7. Understand the difference between banked and crowned roads.

6-3 Braking
8. Describe the proper way to brake under normal conditions.
9. Understand what an antilock braking system is.
10. Describe the three factors that determine stopping distance.

6-4 Speed Limits
11. Describe the purpose of speed limits.
12. Know what the basic speed law is.

KEY TERMS

speed	center of gravity	pumping the brakes
acceleration	loading	locking the brakes
gravity	banked road	reaction distance
traction	crowned roads	braking distance
torque	oversteering	stopping distance
inertia	understeering	speed limits
kinetic energy	antilock braking system (ABS)	basic speed law
force of impact		minimum speed law
centrifugal force	threshold braking	

6–1 PHYSICAL LAWS AFFECTING DRIVING

To properly understand the physical laws that play a crucial role in your ability to control your vehicle, you must first become familiar with some of the basic terms used in discussing the science of driving.

Speed and Acceleration

Speed refers to how fast something is moving—that is, how quickly its position is changing over time. In driving terms, speed is measured in miles or kilometers per hour. **Acceleration** measures the rate of change of an object's speed. When you increase your vehicle's speed over a given period, you accelerate. When you reduce your speed, you decrease acceleration or decelerate.

Gravity

Gravity is an invisible force that massive objects like the Earth exert on other objects. This force attracts or "pulls" other objects toward their center. If you stop a car at the top of a steep hill, put the

Figure 6–1 Gravity pulls objects toward the center of the Earth.

Properly inflated tire

Underinflated tire

Overinflated tire

Figure 6–2 A properly inflated tire provides more contact with the surface of the road, and therefore greater traction.

transmission in NEUTRAL, and take your foot off the brake, what will happen? Not only will the car roll down the hill, it will accelerate as it does so.

The force of gravity exerts a strong pull on all objects, including you and your car. It is most noticeable when climbing or descending a grade. As you travel uphill, it takes more power from your engine to maintain or increase speed because you are moving against the force of gravity. Going downhill takes less power, if any, from your engine to maintain or increase speed because you are moving in the direction of the force of gravity.

Traction

Traction is the friction between your vehicle's wheels and the surface of the road. Traction is necessary to move, change direction, and stop. Without traction, your car's wheels cannot "grip" the road. If you are already moving and lose traction, your car may skid, and you will lose some or all of your ability to steer. You may also be unable to stop your vehicle, at least where you intend to stop it. Many factors influence how much traction your vehicle has at any given time.

The total weight of your vehicle and its load, as well as the manner in which it is distributed over your tires, affects traction. For example, even though a large truck may be extremely heavy, it could still have poor traction if the vehicle has rear-wheel drive and most of the vehicle's weight is put on the front tires.

New tires with deep treads grip the road better than worn tires. Wider tires have better traction than narrower ones. Specialty tires such as snow tires provide peak performance on packed snow but may not handle as well on dry pavement or ice. Overinflated and underinflated tires provide less surface area in which to grip the road and therefore have reduced traction. Tires on vehicles with very "tight" suspension, poor wheel alignment, or badly worn shock absorbers have less contact with the road surface as well, especially at higher speeds or on bumpy roads.

Different road surfaces provide varying degrees of traction. Concrete provides the best traction, followed in turn by blacktop, polished concrete, hard dirt, gravel, and sand. Potholes, bumps, cracks, pavement grooves, and other physical alterations on the roadway can adversely affect traction by diminishing the amount of surface in contact with tires. The presence of loose gravel, water, oil, snow, ice, wet leaves, or any other intervening material will reduce traction. The temperature of the road surface will affect traction as well. For example, heat can make asphalt softer, which increases traction, but it can also release slippery oils that decrease traction.

Torque is the ability of a force to cause an object to rotate. As it applies to driving, it is your engine's ability to apply more force to turn the wheels of your car. As you press down on a car's accelerator, the car's engine applies torque to the transmission. This torque turns the wheels of the car and, provided there is adequate traction, puts the car in motion.

To understand the role that torque plays in affecting traction, you need only be familiar with the term *burning rubber*. When the force applied to turn your vehicle's wheels exceeds the force resisting this motion, the traction between the road and your tires is lost. Your wheels spin more or less freely against the pavement and screech loudly. The resulting rapid rotation of the tires against the road surface produces intense heat. This heat can indeed melt the rubber on the tires of your car, and you can smell the proof.

Inertia

The law of **inertia** states that *an object at rest tends to remain at rest, and an object in motion will continue in motion in a straight line until acted upon by a force.* Anyone who has ever tried to drive a car with a dead battery can confirm the fact that objects at rest tend to remain at rest! Our experience tells us that the more an object weighs, the more difficult it is to move. You have probably experienced the feeling of being pushed back against the seat of an automobile when it is accelerating rapidly. You were merely experiencing your body's tendency to remain at rest—its inertia—even as the car's speed was increasing.

The law of inertia also states that objects in motion tend to remain in motion until acted upon by another force. At first, this may seem to go against what we see in everyday life. If you take your foot off the accelerator, for example, your car will slow down and eventually stop. It will not keep moving endlessly down the road on its own. This is because the force of friction between your car's tires and the road opposes the forward motion of the car.

 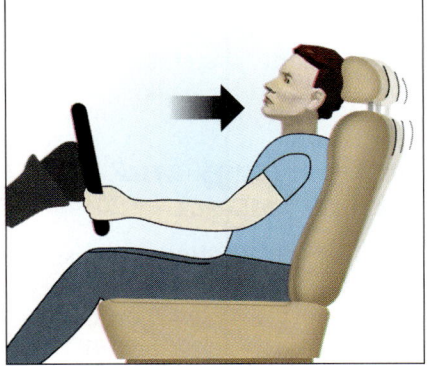

Figure 6–3 The inertia of a passenger at rest "pushes" the passenger back as the car accelerates forward.

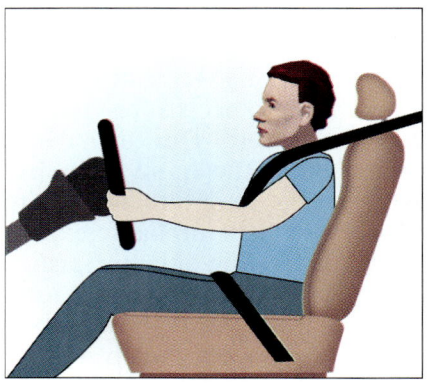 

Figure 6–4 The inertia of a moving passenger makes it seem as if the car is "thrown forward" as it decelerates.

auto accessories

If you have a large trunk or a cargo area in your vehicle, you may want to invest in a trunk organizer, usually made of plastic or canvas, that will hold your supplies and personal items in place when they are subject to the inertia of motion.

The same force that makes our car able to move also means it is necessary to continue to apply the force of the engine to keep it moving!

You experience inertia while driving if you apply the brake forcefully. The car slows, and you are "thrown forward." In fact, your body is merely exhibiting its tendency to stay in motion as the car is slowing. To prevent your body's inertia from taking you straight through your car's windshield during a very sudden stop, vehicles are equipped with occupant restraint systems such as safety belts and air bags.

Kinetic Energy and Force of Impact

All objects in motion possess a measurable quantity of energy called **kinetic energy**. For an object to move, it must acquire kinetic energy. A car's kinetic energy is related directly to its weight and speed.

For a vehicle to stop, it must lose its kinetic energy. This normally happens in one of two ways. The most common method is to apply the vehicle's brakes to reduce its speed. Another is to remove your foot from the accelerator and let the car coast to a stop by allowing the friction of the car's wheels on the roadway to decrease its speed slowly.

A third and much less desirable way for a vehicle to lose kinetic energy is through an impact with a solid object. The less distance over which an object is stopped, the faster it must lose its kinetic energy, and the greater the **force of impact.** The force of impact is determined by both the magnitude of the kinetic energy of the objects colliding and the distance over which the kinetic energy is lost.

Although you can do little to reduce your car's weight, you can reduce its speed. As both common sense and the physics of kinetic energy tell you, decreasing

Figure 6–5 Sand-filled canisters and safety cushions spread out the force of impact.

speed reduces the severity of an impact with any object. Similarly, objects that are softer or have more "give" will spread the force of impact over a greater distance and reduce the severity of the collision. Crashing into a row of bushes will result in much less damage than hitting a concrete wall or telephone pole.

Figure 6–6 The force of impact resulting from a collision with a fixed object can be tremendous.

Precisely for this reason, newly built freeways include steel-beam and cable guardrails to shield objects that cannot be removed or relocated off the roadway. Safety-shaped concrete barriers separate opposing traffic, and "crash cushions," sand-filled plastic barrels or crushable foam cartridges surrounded by sections of guardrail, are sometimes placed at bridge abutments, off-ramp neutral areas, and tollbooths. Also, many road signs and light supports have breakable bases or hinge mechanisms that allow them to break when struck by a vehicle.

Not only are today's roadways designed with the force of impact in mind, but so are vehicles. Unlike the interiors of older cars, which were almost entirely made of metal, the dashboards of modern cars are made of plastic or other soft material that will "give" on impact. Bumpers and bodies are designed with "crush zones" to absorb as much of an impact as possible rather than have the car's occupants experience the full brunt of the force.

6–2 NEGOTIATING CURVES

You have undoubtedly experienced the feeling of being "pulled" outward when rounding a curve in a car. The faster the car goes around the curve, the stronger this force is. This mysterious pull, called **centrifugal force,** is nothing more than the inertia of your body attempting to continue traveling in a straight line as the car's path changes directions. Sharper curves act to change the direction of

Wild Wheels

Figure 6–7 What is the name of this sturdy car, built in 1934, that was able to drive off under its own power after being driven off a cliff?

 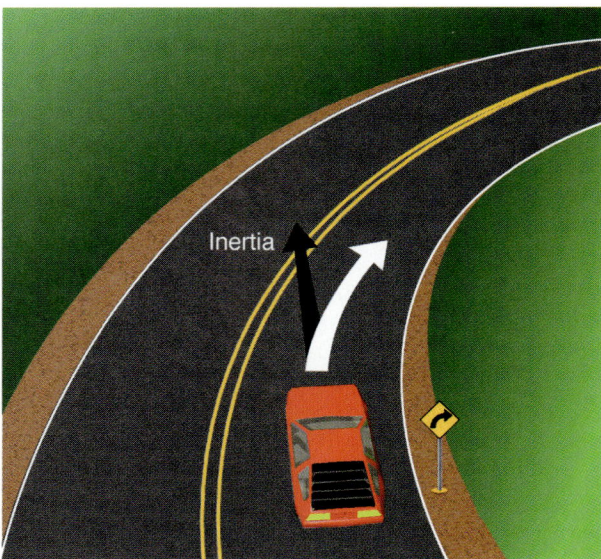

Figure 6–8 A car's inertia exerts force to keep it traveling in a straight line.

your car more quickly. At speeds that are high relative to the angle of the turn, your vehicle's inertia will exceed its ability to follow the curve of the roadway. If your car's inertia exceeds the traction force of your tires, your vehicle may swing wide into oncoming traffic, slide off the road, or roll over.

JACKSON POLLOCK

Born in 1912, Jackson Pollock was the leading painter of a group of artists known as the Abstract Expressionists working in the United States after World War II. Pollock became internationally famous for his "action painting," in which he dripped colored paint onto canvases laid on the floor. On the evening of August 11, 1956, Pollock was speeding to his home in Long Island, New York, with two weekend guests in his 1950 Oldsmobile convertible when he lost control of the vehicle at a curve. The car ran off the road, rebounded from an embankment, swerved to the other side of the road, and turned end over end through underbrush until it collided, upside down, with some trees. Pollock was thrown clear, but he died instantly when his head struck a tree. One of the passengers was also thrown clear and survived with severe injuries, but the other died when her neck was broken as she was pushed back into the trunk by the great force of the impact. Pollock's collision is a gruesome example of how a moving vehicle is affected by physical forces. As a driver, you cannot change the laws of physics, but you can always control your speed.

Center of Gravity and Loading

An object's **center of gravity** is the point about which the object's weight is centered. Vehicles that are wide and low have a relatively low center of gravity. Other vehicles, such as jeeps and sport utility vehicles, have a relatively high center of gravity. When vehicles with a high center of gravity make sharp turns, or attempt to round curves at high speeds, they are much more vulnerable to rolling over.

A vehicle's loading can significantly change its center of gravity. **Loading** means adding weight to your vehicle's weight as measured when it is empty. You "load" your vehicle anytime you or any passengers get into it, or when you attach cargo or pack luggage for a trip. Increased loading will make your engine work harder, which will reduce acceleration. Consider this when calculating the safe distance you will need to make a maneuver such as turning left in front of oncoming traffic. Your engine is much more likely to overheat as it works harder, especially if you are driving in hot weather or up steep inclines. Your fuel economy will also be significantly less than normal.

Adding weight increases your car's kinetic energy, making it harder to stop or change the direction of your vehicle. Depending on the amount of weight being added and the physical location of the added weight, your car's center of gravity can also be altered. This is potentially the most dangerous aspect of loading because you may not fully realize the extent to which the loading has been changed until it is too late. Your car might begin to tip when going around a curve at a speed you normally would think of as safe.

Road Slopes

The slope of the road is another important factor in negotiating a curve. Roads may be either level, banked, or crowned. A **banked road** dips down in one direction, so either the left or right side of the road is higher than the other. A banked road that dips in the direction of a curve reduces the risk of rollover. In most cases, roads are banked in the direction of the

DRIVING TIPS Do Not Drive in the Shadow of a Large Load

Large tractor-trailer trucks often carry loads that can shift, such as tanks containing liquid or propane, large metal pipes, or lumber. Avoid riding next to "big rigs" carrying this kind of load, especially on curves. Depending on the speed of the truck and the road conditions, the load could break loose and land on you, or the tractor-trailer could jackknife and collide with your vehicle. Drop back and allow the truck to take the curve ahead first, maintaining a large following distance in case you need to stop.

Level road   Banked road   Crowned road

Figure 6–9 Roads may be level, banked, or crowned.

curve, but some stretches of rural roadway may actually bank in the wrong direction. This is not done by design, but is often the result of insufficient funding to properly grade lesser-used roads.

Crowned roads are higher in the center than on the sides. Roads are crowned to promote water runoff after rains and to reduce the risk of hydroplaning. The bank of a crowned road can either work with or against you, depending on the direction of travel. If you are going around a right curve, the bank reduces the risk of turnover, but if you are going around a left curve it may actually *increase* the chance of rolling over.

Entering a Curve

Of all the factors that affect negotiating a curve, speed is the only one over which you have a large degree of control. You cannot make a sharp corner less sharp, nor can you change the bank of the curve. You cannot reduce the weight of your vehicle or lower its center of gravity either. The only way to safely negotiate a curve is to reduce your speed *before* entering the curve. Reducing speed by sudden braking or downshifting

USING SAFE — Curves

As you approach a curve in the roadway, remember to use the SAFE process. *Scanning* ahead is especially important on curves, where objects can appear suddenly because of your limited line of sight. Watch for speed advisory signs and special warning signs, chevrons (yellow markers with V-shaped "arrows" pointing in the direction of the curve), and reflectors posted along the edge of the roadway to help guide you around the curve. *Assess* how your vehicle will react to the curve given your speed, the sharpness of the turn, the weight of your vehicle, the condition of the road, and other factors. *Find* an escape route in case an approaching car is straddling the center line or an unexpected hazard appears suddenly. Be especially careful on two-lane, two-way roads and roads with narrow shoulders and drop-offs. *Execute* the best alternative to negotiate the curve and avoid potential dangers smoothly and safely.

during a turn is extremely dangerous and should be considered a measure of last resort. Reducing speed before entering a curve also allows more time to react should an emergency arise.

Once you have adequately reduced your speed before entering the curve, maintain your lane position by avoiding either oversteering or understeering around the curve. **Oversteering** is turning too sharply into the curve. This can cause your rear wheels to slide out, forcing you into a skid. **Understeering** is not turning sufficiently to round the curve. This can cause you to drift into an outside lane, off the road, or into the opposing lane of travel. By properly maintaining your lane position, you will retain solid directional control of your vehicle while rounding a curve.

Under normal circumstances, you should gently accelerate into the curve after entering it and resume a safe speed once you have straightened out again. However, if it is raining or snowing or if the roads are icy, accelerate only *after* you have exited the curve.

6–3 BRAKING

By now, you may be wondering exactly how brakes "remove" kinetic energy from a moving car. When brakes slow a car down, where does this invisible energy

Figure 6–10 Reduce speed *before* entering a curve.

factoid

A braking system exerts up to 1,000 pounds (450 kg) of hydraulic pressure on each of the four brakes.

actually go? When you apply your brake pedal, the force from your foot is transferred through the hydraulic pressure of your braking system to the brake pads or drums against the brake linings on your wheels. Friction increases greatly as the pad or drum comes into contact with the brake linings on the wheel. The friction opposes the force of the rotating wheels, causing them to slow. As the wheels slow, the friction between the rubber of your tires and the surface of the road also increases. If you brake rapidly, the friction between your tires and the road surface increases to the point where traction is lost, resulting in a skid.

The kinetic energy of the car's motion is transformed into heat as the friction of the brakes resists the motion of the wheels. Your brakes and the rubber on your tires get hot. As you may have experienced at one time or another by smelling "burning" brakes, this heat becomes more intense as the brakes are applied continually for long periods or very hard for short periods.

How to Brake

To prepare for those times when you will have to apply your brakes under less than ideal conditions, you should practice proper braking procedures on a dry, flat, firm surface with properly functioning brakes. In time, you will learn to apply just the right amount of pressure to decrease your speed uniformly so that you avoid either stomping on the brake at the last second or stopping far short of your intended point.

Always practice with the particular vehicle in which you will be doing the braking. When you drive a new or unfamiliar car, allow yourself time to "get a feel" for the car's brakes. They may be "mushier" or "tighter" than the

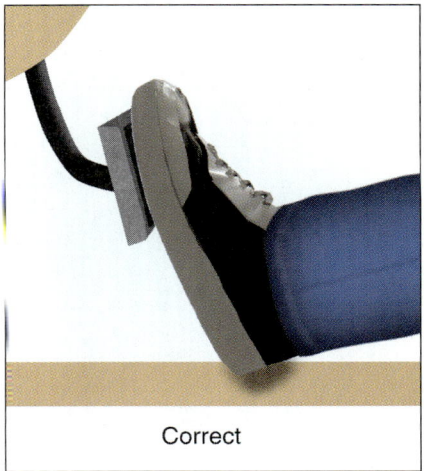
Correct

Incorrect

Figure 6–11 Keep your heel on the floor when braking.

ones in the car you are used to driving. Waiting until you are traveling at a high speed approaching a yellow light is the wrong time to learn.

When braking under normal conditions to reduce speed or stop, begin by lifting your right foot from the accelerator while keeping the heel of your foot resting on the floor of the car. Then, without lifting the heel of your foot off the floor of the car, move your foot to the brake pedal and gradually apply pressure against it with your toes. Do not lift your entire foot off the floor or use the ball of your foot to brake. If you lift your entire foot off the floor, you have to use your more powerful thigh muscle, which makes applying pressure to the pedal more difficult to control.

Antilock Braking Systems

Antilock braking systems (ABS) consist of a sensor mounted on each wheel that is capable of determining whether the wheel locks or stops rotating during braking. When this occurs, the sensor sends a signal to the microprocessor that controls the brakes. The microprocessor then instructs the braking system to release just enough pressure on *that particular wheel* to allow it to resume rotation and avoid a skid. As soon as the wheel begins to rotate, the sensor sends a different signal to the microprocessor that the wheel is no longer locked. The microprocessor then increases pressure on the brake once again.

This cycle is repeated many times a second so that the ABS can stop your car in the shortest possible distance while leaving you with full steering control. There are two primary reasons ABS brakes are more effective than standard brakes. First, because antilock braking systems work *independently* on each wheel, they are more efficient and give you more control than non-ABS brakes. Second, antilock braking systems are controlled by a computer chip that can sense a wheel lock, release it, and then reapply the brakes much faster than a person ever could.

For automobiles with antilock braking systems, you can apply your brakes without worrying whether the brakes will "lock." To properly apply ABS brakes, you need only firmly apply the brake pedal. The ABS computer will do all the remaining work for you. A common mistake made by ABS users is to fully apply the brakes but let off pressure when they feel the "pulsing" sensation from the pedal. This sensation, however, is typical of a properly functioning ABS and is *not* a brake malfunction. Antilock braking systems will prevent you from entering a skid even if the brakes are applied suddenly and forcefully.

Emergency Braking

Even when you do your best to drive defensively, there are times when you must reduce your speed as quickly as possible using a full application of the brakes. If you do *not* have ABS brakes, you can use

Antilock braking systems automatically "pump" the brakes at a rate of eighteen times per second.

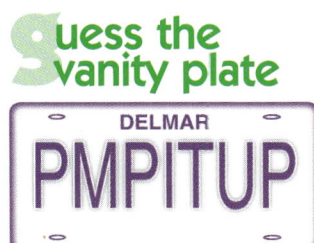

Figure 6–12

factoid

It takes ¾ second to move your foot from the accelerator to the brake pedal.

one of three techniques, depending on how fast you need to stop.

Threshold braking involves using a full and firm application of the brake pedal *up to* the point where the brakes lock and cause the vehicle to enter a skid. When a car begins to skid, the driver loses steering control, making a bad situation worse. When possible, threshold braking should be used to avoid skidding. Skidding not only is dangerous, but also greatly accelerates wear on your tires by burning rubber off them during the skid.

Pumping the brakes involves alternately applying the brakes completely until they lock and then releasing them in rapid succession. In effect, you are attempting to imitate the action of antilock brakes. To avoid losing control of your vehicle if you experience some skidding, you must disengage the vehicle's power train to remove any unbalancing effects it presents. If you are driving a vehicle with an automatic transmission, you disengage the power train by putting the car in NEUTRAL. If you are driving a car with a manual transmission, push and hold in the car's clutch. Because this method involves some degree of skidding, it is a more aggressive technique than threshold braking, but it will also stop your car in a shorter distance.

Never pump the brakes on a vehicle with ABS. This sends false or confusing information to the system's sensors that can actually cause the car to begin skidding.

Figure 6–13 Locking brakes results in a loss of steering control.

Locking the brakes is the most radical braking technique and should be used only when there is no time to react otherwise or use alternative braking methods. To lock your brakes, you firmly press and hold the brake pedal until the vehicle comes to a complete stop. Locking the brakes can stop your car in the shortest possible distance, but you will also go into a skid and lose all or most of your ability to steer.

Stopping Distance

The speed at which you are traveling and the time required for you to stop your vehicle are strongly and directly correlated. Your ability as a driver to stop smoothly, accurately, and rapidly is a function of both reaction distance and braking distance.

CHAPTER 6 The Speed Factor: Negotiating Curves and Braking ◆ 133

> **DRIVING MYTHS** — **Speeding**
>
> Many drivers feel that speeding is worth the risk of a ticket or collision to save precious time on their busy schedule. Speeding, however, will not save as much time as you think. On a 55-mph (90-km/h) highway, driving 10 miles per hour (15 km/h) over the limit will save you only 3 minutes on an average 20-mile (30-km) trip. On city streets, speeding may result in more stops at red lights since traffic signals are often programmed to have consecutive green lights for vehicles traveling at or below the speed limit. The potential costs of speeding are also higher than you might expect. A speeding ticket can cost you up to $1,000 in fines, up to 50 points being put on your driving record, and the loss of your driving privileges for up to a year. Speeding results in lower fuel economy as well, meaning more stops at the gas pump. Finally, speeding significantly increases the amount of pollutants emitted from your engine, which reduces the quality of life for everyone.

Reaction distance is the distance traveled by your vehicle during the time it takes you to identify the need to stop and react to the braking situation. This time translates into actual feet or meters based on the speed of your vehicle, visibility, and on your

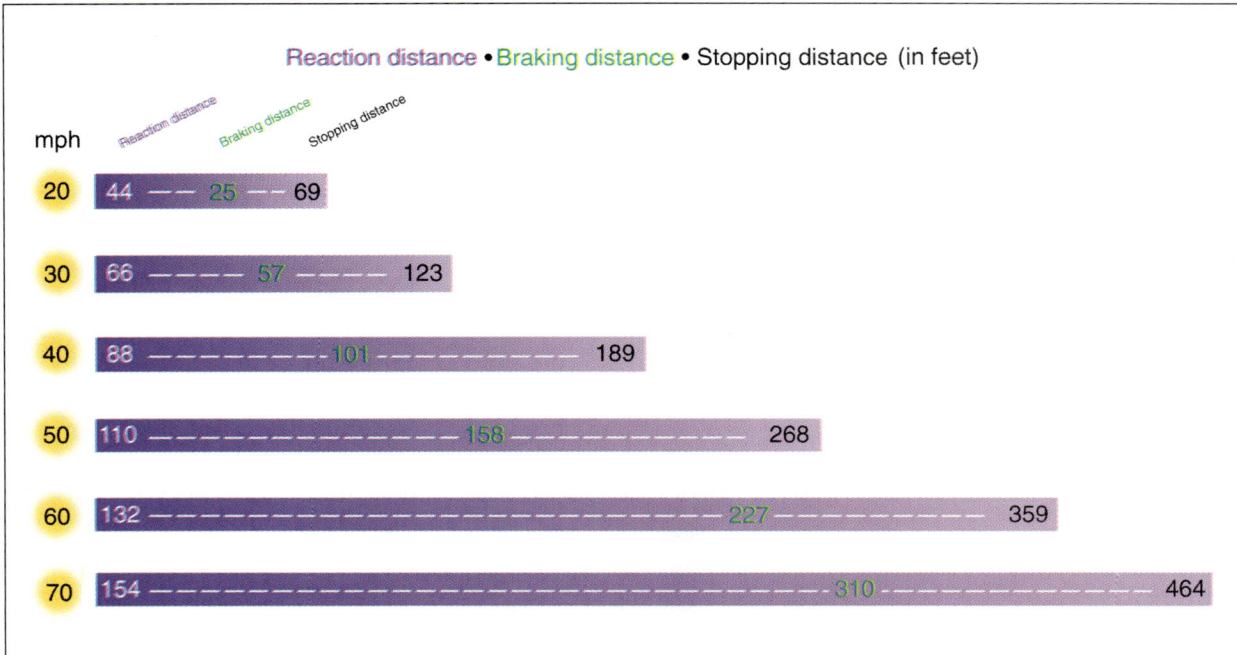

Figure 6–14 The time it takes to come to a complete stop once you recognize the need to apply your brakes increases dramatically at higher speeds.

factoid

The chance of being killed in a collision doubles as speed increases from 45 to 60 miles per hour (70 to 100 km/h), and nearly doubles again at 70 miles per hour (110 km/h).

mental and physical state. The faster you are moving and the harder it is to see, the longer it takes to react. Inattention, fatigue, sleepiness, the presence of alcohol or other drugs in your system, and age can also adversely affect your reaction time.

Braking distance is the time needed to come to a complete stop once the brakes have been applied. Brake effectiveness depends on vehicle type and weight, road and weather conditions, type and wear of tires, and the type and physical condition of your brakes.

Together, the reaction distance and braking distance add up to your **stopping distance**, which is the total distance required to stop from the time you first recognize the need to brake to the time your vehicle is no longer moving. The most important determining factor in your ability to brake is your speed. Simply put, the faster you travel, the longer it takes to stop. Therefore, you should always carefully monitor the one thing over which you have complete control—your speed. Remember that if you ever suspect trouble ahead, slow down!

6–4 SPEED LIMITS

Now that you understand how speed relates to kinetic energy, inertia, traction, force of impact, turning, and braking, you should have a better appreciation for the importance of operating your vehicle at a safe speed. What, exactly, is a "safe" speed? To determine the proper driving speed, you must consider posted speed limits, weather conditions, visibility, the surface condition of the roadway, traffic conditions, lane width, and any special speed laws that might apply to the type of vehicle you are driving, road you are on, or area you are in.

Posted Speed Limits

Both state and local authorities in Illinois post **speed limits** for different roadways. Posted limits are not there to recommend a driving speed, but to advise you of the maximum legal speed, based on considerations of safety and economy, at which to operate your vehicle *under ideal conditions*. Factors considered in setting speed limits include road conditions, traffic volume, pedestrian activity, collision history, public de-

Figure 6–15 Different speed limits exist to take special circumstances into account.

factoid

The Great Fire of 1871 swept through Chicago at the rate of 65 acres per hour.

mand, noise and air pollution restrictions, and parking patterns.

The highest speed allowed anywhere in the state is 65 miles per hour (110 km/h), which is posted on some rural interstate highways. The speed limit on interstate highways near or in major urban areas and on most other highways in Illinois is 55 miles per hour (90 km/h). If no limit is posted on a rural highway, you may not exceed 55 miles per hour. Vehicles towing trailers, most recreational vehicles, and commercial trucks may never go faster than 55 miles per hour, even on rural interstates.

Speed limits on other roadways vary by location. Those in urban areas are generally lower than they are in rural areas. If no limit is posted, you may not exceed 30 miles per hour (50 km/h) in a business or residential district. You are also restricted to a maximum speed of 15 miles per hour (25 km/h) in an alley. School zones have a maximum speed limit of 20 miles per hour (30 km/h) during school days between the hours of 7 AM and 4 PM. Most school zones in Illinois have signs indicating that the limit applies WHEN CHILDREN ARE PRESENT. Many also have mounted yellow lights that flash when you are required to reduce your speed.

Always remain alert to posted limits. It is possible for the same roadway to have different speed limits for each direction of travel. Lower speed limits are commonly posted on dangerous stretches of roadway, such as curves and freeway ramps, on bridges, and in construction zones. Be aware that speeding in a construction zone while workers are present is a separate offense that carries higher-than-average penalties because of the dangers faced by road crews.

The Basic Speed Law

Never blindly follow maximum speed limits. According to Illinois's **basic speed law,** it is illegal to operate a motor vehicle at a speed greater than is reasonable and proper for existing traffic conditions, regardless of posted limits. A "reasonable and proper" speed is one that does not endanger the safety of persons or property. Always take into account both actual and potential hazards—the weather, visibility, the amount of traffic, the surface condition and width of the roadway, and the presence of bicyclists and pedestrians—when determining a safe speed. If you are driving 10 miles per hour (15 km/h) below the speed limit in a pounding rainstorm, you can still be cited for speeding if you are going "too fast for conditions."

Be aware that state law also states that you have a duty to slow down on curves, on narrow or winding highways, and when approaching an intersection or the crest of a hill even if no speed limit sign is posted.

The Minimum Speed Law

Driving too slowly can be just as dangerous as driving too fast. Illinois's **minimum speed law**

Speed is a factor in one-third of all fatal collisions.

Figure 6–16 This facility accelerates subatomic particles to nearly the speed of light.

REALITY CHECK — *Radar Detectors*

In addition to "pacing" (following a driver for a certain period of time to determine his or her speed) and aerial speed measurement, law enforcement in Illinois uses a variety of sophisticated gizmos to crack down on speeders, including hand-held radar, moving radar, and lasers, and not everyone likes it. While radar detectors are permitted in non-commercial vehicles, they are useless—the only thing they will detect is the ticket you are going to get. Instead of investing your hard-earned dollars in an expensive toy, buy something practical like a roadside emergency kit.

factoid

America's first automobile race, which ran from Chicago to Evanston, took place in 1895 and was won by J. Frank Duryea, who attained an average speed of 7½ miles per hour (12 km/h).

prohibits drivers from moving at such a slow speed as to impede or block the normal and reasonable movement of traffic, except when reduced speed is necessary for safe operation or in compliance with the law. Studies have shown that when all vehicles on a roadway are going at about the same speed, the chance of a collision is less than when one vehicle is being driven at a much slower speed than the ones surrounding it. Some roadways in Illinois have minimum posted speed limits, and you can be ticketed for driving below these limits unless road conditions otherwise demand a slower speed.

WHO'S AT FAULT?

1. Driver 1 was on his way to Aurora University in Aurora with a hard storage carrier strapped to the roof of his vehicle. He was driving on Interstate 88 in heavy traffic when Driver 2 ahead of him had to stop suddenly. Driver 1 was able to stop in time to avoid hitting Driver 2 with his car, but the storage container on his roof broke loose and landed on Driver 2's vehicle, smashing the trunk and breaking the rear window. *Who's at fault?*

2. In Peoria, a pedestrian stepped onto Sterling Avenue outside of a crosswalk in front of Driver 1, who was driving a brand-new sport utility vehicle just over the posted speed limit. Driver 1 swerved sharply to avoid the pedestrian and overturned the vehicle, which skidded for some distance before hitting a lamp post. *Who's at fault?*

YOUR TURN

6–1 Physical Laws Affecting Driving

1. What is the importance of gravity as it relates to driving?
2. What are the factors that affect vehicle traction?
3. What is the importance of inertia as it relates to driving?
4. Why is it better to collide with a soft object than a hard one?

6–2 Negotiating Curves

5. What is centrifugal force?
6. What is the importance of center of gravity and loading in negotiating curves?
7. What is the difference between banked and crowned roads?

6–3 Braking

8. What is the proper way to brake under normal conditions?
9. What is an antilock braking system?
10. What three factors determine stopping distance?

6–4 Speed Limits

11. What is the purpose of speed limits?
12. What is the basic speed law?

SELF-TEST

Multiple Choice

1. Which of the following can significantly change a vehicle's center of gravity?
   - a. loading
   - b. force of impact
   - c. antilock brakes
   - d. torque

2. Which of the following statements about antilock braking systems (ABS) is *not* correct?
   - a. ABS works on each wheel independently.
   - b. ABS works best when you "pump" your brakes.
   - c. ABS causes the brake pedal to "pulse."
   - d. Motion sensors and a microprocessor are key components of ABS.

3. Where posted, maximum speed limits indicate:
   a. the suggested speed for driving under normal conditions.
   b. the suggested speed for driving under *most* conditions.
   c. the maximum legal speed permitted at *all* times.
   d. the maximum legal speed permitted under normal driving conditions.
4. The proper way to brake in an emergency situation with non-ABS brakes is to:
   a. apply threshold braking.
   b. "pump" the brakes.
   c. lock the brakes.
   d. all of the above
5. If it were not for _____, your car would be unable to move.
   a. centrifugal force and kinetic energy
   b. centrifugal force and gravity
   c. torque and traction
   d. torque and centrifugal force

Sentence Completion
1. _____ measures the rate at which an object's speed is changing.
2. _____ is the force that "pulls" you to the outside while rounding a curve.
3. The _____ says that a driver should always operate his or her vehicle at a speed that is reasonable and prudent for existing conditions.
4. _____ is the force that pushes you back into your seat when your car accelerates.
5. _____ distance is the distance traveled by your vehicle during the time it takes you to identify the need to stop and react to the braking situation.

Matching
Match the concepts in Column A with examples of the concepts in Column B.

Column A
1. __ Gravity
2. __ Force of impact
3. __ Inertia
4. __ Center of gravity
5. __ Acceleration
6. __ Traction
7. __ Speed
8. __ Torque

Column B
a. How quickly an object's position changes over time
b. The friction between your tires and the road
c. Is zero while maintaining a constant speed
d. "Pulls" objects toward the Earth
e. Continuing in motion until acted upon by another force
f. The force your engine applies to turn your car's wheels
g. Point at which an object's weight is centered
h. Varies with the distance over which kinetic energy is lost when objects collide

Short Answer

1. How should you apply the brakes on a vehicle equipped with an antilock braking system?
2. What are the advantages and disadvantages of crowned roads?
3. In terms of the natural laws that act on cars and drivers, what role do safety belts play?
4. What factors should you consider in deciding how fast to drive?
5. What might happen if you understeer going around a curve?

Critical Thinking

1. Your friend, who usually drives a small compact car, takes you on a "test ride" in a large sport utility vehicle that she is considering purchasing. She begins to drive aggressively to "put it to the test." What are some of the risks you both face even if few cars are on the road? What specific risks may be associated with negotiating curves and braking?
2. You have been "elected" to drive several friends back down the mountain after a full day of skiing. It is late in the afternoon when you begin your trip down the steep, treacherous grade. What factors do you need to consider when deciding how closely to follow a large truck in front of you that you are unable to pass?

PROJECTS

1. Using a local street map, draw a circle around your home that extends for several blocks in all directions. Find all of the speed limit signs that fall within the circled area and indicate their location on a map. How far apart are the signs placed? If different speed limits are posted on the same stretch of roadway, what might account for the increase or decrease? Do you feel that the posted limits are appropriate? Are there places that should have signs but do not?
2. Armed with a pad and pen, visit a nearby amusement park. Select three rides that best demonstrate inertia, the effects of gravity, and centrifugal force. Sketch each and explain to the class how the ride demonstrates one or more of these forces at work.

Welcome to the Road
unit 3

Chapter 7   Signs, Signals, and Roadway Markings
Chapter 8   Intersections and Right-of-Way
Chapter 9   Turning

Signs, Signals, and Roadway Markings

chapter 7

Signs, signals, and roadway markings are all designed to reduce confusion and increase safety. Traffic controls ensure an orderly flow of traffic, regulate where you can drive and park, warn you of hazardous areas or conditions ahead, and provide information about nearby places and services. It is important that you familiarize yourself with the different kinds of traffic controls to protect yourself and other users of the roadway.

Chapter Objectives

Upon completion of this chapter, you should be able to:

7–1 Traffic Signs
1. Understand the meaning of different colors in traffic signs.
2. Identify signs that have their own distinctive shapes.

7–2 Traffic Signals
3. Explain what you do at a red light, yellow light, and green light.
4. Understand how signal arrows control traffic.
5. Understand the purpose of variable message signs.

7–3 Roadway Markings
6. Understand how white and yellow lines are used on roads to control traffic.
7. Identify special lanes with distinctive roadway markings.
8. Understand how different colors are used to regulate parking at curbs and in parking spaces.
9. Understand the purpose of pavement markers.

KEY TERMS

- regulatory signs
- crossbuck
- warning signs
- route markers
- guide signs
- milepost
- construction signs
- service signs
- recreation signs
- traffic signals
- lane-use signals
- ramp signals
- variable message signs
- pedestrian signals
- limit line
- turn lanes
- two-way left-turn lane
- reversible lanes
- bus lanes
- bicycle lanes
- reflective pavement markers
- rumble strips
- speed bumps

7–1 TRAFFIC SIGNS

Nothing about a traffic sign's color, size, shape, or location is an accident. There are very strict and detailed requirements about what signs look like and how and where they are posted. Traffic signs tell you which lanes you can use, how fast you can go, what dangers lie ahead, and where and how long you can park. The use of different combinations of colors, shapes, and symbols allows drivers to recognize signs quickly and from a distance.

Regulatory Signs

Regulatory signs tell you what you can or cannot do at certain times and places. They direct traffic to stop, yield, or go in a particular direction. They also control parking and passing, and restrict pedestrians and drivers from doing certain things. Obey regulatory signs wherever you see them. The next time you feel like ignoring a STOP sign in the parking lot of a shopping mall, motel, theme park, or stadium, keep in mind that signs posted on private property may be just as enforceable as signs on the public roadway.

Regulatory signs come in a variety of shapes and colors. The STOP sign is the only octagonal (eight-sided) sign on the road. The YIELD sign is the only sign in the shape of a downward-pointing triangle. Both of these signs usually appear at intersections of all types.

Most other regulatory signs are white squares or rectangles with black letters, words, or symbols. In general, these signs tell you what to do or set limits on what you can do. Red is reserved exclusively for STOP, YIELD, WRONG WAY, and DO NOT ENTER signs and to prohibit certain movements. Some square-shaped signs have a black symbol—for example, a U-turn or "P"

Figure 7–1 There are specific laws about the shape, color, and placement of signs.

Figure 7–2 Regulatory signs

REALITY CHECK *Sign Vandalism*

On a clear night in February, 1996, an eight-ton truck hit a white Camaro at a rural intersection in Tampa, Florida, instantly killing the three teenagers inside. The STOP sign at the intersection had been pulled from the ground and lay face down by the side of the road. Just days earlier, three young friends heading home from a shopping trip had removed several road signs in the area "for a rush," piling them in the back of their pickup. Although they claimed not to have removed the STOP sign at the site of the fatal crash, the trio was arrested and convicted of manslaughter the following year in a case that received national attention. Although you may think it is fun to knock down signs or have a road sign on your bedroom wall, sign vandalism is a crime, not a prank. As the mother of one of the victims said, "Pranks don't kill."

factoid
The City of Chicago handles more than 42,000 requests to install, repair, or remove traffic signs each year.

for parking—inside a red circle crossed by a diagonal red slash. This type of sign indicates that the action or vehicle shown is not permitted on the roadway.

A special type of regulatory sign called a **crossbuck** is posted at most railroad crossings. It consists of a crossed white sign with the words RAILROAD CROSSING in red or black, and is equivalent to a YIELD sign. Crossbucks are sometimes mounted above a pair of red lights that flash when a train is approaching. If there are multiple tracks at the crossing, the number of tracks may be indicated on signage below the crossbuck.

Parking signs are rectangular in shape with black, green, red, or blue colors on a white background. Green letters usually indicate a time restriction—for example, parking is allowed only for two hours. Parking, "standing" (sitting in your vehicle while parked), or loading is either not allowed at all or prohibited at certain times or on certain days if the letters and/or numbers are red.

Figure 7–3 STOP and YIELD signs have their own shapes.

CHAPTER 7 Signs, Signals, and Roadway Markings ◆ 147

Figure 7–4 Pay careful attention to signs before you park your vehicle.

Figure 7–5 Do not be tempted to park in a space designated for the disabled if you are not qualified to use it.

Figure 7–6 Warning signs

Blue letters—sometimes in combination with green letters and/or a green border—usually mean that the area is designated for use by disabled persons. Parking signs for the disabled often display the symbol of a person in a wheelchair. Disabled parking spots are located closer to building entrances than are other parking spots. You may only park in such a space if you have a special license plate or placard. If you illegally park in a space reserved for the disabled, you will pay a large fine and may get your car towed. More importantly, you will be taking up a space that a disabled person needs much more than you do.

Posted parking regulations can be complicated, so take the time to read all of the signs carefully before you leave your vehicle. Parking on major urban and suburban roads may be restricted during commuting hours and designated times for street sweeping. Also, be aware that some neighborhoods in Illinois require you to have a special permit to park on residential streets.

Warning Signs

Warning signs alert you to possible hazards ahead, a change in road conditions, or an upcoming intersection or pedestrian crossing. If you see a warning sign, slow down and stay alert. Most warning signs are diamond-shaped with a yellow background and black words and/or symbols. Square-shaped advisory speed signs, often

factoid

For every $1 spent on signs, about $20 is saved in expenses associated with collisions.

Figure 7–7 Yellow signs alert drivers to upcoming hazards or dangerous conditions.

reality check

When driving through a school zone, you cannot resume a normal driving zone until you get to the exact spot where the sign is posted indicating the end of the school zone, not when you see the sign.

posted below diamond-shaped signs, recommend a safe speed for traveling on curves, freeway ramps, and other dangerous stretches of roadway. Large square-shaped yellow signs typically combine a symbol with an advisory speed or special warning.

School zone signs have five sides (a pentagon). Most are fluorescent yellow-green in color to better set them off from other warning signs, but some are the traditional yellow. There are two types of pedestrian and school crossing signs, and many motorists confuse their meanings. One type shows a pedestrian or pair of schoolchildren alone, while the other shows them within crosswalk lines. The first type of sign is posted well ahead of an intersection or established crossing, which may or may not be marked with crosswalk lines, where pedestrians or children cross the road. The second type is placed at or just before an actual crosswalk. Be aware that pedestrian and school crossings can sometimes occur in the middle of a block. If you mistake a crosswalk warning sign for an advance crossing warning sign, you might hit someone.

In addition to school zone signs, there are two other types of warning signs with unique shapes. Pennant-shaped signs indicate no-passing zones. Because your view of the right side of the road is partially blocked while trying to pass another vehicle, they are posted on the left side of a two-way road. A round sign with an X and two R's (for "railroad") warns you that you are approaching a railroad crossing.

Route Markers

Route markers come in a variety of colors, shapes, and sizes and are used to identify numbered roadways or roadways designated for special purposes. In Illinois, these

Figure 7–8 Railroad crossing signs

include state routes, U.S. routes, interstate routes, off-interstate routes (spurs and business loops), bike routes, and scenic routes such as the Great River Road and the Lincoln Heritage Trail. Small, rectangular "auxiliary markers" are mounted above and/or below some numbered route markers to provide additional information or guidance. These markers consist of directional arrows or words such as WEST, TO, BUSINESS (business loop), JCT (junction), and DETOUR.

actoid

Before the United States Congress in 1926 approved a plan to number all major roads in the country, "auto trails" were marked by colored bands or symbols painted or posted on telephone poles.

Figure 7–9 Route markers help you navigate the state's extensive network of highways.

Guide Signs

Green-and-white square and rectangular **guide signs** tell you where you are, where you are going, or how to get to a particular place from the roadway you are on. One type of guide sign has directional arrows to indicate which road, lane, or exit to use to get to a destination or to access a route. Another type indicates the distance in miles from certain locations, roadways, or exits ahead of you.

A special type of guide sign, called a **milepost,** appears at one-mile (1.6-km) intervals along sections of interstate highways to indicate your location on the roadway. Mileposts are narrow, vertical rectangular panels located along the outside edge of the shoulder. They have a number, usually accompanied by the word "MILE," that identifies how far you are from the state boundary line. Numbers begin at the southern border for north–south routes and the western border for east–west routes. If you are driving north on Interstate 57 and see a MILE 20 sign, for example, you are 20 miles (32 km) north of the Missouri state boundary.

In addition to helping drivers estimate distances they have traveled, mileposts are used by law-enforcement officers and emergency services to identify the location of a crash, violation, or breakdown. Because of the large volume of traffic in metropolitan Chicago, secondary mileposts mark each tenth of a mile between each regular milepost.

Construction Signs

Construction signs alert you that you are in or about to enter a construction or maintenance zone. Like warning signs, most of these signs are diamond-shaped and have black letters, words, or symbols, but they have an orange instead of a yellow background. Cones, drums, panels, tubes, re-

bumper sticker sightings

> LEAD ME NOT
> INTO TEMPTATION,
> I CAN FIND IT MYSELF

Figure 7–10 Guide signs tell you where you are and where you are going.

CHAPTER 7 *Signs, Signals, and Roadway Markings* ◆ 151

Figure 7–11 Construction signs

Service Signs

Blue-and-white service signs inform the driver of nearby services, including gasoline stations, law-enforcement agencies, roadside motels, restaurants, campgrounds, picnic areas, hospitals, telephones, rest stops, and scenic overlooks. Some service signs include logos provided by specific businesses.

Recreation Signs

Rectangular- and trapezoid-shaped recreation signs inform travelers about public places of cultural interest and recreation such as historic sites, natural landmarks, lakes, forests, reservoirs, museums, and parks. These signs are brown with white or yellow letters and numbers. Within recreation areas, small square-shaped signs with symbols direct travelers to rest

flectors, flashing yellow lights, flares, flags, flashing or sequencing arrow panels, and other channeling devices may also be used to guide drivers. Always reduce your speed when driving through an area where workers are repairing or constructing roadways, even if no special speed limits are posted.

Highway heroes

While preparing to stop traffic on the Kennedy Expressway near one of the North Avenue exits in Chicago, flagger Dyclenn Crenshaw was killed by a dump truck that backed over her on May 5, 1994.

Figure 7–12 Slow down and use extra caution when driving through a construction zone.

Figure 7–13 Service signs

Figure 7–14 Blue service signs often include the names of restaurants and other roadside businesses.

Figure 7–15 Recreation signs direct you to the many points of interest in the "Land of Lincoln."

rooms, hiking trails, bicycle rentals, fishing areas, cross-country skiing or snowmobile trails, swimming areas, boat ramps, and other facilities.

Local Traffic Signs

Cities and towns in Illinois often have their own traffic signs that identify local parks, libraries, parking facilities, public transportation routes, and other services. Communities may post public awareness signs to highlight various activities like neighborhood watch groups. On private streets and parking lots, you may encounter signs that have unfamiliar color schemes or irregular shapes. You should obey local signs even if they do not look exactly like those found on public roadways.

7–2 TRAFFIC SIGNALS

Traffic signals are normally found in intersections and other areas where traffic is heavy or there is a high level of risk that different roadway users will cross paths. They are also used to inform users about the roadway that they are approaching or to let them know that they are entering an area with special dangers or restrictions, such as a school zone, construction zone, or toll plaza.

Traffic signals have two advantages over signs. First, because they are illuminated and mounted higher than most signs, they are easier to see from a distance, especially if it is dark or the weather is bad. Second, because they are often timed, linked to sensors in the roadway, or equipped with overhead motion detectors, they can be programmed to adjust to the changing flow of traffic.

Most traffic signals use a red-yellow-green sequence of lights, arranged vertically or horizontally. Red lights always appear at the top or left, yellow lights in the center, and green lights at the bottom or right. This uniform arrangement allows drivers who cannot distinguish between red and green (a common form of color blindness), or who cannot see the entire signal because of an obstruction like an overhanging tree branch or truck, to correctly interpret the signal. Lights can be solid, they can flash, or they can be illuminated arrows. Signals at complicated intersections may have multiple light sequences, each controlling a different lane. When approaching such a signal, especially where solid lights and arrows appear together, it is always important to identify which lights apply to your lane.

Where am I?

Figure 7–16 This renovated theater, which opened in 1931, was the first building in Illinois outside of Chicago to have air conditioning.

Red Light

You must come to a *full* stop at a red light. Unless you are lawfully making a "right on red" or turning left from a one-way street to another one-way street, you cannot proceed after stopping until the light turns green. A flashing red light works like a STOP sign. Come to a complete stop and then proceed when it is safe to do so and you have the right-of-way.

Yellow Light

A yellow light indicates that the signal is changing from green to red. It warns drivers to slow down and prepare to stop. When you see a solid yellow light, make a complete stop if you can safely do so within the amount of space between your vehicle and the limit line. If you cannot stop safely, proceed through the yellow light. If the light is flashing yellow, slow down, give the right-of-way to other traffic, and proceed with caution.

Green Light

A green light means to proceed if safe. If you have been stopped at a red light, make sure that you first look to your left and right to check for any vehicles or pedestrians that might still be in the intersection. If you are already moving when you approach a green light, continue to move through it with caution. When you see a green light, especially if you have noticed that it has been green for a while, be prepared for the "stale" green light to turn yellow.

Red Arrow

You cannot go in the direction indicated until the light changes to green. *A right or left turn is not allowed on a red arrow, even after coming to a full stop.*

Yellow Arrow

Come to a complete stop if you can do so safely. If the yellow arrow is flashing, slow down, watch for other traffic and hazards, and

---

DRIVING TIPS — *"House" Signals*

Some major intersections in Illinois with left-turn lanes are equipped with a five-light signal known as a "house" signal to improve traffic flow. It consists of a single red light in the top middle above a cluster of four lights. The light on the bottom left is a green arrow, while the light on the bottom right is a solid green. If the green arrow is lit, this indicates that you may make a left turn without the interference of oncoming traffic, which is halted with a red light. When the solid green light is on, you are allowed to make a left turn only after yielding to oncoming traffic. Each green light has a corresponding yellow light above it. When the red light is on, left turns are prohibited. These types of signals are often accompanied by a LEFT TURN YIELD ON GREEN regulatory sign as a reminder. Through and right-turning traffic is usually controlled by standard red-yellow-green signal mounted about 6 feet (2 m) to the right of the signal.

Figure 7–17 Signals regulate traffic more efficiently than signs.

proceed in the direction indicated with caution.

Green Arrow

Proceed in the direction indicated if it is clear and safe to do so. Watch for signs indicating that you may turn *only* on a green arrow.

Lane-Use Signals

Overhead **lane-use signals** are used to indicate whether or not lanes are open to traffic moving in your direction. Though no longer common in Illinois, you will see them in tunnels, on bridges, and at toll plazas throughout the United States. If you see a solid green light or a downward-pointing green arrow, this indicates that the lane below is open. *Never* drive in a lane with a solid red light or red X above it. Lane-use signals are also used on municipal streets with reversible lanes, and in the course of your

CHAPTER 7 Signs, Signals, and Roadway Markings ◆ 155

RED X
You must never drive in a lane under a RED X signal.

GREEN ARROW
You are permitted to drive in a lane under a GREEN ARROW.

YELLOW X
A steady YELLOW X indicates the driver should safely vacate this lane—because it soon will be controlled by a RED X.

FLASHING YELLOW X
A flashing YELLOW X indicates the lane is to be used, with caution, for left-turn movements only.

Figure 7–18 Lane-use signals

travels you may encounter overhead yellow X's and flashing yellow X's. A yellow X indicates that the lane is about to be opened ahead to oncoming traffic and that you should move over as soon as you can do so safely to a lane with a downward-pointing green arrow. A flashing yellow X means that you may use the lane only to make a left turn.

Warning Signals

Some roadways have permanently flashing yellow lights to warn drivers of an approaching curve or exit, a dangerous intersection, a dividing roadway, the beginning of a school zone, or any other place requiring drivers to slow down. Signals posted near fire stations that normally flash yellow may turn to a steady yel-

There are more than 100 ramp signals installed on freeways in the Chicago area.

Figure 7–19 Flashing yellow lights warn you to slow down for a hazardous area ahead.

Figure 7-20 Special signals are used to regulate traffic entering some stretches of freeway in the Chicago area.

low, and then red, when emergency vehicles are exiting the station on a call.

Ramp Signals

In the Chicago area, special signals are placed at the end of some freeway on-ramps to control the number of vehicles merging onto the freeway at one time. Similar in appearance to stop lights, **ramp signals** have alternating red and green lights that allow one vehicle at a time to proceed onto the freeway. A sign posted with the signal indicates where you should stop to wait for the light to turn green before proceeding. The entrances of some on-ramps have a warning sign that reads RAMP SIGNAL AHEAD. Ramp signals generally operate only at peak traffic hours; otherwise they are turned off or have a continuous solid green light to allow cars uninterrupted freeway entry.

Variable Message Signs

Some freeways in Illinois have special electronic "signs" that provide up-to-date information about lane and exit closings, collisions, weather problems, and other warnings about changes in driving conditions. These signs

Figure 7-21 Variable message signs provide current information on road conditions.

The typical pedestrian WALK signal stays on from 4 to 8 seconds.

are usually posted in areas with high-density commuter traffic. Portable **variable message signs** may also be set up by construction or maintenance crews to detour traffic or warn drivers about problems on the roadway ahead.

Pedestrian Signals

Pedestrian signals are designed for intersections with heavy pedestrian traffic. Most are mounted just below traffic signals. Pedestrians are allowed to cross when they face a green or white WALK signal or the symbol of a person walking. When they face a flashing orange or red DON'T WALK signal or the symbol of a raised hand, they must wait at the curb if they have not yet started to cross the street, or they should clear the intersection if they are still in it.

Sometimes pedestrian signals must be activated with special pushbuttons. Do not assume that a pedestrian who approaches a green light at an intersection and is too late to activate the WALK signal will stop. Also, be aware that pedestrians may ignore a DON'T WALK signal or not realize it is on and try to cross in front of you.

Hand Signals for Directing Traffic

Sometimes signal lights malfunction or traffic becomes so heavy at an intersection that police or traffic control officers are called in to direct traffic. This is common at festivals, sports events, parades, concerts, and theme parks where the streets are closed off or where the parking lot entrances and exits merge with the main roadway. School crossing guards and construction workers also use hand signals, as well as handheld signs, paddles, or flags, to temporarily halt or detour traffic. Holding the hand up, palm facing outward, is the signal to stop. Waving one arm forward or in a circular motion is the signal to proceed. Hand signals always take priority over signs and signals.

7–3 ROADWAY MARKINGS

Roadway markings are painted on the pavement and curbs to provide warning and direction. They can be lines, symbols, letters, or words. Roadway markings define where lanes are and how they may be used, regulate traffic and parking, and warn of approaching dangers. Like signs and signals, they have colors that mean different things.

White Traffic Lines

Broken white lines separate traffic moving in the same direction, and may be crossed when changing lanes or merging. Solid white lines are used to mark turn lanes, bicycle lanes, the right edge of the roadway, and areas where changing lanes is either dangerous or illegal, such as near intersections. They also may appear around fixed obstacles on one-way roads, such as bridge supports, and neutral areas (called "gore points")

Figure 7–22 Pedestrian signals

Single white lines were first used as road dividers in 1927.

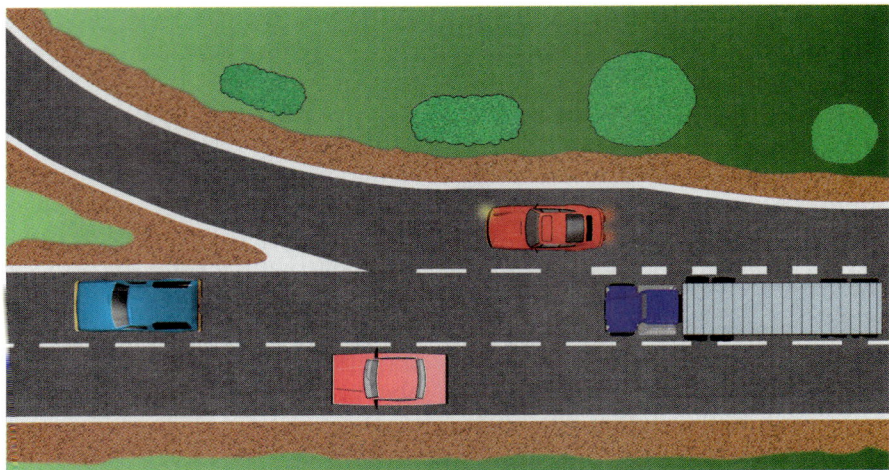

Figure 7–23 White lines may be solid, dashed, or dotted.

between freeway lanes and access ramps. Single solid white lines should not be crossed and, depending on where they appear, may be illegal to cross.

Yellow Traffic Lines

Yellow lines separate traffic going in opposite directions on a two-way road. They may be solid or dashed, or a combination of solid and dashed yellow lines. They may also be single or double lines. If two-way traffic is separated by a median strip or concrete divider, the left side of the roadway is marked with a single yellow line. Solid yellow lines also mark obstacles in the center of a two-way road. Unless you are lawfully passing another vehicle or making a left turn, or are detoured by roadway construction, temporary markers, or law-enforcement or emergency personnel, you must stay to the right of yellow lines.

The pattern of solid and dashed yellow lines used on a roadway indicates whether passing is permitted. No passing is allowed if there are two solid yellow lines or, as may occur on very narrow roads, a single solid yellow center line. If a solid and a dashed yellow line appear together, passing is allowed if there are no oncoming cars and you are next to the dashed line; passing is not allowed

REALITY CHECK *Old Lane Markings*

When roads are repainted or new lanes are added, sometimes the old roadway markings are not completely removed. Watch out for other drivers that move into your lane because they are following the old markings.

Figure 7–24 The pattern of yellow center lines on a roadway determines whether passing is permitted.

RUDY TOMJANOVICH

During his eleven-year career as a player with the Houston Rockets basketball team, Rudy Tomjanovich competed in the NBA playoffs five times, scored 13,000 points, and made five trips to the All-Star Game. Upon his retirement in 1981, the "Rocket Man" became a scout, then assistant coach, and eventually head coach of the Rockets. In July 1994, 3 weeks after his team won the NBA championships, Tomjanovich was arrested on a highway in Houston for crossing a yellow dividing line twice in his jeep. The coach spent the night in jail for this offense, as well as other violations. Straying out of a lane is often an indication of drunk driving, fatigue, or speeding, and is certain to get the attention of law enforcement. If you want to avoid being pulled over by the police, pay close attention to roadway markings and other traffic controls.

Highway heroes

On May 15, 2000, a driver attempting to pass several tractor-trailers in a no-passing zone and in an IDOT construction zone on Illinois 34 north of Eichorn in Hardin County ignored instructions to stop by flagger John Bramlet and collided with a stopped vehicle, which struck and killed Bramlet.

if you are next to the solid line. If there is only one dashed yellow line and no oncoming cars, passing is allowed.

In Illinois, you are permitted to turn left across solid double-yellow lines only when entering or leaving a building, alley, private road, or driveway. It is always illegal to drive across two sets of double-yellow lines. These markings, which designate a "virtual" island, are sometimes used in the center of a street where there is no physical barrier to separate traffic. If in doubt, never turn across double-yellow lines.

Limit Lines

A **limit line** is either a stop line—a white or yellow line, sometimes accompanied by the word "STOP," painted across the roadway—or the nearer of two crosswalk lines. Crosswalk markings are sometimes connected with diagonal lines, or the crosswalk itself may appear as a series of painted bars or raised markers on the road. Sometimes words such as "PED XING" (pedestrian crossing) or "X WALK" (crosswalk) are also painted nearby on the pavement. When you are required to come to a stop at an intersection by a sign or signal, your vehicle must come to a full stop *behind* the limit line. If there is no limit line, you must stop at the sign or, if there is no sign, at a point nearest the intersecting roadway where you have a clear view of cross traffic.

Special Lanes

Special roadway markings are used to separate lanes reserved for certain actions or vehicles from the normal lanes of traffic.

Turn Lanes

Turn lanes are often added near intersections to separate left- or right-turning traffic from through traffic at an intersection. White arrows, often accompanied by the word "ONLY," indicate that you must stay within a designated lane while turning onto the cross street. Some turn lanes have multiple arrows, allowing you to turn

CHAPTER 7 *Signs, Signals, and Roadway Markings* ◆ **161**

Figure 7–25 Two-way left-turn lanes can be used by traffic turning left from both directions.

left or right, or even go straight. To discourage drivers from changing lanes near an intersection, turn lanes are often separated from through traffic lanes by solid white lines.

On some roadways, a **two-way left-turn lane** between two opposing directions of traffic—also called a "center turn lane"—is designated for left turns only from either direction. This lane is marked by parallel solid and dashed yellow lines, sometimes accompanied by white arrows painted on the pavement that alternately

KNOW YOUR NEIGHBOR — HOV Lanes

Portions of freeways in Milwaukee, Wisconsin, as well as many other metropolitan areas in North America have special lanes restricted to buses, motorcycles, and vehicles carrying a minimum of two or sometimes three persons. Often called "high-occupancy vehicle" (HOV) or carpool lanes, they have a diamond symbol and other distinctive roadway markings. They are typically found on the far-left side of the freeway as well as on the left side of multilane on-ramps. Be careful not to stray into an HOV lane by mistake or use one to bypass gridlock if you are not eligible to drive in one. You will get a very heavy fine if you are caught by law enforcement as well as arouse the anger of fellow motorists who are obeying the law. If you are traveling in a carpool lane or next to one, be aware that HOV lanes often appear and disappear suddenly, forcing vehicles in the carpool lane to merge back into regular freeway traffic. Also, be sure to never cross the solid lines that separate HOV lanes from the other lanes. You may enter or exit HOV lanes only at designated places with dashed or dotted lines. To minimize interruptions to the flow of traffic, these breaks do not appear before every freeway exit, so pay attention to signs indicating when you should get out of an HOV lane to reach the off-ramp of your choice.

point left and right. Vehicles moving in either direction can use these lanes to make left turns into another road or driveway.

Reversible Lanes

Many major municipal streets in North America, including Chicago's North Lakeshore Drive, have **reversible lanes** that are open to city-bound drivers in the morning and to commuters returning home in the afternoon and early evening. They are separated from regular traffic lanes by dashed double-yellow lines and/or, in some cases, concrete barriers, cones, or other channeling devices. If you find yourself driving in one of these lanes, pay close attention to all signs and signals indicating when the lane is open to traffic moving in a particular direction. Be especially cautious just before and after the designated times when traffic is reversed.

Temporary Lanes

Temporary lanes are sometimes established during sports or cultural events to create extra lanes of traffic in one direction. Orange cones or other types of dividers may be used to "borrow" one or more lanes from one side of the road and "lend" them to the direction of highest traffic flow until the congestion is cleared up. Temporary lanes may also be created in construction zones. When you encounter these lanes, remember that traffic will be traveling in lanes otherwise used by opposing traffic. The danger is that a driver may get confused or will not be paying attention and cause a deadly head-on collision. Be especially cautious when driving in temporary lanes.

Figure 7-26 Use extra caution when driving in temporary lanes.

Bus Lanes

Champaign-Urbana is one of many metropolitan areas in Illinois that have special lanes on the far-right side of the road reserved for buses. Bus lanes are usually identified by a diamond symbol, words such as "BUS LANE" on the pavement, and regulatory signs indicating when the lanes are reserved. During those hours when they are restricted to buses, you may only enter a bus lane to make a right turn at an upcoming intersection.

Bicycle Lanes

Some roadways in Illinois have a special lane on the far-right side for bicycles. Bicycle lanes are separated from the traffic lanes by a solid white line, which sometimes changes to a dotted line near intersections. Vehicles are permitted to enter a bicycle lane only to park where permitted by the side of the road or, where the line is dotted, to prepare for a turn at an upcoming intersection. Bicycle lanes are typically marked with the logo of a bicycle or words such as "BIKE LANE" or "BIKE ONLY" on the pavement. Green BIKE ROUTE signs and black-and-white BIKE LANE signs may be posted by bicycle lanes as well.

Painted Curbs

Curbs and parking spaces are sometimes painted to regulate parking. If a curb is painted any color, or if there are colored lines on the pavement, it means that there are special parking rules. Illinois has no statewide laws regulating the color of curbs and parking-space lines, but some municipalities do color curbs. Always pay careful attention to nearby

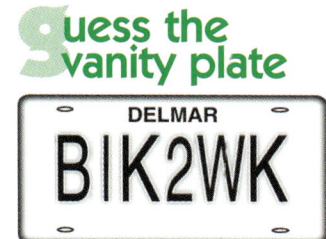

Figure 7–27

Figure 7–28 Bicycle lanes provide a safe alternative to riding on the shoulder of a roadway.

Figure 7–29 A painted curb usually indicates a parking restriction or prohibition.

posted signs or instructions lettered on the curb or pavement if colors are used.

As a rule of thumb, never stop, park, or stand by a curb painted yellow or red. Blue is often used to identify parking spaces for the disabled, sometimes accompanied by the symbol of a wheelchair on the pavement. In parking lots, be careful not to park in the designated areas *next* to disabled parking spaces or you could limit access to and from a disabled driver's vehicle.

Pavement Markers

If you are driving at night or in bad weather, it is often hard to see painted lines on the roadway. It can also be difficult to stay within lines if you are tired or otherwise impaired.

Many roadways in Illinois are equipped with square- or rectangular-shaped **reflective pavement markers** to help drivers stay within lanes. These devices are spaced at intervals between lane-line dashes and, on freeway ramps, along the edge of the roadway next to the lines. In addition to making a noise when your tire strikes them, pavement markers reflect your headlights to help you see the traffic lines in conditions of low visibility. Most are colored yellow or white to match pavement markings, although some have two-way reflectors that shine red as well. If you see red reflectors on the right, you are driving the wrong way! Where the roadway is too narrow to accommodate pavement striping trucks, markers may be placed on the adjoining curb or median.

Rumble strips are narrow grooves or wide depressions cut into the pavement ahead of areas where you are required to reduce your speed such as toll plazas. They are also found alongside the shoulders of many highways, including the interstates. Rumble

Figure 7–30 (a) Reflective pavement markers make it easier to see lane lines at night; **(b)** rumble strips are often used to warn drivers that they are straying off the roadway; **(c)** speed bumps are an effective way to force drivers to stop or slow down.

strips cause your vehicle to vibrate when you drive over them. Those that run across the road serve as an extra warning device to get you to pay attention to nearby signs or signals. Rumble strips placed along the shoulder are designed to warn dozing or distracted drivers that they are drifting off the roadway.

Speed bumps are raised slabs of pavement commonly found in shopping center and motel parking lots, residential neighborhoods, and large condominium or apartment complexes. They are often painted or striped white or yellow to get your attention. Speed bumps force you to reduce speed to about 5 miles per hour (8 km/h) for safety or, in some cases, come to a stop. Always brake before reaching a speed bump, and remember to release the brake pedal before going over it. Some speed bumps are very high, and if you travel over them too fast you can seriously damage your vehicle.

A variation of the speed bump, called a "speed hump," is used on streets in some residential neighborhoods. Wider and lower than speed bumps, speed humps are designed to force drivers to slow to the speed limit in the area.

Reflective pavement markers are encased in metal to prevent damage from snowplows in bad weather.

WHO'S AT FAULT?

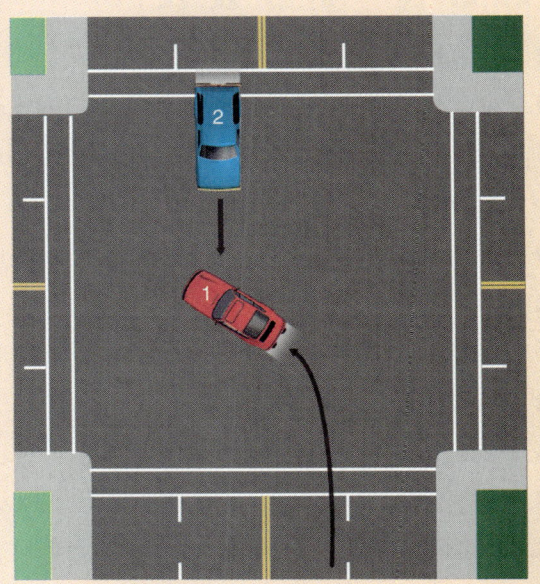

1. Driver 2 was proceeding south on Illinois Street in Carbondale. Observing that the traffic light at the intersection with Grand Street was yellow, he accelerated. Driver 1 entered the intersection northbound on Illinois and, intending to make a left turn on a green light, pulled suddenly into Driver 2's path of travel. Driver 2 was unable to stop and broadsided Driver 1. **Who's at fault?**

2. Driver 2 was eastbound in the two-way left-turn lane of Perryville Road in Rockford for quite some distance in preparation to make a left turn into a driveway. Driver 1 made a left turn out of a strip mall parking lot and also entered the center turn lane to prepare to enter westbound traffic. Driver 2 attempted to veer to his left to avoid impact, but traffic in the westbound lanes prevented him from doing so. Driver 2 then stopped abruptly and was impacted on the front right side by Driver 1, who was apparently unaware of Driver 2's presence in the lane. **Who's at fault?**

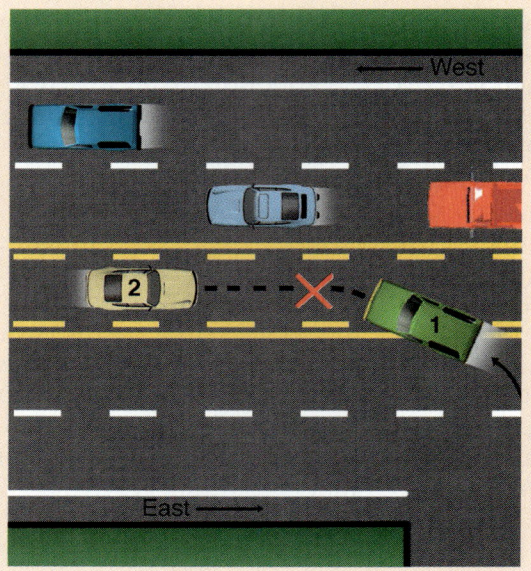

YOUR TURN

7–1 Traffic Signs

1. How are different colors used in traffic signs?
2. Which signs have their own distinctive shapes?

7–2 Traffic Signals

3. What do you do at a red light, yellow light, and green light?
4. How do signal arrows control traffic?
5. What is the purpose of variable message signs?

7–3 Roadway Markings

6. How are white and yellow traffic lines used on roads to control traffic?
7. What special lanes have distinctive roadway markings?
8. How are different colors used to regulate parking at curbs and in parking spaces?
9. What is the purpose of pavement markers?

SELF-TEST

Multiple Choice
1. What are the three main elements of a traffic sign that communicate information?
   a. colors, weight, and symbols
   b. symbols, colors, and reflectiveness
   c. shapes, symbols, and colors
   d. reflectiveness, height, and colors
2. What shape would a FALLING ROCKS sign have?
   a. square
   b. rectangle
   c. circle
   d. diamond
3. Speed bumps are commonly used:
   a. at the edge of roadways.
   b. at dangerous intersections.
   c. between dividing lines.
   d. in parking lots.
4. When are you *not* allowed to temporarily move to the left side of yellow center lines?
   a. when your side of the roadway is blocked
   b. when you are turning into a driveway
   c. when you want to see the roadway ahead of a big truck in front of you
   d. when you are passing another vehicle going in your direction
5. Disabled parking zones are often indicated by curbs and parking-space lines painted:
   a. green.
   b. blue.
   c. orange.
   d. yellow-green.

Sentence Completion
1. The only octagonal traffic sign is a _____ sign.
2. At a railroad crossing, a _____ sign is often accompanied with flashing red lights.
3. _____ signals always take priority over signs and traffic signals.
4. A lane that allows both you and oncoming traffic to make a left turn is called a _____ .
5. HOV stands for _____ .

Matching
Match the concepts in Column A with examples of the concepts in Column B.

Column A	Column B
1. __ Blue sign	a. SHOULDER WORK
2. __ Pedestrian signal	b. INTERSTATE 64
3. __ Pavement marker	c. GALESBURG 10
4. __ Green sign	d. WELDON SPRINGS STATE PARK
5. __ Brown sign	e. Phone symbol
6. __ Orange sign	f. WALK
7. __ Limit line	g. Speed bump
8. __ Route marker	h. "STOP"

Short Answer

1. Where are disabled parking spots located in a parking lot and why?
2. What is the purpose of a milepost?
3. What must you do when you approach an intersection and the light turns yellow?
4. What are reversible lanes?
5. What does a solid yellow line combined with a dashed yellow line mean?

Critical Thinking

1. You are driving late at night in the almost empty parking lot of a large suburban shopping mall, only a few stores away from a 24-hour liquor store. You approach a small STOP sign in the middle of the parking lot, obviously placed there by the shopping mall management and not by the municipal authorities. What should you do?
2. You are driving a large recreational vehicle on a two-lane country road when you notice a small bridge up ahead that is very narrow. The bridge is not sign posted, but it seems that the lanes are smaller than they are on the roadway on either side of the bridge. The way looks clear, but there is a curve beyond the bridge along with a lot of trees blocking your view of any oncoming traffic. What should you do?

PROJECTS

1. As a passenger, take a notebook and jot down as many different types of traffic signs as you can see. When you get back home divide the signs into their appropriate categories of regulatory signs, warning signs, route markers, guide signs, construction signs, service signs, recreation signs, and local traffic signs. Total them up and see which category has the most signs. Compare your results with those of other students in class.

2. Contact the Illinois Department of Transportation (IDOT) to find out what new transportation technologies it is implementing in terms of signs, signals, and pavement markers to make the roadways safer and more efficient.

8 chapter

Intersections and Right-of-Way

Intersections are places where different roadways meet or cross. Because pedestrians and vehicles of different sizes and speeds must cross one another's paths, intersections are extremely dangerous places. Studies indicate that one-third to nearly one-half of all collisions occur in and around intersections. Your chance of having a collision at an intersection is greater than at any other place on the roadway.

Chapter Objectives

Upon completion of this chapter, you should be able to:

8–1 Intersections

1. Understand the difference between a controlled and semicontrolled intersection.
2. Understand the proper way to approach an uncontrolled intersection.
3. Describe some of the hazards associated with alleys, traffic circles, and parking lots.

8–2 Right-of-Way

4. Understand what right-of-way is.
5. Understand what the right-hand rule is.

8–3 Railroad Crossings

6. Understand the difference between a controlled and uncontrolled railroad crossing.
7. Understand the proper procedure for crossing railroad tracks.

KEY TERMS

intersections
controlled intersections
semicontrolled intersections
uncontrolled intersections
"stale" green light
"fresh" green light
traffic circles
right-of-way
right-hand rule
controlled railroad crossings
uncontrolled railroad crossing

CHAPTER 8 Intersections and Right-of-Way ◆ 171

Figure 8–1 Traffic controls allow drivers to navigate intersections safely and efficiently.

8–1 INTERSECTIONS

There are three basic types of intersections. **Controlled intersections** use some form of sign or signal to direct the flow of traffic. **Semicontrolled intersections** have either signs or signals on one or several approaches to the intersection, but not on every approach. This type of intersection is very dangerous precisely because it does not tell every driver exactly what to do. **Uncontrolled intersections** lack any form of control, such as traffic signs or signals, to regulate traffic.

All addresses in Chicago originate from the intersection of Madison and State streets.

Figure 8–2 Signs rather than signals are used at intersections with a low volume of traffic.

factoid

Converting two-way stops to four-way stops reduces all crashes by up to 60% and injury crashes by up to 80%.

famous collisions

On July 12, 1987, a woman eight months' pregnant driving south on Eola Road in Aurora was killed along with her unborn child when another driver ran a STOP sign at the Liberty Street intersection and slammed into her car.

They are typically found in rural and residential areas with little traffic congestion.

Approaching a Controlled Intersection

As you approach an intersection, scan for signs, signals, special pavement markings (such as "STOP AHEAD") and other clues such as bus stop benches, rows of mailboxes, newspaper vending machines, parked cars, and pedestrian activity to help you identify a controlled intersection as soon as possible. Stay far enough behind the vehicles in front of you so that your field of vision is not blocked. The earlier you can spot an intersection, the more time you have to slow down, stop, move into the proper lane, or signal to others that there is a halt in traffic ahead by lightly tapping your brakes.

Many lanes change upon approach to an intersection. For example, a through-traffic lane can quickly turn into a left-turn-only lane, forcing you to go somewhere you do not want to go. Make sure that you leave yourself plenty of time to get in the proper lane before arriving at an intersection. Changes in the number and type of lanes are normally indicated with a sign and/or roadway markings. If you pay attention, you will be able to move out of a lane that does not suit your travel plans well in advance of the intersection, where it will be too late to switch.

Red Lights and STOP Signs

If you are approaching a red light or STOP sign, reduce your speed to stop. Even if it is the middle of the night and no cross traffic is in sight, you must make a *full* stop behind the limit line. In urban areas, a stop line is often painted on the pavement 4 feet (1 m) ahead of the nearer of two crosswalk lines. At these types of intersections, you must stop before any part of your vehicle crosses the stop line. If there is no stop line, you must stop before the nearest crosswalk line.

Figure 8–3 What is dangerous about this kind of intersection?

TRINIDAD SILVA, JR.

Actor Trinidad Silva, Jr., moved from his native Texas to Los Angeles to pursue his dream. He worked his way up in Hollywood the hard way, winning praise for his work on stage and in independent films. Silva got his big break in 1981 with the role of gang leader-turned-lawyer Jesus Martinez in the TV cop show *Hill Street Blues*. This popular series aired for seven years and won twenty-six Emmy awards. After the last episode in 1987, Silva appeared in several critically praised movies, including *The Milagro Beanfield War* and *Colors*, and formed his own television production company. On July 31, 1988, Silva's promising career abruptly ended when a driver ran a red light and broadsided his pickup at an intersection in Whittier, California. The thirty-eight-year-old Silva was thrown 100 feet (30 m) from his vehicle and died instantly. His wife and two-year-old daughter were also injured. The next time you think about racing through an intersection to "beat" a red light, remember that you are risking other lives, as well as your own.

If the intersection is controlled with lights and there is no limit line, you must stop at a point nearest the intersecting roadway where you have a clear view of approaching traffic. If it is controlled by STOP signs and there is no limit line, you should stop behind the sign. At sign-controlled intersections where the limit line or STOP sign appears well in advance of the actual intersection, you should stop again at the corner to make sure that the way is clear.

Many drivers fail to come to a complete stop at STOP signs or before turning right on a red light. Not only is "rolling" through a stop dangerous, but it also is an easy way to get a ticket in Illinois.

Some drivers like to play a game with traffic signals to save time by trying to enter the intersection at full speed just as the red light turns green. However, if you guess wrong and the light does not turn green as fast as you think it will, you may be forced to come to a quick stop, or worse, enter the intersection on a red light. Never try to "beat" a red traffic signal by attempting to anticipate the light change. Worse yet, never react to the cross traffic's signal. You do not know the timing of that particular signal, and there may be special left- or right-turn arrows you are unaware of that will delay the green light you expected. Be aware that traffic signals at some intersections are programmed in such a way as to momentarily stop traffic from *all* directions. Even after the cross-traffic light turns red, your light may briefly remain red before changing to green.

As you approach a red signal, pay attention to posted signs. A right turn on a red light after a full stop is generally permitted everywhere in the state, but it may be prohibited at certain intersections for reasons of safety.

Yellow Lights and YIELD Signs

If the light ahead is solid yellow, the general rule of thumb is to

According to a recent survey, more than half of all Americans admit to running red lights.

factoid

Chicago has nearly 2,800 signalized intersections, about 12 times as many as Rockford, the second largest city in Illinois.

factoid

With an average of nearly 250 collisions each year, the intersection of Golf and Roselle roads in Schaumburg is the most dangerous in Illinios and the third-most dangerous in the country.

stop safely before entering the intersection. If you are already in the intersection, or cannot stop safely, proceed through it at a constant speed, watching carefully for cross traffic that might enter the intersection before you have cleared it. If you are near the intersection but think that you will be able to stop safely before entering it, check the traffic to your rear at this critical decision point. If it appears that a car behind you is attempting to make it through the intersection on the yellow light, it is better to proceed if you can do so safely rather than risk getting rear-ended.

If you can see one or two lights down the road, you may be able to discover whether the light ahead is a "long" or "short" yellow. In many cases, how long a light stays yellow depends on the time of day or night and the volume of traffic. If you observe that the yellow light is short, you can be better prepared to stop at the intersection rather than have to worry about making a sudden stop. If you know that the yellow light ahead is long, you will be better able to judge whether you can make it through the intersection without having to accelerate at the last second.

If you are approaching a flashing yellow or YIELD sign at an intersection, reduce your speed and scan in both directions of the cross street for oncoming traffic. Look left, right, ahead, and left again. If there are pedestrians or vehicles crossing the intersecting roadway, you must stop. If the way is clear, proceed cautiously through the intersection or turn onto the cross street and gradually increase your speed.

Green Lights

Scanning ahead will help you determine whether a green light is "stale." When you approach an intersection with a **"stale" green light,** a light that has been green for a long time or ever since you first noticed it, anticipate that it will turn yellow and slow down. A flashing red pedestrian DON'T WALK sign is a good indication that a stale green light is about to turn yellow. As you approach a stale green light, keep an eye out for drivers or pedestrians in cross traffic who may try to "get a jump" on the light and proceed before their light turns green.

If the light is a **"fresh" green light,** or has just changed from red to green, proceed when cross traffic has safely cleared. Always remember to look left first because on a two-way road cross traffic will approach you first from this side as you enter the intersection. Then look right, ahead, and left again. Watch for drivers who may have tried to "run" a yellow light just as it was changing to red. If you are turning left, make sure that you know whether you have to wait for a green arrow or can go as soon as there is a sufficient break in oncoming traffic.

Do not treat your initial forward movement like a drag race. Pedestrians or other hazards may come into view as you accelerate, and you will not have time to stop. At the same time, you risk getting rear-ended if you move too slowly or stop once you begin your movement. Do not enter an intersection unless you are sure that you can get all the way across before the light changes to red. In Illinois, you can be ticketed for blocking an intersection. Even if the driver behind you is furiously honking at you to move, do not endanger your life or other lives by getting stuck in the middle of an intersection. Black-and-white DO NOT BLOCK INTERSECTION or STOP ON RED HERE signs are sometimes posted near traffic signals as an added caution, but remember that even if there is no sign it is illegal to block cross traffic.

Automated Signals

Some signals at intersections are controlled with sensors in the roadway that trigger a light change in response to traffic flow and congestion. Other signals are timed to go off at certain intervals that vary with the day and time. Newer signals have video motion sensors on the signal head to monitor traffic flow. As a driver, you can adjust your speed to the timing of traffic lights to "make" all the greens and save both gas and wear and tear on your brakes.

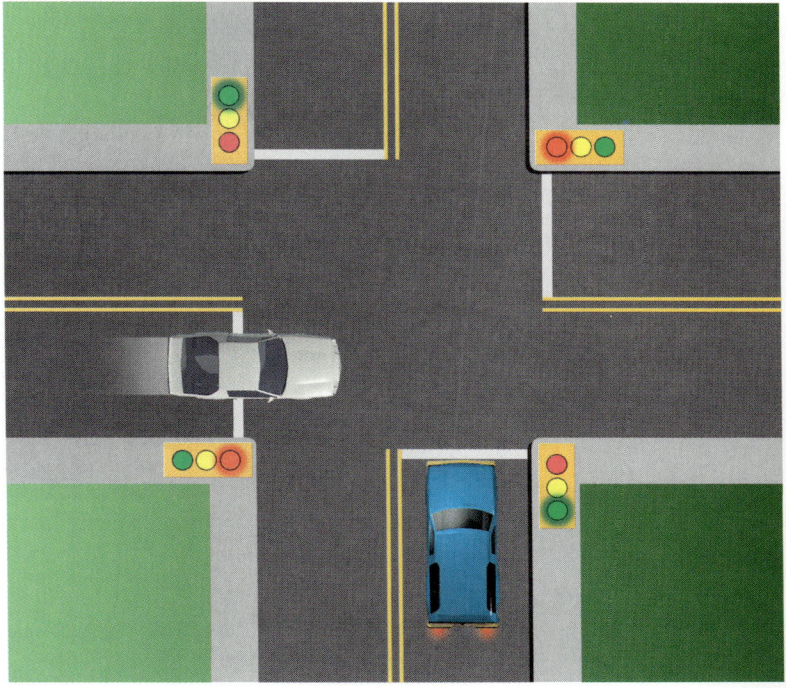

Figure 8–4 Just because your light is green does not mean it is safe to proceed.

However, you should do this only if you can stay within the speed limit, maintain a safe space cushion around your vehicle, and not take your concentration off other vehicles on the roadway.

REALITY CHECK   *The Right-Hand "Quick Starter"*

Rather than wait behind other cars at a red light, impatient drivers will sometimes leave the regular stream of traffic and pull over into the far-right lane to get a "line position" for the next green. The "lane" they use is often a bicycle lane or a lane filled with parked cars. These racers will then creep up to the crosswalk so that once the light turns green they can floor the accelerator of their cars and rapidly swerve back into the regular stream of traffic, missing parked cars or other obstacles ahead by a hair. If this happens to you, do not be tempted to race the offending driver because he or she has gotten in front of you. Reduce your speed and allow the other driver to get ahead of you. It is not worth a crash to prove a point.

Figure 8–5 Be careful not to get stuck in the middle of an intersection.

Approaching an Uncontrolled Intersection

Use extra caution when approaching an uncontrolled intersection. Because there are no signs or signals, drivers often do not pay attention to cross traffic, or they just assume that they are the only ones on the road and race right through. Scanning ahead is especially critical at uncontrolled intersections. Because many are in residential areas, watch out for children, bicyclists,

DRIVING TIPS   *Intersections*

- Look carefully for pedestrians and other vehicles in an intersection before you proceed either on a green light or after you have stopped at a STOP sign or uncontrolled intersection.
- When entering a major roadway from a side roadway, always stop at the intersection, even if there is no limit line or no STOP sign is posted.
- If you intend to park on the other side of an intersection, slow down, get into the far-right lane, and indicate your move toward a parking spot by signaling only *after* you have entered the intersection. If you put on your turn signal too early, drivers around you may assume that you are making a right turn at the intersection.
- When driving beside slowing or stopped vehicles in or near intersections, use extreme caution—the other drivers may have spotted something, or someone, in the street that you cannot see. Pedestrians are often hidden from view by parked cars or other vehicles in an intersection. Be patient and avoid hitting the unseen pedestrian or vehicle.

DRIVING MYTHS — Timed Signals

Many signal-controlled intersections have what is known as a "loop detector" in the road that detects the presence of a vehicle at the limit line. When the electrical field generated by the wire is interrupted by a large object, the signal will provide a green light. The signal change is not automatic, but the time you have to wait for a green light is reduced. Some drivers get frustrated and believe that by backing up and driving forward again they can get the signal to change faster. This, however, is as ineffective as repeatedly pushing an UP or DOWN elevator button. In fact, if you roll past the limit line and the sensor, the signal's timing might be thrown off, causing the red light to stay on even longer! Another myth is that you can get a timed signal to turn green faster by jumping out of your car and pushing the pedestrian push-button. This will extend the length of the green light, but it will not make it come on sooner. Backing up or getting out of your vehicle at an intersection are both dangerous and illegal, so be patient and wait your turn.

skateboarders and rollerbladers, balls flying into the road, dogs without leashes, and so on. Keep in mind that small children can easily be hidden by parked cars or bushes. Always reduce your speed when approaching an uncontrolled intersection and prepare to stop your vehicle.

Other Types of Intersections

When approaching an intersection that you have trouble identifying, slow down to give yourself extra time to recognize what you are facing and to react to any potential obstacles or problems.

Alley Intersections

Because drivers on a roadway have a hard time even seeing alleys, especially in urban environments, this type of intersection can be very dangerous. The driver in the alley who is watching for pedestrians on the sidewalk and vehicles in the roadway must be more patient than usual while

Figure 8–6 This mansion, the oldest brick house now standing in Illinois, is located near the intersection of two state highways.

factoid

It can cost up to $100,000 to install a single traffic signal at an intersection.

Figure 8–7 Uncontrolled intersections are most common in residential areas.

> ### AUTO ACCESSORIES — "Round-the-Corner" Mirrors
>
> The next time you drive out of an alley or parking garage where it is hard to see cross traffic, look for a small, round mirror mounted nearby on a wall, post, or overhang. These mirrors are designed to allow drivers and pedestrians to see one another at dangerous narrow intersections where the driver's line of sight is blocked, often by a building.

waiting for a safe opportunity to join the through traffic. Make sure that it is the *best* opportunity, not just the first opportunity, and never assume that you have a clear path.

When driving in an alley you must be extra cautious. Alleys are not designed for high-speed travel. They are typically narrow and have limited views because of the existence of obstacles such as the walls of the buildings through which the alley runs, parked trucks, and trash dumpsters. Keep your speed under 15 miles per hour (25 km/h), and watch out for vehicles exiting parking lots and garages.

Traffic Circles

Traffic circles, also called rotaries, are circular roadways where traffic from many different directions intersects. There are three basic types of traffic circle. One type uses STOP signs and/or signals to control access, and some or all of the roads may intersect the circle at a 90-degree angle. Another type of traffic circle controls traffic with YIELD signs, and forces drivers

Figure 8–8 You may have to go around a traffic circle more than once to reach your exit safely.

USING SAFE — Large Parking Lots

Parking lots may seem like one of the easiest places to drive, but they can actually be very dangerous. Pedestrians and vehicles mix directly on the roadway, cars continually pull in and out of spaces, and drivers often ignore signs and roadway markings. When navigating through a parking lot, you must *scan* aggressively from side to side for pedestrians, especially small children, who may dart from between cars into your path at any moment; drivers who back out of spaces without first looking behind them; and drivers ahead of you who slam on their brakes when they see an opening. *Assess* what may happen. Can that shopper loaded down with merchandise see you coming? Is the driver in front of you looking for a parking space or trying to get out of the lot? *Find* an escape route by maintaining a large following distance in front of you. *Execute* your decision to avoid a potential collision with a pedestrian or another car by tapping your horn or flashing your lights to communicate your presence.

---

to enter at an angle. This type of traffic circle, which is common in Europe, is also called a roundabout. A third type of traffic circle is found in some residential areas in Illinois, and consists of a circular barrier placed within an otherwise uncontrolled intersection.

Although traffic circles reduce the *severity* of potential collisions, they can be confusing to drivers, thereby increasing the *risk* of a crash. Always stay to the right when entering a traffic circle. As you progress through the circle, watch out for vehicles on both of your sides. Some traffic circles have more than one lane of travel. As a general rule, the inner lane(s) of the circle should be used for through traffic, and the outer lane should be used by vehicles preparing to exit the circle. Moving to the outer lane can be a tricky maneuver because you must simultaneously follow the turn and watch for other vehicles that are trying to do the same thing you are. Do not cross lanes too early to reach your exit. You can always continue to go around the circle, and next time you will be more prepared to exit at the appropriate place.

Large Parking Lots

Safety is often compromised in parking lots where large numbers of cars and pedestrians heading every which way are in close proximity. The excitement associated with shopping, concerts, and sporting events makes it more difficult for drivers to concentrate, especially at night. Pedestrians are hard to spot among parked cars, and when traveling in groups, often assume that there is "safety in numbers."

Drivers often go too fast within large parking lots, treating the lot as an extension of the surrounding street scene. In crowded lots, some drivers compete fiercely for scarce parking spots, ignoring intersections within the parking lot or between the lot and the roadway. Congestion naturally leads

Highway heroes

While responding to a call, Officer William M. Morrison, Jr., of the Chicago Police Department was killed when his vehicle was broadsided by a drunk driver at the intersection of West 15th Street and South Ashland Avenue on the evening of September 4, 1987.

Figure 8–9 Dangers lurk in every direction in crowded parking lots.

to frustration and impatience. In these situations you must slow down, stay calm, and remain at peak alertness. This is not a time to change radio stations or argue with passengers.

Stay in the lanes designated for traffic movement, and do not cut across rows of empty parking spaces even if no one appears to be around. When leaving a head-in parking space, avoid the temptation to drive forward if the space directly ahead of you is vacant. Not only might you collide with another driver entering the spot from the other side, but you could damage your vehicle if you drive over a parking curb you might have forgotten was there.

8–2 RIGHT-OF-WAY

Right-of-way is the right to use a certain part of a roadway when two or more users of the roadway want to use it at the same time. When you give the right-of-way, or "yield," to another driver or pedestrian, you are giving that person first use of the road. When someone gives the right-of-way, or "yields," to you, that person is giving you first use of that stretch of roadway.

Understanding and obeying right-of-way laws is one of the most important components of safe driving. Illinois has such laws to increase safety for motorists, bicyclists, and pedestrians alike.

guess the vanity plate

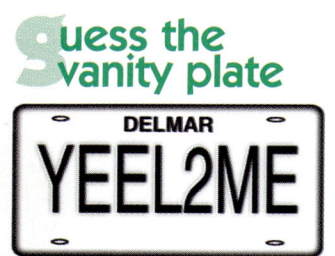

Figure 8–10

DRIVING TIPS — "Don't Be There, Be Square"

You may have heard the expression, "Be there or be square." At intersections, however, you should do the opposite. Have you ever been stopped in the left lane at a traffic light and been nearly clipped by a driver on the cross street turning left across your bow? You can decrease the risk of a collision by "squaring" the intersection—that is, by giving left-turning drivers on the cross street more room to make their turn. Instead of pulling all the way up to the limit line, stop well behind the vehicle in the adjoining lane to your right so that you are looking at the driver's back seat, not at him or her.

They are founded on common sense and rely on drivers' patience and courtesy to be effective. In a certain situation, you may want to give the right-of-way to someone else even if the law says you have the right-of-way! Why? To contribute to an orderly flow of traffic, to avoid collisions, or even just to be nice. In fact, the law *requires* you to give up your right-of-way if you have the last clear chance of avoiding a crash.

Never assume that others will give the right-of-way to you, even if you legally have the right-of-way. Giving up your right-of-way is a small price to pay to avoid collisions. Equally important, *never demand or insist on taking the right-of-way.* If you are the type of person who gets mad at someone for cutting you off, ask yourself if proving a point to a total stranger is really worth a crash. The answer is no! Let the other driver go and forget the whole thing.

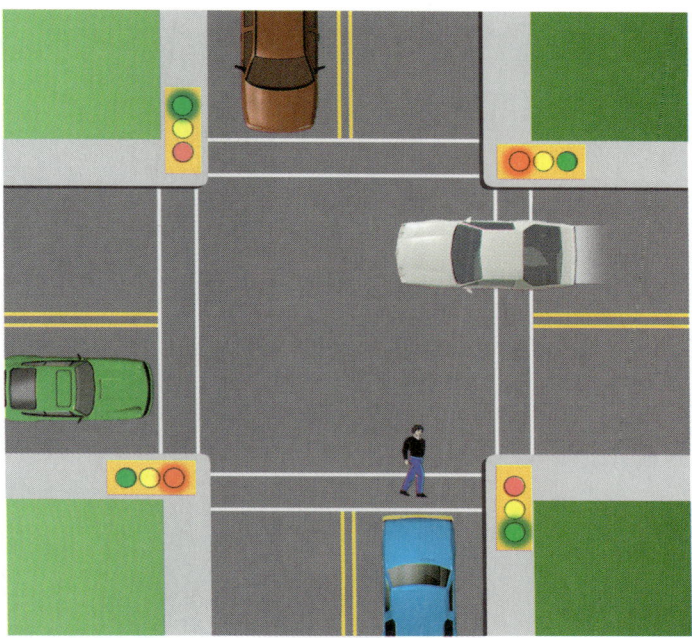
Figure 8–11 Signal-controlled intersections

Right-of-Way at Intersections

At uncontrolled intersections, intersections with STOP signs on all

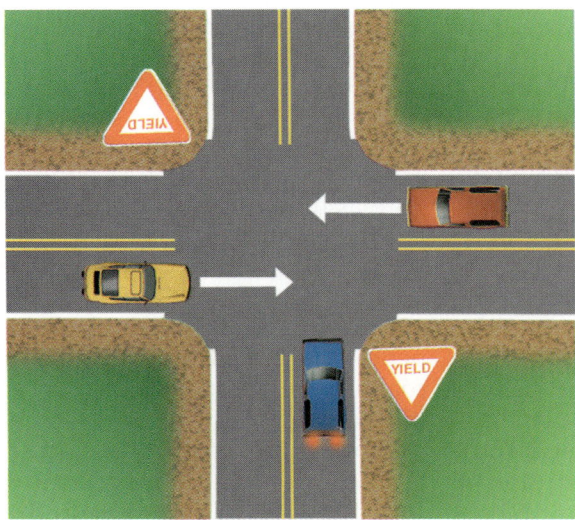
Figure 8–12 Semicontrolled intersections

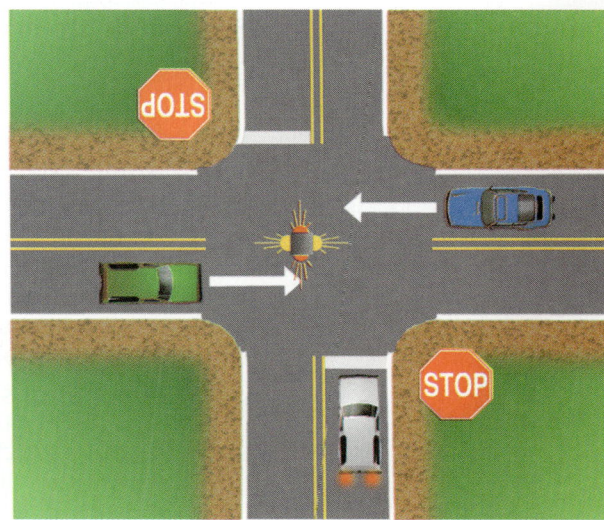

182 ◆ UNIT 3 *Welcome to the Road*

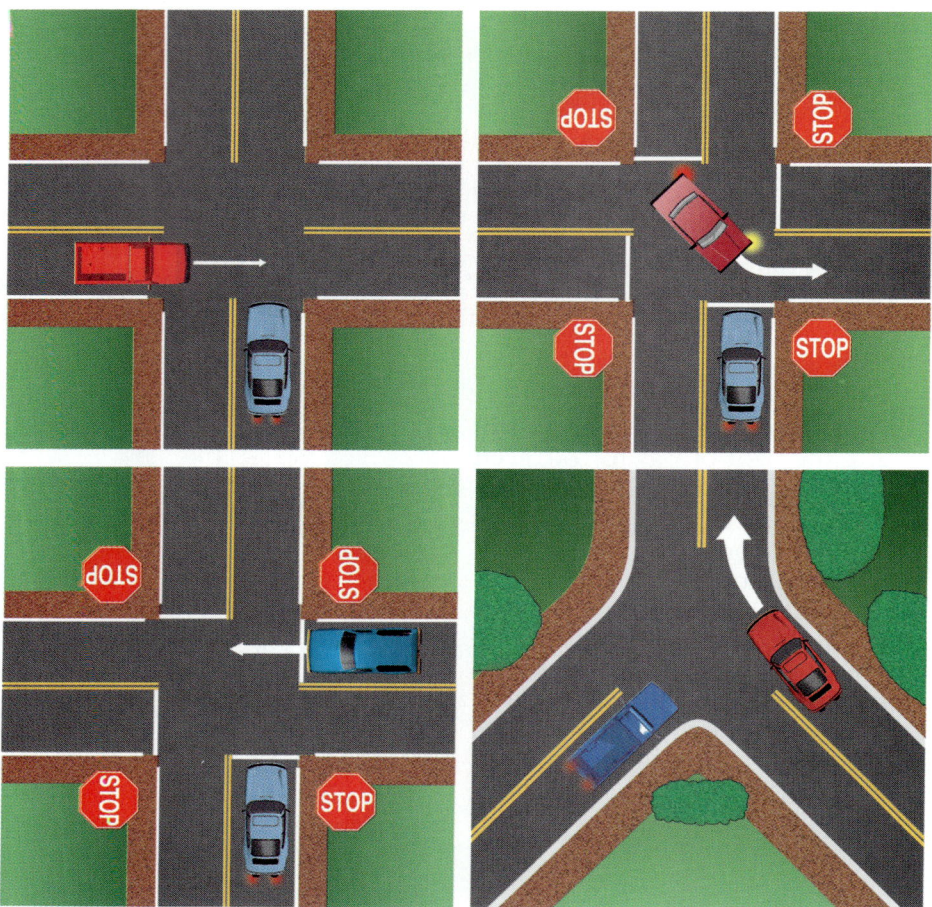

Figure 8–13 Uncontrolled intersections or intersections with STOP signs on all the corners

the corners, and intersections with flashing red or broken signal lights, the first vehicle approaching or entering the intersection has the right-of-way. If two vehicles reach the intersection at the same time, keep in mind the **right-hand rule:** Drivers on the left should always give the right-of-way to drivers on the right. If three or four vehicles reach the intersection from different directions at the same time, each driver should wait to take his or her turn to cross the intersection. This situation calls for caution and clear communication among all drivers as to who will go first, second, and so on.

If you face a YIELD sign at an intersection, all vehicles and pedestrians legally crossing the roadway on which you are driving have the right-of-way.

Always give the right-of-way to pedestrians before coming to a sidewalk and all vehicles in the roadway when approaching

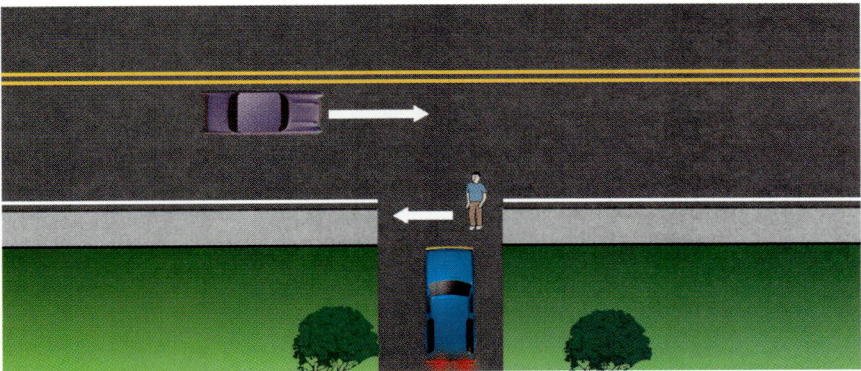

Figure 8–14 Alleys, driveways, and private roads

from an alley, private driveway, or private road. This means you will have to stop *twice:* once for pedestrian traffic on the sidewalk and once before entering the roadway.

The driver of a vehicle entering a traffic circle must give the right-of-way to vehicles already in the circle. The driver of a vehicle within the circle must give the right-of-way to other vehicles that wish to exit the circle. Always be extra alert when entering or exiting a traffic circle, and assume that those around you do not know what they are doing.

If you are making a left turn, all approaching vehicles have the right-of-way. Wait until the road is clear of oncoming traffic to make your turn. If you are in an intersection when your light turns yellow, wait until oncoming traffic has stopped, then proceed. If the light turns red while you are waiting to make your left turn, stay calm and proceed cautiously only when oncoming traffic has stopped. Cross traffic must give the right-of-way to you even though they now have a green light.

Other Right-of-Way Situations

Negotiating different types of intersections is not the only situation in which right-of-way comes into play.

Passing

On a multilane roadway, always give the right-of-way to any other vehicle passing you and to any vehicle close enough to prevent you from passing safely.

Figure 8–15 Traffic circles

DRIVING TIPS — Right-of-Way

- Always give up the right-of-way, even if you think you have it, to avoid a collision.
- If you force another vehicle to slow down or stop, you have not given up the right-of-way.
- Although traffic signals and signs determine who has the right-of-way, other drivers will not necessarily observe them. Watch out for the driver who rolls through a STOP sign, ignores a YIELD sign, or runs a red light. Even if you have the clear right-of-way, always keep alert.
- In case of doubt, always yield the right-of-way. This will contribute to greater safety on the roadway, and other drivers will signal their appreciation of your courtesy.

Figure 8–16 Left turns

Merging Lanes

If the lane in which you are traveling ends, either because the road narrows or because an obstacle is in your lane, you will be forced to merge into another lane. When this occurs, you must give the right-of-way to vehicles passing you in this other lane.

Roadway Parking

A car parked on the side of a road that is pulling into the roadway must give the right-of-way to through traffic on the road.

Figure 8–17 Drivers in merging lanes must give the right-of-way to through traffic.

Figure 8–18 Watch out for pedestrians who illegally "take" the right-of-way.

Pedestrians

Pedestrians crossing a roadway within a marked crosswalk have the right-of-way. If there is no painted crosswalk at an intersection, which often occurs in residential and rural areas, pedestrians have the right-of-way as long as they stay within the space where crosswalk lines would normally appear—for example, between the stop line and the corner or where the sidewalk would appear if it continued across the road you are on.

Some pedestrians, including the elderly and the disabled, may need more time than others to cross a street. Those who start to cross legally but are not able to complete the crossing before the light changes to red still have the right-of-way. Even if you get a green light, you must remain stopped until such persons have reached the other side of the street or a point of safety such as a median or island specifically designed to accommodate pedestrians (these are often equipped with their own pushbuttons).

Be prepared at all times for a pedestrian to illegally "take" the right-of-way from you by crossing outside of a marked or unmarked crosswalk, or by ignoring traffic controls. Take all measures to avoid hitting the pedestrian and to warn him or her of your presence, such as honking your horn or flashing your lights. All drivers have a duty to protect pedestrians even if the pedestrians are breaking the right-of-way laws.

Pedestrians with clearly visible disabilities, include those who are blind or deaf, have the right-of-way at all times no matter where they are in the street.

Emergency Vehicles

Emergency vehicles include police and sheriff's department cars, ambulances, paramedic vans, fire department vehicles, and vehicles used by various volunteer units such as first-aid workers and rescue squads. When responding to an emergency, they will display flashing red, blue, or green lights and activate their siren and/or air-horn.

factoid

A system called "preemption" turns all traffic lights to green along a rapid response vehicle's route to an emergency.

Figure 8–19 Emergency vehicles have the right-of-way in *every* situation.

famous collisions

On the night of March 15, 1999, eleven people on board an Amtrak train were killed after it slammed into a truck whose driver had tried to maneuver around lowered gates at a rural crossing in Bourbonnais, 50 miles (80 km) south of Chicago.

You must give the right-of-way to an in-service emergency vehicle approaching from either direction on an undivided roadway by moving to the far-right side of the road, as near to the curb as possible, and coming to a complete stop. On roadways divided by any form of *physical* barrier such as a median strip, guardrail, or concrete island (but not including a "virtual" island), you must pull over and stop if you are on the side of the road on which the emergency vehicle is traveling.

If it is unsafe or impractical to move to the right, clear a path for the emergency vehicle to get by you and stop where you are. Never stop in an intersection unless specifically directed to do so by law enforcement or emergency personnel.

Wait for emergency vehicles to fully pass you before you signal and turn back into the normal stream of traffic. Illinois law requires drivers to stay at least 500 feet (150 m) behind moving in-service emergency vehicles. If you encounter emergency vehicles at the scene of a fire, collision, or other emergency situation, you may not park within 500 feet of where they are stopped. Never drive over any unprotected fire or chemical hoses laid down on a street or driveway.

Funeral Processions and Motorcades

A group of vehicles forming part of a funeral procession or motorcade has the right-of-way over other vehicles. These vehicles are generally easy to identify because they travel at a slow speed; have signs, flags, or other markings on their cars; and are often escorted by police. Each car in the motorcade must have its headlights on. The lead vehicle may also have a flashing yellow light in operation.

The driver of the lead vehicle in a funeral procession or motorcade must obey all traffic signs and signals, but other participating drivers may follow this lead vehicle

Figure 8–20 More than half of all train–car collisions are caused by drivers trying to go around lowered gates or ignoring flashing lights.

without stopping, as if they were an extension of the lead vehicle. Cutting through or otherwise disrupting a funeral procession or motorcade is prohibited unless you are authorized to do so by a law-enforcement officer.

8–3 RAILROAD CROSSINGS

Illinois is crisscrossed by more than 10,000 miles (16,100 km) of railroad track. Although rail traffic is not as heavy as it once was, Chicago is still considered the rail capital of the world. The number of train–car collisions in the Prairie State has declined by nearly 50% in the last decade, but railroad crossings are still one of the deadliest types of intersections. Most of the nearly 200 crashes that occur at these locations each year are caused by human error: drivers who do not obey warning devices, try to outrun a train, or ignore crossing barriers.

It takes a 150-car freight train going 30 miles per hour (50 km/h), the speed under which most highway–rail crashes occur, two-thirds of a mile (1 km) to stop. An eight-car passenger train going 60 miles per hour (100 km/m) needs an equal distance to brake. Even if the engineer spots your car on the tracks ahead, it is next to impossible for him or her to stop the train in time to avoid

Illinois is ranked third in the nation in the number of train–car collisions.

A motorist is forty times more likely to die in a collision with a train than with another car.

REALITY CHECK   Playing "Chicken" with a Train

If you live in a rural area where weekend nights are slow and there is not much to do, you and your friends might be tempted to head for a remote stretch of railroad track to play an old and dangerous car game of "chicken." When a train approaches, you wait until the last minute before you hit the gas and try to cross the track only a few feet before the train hits your car. Sounds fun, right? The only problem is that in your desire to look cool in front of your friends you have forgotten that you are about to compete with a piece of metal weighing thousands of times more than your car. Talk about the odds being against you! Playing chicken with another driver is stupid enough, but playing chicken with a train is certain suicide.

About 70% of all train–car collisions in Illinois occur at crossings with active warning devices.

hitting your vehicle. Train–car collisions are not a pretty sight. The weight difference between a train, which can weigh more than 10,000 tons, and a car is like that between a car and an aluminum soda pop can. Think about that the next time you *flatten* a littered can on the road.

Railroad Crossing Controls

There are nearly 9,000 public railroad crossings in Illinois, and almost all of them have at least a crossbuck on each approach. Most controlled railroad crossings also include advance warning

DRIVING TIPS *Railroad Crossings*

- Never rely on crossing signals or gates to warn you of an approaching train. Just like anything else, they can break. There is no substitute for slowing down, looking, and listening.
- Always wait for the vehicle in front of you to clear the tracks before you proceed. You do not want to be stuck behind a car or truck if it stalls in the middle of the crossing and exposes you to potential danger.
- Never pass another vehicle, no matter how slow it is going, at a railroad crossing.
- Be extra careful when following motorcyclists or bicyclists across railroad tracks. They may lose traction on the slippery tracks or lose control if their tires get caught in the grooves of the rails. If the tracks cross the roadway at an angle, be prepared for the cyclist to swerve to cross the tracks at a right angle.
- When you are following buses, trucks, or vehicles carrying flammable or other hazardous cargo, always increase your following distance. Be prepared to stop behind them when they get to any railroad crossing, not just ones with warning signs and signals.
- Never stop your vehicle on railroad tracks. If your car stalls on the tracks and you cannot restart it immediately, get clear of your vehicle. Do not try to push it out of the way. Instead, seek immediate help from a towing service and/or local authorities. If you escape from your vehicle just before a train makes impact with it, run away from the tracks in the direction of the approaching train to avoid being hit by debris caused by the collision.
- Trains travel at all hours of the day and night. It is never too late or too early for a train to be running, so always use the same care when checking for the presence of trains. Be especially careful at night at uncontrolled railroad crossings. If the train is moving slowly enough or if the weather limits your ability to see or hear, you may not realize that it is there.
- If you happen to cross railroad tracks frequently and are used to never seeing train traffic on them, do not assume that the tracks are never used just because they appear to be in bad condition and overgrown with weeds. Even if certain stretches of track are not often used, railroad companies may run maintenance trains down these tracks from time to time.

Figure 8–21 Approaching a railroad crossing

Wild Wheels

Figure 8–22 What is the maximum speed of a passenger vehicle when it is adapted to ride on railroad tracks with this accessory?

signs and pavement markings—typically an "R X R" before the crossing and a limit line at the crossing itself. There may also be a STOP or YIELD sign at the crossing. More than half of all public railroad crossings also have some combination of crossing gates, flashing lights, or traffic signals that are triggered by a train as it approaches.

Many of the state's 5,000 private railroad crossings, located primarily in industrial and rural areas, have no controls at all. If you ever encounter an **uncontrolled railroad crossing,** treat it as if you were approaching a YIELD sign at a regular intersection.

Approaching a Railroad Crossing

Follow these steps when approaching a railroad crossing:

1. Reduce your speed. Signal to any drivers behind you that you are slowing down by using several light taps on your brakes. Turn off the stereo, stop talking on the car phone or with passengers, turn off the air conditioner or heater, and lower the windows so that you can hear if any trains are coming. Do not assume that the engineer will sound a warning or that you will be able to detect the distinctive *clackety-clack* sound of an approaching train, which is harder to hear on today's more smoothly welded rails.
2. As you approach the crossing, look and listen in both directions—left, right, and left again. Never try to beat a train to a crossing. Your ability to guess the speed of the train is not as good as you think it is. A typical locomotive is nearly 20 feet (6 m) high and 10 feet (3 m) wide. Its large size makes it appear to be traveling slower than it really is. The parallel lines of railroad tracks, which converge toward the horizon, add to the illusion by making it appear that a train is farther away than it really is.
3. If you have determined that no train is approaching from either direction, and the crossing is uncontrolled or controlled only by a crossbuck, proceed cautiously across the tracks. If a train is approaching and there is no gate or pavement marking to indicate a stopping point, you are required by law to stop no closer than 15 feet (5 m) away from the nearest rail of the tracks. You must be far enough away from the tracks not only to let the train clear safely, but to avoid being hit by any cargo that might be overhanging the sides of freight cars. Observe all railroad crossing controls. It is dangerous as well as illegal to cross railroad tracks if bells are sounding or lights are flashing, or to try to drive around, under, or through lowered crossing gates even if no train is visible.
4. Be patient. Depending on its length and speed, it can take up to 10 minutes for a freight

train to clear a crossing. Once you have waited for a train to pass, never rush forward just as soon as it passes you. Another train may be behind it, or one may be coming from the other direction that you can neither see nor hear because of the presence of the first train. Once the gates are raised and the lights have stopped flashing, and you are sure that no other trains are crossing at the same time, move across the tracks quickly and with enough momentum to coast to safety if your vehicle stalls. If you are driving a vehicle with a manual transmission, shift to a gear that will allow you to cross without having to shift again on the tracks.

"Light Rail" Crossings

Metropolitan Chicago's famous "light rail" public transportation system, called the "El" (for "elevated"), serves the city and 38 of its suburbs. The more than 1,000 light-rail vehicles that travel over seven different lines are smaller and shorter than cars found on passenger and freight trains, but they sometimes pass through crowded urban areas.

Light-rail crossings generally have fewer safeguards than standard railroad crossings, and light-rail vehicles can be difficult to hear in city traffic. If you encounter a light-rail crossing, exercise the same caution you would at a regular railroad crossing. Check for signs, signals, roadway markings, rails in the pavement, overhead electric wires, passenger stops, and other clues about the light-rail route and whether any vehicles are approaching. Observe all traffic controls, and check both directions to make sure the way is clear before proceeding across the tracks.

WHO'S AT FAULT?

1. Taking a local shortcut to avoid traffic, Driver 2 had just made a quick right turn off Capitol Avenue in Springfield and was proceeding eastbound down the alley at approximately 30 miles per hour (50 km/h). Driver 1 had just finished a meal at the restaurant on the corner and was pulling into the alley. Driver 1's view was limited because of a high fence at the edge of the parking lot. Driver 2 ran into Driver 1's front end. **Who's at fault?**

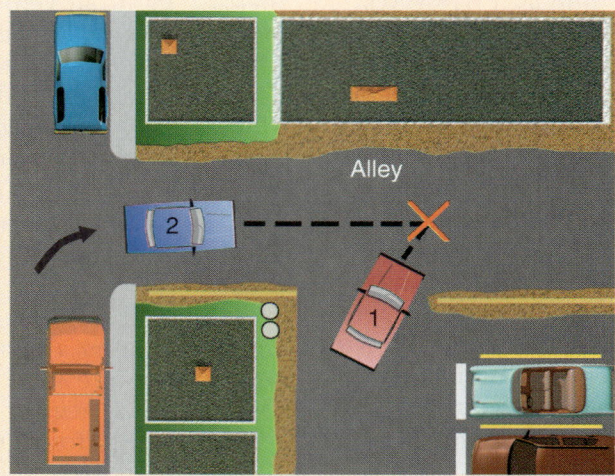

2. Driver 1 signaled in preparation for a left turn at the intersection of North Second Street and Ethel Avenue in Rockford. Upon seeing Driver 2 approaching the intersection, Driver 1 judged that he had just enough time to make it through the intersection and proceed rapidly with the turn. As Driver 1 was completing his turn, he sighted a bicyclist crossing his path from the left. Driver 1 was unable to stop in time and struck the bicyclist. **Who's at fault?**

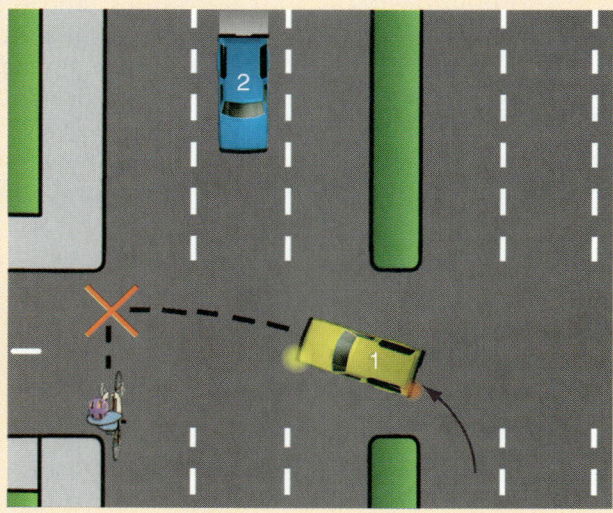

YOUR TURN

8–1 Intersections

1. What is the difference between a controlled and semicontrolled intersection?
2. What is the proper way to approach an uncontrolled intersection?
3. What are some of the hazards associated with alleys, traffic circles, and parking lots?

8–2 Right-of-Way

4. What is right-of-way?
5. What is the right-hand rule?

8–3 Railroad Crossings

6. What is the difference between a controlled and uncontrolled railroad crossing?
7. What is the proper procedure for crossing railroad tracks?

SELF-TEST

Multiple Choice

1. When approaching an intersection, always look:
   a. right, left, ahead, and right again.
   b. left, right, ahead, and left again.
   c. ahead, left, right, and left again.
   d. left, ahead, right, and left again.
2. A semicontrolled intersection is:
   a. an intersection with only STOP signs on all approaches.
   b. an intersection with both STOP and YIELD signs on all approaches.
   c. an intersection with traffic controls on one or more approaches, but not all approaches.
   d. an intersection with traffic controls on only two approaches.
3. When can you stop in an intersection?
   a. when directed to do so by a firefighter
   b. when you think it will help an emergency vehicle
   c. when all the signal lights are not working
   d. when you are waiting for a parking space on the other side
4. You should slow down when approaching a:
   a. "fresh" green light.
   b. green arrow.
   c. "stale" green light.
   d. flashing green light.
5. If you are in an intersection making a left turn when the light turns red, you should:
   a. back up.
   b. proceed cautiously when oncoming traffic has stopped.
   c. drive straight through the intersection.
   d. stay where you are and wait for the next green light.

Sentence Completion

1. You must give the right-of-way to any funeral procession or police-escorted _____ .
2. Never cross railroad tracks if warning lights are _____ .
3. A _____ sign means "give the right-of-way."
4. A _____ green light is one you have noticed has been green for a long time.
5. You can be ticketed for _____ an intersection.

Matching
Match the concepts in Column A with examples of the concepts in Column B.

Column A
1. __ Clue to an approaching intersection
2. __ Controlled railroad crossing
3. __ "Give the right-of-way"
4. __ Uncontrolled intersection
5. __ Semicontrolled intersection
6. __ Roads intersect from many directions
7. __ Controlled intersection
8. __ Always has the right-of-way

Column B
a. Two STOP signs
b. Blind pedestrian
c. YIELD sign
d. Four STOP signs
e. No signs or signals
f. Crossbuck
g. Bus stop benches
h. Traffic circle

Short Answer
1. What would you do if a pedestrian suddenly crossed in front of you outside of a crosswalk?
2. How many times do you have to stop as you move from an alley into a roadway?
3. When you give the right-of-way to another driver or to a pedestrian, what are you actually giving that person?
4. What does it mean to "roll" through a stop?
5. Who has the right-of-way in a traffic circle?

Critical Thinking
1. You approach an uncontrolled intersection at the same time as a vehicle to your right. You stop and let the vehicle to the right proceed. The other driver waves you to go first. Just then you hear a siren behind you. What should you do?
2. You are stopped at a red light at a fully controlled intersection and you are the first car behind the crosswalk line, in the right-hand lane. A teenage driver pulls up on your right in a souped-up hot rod, squeaking between your car and a row of parked cars along the curb. The teenager revs his engine as he pulls up even with you, shooting you a dirty look. He pulls up into the crosswalk, looking at the cross traffic's signal light. What should you do?

PROJECTS

1. Go out into your neighborhood and try to find an example of every type of intersection: a controlled intersection, a semicontrolled intersection, an uncontrolled intersection, a railroad crossing, an alley intersection, a traffic circle, and a parking lot intersection. Take a photograph or make a sketch of each one and show them to the rest of your class.

2. Find a stretch of railroad tracks in your community and see how many times roads and pedestrian paths intersect the tracks. What types of controls do they have? Are any uncontrolled? Share the results of your investigation with your class.

9 chapter

Turning

Turning is more than simply changing direction. Whether you turn left, right, or completely around, you move in an instant from one flow of traffic to another. Maintaining peak alertness and scanning constantly are essential to safely executing a turn. You must visualize the turn you are about to make, plan your entry point, control the steering wheel and speed of your car, and maintain a comfortable cushion of space between yourself and other vehicles or pedestrians that can come at you from many different directions.

CHAPTER OBJECTIVES

Upon completion of this chapter, you should be able to:

9–1 Turning Basics
1. Understand the difference between a protected and unprotected turn.
2. Understand the proper way to prepare for a left or right turn.

9–2 Left Turns
3. Understand the proper way to execute a left turn.
4. Describe some of the special dangers of left turns.

9–3 Right Turns
5. Understand the proper way to execute a right turn.
6. Describe some of the special dangers of right turns.

9–4 Reversing Your Direction
7. Understand the proper way to execute a two-point turn.
8. Understand the proper way to execute a U-turn.
9. Understand the proper way to execute a three-point turn.

KEY TERMS

protected turn
fully protected turn
semiprotected turn
unprotected turn
two-point turn
U-turn
three-point turn

Figure 9–1 At some intersections, you may turn only when you have a green arrow.

9–1 TURNING BASICS

A **protected turn** is one made from a turn lane posted with signs, road marked with arrows, and accompanied by a traffic signal arrow. A green arrow allows you to turn while oncoming traffic is halted with a red light. If the signal allows you to make a turn only when you have a green arrow (the green arrow turns to a red arrow or red light), it is a **fully protected turn**.

A **semiprotected turn** is one made from a turn lane but *not* accompanied by a special traffic signal that directs your turn with a green arrow. You are allowed to make such a turn only while you have a regular green light—the green arrow is replaced by a solid green or there is one green light for all traffic moving in your

Figure 9–2 When making a semiprotected turn, you may turn on a solid green light once a sufficient gap in traffic exists.

Figure 9–3 Be especially careful turning at intersections without arrows or turn lanes.

direction—and a sufficient gap in traffic occurs.

An **unprotected turn** is one not made from a turn lane at an intersection and where there are no arrows—whether designated by a signpost or road markings, or signaled by a traffic control—to guide your turn. It means that you have to make a turn either from the far-right lane (right turn) or from the far-left lane (left turn).

Preparing for a left or right turn and positioning your vehicle in the correct lane should begin well before you reach the intersection where you intend to make the turn. Scan the roadway in front of you, to your sides, and to your rear. Your eyes should be roving in all directions to make sure that no hazards or obstacles will interfere with your turn. In particular, be on the lookout for bicyclists,

Figure 9–4 By properly scanning before making your turn, you can avoid this kind of situation.

> **DRIVING TIPS** **Avoid the "Pickle"**
>
> Imagine a base runner in baseball who has been caught trying to steal a base halfway between two opposing players moving toward him to tag him out. If you do not scan the sidewalk far enough up and down the side of the street you are about to turn across, you may find yourself in the same jam. Right in the middle of your turn, a fast-moving pedestrian—often a skateboarder, rollerblader, or jogger—crosses the street right in front of you, leaving you nowhere to go! Avoid getting caught in this "pickle" by scanning the sidewalk to either side of the intersection for at least half a block before starting your turn.

pedestrians, and other vehicles in your intended path. Make sure that you first look to see whether there are any signs that restrict or otherwise prohibit the making of your turn. If there is a light at the intersection, make sure that it allows you to make the turn you intend.

Once you have scanned the roadway, select the correct lane and move into it, at the same time reducing your speed. The sharper the turn, the slower you should be driving. Tap your brake pedal to warn vehicles behind you. If you are making a left turn, move just to the right of the center line or into another designated left-turning lane on the left side of the roadway. If you are on a one-way street, move just to the right of the left curbside of the far-left lane. If you are making a right turn, move to the right side of the far-right lane.

A good rule of thumb is to signal a turn 200 feet (60 m) before an intersection in urban areas and 300 feet (90 m) before an intersection in rural areas. Remember, however, not to signal your turning intentions too soon. If another turn is possible before you reach your intended turn, you will confuse the drivers behind you. Keep your signal on throughout the time that you are preparing for and making a turn.

Over time, you will learn to judge when to begin a turn. Your speed, the width of the traffic lanes, the turning radius of your car, and the presence of obstacles such as center dividers and parked cars are all factors to consider. If you start a turn too early or late, you may drift into an adjoining lane of traffic. As you proceed through a turn, keep your speed constant. Wait until you begin to enter the new lane before gradually accelerating to match the speed of traffic.

9–2 LEFT TURNS

Left turns can be one of the most hazardous maneuvers a driver can perform. For new drivers, judging when to cross in front of multiple lanes of oncoming traffic can be a nerve-rattling experience. Often your view of

bumper sticker sightings

DON'T WORRY ABOUT WORLD PEACE. WORRY ABOUT YOUR TURN SIGNAL.

Figure 9–5 Wait until you have a clear view of oncoming traffic before turning left.

oncoming traffic is completely or partially blocked by a line of left-turning cars facing you or by a large vehicle ahead of you. Do not make the turn if you cannot see oncoming traffic clearly. Wait until the light switches to red and cars clear the intersection to give you a clear view on the next green light before making your left turn.

Do not forget to watch out for drivers who may race through a yellow or red light at the last second, and do not be pressured into turning left if you are not ready. Remember that *you* must decide when it is safe to proceed. If drivers to your rear honk at you because they think you are taking too long or are waiting for too large a gap to turn, ignore them. Do not risk a collision because of someone else's impatience.

When you move into an intersection to begin a left turn, you may face an oncoming driver whose real turning intentions are unclear. For example, the other driver may have activated the left-turn signal but now appears to be moving to the right to proceed straight. Or the driver may not have activated a signal but has slowed and appears ready to make a rapid left. To avoid any

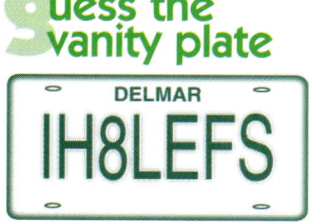

Figure 9–6

Figure 9–7 When turning left, make sure that you leave enough room for oncoming traffic to turn left as well.

possible confusion, leave enough space for both vehicles to turn in their respective directions.

Driver miscommunication is a major factor in many crashes. Motorists unfamiliar with an area or in a hurry can become indecisive and frustrated, changing their minds about a turn at the last second. They may be trapped in a lane that has changed unexpectedly from a regular traffic lane into a left-turn-only lane when they want to go straight. Whatever the reason, *never second-guess other drivers' intentions.* Wait until they make their move, and then proceed.

Types of Left Turns

The steps for making a left turn depend on what type of roadway you are turning from and what type of roadway you are turning onto. When turning left from a two-way street onto another two-way street, always turn from the far-left lane unless there is a road marking, sign, or signal that allows you to turn from one or more adjoining lanes. If there is only one lane from which to turn left, complete your turn in the far-left lane of the roadway you are turning onto.

On two-way streets, if there is only one lane from which to turn left and no signal arrow controlling your turn, follow these steps (see Figure 9–8):

1. Allow for a gap in oncoming traffic of at least 10 seconds on city streets and at least 15 seconds on higher-speed roadways.
2. Move slightly into the intersection to prepare for opposing traffic to clear, allowing you to make your left. Remember to first give the right-of-way to pedestrians or other vehicles in the intersection. Keep your wheels straight so that in case you get rear-ended you will not be pushed into oncoming traffic. Remember that a left-turning driver must give the right-of-way to any and all oncoming traffic and to pedestrians in the path of the turn.
3. Gaze down the path of your intended left turn as it curves into the far-left lane of the cross street you are turning onto. Begin turning the steering wheel as you slowly accelerate. Follow the curvature of the imaginary turn you visualized seconds earlier, ending your turn in the far-left lane of the cross-traffic roadway.
4. Finish the turn by correcting your steering wheel to proceed straight in the far-left lane of the cross-traffic roadway. Make sure that your turn signal is shut off.

When turning left from a one-way street to another one-way street or two-way street, always begin your turn from the far-left lane and complete your turn in the far-left lane of the road you are entering.

Left-Turn Lanes

Often a left-turn lane is created out of a center turn lane, a concrete dividing area (island), or an

driving tips

If you are approaching a driver who is preparing to make a left turn across your path and you think that the driver may not be paying close attention to the driving task, let him or her know that you intend to proceed through the intersection by flashing your headlights, day or night.

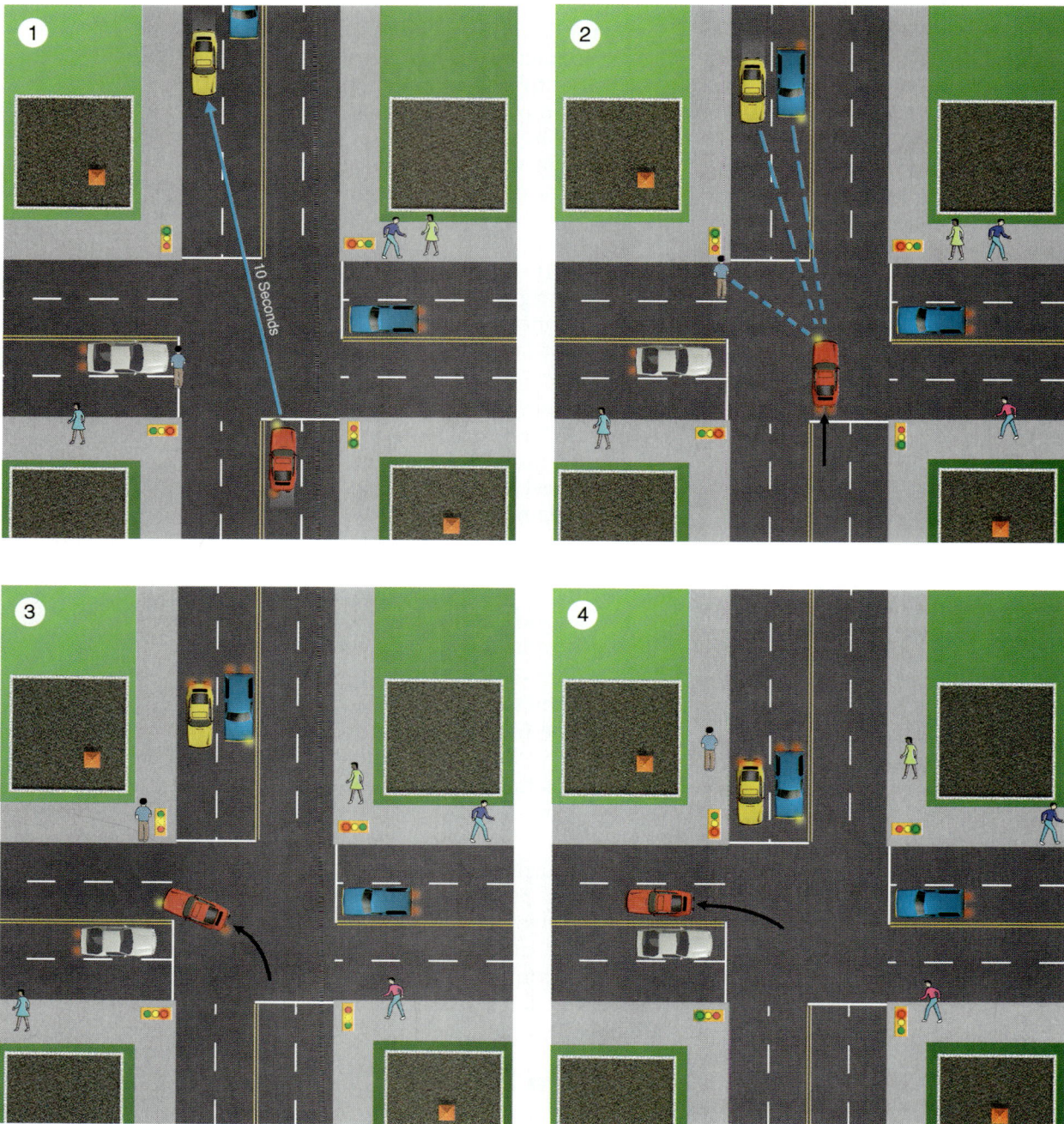

Figure 9–8 Making a left turn

off-limits area set off by two sets of double-yellow lines ("virtual island"). These left-turn lanes have specified entry points usually just before the space allowed for the actual turn.

Some drivers illegally drive across a virtual island or use the center turn lane to avoid having to wait to enter a left-turn lane. This can result in a collision if another driver who has waited correctly for the opportunity to move into the left lane merges without checking traffic on his or her left.

Always look in your rearview and sideview mirrors, and turn your head to make sure that you can enter the left-turn lane safely. Assume that the lane change you are making will involve traffic in the lane next to you. Never enter a left-turn lane early by crossing double-yellow lines or entering a lane that has opposing traffic. You risk a serious crash for which you will be completely at fault.

Just as dangerous as those drivers who enter a turn lane too early are those who enter it too late. This happens when a driver decides to make a left turn after already missing the opening to the turn lane and has to cross the solid white line separating the turn lane from the adjoining through-traffic lane. This kind of last-second maneuver is both dangerous and illegal. Odds are, there will be a driver right behind you already in the turn lane preparing to turn, and you will drive right into him or her or cause a crash trying to avoid hit-

Figure 9–9 When turning left, always start and end your turn in the far-left lane.

Wild Wheels

Figure 9–10 What is the name of this 1961 vehicle, whose driver had to use a joystick rather than a steering wheel to make turns?

ting that car. Instead of risking a collision, turn at the next intersection and retrace your steps.

Center Turn Lanes

If you wish to make a left turn on a street that has a center turn lane, you may only do so from this lane. Move completely within the lane so that through traffic is not blocked. Do not turn into the lane too early, or you may find yourself facing a left-turning vehicle moving in the opposite direction that is in the way of you making your left. If you get caught in this situation, it is always best to let the other driver go first.

Center turn lanes are also a convenient way to make a left turn from a driveway or side street onto a main road. For example, if there is a lot of traffic or it is hard to see, it may be difficult for you to make the complete left turn in one go. First, turn from the driveway or side street into the center turn lane. Second, let vehicles on the roadway going in your direction pass and then enter the far-left traffic lane. Remember that drivers already on the road have the right-of-way and will be traveling at relatively high speeds, so use extra caution before merging. When driving on a road with a center turn lane, watch out for other vehicles making this maneuver that may cut in front of you.

"Stealing" Left Turns

Motorists making semiprotected or unprotected left turns some-

Figure 9–11 You can only make left turns from a center turn lane.

GENERAL GEORGE S. PATTON, JR.

United States General George S. Patton, Jr., performed military miracle after miracle in North Africa and Europe during World War II. "Old Blood and Guts," as the general was known by his soldiers, was famous for taking great chances with his personal safety in the front lines of battle. Ironically, rather than dying gloriously in combat, General Patton was killed several months after the war ended in Europe in a slow-speed automobile crash. The sedan in which he was traveling collided into a truck making a left turn into a driveway. The speed of the truck was estimated at approximately 10 miles per hour (15 km/h), and the speed of the car General Patton was riding in was estimated at less than 30 miles per hour (50 km/h). Patton's tragic death illustrates the fact that a poorly planned turn can kill even at slow speeds.

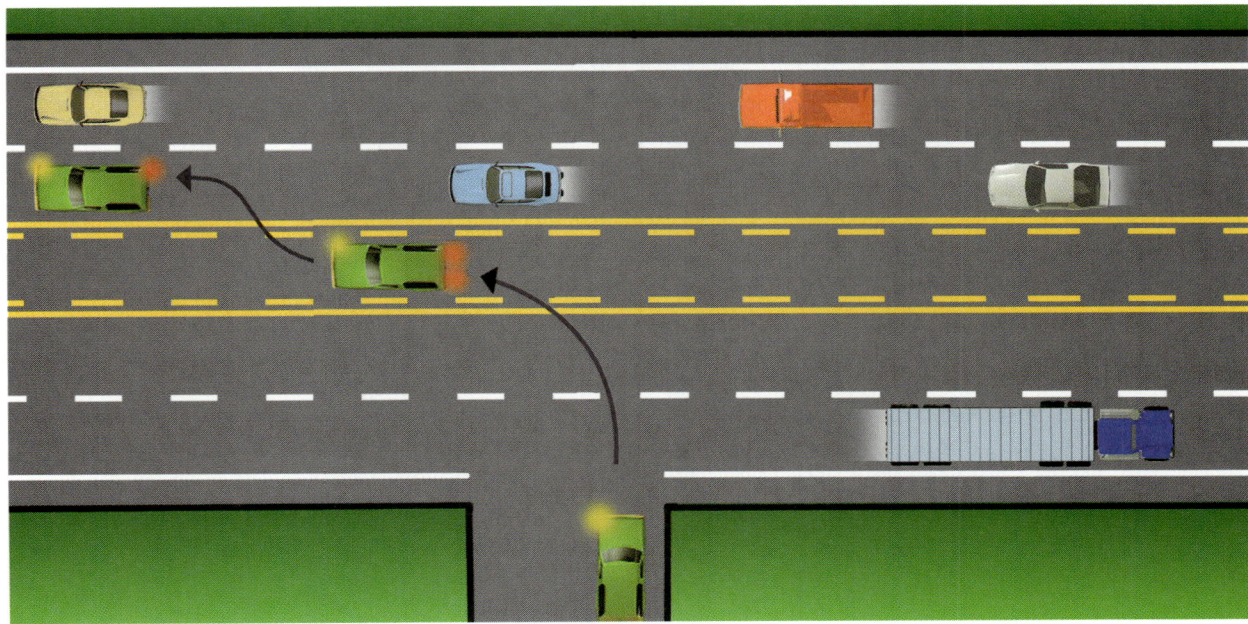

Figure 9–12 Vehicles turning onto a center turn lane from a side street must give the right-of-way to traffic already on the road.

times have to wait to complete their turn until the light turns yellow or even red and opposing traffic has stopped. In these situations, only the one or two drivers that have partly entered the intersection can legally turn left at each signal sequence. However, many drivers are too impatient to wait for a green light when they are second or third in line to turn. They try to "steal" left turns at intersections behind other cars *after* the light has turned red. By the time these drivers are completing their left turns, the light has been red for several seconds and cross traffic is already well within the intersection.

Stealing a left turn is both illegal and extremely dangerous. If you are stuck in heavy traffic, remember that *three rights make a left*. You can get to the same point of your intended left turn by going up a block, making a right, turning right at the next intersection, and then making a right to get back to the original street where you intended to make your left. This may sound like a lot of trouble, but in the end, you will actually get where you want to go faster and more safely.

Heavy Traffic Situations

Turning left in the middle of a block to enter a driveway or alley in heavy traffic can be tricky. Because there are no intersection lines for the opposite traffic to stop behind, there is nothing to

USING SAFE — "Stealing" Left Turns

You are stopped at a red light at an intersection. Two left-turning drivers in oncoming traffic have just completed their turns, and your light changes to green. You accelerate into the intersection and nearly collide with a third driver trying to "steal" a left across your path. This is an all-too-common situation that you can avoid by using the SAFE process. While waiting at the light, use the opportunity to thoroughly *scan* the intersection. Note whether left turns are protected or semiprotected and if there are DO NOT BLOCK INTERSECTION signs. *Assess* whether there are any drivers who may try to sneak a left. Is traffic heavy, causing impatient drivers to take risks rather than wait for the next light? Are oncoming drivers queued up in the left-turn lane, displaying signs of nervousness, anxiety, or frustration? Are they creeping forward into the intersection? Are drivers in lanes adjoining yours showing signs of racing into the intersection, leading to a collision that could spill over into your lane? *Find* an "out" by checking the lanes on either side of you and the distance between your vehicle and oncoming left-turn drivers. *Execute* your decision to avoid a potential crash by starting your forward movement slowly and being prepared to tap your brakes to communicate the danger to drivers behind you.

help create a gap in traffic to allow you to turn left. This situation calls for using the "Four C's": *crawling, communication, confidence,* and *caution.*

Before you turn your wheels to move across the stream of oncoming traffic, make sure that the traffic is coming to a gradual halt, or *crawling,* at less than 5 miles per hour (8 km/h), usually in response to a red light or STOP sign further down the road.

Once you have determined that the traffic is moving slowly enough for the other drivers to notice you, get the attention of the driver in the lane closest to you by means of hand signals. You may also use a light tap on your horn or a quick

DRIVING TIPS — Hesitant Left-Turners

Everyone is familiar with the driver who is afraid to move toward the center of an intersection to prepare for a left turn. In heavy traffic, this failure to "move up" will probably mean that you will miss your chance to turn left, making you wait for the next green light. Be patient. Whatever you do, do not pressure the driver to move by honking your horn, flashing your lights, or gesturing in any way. It is not worth scaring the driver and causing a collision. Wait until you can safely advance into the intersection and then make your turn.

> **AUTO ACCESSORIES** *Steering Wheel Grip*
>
> You may have experienced problems turning your vehicle on a hot day because sweat on your hands causes your steering wheel to slip. To correct this problem, try putting on a vinyl or leather steering wheel grip. This accessory is inexpensive, easy to install, and makes turning safer. This type of grip is usually standard on expensive cars, so it will also give your vehicle a touch of class!

flash of your headlights to *communicate* your desire to cross in front of oncoming drivers.

If there is more than one lane of traffic to cross, the first driver who agrees to let you in may help you communicate with other drivers in lanes to his or her right. However, assume that drivers in lanes farther from you do not see you at all. Creep forward just enough so that your bumper is visible and double-check that drivers in these other lanes are stopping.

Unfortunately, some drivers will ignore you or intentionally close any gap in traffic and make it impossible for you to make a left. In these cases, do not attempt to force the gap. If you muscle your way in by sticking part of your car in front of an uncooperative driver, you will either create gridlock when traffic starts to move again or confuse other drivers into dangerous evasive maneuvers that may cause a crash. It is better to take a deep breath and wait it out. Chances are, the next person will let you make your left and you will be safely on your way!

Once you have established eye contact with the driver in the lane closest to you and have gotten his or her agreement to go through, proceed with *confidence* across the lane. Proceed across while exercising constant *caution*. Because these lanes of traffic are typically slowing as well, it is usually not a problem to get other motorists to stop for you, especially because they realize you have already committed yourself to the turn.

9–3 RIGHT TURNS

Making a right turn is less complicated than making a left. Unless otherwise indicated with pavement markings or street signs at the intersection, a right turn always begins with your vehicle in the far-right lane, the one nearest the curb. As you slow down to prepare for your right turn, remember to tap your brakes to indicate to drivers behind you that you are preparing to turn. Keep in mind that through traffic may approach from the rear at high speeds, especially if the signal is turning yellow.

The steps for making a right turn are the same whether you are turning onto a one-way or a two-way street (see Figure 9–13).

1. After ensuring that the intersection is clear of pedestrians, find a gap in the cross traffic to your left. Allow for a gap in the traffic of 8 seconds on city streets and at least 10 seconds on higher-speed roadways.
2. Align your tires up to the point of the curb where it curves or comes to a right angle with the roadway of cross traffic. Just before turning, scan the intersection again for pedestrians that may have entered the crosswalk since the last time you looked.
3. Gaze around the curb along your path of turning and begin turning the steering wheel as you slowly accelerate. Keep in mind that your right rear wheel will have a smaller turning radius than your right front wheel. You will need to compensate for this as you turn your vehicle to avoid striking the curb. Follow the curvature of the curb as you visualized it seconds earlier, staying in the right lane of the cross-traffic roadway.
4. Finish the turn by correcting your steering wheel to proceed straight in the far-right lane of the cross-traffic roadway. Make sure that your turn signal is shut off.

Danger on Your Right Side

As you move into the right-hand side of the far-right lane before your upcoming turn, look for any bicyclists that might be riding either on the side of the road or in a designated bicycle lane that you will need to enter to make your right. Before moving into the bike lane, make sure that the area is clear of bicyclists. If any are nearby, take care that they know your intentions. Signal early and alert them of your presence, if necessary using a friendly tap of your horn.

Be aware that motorcyclists may be in your right-hand-side blind spot or accelerating to pass you on the right just when you are about to prepare for a right turn. Turn your head and look back to make sure that your blind spot is vacant. Do not rely on your rearview or sideview mirrors.

Danger on Your Left Side

Do not swing wide to your left to make a right turn. You will get in

REALITY CHECK Do Not Cut Corners

The next time you are tempted to cut through a gas station to avoid waiting in line to make your right turn, keep in mind that it is illegal as well as dangerous to drive on or through any private property, road, or driveway to avoid obeying traffic rules or traffic control devices.

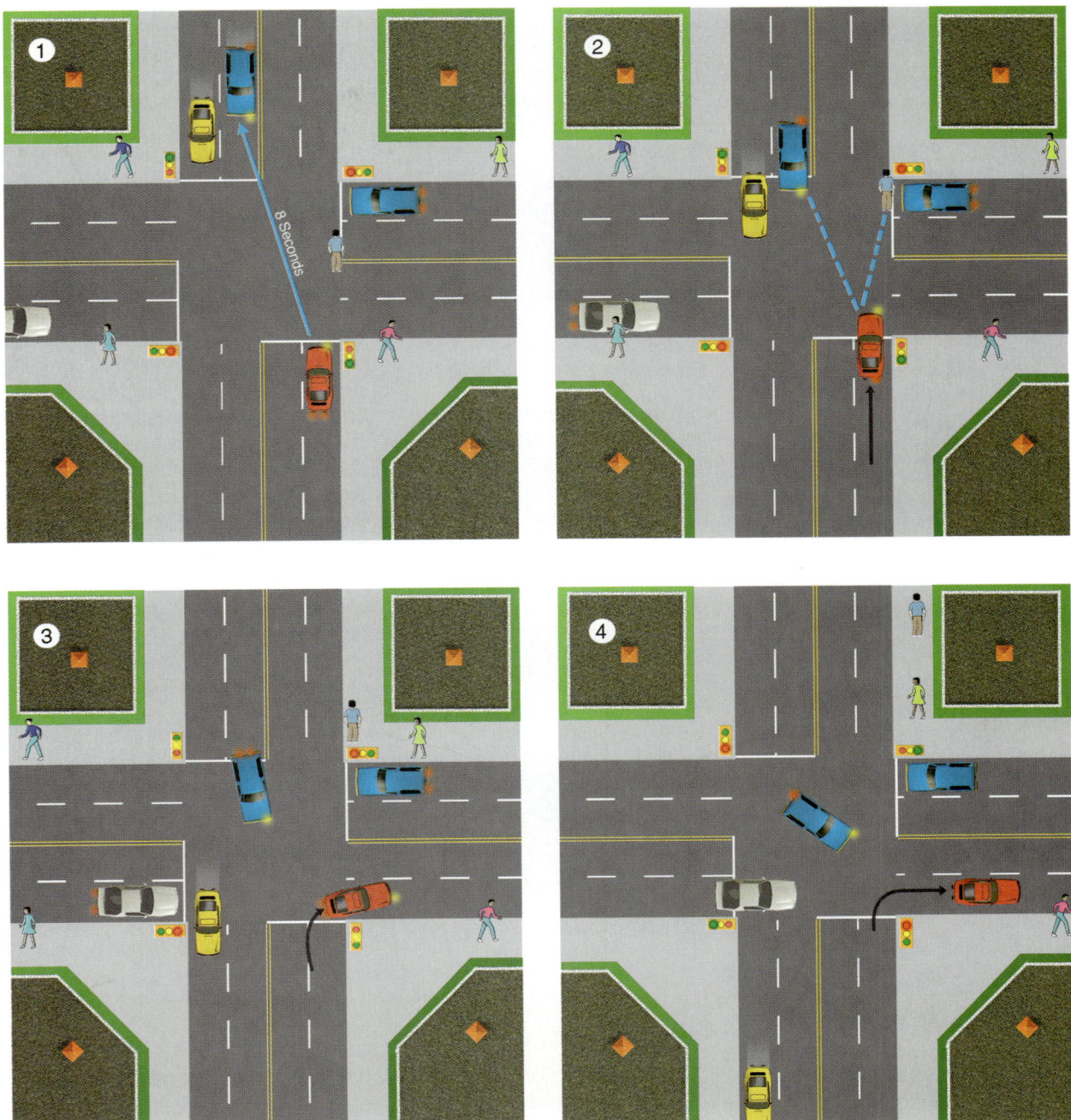

Figure 9-13 Making a right turn

Figure 9-14 Be careful making right turns on roads with bike lanes.

the way of traffic in the lane to your left, which can either be oncoming traffic or traffic going in your direction depending on the type of road you are on. If the type of vehicle you are driving forces you to swing wide left to clear the right-hand curb—for example, if you are towing a trailer or are in a recreational vehicle—signal your intentions to other drivers by using your hazard lights and hand signals.

Danger Ahead

Remember that you must first stop at the stop line or nearer crosswalk line before turning right. Look for pedestrians coming from both directions. As you turn, slow down for any pedestrian traffic on the crosswalk of the street you are turning onto that may dart onto the road from your right side. Also, look for pedestrians who are crossing the street opposite from you on your left.

Figure 9-15 Large vehicles must often swing left first to turn right.

DRIVING TIPS Follow Your "Blocker"

At times when you are preparing to make a right turn, your view of cross traffic from the left may be obstructed by a large parked vehicle. In this case, you can use a "blocker," just as a running back in football uses blockers to clear the way. Your "blocker" is a driver turning left from the cross street onto your street because that driver will have a better view of oncoming traffic than you do. When he or she executes the left turn, the path of cross traffic will be blocked, allowing you to execute your right turn at the same time.

CHAPTER 9 Turning ◆ **211**

Figure 9–16 When making a right turn at an intersection, always look for pedestrians on *both* streets.

Do not forget about oncoming vehicles that will be turning left into either your lane or a nearby lane. Use courtesy and common sense to determine which car should end up in which lane. If both of you will be turning into only one open lane, always be ready to give the right-of-way to the left-turning driver. He or she is cutting across oncoming traffic and may need to accelerate into the lane to avoid getting broadsided by another vehicle.

Figure 9–17 When turning right, always give the right-of-way to left-turning drivers.

Right Turns on a Red Light

When turning right on a red light, which is generally permitted in every jurisdiction in North America except New York City and Quebec, always double-check for pedestrians, bicyclists, and any vehicles that might be either crossing the intersection on their green or making a left turn into your intended lane of travel. Remember that both pedestrians and vehicles approaching the intersection or in the intersection have the right-of-way over you. Keep in mind that you are not required to make the right if you feel safer waiting until the light is green. If you get behind someone who will not make the right on a red, be patient and wait for him or her to turn right when the light turns green.

9–4 REVERSING YOUR DIRECTION

Sometimes it is necessary to reverse the direction in which you are going. For example, if you missed a street that you needed to turn onto and there is no way to go around the block to get to it, you may have to turn around and go back in the opposite way. Because you will have to cross or back into one or more lanes of traffic, this can be dangerous. In many places and situations, it is illegal. Depending on the circumstances and road conditions, there are several ways to turn around.

Two-Point Turns

Two-point turns require the use of a driveway either on your side of the roadway or on the opposite side. Each has important safety implications that you must keep in mind.

Reverse, Then Forward

It is safest to execute a two-point turn from a driveway on your side (the right side) of the street (see Figure 9–18).

1. Reduce your speed as you approach the driveway, tapping lightly on your brake pedal to warn drivers behind you that you are slowing down. Make sure that the driveway is clear of vehicles, pedestrians, and other obstacles.

2. Go past the driveway, lining up your rear bumper with one end of the driveway and staying within a few feet (about 1 m) of the right-hand curb. Check again for any pedestrians or vehicles coming from behind you that may interrupt your maneuver.

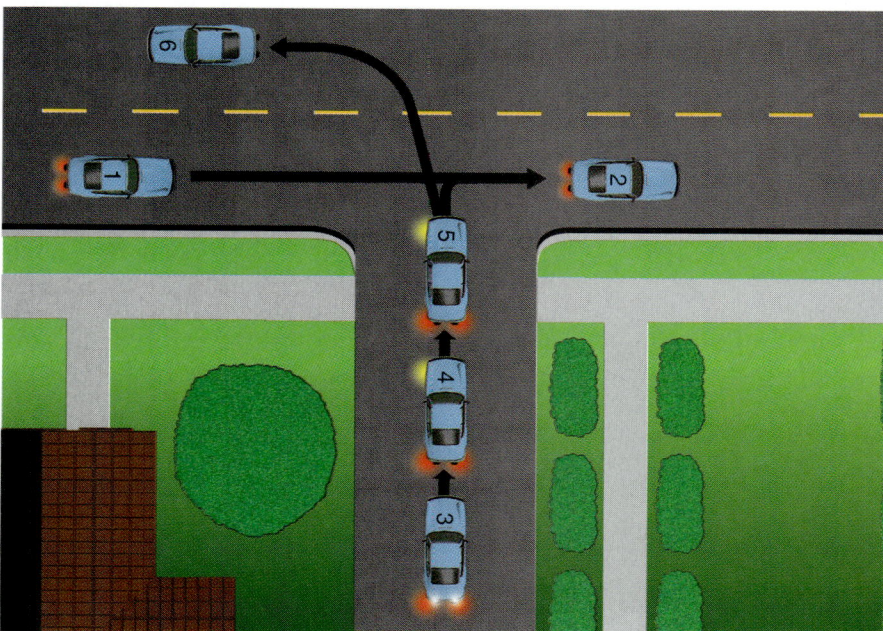

Figure 9–18 Making a two-point turn: "reverse, then forward"

3. Switch gears to REVERSE, turn your body to look out the rear and side windows, and slowly begin backing into the driveway, turning your steering wheel to the right. Be sure to straighten out the steering wheel as you complete your turn backwards into the driveway. Stop the car once you are fully in the driveway.
4. Switch gears to DRIVE or FIRST, and signal a left turn. Creep slowly up to the sidewalk and stop. Make sure that no pedestrians are in your path.
5. Creep up to where the driveway meets the street and stop.
6. Scan traffic in both directions, and when you have a large enough opening to allow you to smoothly accelerate into the lane, make your left turn and rejoin traffic in the direction opposite that which you were going moments earlier. Make sure that your turn signal is shut off.

Forward, Then Reverse

It is best to use a driveway on the opposite side of the street to reverse your direction if you want to go forward first, then reverse (see Figure 9–19).

1. Choose a driveway with a clear view of traffic in both directions. Make sure that no obstacles, such as parked cars, are on the road very close to the driveway that may block your view. Move into the far-left lane of traffic going in your direction. Reduce your speed, tapping lightly on your brake pedal to warn

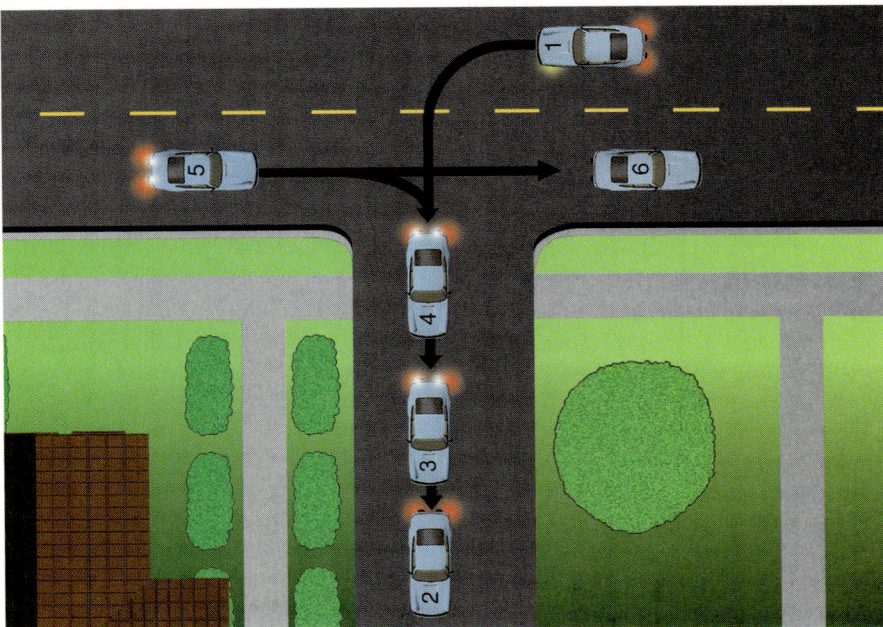

Figure 9–19 Making a two-point turn: "forward, then reverse"

drivers behind you that you are slowing down. Signal a left turn after making sure that the driveway is clear of vehicles, pedestrians, and other obstacles.

2. Turn into the driveway, straightening the wheel out as you fully enter it. Come to a complete stop. Make sure that your turn signal is shut off.

3. Shift your transmission into REVERSE, and go *very slowly* out of the driveway. Look over your shoulder and use your rearview and sideview mirrors to make sure that no pedestrians are in the way or vehicles are approaching from your right. Stop first before reaching the sidewalk.

4. If clear of pedestrians, proceed across the sidewalk and stop a second time before reaching the roadway.

5. When you are sure that traffic is clear in the far-right lane of oncoming traffic going in the direction you want to go in, sharply turn the steering wheel while reversing slowly out into the roadway. Look over your shoulder *the whole time*. This will mean steering the wheel with one hand while your other is supported against the back of the passenger seat, a maneuver you should practice many times before actually using it on the streets. Do not try to back across one lane of traffic into another lane farther away from the curb.

6. As you enter into the far-right lane of travel, straighten out your steering wheel and come to a stop, switching gears to DRIVE or FIRST, and proceed forward.

It is possible to pull forward into a driveway on the right side and then back into the opposite lane to reverse your direction. However, this is extremely dangerous and should be avoided at all costs. You will be backing up across *two* lanes of traffic, a risky move even in the quietest residential areas.

U-Turns

When a vehicle makes a U-turn, it actually traces a path in the shape of a large "U." U-turns are another way to reverse your direction of travel. They are often dangerous, and in many situations and places, they are illegal, so a good rule of thumb is to avoid performing a U-turn if you are unsure whether it is allowed (see Figure 9–21).

1. The first step in making any U-turn is to check for any signs that tell you U-turns are not permitted. If you do not see any signs, then think about where you are. Unlike other kinds of turns, whether a U-turn is legal or not can depend on your location. Often no signs will be present to indicate specifically whether a U-turn is legal or illegal. It is up to *you* to know the difference.

2. Allow at least 500 feet (150 m) visibility in both directions. Make sure that nothing is

Figure 9–20 Before making a U-turn, ask yourself if it is legal where you are.

blocking your view, such as buildings, blind curves, and hills, and that you have enough time and space to complete the turn on the roadway. Consider the turning radius of your vehicle and get over as far to the right as necessary. Vehicles with long wheelbases, such as trucks and sedans, need more room to maneuver. Allow yourself a gap in traffic of at least 20 to 30 seconds if you are unsure whether your vehicle can fully make the U-turn on a narrow road. Check traffic in both directions and signal a left turn.

3. Turn the wheel to the left completely, moving forward at a slow speed. If it looks like you can make the turn without hitting the other curb or a parked car, complete the turn until your vehicle is facing in the opposite direction. As you complete the turn, straighten out your wheels and proceed forward. Make sure that your turn signal is shut off.

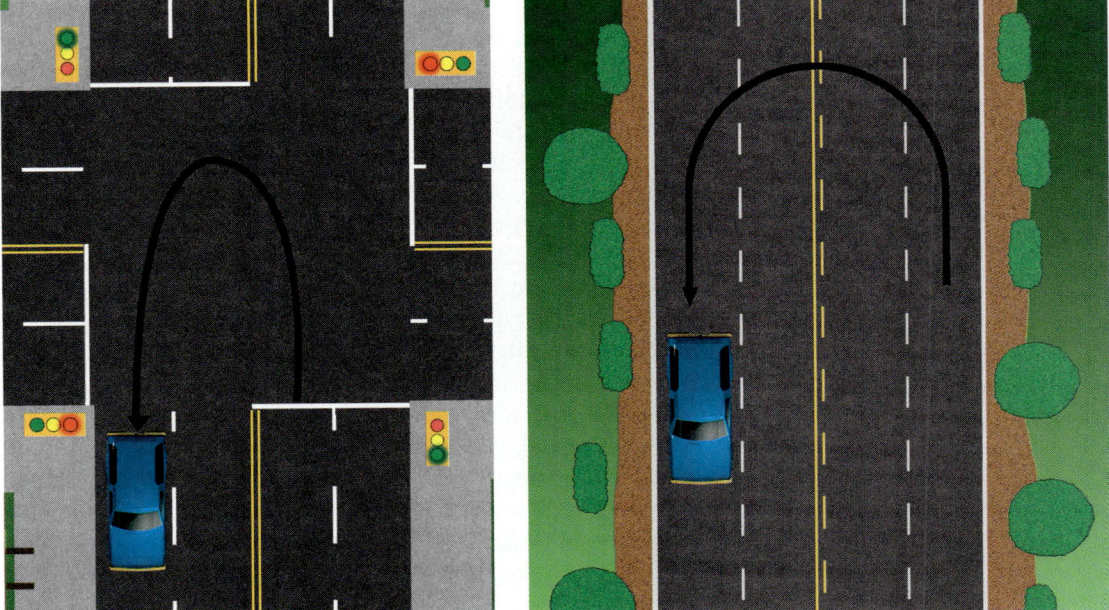

Figure 9–21 Make sure that you have plenty of room and time to make a U-turn.

Three-Point Turns

If no driveways are available for you to make a two-point turn and circumstances prevent you from making a U-turn, you can make a **three-point turn** to reverse your direction. This is a difficult maneuver and should not be tried on the streets until you have practiced it a lot in a safe area away from traffic, like an empty parking lot. You should limit your use of this maneuver to rural roads, residential areas with no driveways, dead-end streets, and emergencies (see Figure 9–22).

1. Make sure that no traffic is approaching you either from the rear or from the front for at least 500 feet (150 m). Stop your vehicle close to the right curb. Double-check that traffic is clear of other vehicles and pedestrians in both directions.
2. Activate your left-turn signal, and turn your car's steering wheel sharply to the left while you start going forward until you get into the opposite lane of traffic (at an angle). Approach the opposite curb within a few feet (about 1 m). Stop your car and turn the steering wheel to the right. Make sure that your turn signal is shut off.
3. Check for traffic again by looking over your shoulder. If it is still clear, put your transmission into REVERSE, and back up into the far-right lane of traffic going in your new direction of travel. Come to a complete stop.

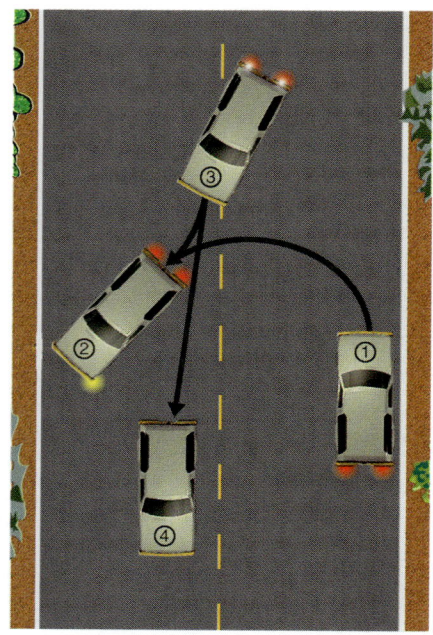

Figure 9–22 Making a three-point turn

4. Switch gears to DRIVE or FIRST and proceed forward.

Other Ways to Reverse Your Direction

The easiest and safest way to reverse your direction is to drive around the block. If all the roads are low-speed roads or roads with signal lights, make three rights and a left. If you are on a high-speed roadway intersected by smaller streets with no signal lights, make three lefts and a right. It is always easier to make your left off of a high-speed street onto a side street if there is no signal light than onto a high-speed road from a side street.

CHAPTER 9 Turning ◆ **217**

Another alternative is to turn around within the "alleys" of a large parking lot, making sure to slow your speed as you move across the lot. Watch out for people getting in and out of cars and drivers that may be busy trying to park. Exit the lot by making a turn in the reverse direction from which you were originally going.

WHO'S AT FAULT?

1. Driver 1 was proceeding northbound on Avenue A and turning right (eastbound) on a green signal at the intersection. Driver 2 was southbound on Avenue A and had pulled into the intersection on the green light to make a left when oncoming traffic was clear. Driver 2 made her left moving toward the lane closest to the curb. In doing so she struck Driver 1, who was making a right and attempting to move immediately into the same lane. *Who's at fault?*

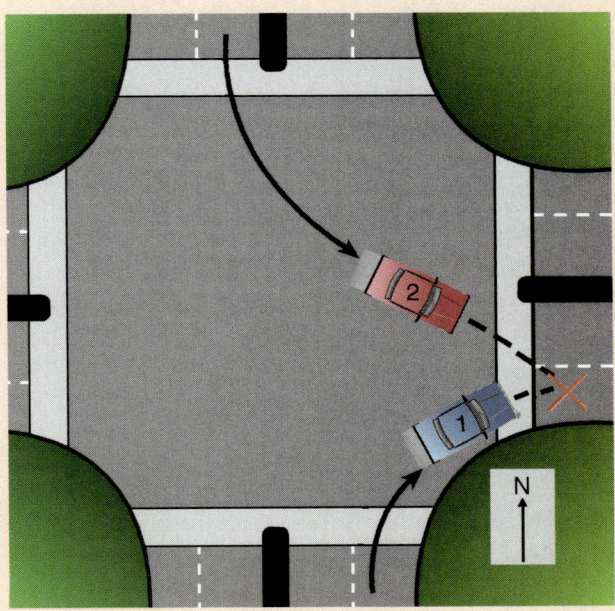

2. All vehicles were stopped at red lights waiting to make left turns or proceed straight once the lights turned green. Lane A is a left-turn-only lane. Lane B is a left-turn or straight lane. When the left-turn green arrow activated, Driver 4 followed Driver 5 in making a controlled left turn from eastbound Central Boulevard onto northbound Third Street. Driver 6 followed Driver 3 in making a controlled left turn from westbound Central onto southbound Third.

Driver 2 was in Lane B waiting for the green light to proceed straight through the intersection on Central. Driver 1, wishing to make a left turn as permitted in Lane B, honked his horn at Driver 2, who could not legally proceed straight. Driver 2, being unfamiliar with the intersection, became nervous and darted forward to clear the way for Driver 1, at which time he was struck on the right front by Driver 3. ***Who's at fault?***

YOUR TURN

9–1 Turning Basics
1. What is the difference between a protected and unprotected turn?
2. What is the proper way to prepare for a left or right turn?

9–2 Left Turns
3. What is the proper way to execute a left turn?
4. What are some of the special dangers of left turns?

9–3 Right Turns
5. What is the proper way to execute a right turn?
6. What are some of the special dangers of right turns?

9–4 Reversing Your Direction
7. What is the proper way to execute a two-point turn?
8. What is the proper way to execute a U-turn?
9. What is the proper way to execute a three-point turn?

SELF-TEST

Multiple Choice

1. A semiprotected turn is made:
   a. from a turn lane but with no special traffic signal.
   b. from any lane but with a special traffic signal.
   c. with help from a traffic officer.
   d. from a turn lane with a special traffic signal.
2. You should signal a turn in a city at least _____ from the intersection.
   a. 100 feet (30 m)
   b. 200 feet (60 m)
   c. 300 feet (90 m)
   d. 500 feet (150 m)
3. You should signal a turn in a rural area at least _____ from the intersection.
   a. 100 feet (30 m)
   b. 200 feet (60 m)
   c. 300 feet (90 m)
   d. 500 feet (150 m)

4. Making a left turn in heavy traffic calls for:
   a. communication, consideration, caution, and coexistence.
   b. crawling, caution, charisma, and communication.
   c. consideration, coexistence, crawling, and cunning.
   d. confidence, caution, crawling, and communication.
5. "Forward, then reverse" is a technique used to make a:
   a. two-point turn on the right side of the street.   c. two-point turn on the left side of the street.
   b. U-turn.   d. three-point turn.

Sentence Completion
1. When turning left, move about 10 feet (3 m) from the _____ of the intersection.
2. Allow at least 500 feet (150 m) visibility in both directions when making a _____ or a _____.
3. If you enter a _____ too early, another left-turning driver may block your way.
4. You are not required to turn _____ on a red light if you do not feel it is safe to do so.
5. It is safest to execute a two-point turn from the _____ side of the street.

Matching
Match the concepts in Column A with examples of the concepts in Column B.

Column A
1. __ Danger ahead for a right turn
2. __ Two-point turn
3. __ Left turn on a city street
4. __ Entering a turn lane too soon
5. __ Answer to a difficult left turn
6. __ Three-point turn
7. __ U-turn
8. __ Safest way to reverse direction

Column B
a. Reverse, then forward
b. Illegal in most business districts
c. Drive around the block
d. Narrow dead-end streets
e. Gap of 10 seconds
f. Crossing solid double-yellow lines
g. Three rights make a left
h. Pedestrians

Short Answer
1. What is a fully protected turn?
2. What are the dangers of entering a turn lane too late?
3. Three-point turns should be limited to which situations?
4. Which lane do you turn into when turning left or right?
5. What does it mean to "steal" a left turn at a red light?

Critical Thinking

1. You are lost on a narrow street in an unfamiliar industrial zone with moderate traffic at night. The area is not well lit and it is hard to tell where the few side streets lead to, if they lead anywhere at all. You need to reverse your direction and head back the way you came. What should you do?

2. You are preparing to make a left turn in a high-traffic business area and have properly waited for a sufficient gap in oncoming traffic to start your left. You commit to the maneuver, beginning to cross before more oncoming traffic approaches, when at the last second a rollerblader comes out of nowhere, zipping across the crosswalk at a blinding speed directly in your path of travel. Oncoming traffic does not appear to be slowing down for you and may soon broadside your vehicle. What should you do?

PROJECTS

1. Knowing the turning radius of your vehicle is an important factor in making a U-turn. Go to an empty parking lot and empty a pail of water on the ground. Drive through the water and turn as sharply to the left as you can. When you have made a complete U-turn, stop and measure your vehicle's turning radius with measuring tape. Compare your results with those of others in your class. List the vehicles in order from shortest to longest turning radius to see how they vary by type of vehicle.

2. Take a clipboard, paper, and pen out to a busy intersection in your town and note how many drivers try to "steal" a left turn. Share your results in class. Appoint a representative to discuss your findings with the traffic section of your local police department and see if they are aware of the size of the problem.

The Driving Environment
unit 4

Chapter 10   City Driving
Chapter 11   Highway and Rural Driving
Chapter 12   Freeway Driving
Chapter 13   Sharing the Road

10 chapter

City Driving

On city streets, you face a variety of complex traffic situations. Intersections occur frequently, pedestrians and bicyclists cross your path from every direction, and business advertisements compete with traffic controls for your attention. Parked cars, buses and delivery trucks, crowds of people, and buildings block your surrounding view. Maneuvering space and escape routes are limited. Urban parking spaces can be hard to find and are typically more dangerous to enter and exit. Stop-and-go traffic, detours for road repairs, and turning restrictions can make city driving frustrating.

Chapter Objectives

Upon completion of this chapter, you should be able to:

10–1 The City Driving Environment
1. Understand the importance of scanning ahead and reducing speed in city driving.
2. Understand the difference between riding and covering the brake.

10–2 Lane Positioning
3. Understand why lane positioning is important in city driving.
4. Understand the proper way to change lanes.
5. Describe the advantages and risks of using one-way streets.

10–3 Passing
6. Understand the proper way to pass another vehicle.
7. Describe some of the dangers of passing on urban roadways.

KEY TERMS

covering the brake
riding the brake
gridlock

one-way streets
pass

10–1 THE CITY DRIVING ENVIRONMENT

The dangers you face on city streets depend on where you are driving. Typically, the streets in residential districts are not very congested, the pace of traffic is slow, and there are fewer controlled intersections. Although this might seem like a safe place to drive, many risks are hidden from view. Streets are not always brightly lit at night, and because residents are familiar with the streets and flow of traffic, some may disregard STOP or YIELD signs or pay less attention than they should while driving.

Business districts have their own hazards. Although roadways are usually wider, have more traffic controls, and are better lit than those in residential areas, they are often packed with traffic. Vehicles often travel at much faster speeds in a business district than in a residential one, making the chances of a serious collision greater. Rows of parked cars, big delivery trucks, and blind alleys often conceal pedestrians or other vehicles that you may only see clearly when it is too late. Drivers competing for a lane opening or parking space can be rude and aggressive.

When driving on crowded urban roadways, every second counts. You cannot afford to look too long at business signs, street addresses, and other distractions. If a driver ahead of you hits his or her brakes in heavy traffic and you are not paying attention, you could rear-end the other vehicle before you even know what happened. Because you have such a small margin for error, you must make an extra effort to stay focused and continually scan ahead in urban environments. To avoid the danger of "information overload" when traveling in an unfamiliar city, consult a local map to familiarize yourself with the layout of the city and the major through streets.

Expect pedestrians and bicyclists to be *everywhere*. People go to and from office buildings, stores, hotels, and restaurants; get

More than 80% of all crashes in Illinois occur on urban roadways.

factoid

Electric lights replaced gas jets on Chicago's Michigan Avenue between 14th and 39th streets on September 1, 1897.

Figure 10–1 Maintaining a space cushion is difficult on crowded urban streets.

Highway heroes

Soon after midnight on August 25, 1997, Officer David C. Evans of the Chicago Police Department was killed when his partner, responding to a call for assistance, swerved to avoid pedestrians crossing West 79th Street and skidded off an overpass, plunging 25 feet (8 m) to the Dan Ryan Expressway below.

in and out of parked cars, buses, and taxis; or just "cruise" the streets. Delivery drivers race to and from their trucks. Couriers weave in and out of traffic on bicycles. Pedestrians in crowds are more likely to cross streets illegally, and some may not pay any attention at all to traffic.

Many residential streets are used by local children as their playground. Kids may chase a ball into the street from a park, yard, or driveway without looking. Children on scooters, bicycles, skateboards, rollerblades, pedal-powered toys, and other recreational equipment can be right around the corner. As you drive through neighborhoods, be thinking of what defensive measures you can take with your vehicle to avoid a tragic collision.

Because of the number of potential hazards and limited space available on city streets, you must make a conscious effort to watch your speed. Cars can and do kill people even at relatively slow speeds. Slowing down will give you more time to assess a dangerous situation on the roadway and respond safely to it. For example, if a car suddenly darts out of an alley ahead of you into traffic, you will be able to apply your brakes early with less risk of causing a crash. If a bicyclist is about to jump the curb and move onto the street directly in front of you, you will have time to warn him or her of your presence with a polite toot of your horn.

Covering the Brake

In city driving, you must be prepared to stop or slow suddenly. It takes time to move your right foot from the accelerator to the brake, and it is not uncommon for people to panic in an emergency situation and completely miss the brake pedal. To avoid this problem, you can use a technique called **covering the brake**. This involves taking your right foot off of the accelerator and holding it over the brake pedal as you cruise forward on your car's momentum. If you have to stop quickly, your foot is already in position to immediately hit the brake.

When covering the brake, be careful not to rest your right foot on the pedal. This is known as **riding the brake**, and it will both add wear and tear to your brakes and confuse other drivers by constantly activating your brake lights. If people behind you think you are going to stop or slow down and then see you continue

Figure 10–2 For many pedestrians, the shortest distance between two places is a straight line.

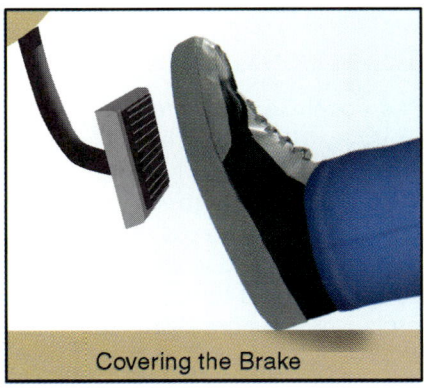 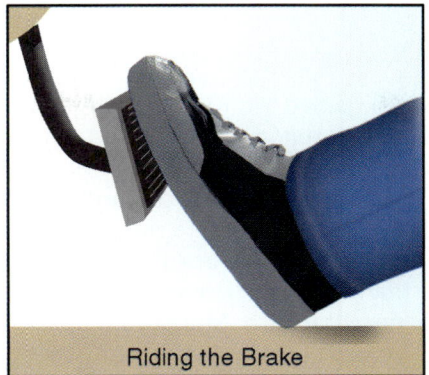

Figure 10–3 Covering versus riding the brake

moving forward, they may not stop as quickly when you really do need to use your brakes.

Parked Cars

When driving in cities, you must constantly watch for cars entering and exiting parking spots. On busy streets, especially in downtown areas, parking spaces can be rare. Drivers may circle a block a couple of times looking for a vacant spot or one about to be vacated. This can lead to dangerous, erratic driving behavior, especially if these drivers see clues that someone is preparing to leave—for example, a person unlocking the door of a parked car or a vehicle's back-up lights. Sometimes these drivers swerve across several lanes or slam on their brakes after passing what looks like a departing vehicle in the hope that they might get that space.

If you are the one looking for a spot, resist the urge to drive dangerously when you see an opening. This is an invitation to get rear-ended. Instead, reduce your speed and scan far ahead so that you will have plenty of time to identify a possible parking space, indicate your intention to park with a turn signal, and come to a gradual stop near the space.

factoid

Crashes at speeds as low as 12 miles per hour (20 km/h) can result in fatal injury to occupants of vehicles.

REALITY CHECK **Do Not Get "Booted"**

To combat the problem of unpaid parking tickets, Chicago has a "boot and tow" program. If you have five or more unpaid parking tickets on your driving record at the time you are cited for a parking violation, the police are authorized to place a mechanical device called a "boot" on the vehicle that prevents it from being driven away. If you do not satisfy the outstanding debt that same day—usually within 24 hours—the car will be towed to prevent vandalism. Until all fines and fees are paid, your vehicle will be impounded, which will add considerable expense to what you already owe.

MARGARET MITCHELL

In 1939, Margaret Mitchell gained world fame with her Civil War novel *Gone with the Wind*, which won a Pulitzer Prize and was made into an Academy Award–winning motion picture. On the evening of August 11, 1949, Mitchell and her husband John Marsh drove into downtown Atlanta, parked their car, and walked together across the street toward a theater where they planned to see a movie. When they were halfway across the street, a driver traveling at a very high speed came around a curve and braked violently when he saw the pedestrians. His car skidded nearly 70 feet (20 m), swerved toward Mitchell, and struck her near the curb. She never regained consciousness before dying at a local hospital almost five days later. When asked why he was exceeding the 25 miles per hour (40 km/h) speed limit, the driver said that "everybody does it." Because city streets are more congested with traffic and have limited visibility, it is essential that you drive at or below the posted speed limit to safely navigate around all potential hazards.

If you are *not* looking for a parking space, be aware that other drivers may be. Be on the look-out for cars entering and exiting parking spaces. Because vehicles must often back up into traffic to squeeze into a space when parallel parking, they can cause an entire lane to quickly stop. Look for wheels turned out, exhaust, brake lights, and other signs of cars leaving parking spaces.

When driving past a row of parked cars, keep your brake covered. Look at the sideview mirrors and through the rear windows of the parked cars to see if any drivers are inside. Be prepared for doors to open suddenly. If your lane is wide enough, stay at least a door's width away from parked cars. If necessary, tap your horn to warn a person getting out of a vehicle of your presence. Keep an eye on the lane next to yours. You should always be prepared to stop behind the door of a parked car, but if you cannot stop, you at least want to avoid swerving into the path of another vehicle to your right.

AUTO ACCESSORIES *Fuel Additives*

Because of constant stopping and starting, city driving can be brutal on your engine. A good way to lessen some of this harsh treatment is to use a fuel additive when you fill up your car. A number of additives are available on the market that can keep your fuel system clean, increase gas mileage, lower engine temperature, reduce emissions, cut down on engine knocking and pinging, extend oil life, and generally increase performance. When correctly used, fuel additives are a small investment that can extend the life of your car.

Gridlock

On city streets in and around urban areas, traffic can get so heavy that vehicles come to a virtual stop. This situation, called **gridlock,** is most common during rush-hour periods and special events such as concerts, athletic contests, and parades. Collisions, road work, police officers issuing tickets, and other distractions can contribute to gridlock and make urban driving a nightmare.

When you know that you will encounter heavy traffic, plan alternative routes. Keep a map in your car for easy reference. Use side streets to get past logjams on through streets. If none are available, try to rearrange your schedule to travel at a different time. If you drive into the city to work, talk to your employer about working earlier or later. Familiarize yourself with turning prohibitions or other special driving restrictions on city streets. Watch TV, listen to the radio, and check traffic-information websites for updates on roadway construction, closures, and other problems. By avoiding heavy traffic, you will save yourself unnecessary headaches.

10–2 LANE POSITIONING

Defensive driving requires you to position your vehicle within a stream of traffic so that you have the greatest possible space between your vehicle and any potential dangers. This is especially true in an urban environment where careful lane positioning is the key to avoiding many of the hazards associated with city driving.

Choosing a Lane

Most people either pick a lane to travel in out of habit or do not think about it at all. However, certain lanes are more appropriate than others, depending on the circumstances. The amount of congestion, the direction of cross traffic, and the availability of turn lanes determine the best lane of travel, as will your immediate goal on the road—for example, to continue straight ahead, park on the street, enter the driveway of a strip mall, or turn at the next intersection. By waiting until the last minute to get into the proper lane before turning or trying to park, drivers needlessly risk causing a crash.

Wild Wheels

Figure 10–4 What are these accordion-like buses, which require special skill to navigate on crowded city streets, called?

REALITY CHECK "Squeaking" By

In city traffic, it is common to get stuck behind a stopped bus, delivery truck, or double-parked car. Avoid the temptation to "squeak" by the vehicle by intruding into the occupied adjoining lane. You could get sideswiped by a vehicle in that lane or, worse, cause it to swerve into oncoming traffic, resulting in a serious collision. You might also get caught between that vehicle and the one you are trying to pass if it suddenly starts forward. Instead of rolling the dice, wait until the vehicle in front of you moves or another lane becomes free and you can pass safely.

Figure 10–5 "Squeaking" by a delivery truck or bus will decrease your space cushion and create problems for everyone.

As a general goal for city driving, you should choose a lane that is less congested than the other lanes. That does not mean you should dart in and out of lanes to get into the one with the "shortest" line in heavy traffic situations. This only slows everyone else down and creates a dangerous situation for other drivers. When you pick a lane, stay there until you need to turn off the road, pass another vehicle, or avoid a traffic hazard. By selecting the less-traveled lane, you increase your distance from other cars and expand your safety cushion.

Two-Lane Roads

On two-lane, two-way roads, traffic often bunches up along the

REALITY CHECK — *Lagging Left-Turn Arrows*

At intersections with fully protected left turns, the left-turn arrow usually appears before the green light for through traffic. This arrangement can lead to backups at closely spaced intersections or those with a large left-turn volume. Divers wishing to turn left often cannot get into the left-turn lane until it is too late because vehicles stopped at the through lane adjoining the left-turn lane block their entry. As a result, they are tempted to crowd the intersection at the end of the green signal to "make" the light and block through traffic when the green arrow is canceled. To address this problem, some cities in Illinois "lag" left-turn arrows so that they are activated *after* the through green light. The danger of this type of signal is that fast-moving through traffic may still be in the intersection when left-turning drivers get a green arrow. If you are the first driver turning left at an intersection with a lagging left-turn arrow, make sure it is clear of oncoming traffic even though you have a fully protected turn.

Figure 10–6 On city streets with lots of signals, the left lane can become very congested.

right-hand side of the road. Buses and taxis picking up or dropping off passengers, vehicles double parking or backing up into parallel parking spaces, slowly creeping traffic-enforcement vehicles checking parking meters, and delivery trucks making their "stop and go" rounds can effectively block an entire lane.

Avoid the temptation to "creep" over the center dividing lines of the road and quickly swerve by a problem. Not only will you be moving into the lane of opposing traffic, but you also will be increasing your speed at the same time. This maneuver is guaranteed to eventually cause a crash. The next time you encounter this situation as a driver, relax, slow down behind the obstacle, and wait patiently for it to move. If it has not moved within a reasonable amount of time, wait until the roadway is completely clear of opposing traffic before *slowly* moving out over the center line to bypass the obstacle.

Multilane Roads

On streets with more than one lane going in the same direction, the left lane is usually the fastest lane. That does not mean, however, that it is the smoothest lane of travel. Traffic can be held up in the left lane by drivers waiting to turn left. This is especially true

Figure 10–7 Congestion can also build up in the right lane at urban intersections.

on streets with intersections that have no left-turn lanes to keep the through lanes clear. Also, you must make room for vehicles that "spill over" into your lane when the right lane ends or when it is filled with parked cars, forcing right-lane traffic into your lane. On crowded multilane streets, the right lane may actually get you to your destination more quickly.

Traveling in the left lane is also more dangerous. Because city streets can be narrow and crowded,

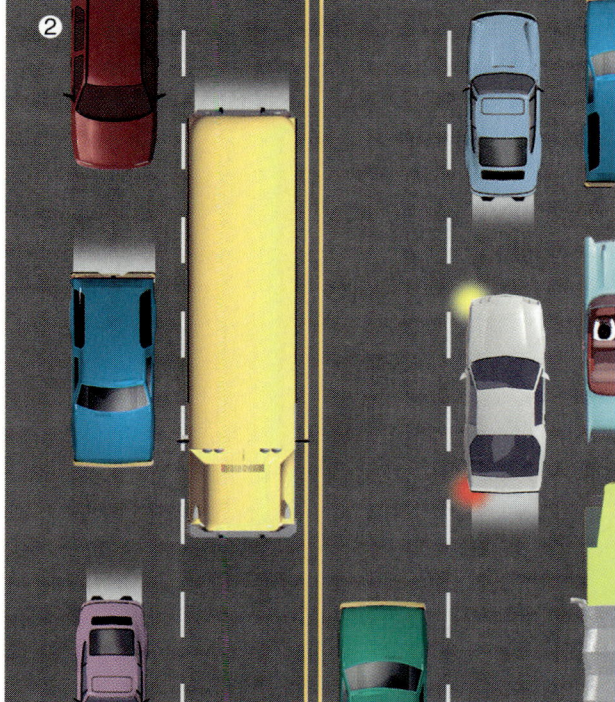

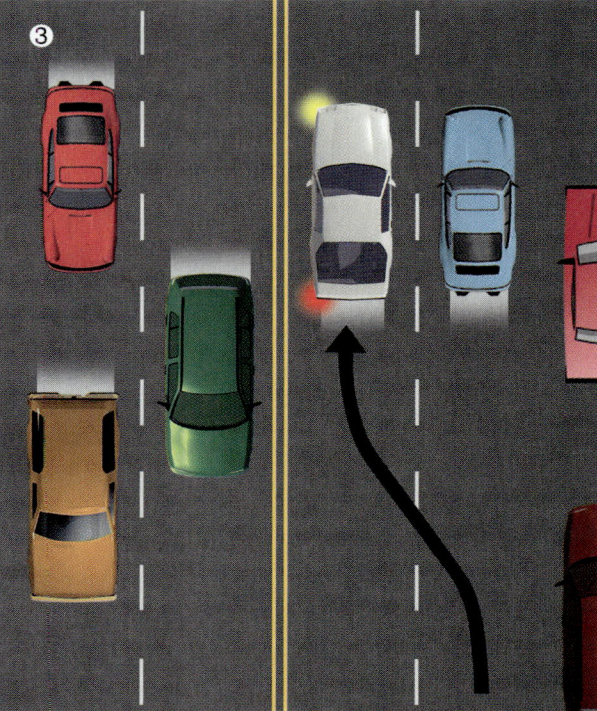

Figure 10–8 Changing lanes

REALITY CHECK — Do Not Let the "Windy City" Blow You Away!

Driving in Chicago can be a nightmare for motorists, especially for visitors, so be prepared before you take to the streets. Many streets are one-way, and a high percentage of the two-way streets are snarled with traffic. In many parts of the city, it is *always* "rush hour." Traffic lights are longer than average, and jaywalking is common. You have to share the road with a wide range of vehicles, including buses, emergency vehicles, bicycles, trucks of all sizes, taxis, mopeds, and horse-driven carriages. In some areas you will even encounter mounted police. It is easy to get lost in the city's maze of bridges and expressways, and at any given time there are lanes or entire roads closed for road work or repairs to utilities. Parking is often impossible to find, prohibited, or ridiculously expensive. The frustrating driving conditions result in a high incidence of road rage and red-light running. When negotiating the tough streets of Chicago, be extra alert and patient. Always expect the unexpected.

there is an increased risk that vehicles will cross over the center lines into your direction of traffic. They might swerve into your lane to pass or avoid a pedestrian or obstacle. They might make left turns in front of you to enter a driveway or parking lot, or they might make a wide right turn that causes them to cross right into your path.

To avoid possible conflicts with oncoming traffic, it is safer to stay out of the left lane unless you are planning to turn left soon or are passing another vehicle. If you have to travel in the left lane, stay to the right side of the lane. If possible, keep an escape route open on your right in case you have to swerve to avoid a car coming at you. If you do face an oncoming car in your lane, slow down, flash your headlights, sound your horn, and give the other driver as much room as possible to clear your vehicle. If the adjoining lane is free, move into the lane only after first checking that it is clear.

Although it is generally safer and quicker, the right lane has its own hazards. You are farther from oncoming traffic, but you are closer to any parked cars. Although right turns are typically easier and smoother than lefts, special right-turn lanes are not as common as left-turn lanes. Sometimes the right lane is wide enough at intersections to allow both through traffic and right-turning vehicles, but this is not always the case. Also, the extra area available to drivers to make right turns at intersections is typically shorter than a left-turn lane, so in some places congestion can actually be worse in the right lane.

On two-way streets with three or more lanes going in the same direction, choose the center lane(s) for the smoothest flow of traffic. This will allow you to avoid both left- and right-turning vehicles *and* parked cars.

Changing Lanes

Changing your lane of travel should never be done too often or without taking the proper precau-

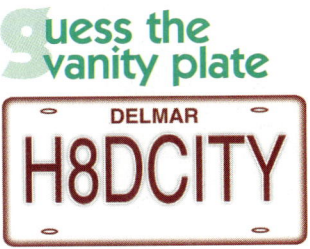

Figure 10–9

factoid

In 1898, Charles Duryea, co-inventor of America's first gas-powered automobile, used Peoria's Barker Avenue as a test track.

factoid

There are 121 cities in 37 states with the name of Springfield, not counting the hometown of television's animated family, the Simpsons.

Figure 10–10 One-way streets are a common feature of the urban driving environment.

tions. Never try to change lanes in or near an intersection. Make sure that you have a clear view of the road ahead and no obstacles are in either lane. When changing lanes, follow these steps (see Figure 10–8):

1. Check your rearview and sideview mirrors for traffic behind you and in the lane you wish to move into. Turn your head and use both eyes to make sure that no vehicles are in your blind spots.
2. When traffic is clear, signal your intention to change lanes. Allow other drivers plenty of time to see your signal. In Illinois, you are required by law to signal at least 100 feet (30 m) before making a lane change in a business or residential area. While signaling, check your mirrors and blind spots *again* to make sure that you have enough room to change lanes safely. For example, if traffic in the other lane will not create space to "let you in," avoid the lane change.
3. Move into the other lane while maintaining a constant rate of speed. Change lanes as smoothly as possible. If you turn the wheel too sharply, you will oversteer and risk losing control of your vehicle.
4. Cancel your turn signal, if necessary. Check the rearview and sideview mirrors again to get oriented to your new position in the flow of traffic.

One-Way Streets

One-way streets are common in urban environments because they can handle a heavier volume of traffic in one direction than two-way streets. Because you do not have to face oncoming traffic, they are also safer. However, when you are unfamiliar with an urban area, it is easy to get lost in a maze of one-way streets that all seem to lead you away from your destination.

DRIVING TIPS The "Herd" Instinct

In most situations, we like to be surrounded by other people. Whether we are at a restaurant, party, concert, or sports event, the presence of others makes us feel more comfortable and secure. For the same reason that animals travel in herds, we instinctively drive near other cars. Resist that instinct and spread that herd out along the roadway! When you can, use the empty space between groups of vehicles to increase the safety factor.

Figure 10–11 Changing lanes at the last second near a stop light is an easy way to get into a collision.

When driving on one-way streets with two or more lanes, always choose the lane with the fewest hazards. If there is a middle lane, it usually has the smoothest traffic flow. On two-lane roadways, use the right-hand lane unless you have to make a left turn off the street soon or are passing another vehicle. If parking is permitted on only one side of the roadway, choose a lane that is not next to parked cars.

Before turning onto a one-way street, make sure that you have correctly identified it as a one-way street, and that you are turning in the proper direction. Watch for posted ONE WAY, DO NOT ENTER, WRONG WAY, and NO LEFT TURN or NO RIGHT TURN signs. Even if there are no signs, other clues can help you identify a one-way street. If parking is allowed on both sides of a one-way street, all vehicles will be pointing in the direction of traffic. Also, one-way streets do not have yellow lines. Once you have correctly identified the one-way street, move into the appropriate lane at least a block ahead of time, and turn into the lane nearest you.

When exiting a one-way street, be sure to correctly identify the street that you will be turning onto. Watch for signs and roadway

REALITY CHECK *The Constant Lane-Changer*

Beware of the driver who makes frequent, unnecessary lane changes. This type of driver impatiently changes lanes to get into the "faster" one, and then when other traffic catches up, changes into the new "faster" lane. Constant lane-changers often fail to signal, and they do not check long enough to make sure that the way is clear before making a lane change. If you are not careful, they may change lanes right into the side of your car! Stay clear of these drivers.

factoid

Chicago is the fifth most congested city in the United States, behind only Los Angeles, Seattle, San Francisco–Oakland, and Washington, D.C.

markings that tell you whether you are turning onto another one-way street or a two-way street. In general, use the left lane for left turns and the right lane for right turns. If there are three lanes, the middle lane should be reserved for driving straight ahead. Some one-way streets have special lanes for turning left or right only. If you are turning onto a multilane street, you may be able to turn from more than one lane. Also, do not be surprised if a one-way street turns into a two-way street, which is usually indicated by special signs or signals.

While driving on one-way streets, be prepared for a driver unfamiliar with the area to be driving the wrong way down a one-way street, heading directly for you in your lane of traffic. Do not panic! Slow down and try to get the driver's attention by flashing your headlights or honking your horn, loudly and continuously if necessary. If this does not work, pull over to the right side of the road and avoid the driver, putting as much space as possible between your two vehicles.

If you mistakenly go the wrong direction on a one-way road, slow down to a near stop, turn on both your headlights and hazard lights, and honk your horn as necessary to get the attention of drivers who may be heading toward you. If you see a nearby driveway, pull into it as soon as you can safely do so. If no driveway is available, slowly make your way to the first available street or alley. If no street or alley is nearby, try to make a U-turn. However, if there is not enough space available to make a U-turn, carefully execute a three-point turn when traffic clears, and reverse your direction. Once you are going in the proper direction, do not forget to turn off your headlights and hazard lights.

Figure 10–12 When driving in cities, allow extra time for unexpected detours.

REALITY CHECK Detours

Detours are common on city streets, usually as a result of street construction, maintenance, or some special event or emergency. When confronted with a detour, always follow the detour signs set up by law enforcement or work crews to keep traffic flowing. Do not stop your vehicle to ask a worker or traffic control officer what the reason for the detour is or how long it will last. You will hold up traffic and contribute to making the detour situation worse than it already is.

Traffic Flow

When only one vehicle drives slower than the rest, it creates problems for all the drivers nearby, who must alter their speeds and paths of travel to deal with the slow driver. Studies indicate that a crash between two vehicles going in the same direction is four times as likely when their speeds vary by 10 miles per hour (15 km/h) than it is when they are traveling at the same speed; if the speed differential is 20 miles per hour (30 km/h), the chance of a crash is *sixteen* times greater! Try to stay with the traffic flow and avoid driving too fast or too slowly. This reduces the chance of you causing a collision by crowding traffic ahead or causing vehicles behind you to pass.

At the same time, "going with the flow" is *not* an excuse for speeding. Even if everyone else is exceeding the speed limit, you must obey the law. If you are in the far-right lane and still feel pressured to speed by drivers tailgating you, be patient and let them pass you.

Traffic flow is often irregular on city streets. When possible, avoid driving next to other cars in adjoining lanes. Side-by-side driving reduces your space cushion and blocks your view of the surrounding area, which increases the chance of a collision. In congested areas, you often have no choice, but as soon as traffic starts to clear up, make sure that you get plenty of separation from other vehicles on the road.

When traffic in your lane is moving but traffic in an adjoining lane is stopped, watch out for impatient drivers who will dart out to escape the holdup. Often these drivers do not adequately check for traffic in your lane before suddenly swerving over. If *you* get stopped in a long line of cars, think twice about pulling out to save a few seconds of waiting in traffic. Drivers in your lane probably are not just testing their brakes. It is likely that they have stopped to allow a bicyclist or pedestrian to cross. If you pull out of your lane and hit the gas without checking first, you stand a good chance of running someone down.

REALITY CHECK *Know Who Not to Pass*

Almost everyone has had the experience of trying to pass a driver who refuses to let you by. This driver speeds up, slows down, swerves, or otherwise obstructs the completion of your pass. Whether this driver feels that he or she was "cut off" by you a few miles/kilometers back, does not like the way you look, or is simply looking for someone to pick a fight with does not matter. What does matter is how you choose to deal with the situation. With road rage a reality of driving today, the best thing you can do is back off and let the other driver feel that he or she has won the "battle." If necessary, slow down and pull over to let the problem driver get far away from you. Take a few moments to relax and then re-enter the roadway and continue driving.

USING SAFE *Passing*

To pass safely, first ask yourself whether the pass is necessary and legal. Use the *SAFE* process. *Scan* for oncoming vehicles, vehicles approaching from the rear, merging vehicles, and any activity on the side of the road. Taking into account the limits of your vision and depth perception, determine how long it will take to make the pass and whether you are able to correctly judge the distance necessary to do it safely. The faster that your vehicle and the vehicle you are trying to pass are traveling, the more time and room you will need to safely complete the pass. *Assess* the driving environment and road conditions, including traction, weather, visibility, width of the road, obstacles, and amount of traffic. Make sure that pedestrians, bicyclists, or anybody else will not be placed in jeopardy by your pass. *Find* safe points ahead to begin and complete your pass. Allow yourself a reasonable margin of safety. Once you have determined that it is safe, you are ready to *execute* the pass.

Where am I?

Figure 10–13 This home was a gift from the town's citizens to a famous general upon his return from the Civil War.

Keep in mind that vehicles in the "open" lane of traffic that you are trying to get into will be approaching from your rear at relatively high rates of speed. They may be unable to stop if you suddenly enter their lane. Also, chances are that you are not the only driver who is tempted to move into the open lane to pass. While you are watching for everyone else, make sure that you are not entering the path of a car pulling out behind you or driving forward into a vehicle pulling into the lane ahead of you.

10–3 PASSING

One skill you will need to develop as a driver is how to overtake or **pass** another vehicle on the road. Passing requires practice and maturity. You must learn how to judge the relative speeds and distances of vehicles, including vehicles going in opposite directions. Just as important, you must be able to identify whether passing is appropriate in different situations. Even if a passing maneuver is legal, it may not be safe.

General Passing Rules

In most situations, you should attempt to pass on the *left*, and only when it is safe to do so. In Illinois, passing on the right is generally allowed only under the following circumstances:

- If the vehicle you are passing is making or about to make a left turn
- If you are on a roadway with unobstructed pavement not occupied by parked vehicles and wide enough for two or more lanes of vehicles moving in the same direction
- If you are on a one-way street, or on any roadway on which traffic is restricted to one direction of movement, that is free from obstructions and wide enough for two or more lanes of traffic

When Not to Pass

There are many situations when passing is either forbidden by law or clearly unsafe and should not be attempted. Illinois law prohibits passing another vehicle within 100 feet (30 m) of an intersection, railroad crossing, or any bridge, tunnel, or viaduct where your view is blocked. Similarly, you are not allowed to pass another vehicle in a school zone or if you are coming to the crest of a hill or a curve where you cannot see enough of the road ahead to be certain of completing your pass safely.

Never try to pass more than one vehicle at a time. It is extremely difficult to look over, through, or around more than one car at a time ahead of you. This is especially true in city driving, where your view of oncoming traffic is severely limited. You simply cannot see far enough ahead of you to determine whether you have enough room to pass.

Never try to pass a vehicle going at or near the legal speed limit. Not only will you have to travel at an illegal speed to pass it, but your higher speed will make it easier to misjudge the gap in oncoming traffic. Passing is also forbidden if it would require driving off the paved or main portion of the roadway or crossing a solid-white-line lane divider. Even if you are in a legal passing section of a two-lane roadway, it is against the law to pass if the visibility of oncoming traffic is limited or not distant enough to allow you to complete your pass safely.

Avoid trying to pass near stopped or slow-moving roadway users, such as bicyclists, pedestrians, and certain vehicles. If the unexpected happens—a driver opens a door of a parked car or a slow-moving car makes a quick U-turn right in front of you—you will have almost no time to avoid a serious collision. On narrow city streets, buses, recreational vehicles, trucks, and vehicles towing trailers preparing to turn sometimes cross two or even three lanes of traffic to make it around obstacles and corners and may run right into you. It is also a good idea to avoid passing other vehicles moving downhill, especially trucks, because they tend to pick up speed on downgrades. Never try to pass if obstacles that would interfere with your maneuver are on the road.

Finally, never pass if you have any doubts about your ability to legally do so or if there is any doubt as to the intentions of the driver ahead of you. A slowing car could be stopping or preparing to turn. If you are about to pass on the left, do not rely on a right-turn signal to tell you it is safe to do so. The driver may have left the signal on from a previous turn, moved the turn-signal lever in the wrong direction, or changed his or her mind.

Making the Pass

Follow these steps to properly pass another vehicle on a road in a business or residential district (see Figure 10–14):

1. Make sure that you are at least at a safe 3-second following distance behind the vehicle you wish to pass. Check your

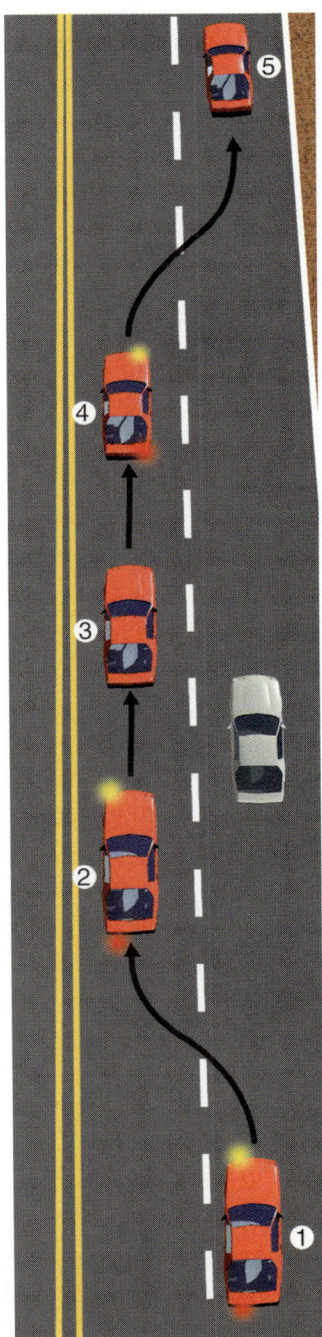

Figure 10–14 Passing another vehicle

Figure 10–15 If you are tailgating a driver who stops abruptly, you will be unable to avoid a crash.

famous collisions

On the morning of July 15, 1994, a teenager lost control of his car on Woodward Avenue in Woodridge (a suburb of Chicago) while swerving to avoid the vehicle he was tailgating and collided with an oncoming jeep, killing himself and injuring his three passengers as well as the driver of the jeep.

rearview and sideview mirrors for traffic behind you and to your sides. Turn your head and use both eyes to make sure that no vehicles are in your blind spots—again, at least 100 feet (30 m) before changing lanes. Signal your intention to pass to other drivers. Tap your horn or, if necessary, flash your headlights to get their attention. While signaling, check your mirrors and blind spots *again*.

2. Move into passing position and obtain a speed advantage over the other vehicle. Do not linger in the other vehicle's blind spots. Overtake the vehicle, passing with a sufficient clearance of both the passed vehicle and of any oncoming traffic ahead of you. Be aware that Illinois law requires you to return to your lane of travel before you get within 200 feet (60 m) of any oncoming vehicle.

3. Recheck conditions ahead of you and to the sides. Create a return space by making sure that you can see the front of the vehicle you have just passed in your rearview mirror. Do not count on other cars making room for your return.

REALITY CHECK *Let the Fast Pass*

Clear out of the faster, left-hand lanes when other, faster drivers appear behind you. Even if they act rudely and flash their lights, honk, or tailgate, do not "up the ante" by staying in your lane. Even worse, do not slow down or hit your brakes to warn these drivers to "back off." Maintain your speed and move to the right as soon as it is safe to do so. If they are speeding, you may later have the satisfaction of seeing them pulled over and cited by a police officer down the road.

If you have to, tap your horn to communicate to other drivers that you need to move back into the line of traffic.
4. Signal your return to the lane. While signaling, check your mirrors and blind spots *again* with both eyes.
5. Move into your return space, creating following-distance space for the vehicle you just passed. Resume a normal driving speed.

Being Passed

As soon as you are aware that another driver is attempting to pass you, maintain or reduce your speed to make it easier for that driver to execute the pass. In most cases, you can accomplish this by simply taking your foot off the accelerator rather than having to brake. Move to the right side of the lane to give him or her more room to pass and a better view of the road ahead. *Never speed up or otherwise block the passing vehicle.* This is both illegal and dangerous.

If you see a hazard ahead that would make another driver's pass of your vehicle dangerous, tap your brakes as a warning not to pass. If necessary, open your window and add the hand signal for slowing/stopping. If he or she does not see your warning or ignores it, reduce your speed as the other driver passes you to increase the space available for that car to re-enter the lane and slow or stop before reaching the obstacle. Always have an escape route ready just in case.

WHO'S AT FAULT?

1. Driver 1 was proceeding northbound on Vermilion Street in Danville at approximately 20 miles per hour (30 km/h). Driver 3 had just completed parallel parking on the street ahead of Driver 1. Driver 2 was traveling southbound and had slowed to approximately 15 miles per hour (25 km/h). She had moved well toward the center of the street to avoid the parked cars on the west side of the street. Driver 1, estimating that he had adequate space to maneuver between Driver 2 and Driver 3, continued northbound in the only lane available and was unable to stop when Driver 3 opened the door to exit his vehicle. Driver 1 applied his brakes but still knocked Driver 3's left front door completely off. **Who's at fault?**

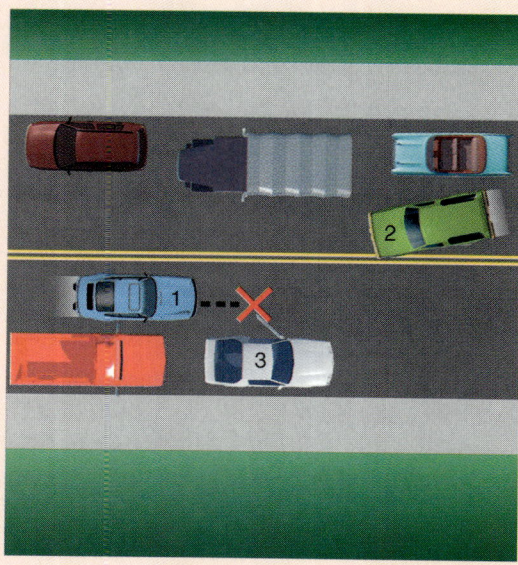

2. Driver 1 was proceeding northbound on 12th Street in Moline approaching the signal-controlled intersection with 19th Avenue at approximately 35 miles per hour (55 km/h) in the right lane. Drivers 3 and 4 were stopped at the red light ahead. Driver 2, who was northbound in the left lane immediately next to Driver 1, was slowing in preparation to stop. Driver 1 had just purchased a hamburger at a fast food restaurant and was eating as he drove. While approaching the intersection he took a bite out of his burger, being careful not to spill it on his clothes. As he looked up, Driver 1 realized he was about to rear-end Driver 3. He simultaneously swerved to his left and stomped on the brake. Driver 1 managed to avoid hitting either Driver 3 or 4, but he slammed into Driver 2, who collided with the center median. **Who's at fault?**

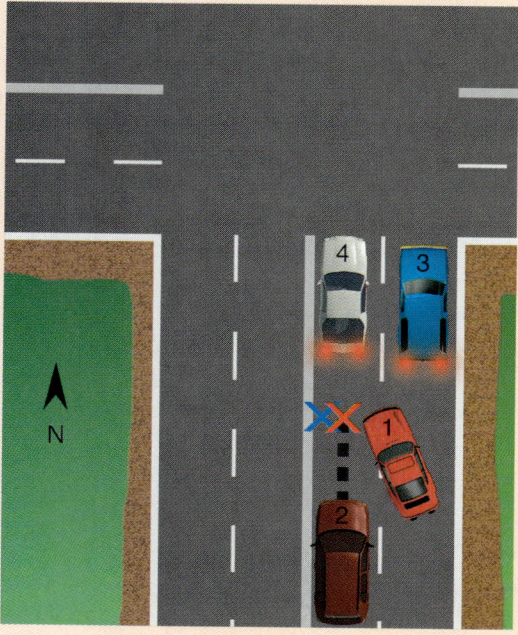

YOUR TURN

10–1 The City Driving Environment
1. Why is scanning ahead and reducing speed important in city driving?
2. What is the difference between riding and covering the brake?

10–2 Lane Positioning
3. Why is lane positioning important in city driving?
4. What is the proper way to change lanes?
5. What are the advantages and risks of using one-way streets?

10–3 Passing
6. What is the proper way to pass another vehicle?
7. What are some of the dangers of passing on urban roadways?

SELF-TEST

Multiple Choice
1. Holding your right foot over the brake pedal as you cruise forward on your car's momentum is called:
   a. riding the brake.
   b. readying the brake.
   c. covering the brake.
   d. holding the brake.
2. When driving past parked cars, stay at least _____ away.
   a. 10 feet (3 m)
   b. a door's width
   c. a car's width
   d. 3 feet (1 m)
3. When traffic gets so heavy that vehicles come to a virtual stop, the result is:
   a. chaos.
   b. signal failure.
   c. road closure.
   d. gridlock.
4. Which determines the best travel lane?
   a. the amount of congestion
   b. the direction of cross traffic
   c. the availability of turn lanes
   d. all of the above
5. On two-way streets with three or more lanes going in the same direction, you should drive in the _____ lane.
   a. left
   b. center
   c. right
   d. turn

Sentence Completion
1. _____ streets are common in cities because they can handle a heavier volume of traffic.
2. Never try to _____ more than one vehicle at a time.
3. You should maintain a _____ following distance behind a vehicle you wish to pass.

4. When possible, it is best to pass on the _____ side of the vehicle ahead of you.
5. When driving in the left lane, avoid the temptation to "creep" over the _____ to swerve by an obstacle.

Matching
Match the concepts in Column A with examples of the concepts in Column B.

Column A
1. __ Illegal pass
2. __ One-way street
3. __ Hazard in the left lane
4. __ "Going with the flow"
5. __ Road closed
6. __ Gridlock
7. __ Hazard in the right lane
8. __ Best travel lane

Column B
a. Parked cars
b. Bumper-to-bumper traffic
c. Traveling at the same pace as other traffic
d. Within 100 feet (30 m) of an intersection
e. WRONG WAY
f. Oncoming traffic
g. DETOUR
h. Least congestion

Short Answer
1. Why should you avoid side-by-side driving?
2. Why is riding the brake dangerous?
3. How can you identify a one-way street?
4. What dangers might you face on residential streets?
5. What should you do if traffic in your lane is stopped but is moving in an adjoining lane?

Critical Thinking
1. As you are turning left onto a two-lane street, you notice that the lines dividing the street lanes are all white across the entire street and that cars parked on both sides of the road are pointing in the direction opposite yours. A car is traveling toward you in the right-hand lane. What should you do?
2. You are traveling uphill behind a slow-moving truck in the right lane of a busy four-lane street. You drop back to see if the left lane is clear for passing and then move into the left lane. As you are passing the truck at the legal speed limit, a pickup speeds up close behind you honking and flashing its lights. What should you do?

PROJECTS

1. Interview a professional bus, limousine, or taxi driver about the particular hazards he or she faces every day while driving in the city. Share the results with your class.
2. Look at a map of Chicago with your classmates and discuss which areas may be more congested at rush-hour times. Find some alternative routes that you could use to bypass the areas that have heavy traffic or gridlock problems.

11 chapter

Highway and Rural Driving

In some ways, rural roadways are safer than city streets. They are less congested, they have fewer distractions, and there is more room to make evasive maneuvers in an emergency. Because they are less crowded, however, rural roads have higher speed limits than city streets. You face *fewer* hazards on a highway than you do on an urban road, but you reach them more *quickly*. Rural roads generally are not cleaned or repaired as often as city streets, and they have fewer warning signs, sharper curves, steeper hills, and fewer controlled intersections. It also usually takes longer for medical help to arrive after a collision.

Chapter Objectives

Upon completion of this chapter, you should be able to:

11–1 The Rural Driving Environment

1. Understand what makes rural roads and intersections dangerous.
2. Understand the proper way to enter, cross, and exit a divided highway.

11–2 Defensive Driving on Highways

3. Understand the importance of scanning ahead and reducing speed in rural driving.
4. Describe some of the dangers of passing on highways.

11–3 Other Dangers of Highway and Rural Driving

5. Understand what velocitation and highway hypnosis are.
6. Understand what you should do if you encounter a slow-moving vehicle or wild animal on the road.

KEY TERMS

- highway
- divided highways
- median strip
- T-intersections
- Y-intersections
- crossovers
- velocitation
- highway hypnosis
- slow-moving vehicles (SMVs)
- all-terrain vehicles (ATVs)
- off-highway motorcycle
- snowmobiles

CHAPTER 11 Highway and Rural Driving ◆ **249**

Figure 11–1 Opposite directions of travel are separated on a divided highway.

factoid

Nearly half of all driver fatalities in Illinois occur on rural roadways.

11–1 THE RURAL DRIVING ENVIRONMENT

Most rural roadways in Illinois serve small communities or link local roads with major roadways like interstates. The most heavily traveled roads in rural areas are highways. A **highway** is a main public roadway designed to carry traffic for long, uninterrupted periods at medium-to-high speeds. Unlike freeways, where access is limited to on-ramps and off-ramps, highways can have intersections with cross traffic, side roads, driveways, and other means of direct access.

Divided Highways

The typical county, state, or U.S. highway in Illinois is a two-way, two-lane road separated by a dashed yellow line or double-yellow lines. Major rural highways may have four lanes. Many multi-lane highways are **divided highways** in which opposing directions of travel are separated by a fixed barrier or area of space called a **median strip.** Open areas may be paved, dirt, or landscaped. The widths of median strips vary depending on the location of the highway and the surrounding environment. The more open the

Figure 11–2 Narrow roadways are a common feature of the rural driving environment.

factoid

Illinois has five national "byways," highways that have been officially recognized as having special scenic or historic value.

area is, the wider median strips tend to be. Median strips may be even with the road, but they are usually angled downward to help channel off water during rainstorms. In dangerous high-traffic areas, narrow median strips may have protective walls or temporary concrete barriers.

Rural Roads

Most rural roads lack the engineering advances featured on modern city streets and freeways. Many are old and were originally designed for slower speeds. They typically have narrow lanes, no shoulders or passing lanes, short sight distances, few traffic controls or pavement markings, and inadequate crash protection devices. They may be paved with a patchwork of materials, resulting in inconsistent traction, or they may not even be paved at all.

Rural roads are more likely than other roads to have "washboard" surfaces with uneven sections, leaving ridges that can cause your tires to lose their grip as you move from one section to another. Potholes and other damaged sections of the roadway may be left unrepaired for long periods of time. Rural roads are not cleaned regularly, and sand, gravel, oil, and leaves can accumulate in slippery patches. Collision debris, tire fragments, and dead animals are also familiar obstacles.

Do not rely on warning signs to indicate rough conditions on rural roads. Pay attention to the sounds your tires make as you drive over different surfaces. Reduce your speed when you suspect any difficulties with traction. Although you cannot expect to detect every potential problem, you will be better prepared for a loss of traction by maintaining a safe speed. Be especially careful when moving from a well-paved highway onto a dirt or gravel road, in which case the sudden reduction in traction could cause you to lose control of your vehicle.

Rural Intersections

Many rural intersections have only signs or no traffic controls at all. **T-intersections,** in which side roads join a main road with through traffic at a right angle, are especially common and can be dangerous. Drivers on the through road may have difficulty seeing vehicles entering from the side road. There may not be warning signs for through traffic, and STOP or YIELD signs on the side road could be blocked from the view of drivers on the main road by trees or bushes. If the intersection is on a

---

AUTO ACCESSORIES Mud Guards

When driving in rural areas, it is common to find yourself on a dirt road. Add some rain to the equation and soon the dirt road becomes a muddy road. Be a courteous driver and install mud guards behind your tires so that drivers behind you do not get hit by mud flying off your vehicle.

Figure 11–3 Many rural roadways have no traffic controls.

Highway heroes

While retrieving a piece of lumber on the shoulder of Illinois 15 in Belleville, IDOT worker Herman Dierkes was killed by a pickup that hit him from behind and threw him 100 feet (30 m) down a steep embankment onto South Belt West.

curve or hillside and visibility is limited, drivers on the side road may misjudge the time it takes to accelerate to keep ahead of approaching traffic. If they are turning from a gravel or dirt road, they may temporarily lose traction in an attempt to jump quickly onto the main road, causing a collision with an approaching vehicle.

In rural areas, roads sometimes meet at unusual angles called **Y-intersections.** These intersections, which commonly form a "Y" shape, are often the site of junctions of minor and major roads. Y-intersections can be confusing, especially to beginning drivers. Drivers may either fail to stop or stop when they are not supposed to. Watch for posted warning signs when approaching Y-intersections and be ready to swerve or stop to avoid hitting confused drivers.

Entering or Crossing a Divided Highway

Roads that intersect divided highways often continue across a median strip. Those vehicles turning left onto the highway from the road, turning left onto the road from the highway, or driving straight across the highway must use a special paved area of the median. Because these types of intersections are usually controlled by signs, not signals, traffic can build up on the median strip. If the median is "full," cars turning left off the highway may have no place to go. There is also the risk that too many cars might try to squeeze onto the median, and the last car might be caught sticking partially onto the highway. Given the high speed limits on most highways, this is extremely dangerous.

To turn left onto a divided highway or proceed straight across it, follow these steps (see Figure 11–4):

1. Come to a complete stop at the intersection. Stay well clear of the right lane of the highway. Remember that cars on the highway will be driving *extremely* fast.

2. Look ahead at the median strip to determine how wide it is and how many vehicles can

252 ◆ UNIT 4 *The Driving Environment*

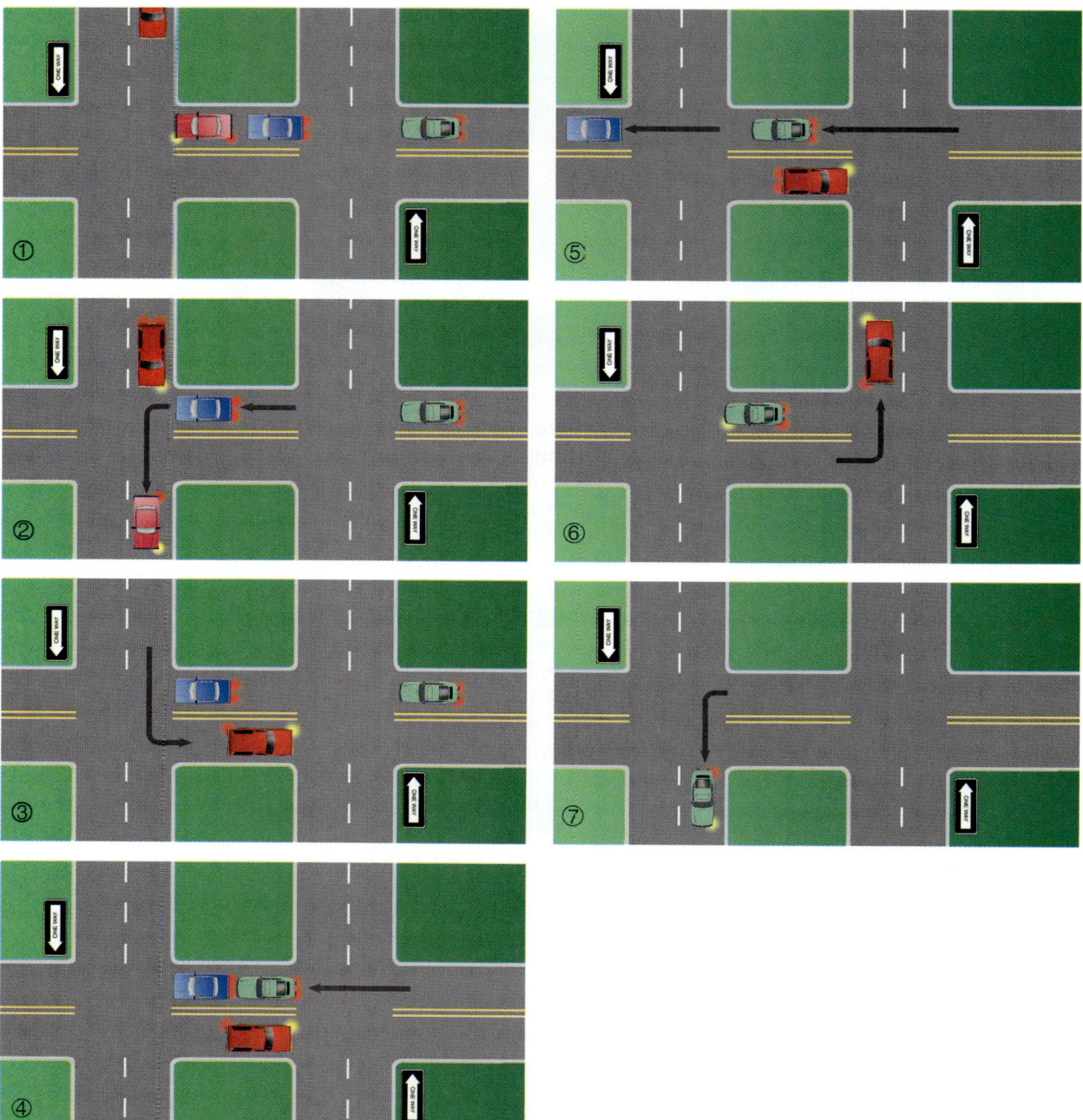

Figure 11–4 Crossing a divided highway

safely occupy it with at least a car length of space to spare. If the median is full, patiently wait until one of the cars on it has turned onto the highway or proceeded across it.

3. Look left at oncoming traffic on the highway. Watch for any vehicles turning left off the highway onto the median strip that might fill the gap you are waiting for. Also, watch for any oncoming vehicles on the median that are turning left across your path onto the side of the highway that you want to cross. Although you have the right-of-way, do not assume that drivers turning onto the highway will wait for you.

4. Once the way is completely clear of any approaching vehicles and there is enough room for you to safely occupy a place on the median, look left again and then proceed cautiously across the highway.

5. Wait for any vehicles in front of you to turn left or proceed straight across the other side of the highway. When you are "next," move to a safe spot on the median to wait for a gap in traffic. If there is no STOP sign or limit line to help you position your vehicle, move to within about half a car length of the edge of the median. Come to a complete stop.

6. If you are turning left, activate your left-turn signal. Look across the highway for oncoming vehicles that will go straight onto the median to drive across or turn onto the other side of the highway. Look to your right for a gap in traffic on the highway.

7. If you are driving across the highway, proceed across when the way is clear. Watch for vehicles exiting the highway onto the road you are taking. If you are turning left onto the highway, watch for oncoming vehicles turning right onto the same side of the highway. Turn left into the left lane of the highway just as you would on a two-lane, one-way street. Make sure that your turn signal is off. Because you will be entering the passing lane, immediately activate your right turn signal and change into the right lane as soon as it is safe to do so.

Turning Left off a Divided Highway

When turning left off a divided highway to turn onto an intersecting road or make a U-turn, you must first turn onto the median strip to wait for a gap in oncoming traffic. Keep in mind that you will be slowing down in what is normally the passing lane. To avoid holding up traffic, get into the left lane and signal your intention to turn at least 200 feet (60 m) ahead of the intersection. As you approach the intersection to prepare to turn, frequently check your rearview mirror for fast-moving cars behind you and look for an escape path on the median in case you risk being rear-ended.

famous collisions

While turning left from the northbound lanes of Illinois Highway 59 onto Elmwood Avenue in West Chicago, two teenagers were struck and killed by a drunk driver traveling at a high rate of speed just before midnight on October 14, 1994.

Figure 11–5 Driveways and side streets may be hard to spot on rural highways.

Some intersections of divided highways have turn lanes. If the median is already occupied by cross traffic or other cars turning left, turn lanes provide a safe place for you to wait for an opening on the median. If there is no turn lane, you may have to stop in the left lane. Keep your wheels straight while waiting to turn. If you get hit from the rear, it is better to be pushed straight ahead than into other vehicles on the median or oncoming traffic on the other side of the highway. Keeping your wheels straight also gives you the opportunity to rapidly accelerate. You may not be able to avoid a rear-end collision, but the faster you are traveling when struck from behind the better chance you have of avoiding a serious injury.

Unmarked Driveways

Driveways to farms and ranches appear frequently on rural stretches of highways. Some of these unmarked roads are hidden by bushes, crops, or trees. Watch for mail and newspaper boxes, gates, openings in fences or shrubbery, tire tracks turning off and onto the highway, and other signs of a property entrance. Reduce your speed as you approach these areas

REALITY CHECK   *Mailbox Bashing*

A favorite pastime of some teens is to drive around and smash mailboxes along rural roads and highways. One person drives, while the passengers take turns hanging out of the windows to take swings at mailboxes. This "prank" is illegal and extremely dangerous. Not only does the driver risk losing control of the car and having a collision, but also the "basher" may lose his or her balance and fall out. If you are caught mailbox bashing, which is a federal offense, you can be arrested for vandalism as well as cited for criminal damage to property and reckless driving.

REALITY CHECK Do Not "Trash" Your Environment

The Illinois Department of Transportation's Adopt-A-Highway Program provides volunteers from more than 1,000 groups across the state with training, signs, vests, and trash bags to clean up designated portions of non-interstate highways, but their efforts to maintain the Prairie State's scenic beauty depend on your cooperation. Do not contribute to the problem of roadside litter by tossing bottles, cans, rags, plastic six-pack covers, clothing, fast-food wrappers, and other types of nonbiodegradable trash out the window of your vehicle. Many of these items can last up to 500 years or more. If that is not enough to motivate you, be aware that littering can cost you up to $500 if you are caught.

and be prepared for a vehicle to dart out into your path.

Entry/Exit Problems

Many highways either do not have acceleration lanes or have very short acceleration lanes, so pay close attention to YIELD or MERGE signs before entering a highway. Wait for a long gap in traffic before you accelerate and merge with through traffic. If you are on the highway, keep an eye out for other vehicles that may quickly enter the roadway. You may have to reduce your speed or, if you are on a multilane highway, change lanes to allow the other driver to merge smoothly. When you see signs warning you of merging traffic ahead, check your rearview and sideview mirrors for vehicles behind you and try to spot any drivers on a side road or at an intersection ahead that might be preparing to merge.

There are often no deceleration lanes on highways either, so remember to slow down more quickly than you would on a freeway as you approach highway exits. You are required by law to signal at least 200 feet (60 m) before turning off any road that is not in a business or residential district to give drivers behind you enough time to reduce their speed. If you are on a multilane highway, make sure that you get in the far-right lane well before you reach the exit.

Crossovers

Many divided highways have **crossovers,** special areas where vehicles can turn around to go in the opposite direction. Unpaved crossovers are generally restricted to emergency vehicles. Warning signs may be placed near the crossovers indicating that they are reserved for official use, but do not assume that you can use a crossover if you do *not* see a warning sign.

Roadside Stands

Roadside produce and souvenir stands are sometimes located very close to rural roadways. Because many highways were built during a time of slower-speed driving, today's greater speeds pose a great degree of danger. When you pass

factoid

U.S. Highway 30, also known as the Lincoln Highway, was the first coast-to-coast paved highway in the United States.

famous collisions

On the morning of August 13, 1997, a pregnant woman who suddenly stopped and attempted to turn left on County Farm Road in Hanover Park (a suburb of Chicago) was killed along with her unborn child when a sanitation truck rear-ended her vehicle, pushing her into oncoming traffic.

DRIVING TIPS *Meeting a Line of Cars*

On open highways with only one lane of traffic going in each direction, a collision, an animal crossing, roadway construction, or some other obstacle can tie up traffic for hours. As you approach a long line of cars, immediately reduce your speed and tap on your brakes to warn others behind you that they must stop ahead. If necessary, switch on your emergency flashers to get their attention.

famous collisions

On the night of October 10, 1999, while heading southbound on State Highway 21 in Libertyville, the father of a former Illini basketball star crossed the center line and collided head-on with a northbound van, killing himself, his son, and the married couple in the van.

one of these places, reduce your speed, increase your following distance, and watch for pedestrians or cars pulled over to the side of the road.

11–2 DEFENSIVE DRIVING ON HIGHWAYS

Speed limits on rural highways can be as high as 55 miles per hour (90 km/h) in Illinois. Sharp curves, detours, lane endings, areas with gusting winds, and other hazards on major rural roads are usually accompanied by warning signs, but do not rely on such signs to always be there. Remember the basic speed law: *Do not drive faster than conditions allow.* The faster you drive, the harder it is to control your vehicle, the less time you have to swerve, the more room you need to stop, and the higher the risk of a serious injury or damage to your car in a collision. Almost two of every three traffic fatalities occur in rural areas, and excessive speed is usually the reason.

Reduced Visibility

Your line of sight on rural roads is often more limited than it is on city streets or freeways. Trees, bushes, crops, billboards, fences, and other obstructions close to the road can reduce your ability to see oncoming vehicles, cross

Figure 11–6 Visibility on rural roads is often limited.

Figure 11–7 It can be difficult observing details at highway speeds.

traffic, signs, or signals. The higher speeds of highways make it even more difficult to recognize details. Your side vision can be reduced to a virtual blur, sometimes called "speed smear."

Traffic safety engineers try to compensate for the reduced visibility of high-speed travel when they design signals, post signs, and paint roadway markings on rural highways. Signs are generally placed farther ahead of hazards, signals may be larger than those in cities, and roadway markings are painted well in advance of intersections and crossings. Rural intersections may have unusual signals or oversize signs to get your attention early. For example, you might see a yellow signal ahead of an intersection concealed by a hill, sharp curve, or trees that flashes only when the light at that intersection is red.

Make sure that you remember to look off-road, as well as directly ahead of you. A dust cloud off to the side could mean that a vehicle is just about to enter the highway, a herd of livestock is approaching a crossing, or a tractor is cruising along the edge of the road on the dirt shoulder. Fenced-in property is bound to have occasional outlets or driveways. Water towers, power lines, telephone poles, or other structures may indicate a nearby road or railway that will cross your path.

Look for patterns in traffic controls and roadway surfaces to better anticipate dangers or changing conditions ahead. While scanning, gather information about your environment. Keep a mental note of roads or service stations you pass, safety features like shoulders and medians, how often and where you see patrol cars, and other information that could be of use to you in the event of an emergency.

Increasing Following Distance

On city streets, a following distance of 3 seconds is safe under normal

factoid

Also known as "The Mother Road," "The Main Street of America," and "The Will Rogers Highway," Route 66 runs for 2,448 miles (3,917 km) from Chicago, Illinois, to Santa Monica, California, and crosses eight states and three time zones.

DRIVING TIPS **Solo Signaling**

Even if you are the only one on the highway, do not fall out of the habit of signaling before turning, pulling over, or any other maneuver requiring you to signal. The next time you are turning or passing another vehicle you might cause a collision by forgetting to signal properly.

factoid

You are more likely to be killed in a passing collision than any other type of crash on a two-lane road.

driving conditions. At highway speeds greater than 50 miles per hour (80 km/h), however, you need at least 6 seconds to stop or avoid an obstacle. Maintaining a long following distance will also help you save fuel and wear on your brakes. Drivers ahead of you may momentarily use their brakes to slow down—for example, because they see a warning sign, an upcoming curve, or a law-enforcement officer—and then speed up again when the danger or threat is past. When you see brake lights ahead, take your foot off of the accelerator and cover the brake. Let your car's speed decrease naturally to avoid having to use your brakes. If you *do* have to use your brakes, you will not have to push them as hard.

Highway Passing

The higher speeds of highway travel and the dangers associated with rural environments like low traction and inadequate lighting at night mean that you must take extra precautions when passing on highways. On multilane highways, you should stay in the right lane unless you are passing another vehicle or preparing to turn.

When passing on a two-lane highway, never move to the left of the yellow center line(s) unless you can see far enough ahead to know whether you can pass safely. Make sure that you have enough room to return to the right side of the roadway without affecting the safe movement of oncoming traffic or the vehicle that you are passing. Consider the width and condition of any shoulder before passing. The wider your escape route, the better. Check for warning signs of upcoming intersections, sharp curves, merging traffic, railroad crossings, bridges, underpasses, no-passing zones, and other places where passing is prohibited or dangerous.

Pass only on long, straight sections of the road where you can see clearly ahead of you. Never pass on or near curves or hills. Especially

MICHAEL HEDGES

Born on December 31, 1953, acoustic guitarist Michael Hedges helped establish the "New Age" Windham Hill record label in the early 1980s. Hedges was known for his unusual tunings and his two-handed picking style. He described his own music as "heavy metal" and "acoustic thrash." Others compared his skill and intensity with that of rock guitarist Eddie Van Halen. On December 2, 1997, a work crew discovered Hedges's body in his crashed BMW at the bottom of an embankment on California Route 128, in rural Mendocino County near San Francisco. The California Highway Patrol determined that Hedges had died several days earlier when his car skidded off the road as he was going around a curve. Rural roads can be deadly if you do not remain vigilant about the hazards they present. When dealing with the sharp curves and steep hills that you will encounter on rural roads, the best way to keep control of your car in a dangerous situation is to slow down.

Figure 11–8 Make sure that you finish your pass *before* you reach a no-passing zone.

U.S. Highway 40 traces the course of the National Road, which was built in the early nineteenth century and stretched from Maryland to Illinois; for a generation, it was the only hard-surfaced road in what was then the "West."

avoid passing while ascending a grade. Use your low-beam headlights to improve your visibility as you pass another vehicle. Never try to pass if your sight distance is limited by rain, snow, fog, blowing dust, or other severe weather.

Keep in mind that until an approaching vehicle is close to you, it is almost impossible to estimate its rate of speed. Just because you are going the speed limit does not mean that the oncoming driver is. Do not take chances. To be on the safe side, make sure that oncoming vehicles are at least a half a mile ahead (1 km) of you before you attempt a pass.

If you are passing another vehicle on a multilane highway, first check *all* lanes of traffic going in your direction. If you are in the center lane, make sure that no vehicles are in either the left or right lanes that might interfere with your pass. Watch for any vehicles that might enter your passing lane. Do not rely only on your

USING SAFE *Rural Roads*

Use the SAFE method to reduce the dangers of driving on rural roads. *Scan* ahead for traffic controls, debris on the roadway, dust clouds, animals, brake lights, vehicles jutting into the road from hidden driveways, slow-moving vehicles, and pedestrians and parked cars by roadside stands. In conditions of low light or bad weather, watch for headlights of oncoming cars, taillights of cars ahead of you, and pedestrians or animals that reflect the light from your headlights. *Assess* the width and condition of the roadway and shoulder, the visibility of oncoming drivers, the following distance of traffic behind you, and the off-road environment. *Find* an escape route in case an oncoming car veers into your lane, an animal or pedestrian darts across your path, or you suddenly come across debris blocking the road. Watch your speed, especially on curves, and be prepared to *execute* your decision to avoid a collision by moving off the road, by changing lanes, or by emergency braking.

famous collisions

Three children and their mother were killed on the night of August 21, 1999, when their van, heading west on State Highway 120 in McHenry County, was hit head-on by a driver who veered over the center line near Liberty Lake Road.

Where am I?

Figure 11–9 It was here that the first successful steel plow was developed by a young blacksmith whose company, founded in 1837, is now the world's leading producer of agricultural equipment.

mirrors, even if you have been "tracking" other cars for a while. Signal early, and look over your shoulder to check that your blind spots are clear with both eyes before making your pass. Always avoid passing in the presence of oncoming traffic. Good defensive driving calls for allowing yourself the widest possible margin for error in case of an emergency.

Always pass on the left when possible. Sometimes, however, you have to pass on the right on a multilane highway—for example, if a driver is preparing for a left turn or is not keeping up with the flow of traffic. Wait until the lane is clear, activate your right-turn signal, double-check for vehicles in your blind spots by looking over your shoulder, and enter the lane smoothly. Watch for other vehicles ahead of you that might be preparing to turn right.

Remember that you can enter the lane of oncoming traffic to pass another vehicle only if the yellow line on your side of the road is dashed. On some highways, especially where curves and hills prevent normal passing, extra lanes are added at intervals to each side of the roadway to make passing safer and easier.

11–3 OTHER DANGERS OF HIGHWAY AND RURAL DRIVING

When driving on highways and rural roadways, be aware of some of the unique hazards and challenges that you will face.

Velocitation and Highway Hypnosis

Driving on a roadway that is extremely familiar or that is long, straight, or boring can cause you to take your mind off what you are doing. Two common psychological effects of open highway driving are velocitation and highway hypnosis. **Velocitation** occurs when you find yourself unconsciously driving much faster than you intended. (The root word of velocitation is "velocity," which means "speed.") **Highway hypnosis** results when you literally become hypnotized by the road. After a few seconds in this hypnotic state, you will feel as if you just awakened from a dream, leaving you with no recollection of the last few seconds you spent driving down the road.

To prevent either problem from occurring, drive only when you are rested and mentally fresh. Break up long journeys into short segments, and avoid trying to make the trip in one long haul. Stop frequently for food and rest. Break up the monotony of open highway driving by talking to passengers and changing drivers. When practical, take "the road less traveled." By taking time to explore and enjoy your route, you may enjoy the trip more than the destination.

Slow-Moving Vehicles

Various types of **slow-moving vehicles (SMVs)** are common on highways. "Wide-load" vehicles carrying homes, trailers, and

CHAPTER 11 Highway and Rural Driving ◆ **261**

Figure 11–10 Approach slow-moving vehicles with caution.

heavy equipment are usually accompanied by lead and trailing vehicles with warning lights or signals. Construction zones may have pavers, bulldozers, and a variety of other heavy equipment going to and from the job site. Near farms, you are likely to encounter combines, tractors, harvesters, and cultivators, especially in late summer or fall. During or after a snowstorm you may encounter snowplows and deicing trucks.

SMVs present two major dangers to motorists. First, because they are not designed for high-speed travel—many cannot go more than 25 miles per hour (40 km/h)—they can cause rear-end collisions or traffic backups on highways. Second, because they are wider than most vehicles or have equipment that projects into the roadway in a dangerous manner, they are difficult to pass. In some cases you may not be able to see cables, booms, cranes, or other pieces of equipment that stick out. Some SMVs can take up the entire road. If the vehicle is on an unpaved road or riding partially on the shoulder, it can send rocks careening into your vehicle or kick up clouds of dust that reduce visibility.

When you approach an SMV, reduce your speed and take a moment to identify the vehicle and its potential dangers to you. Look for warning signs, flags, reflectors, signals, or orange-and-red, triangular-shaped SMV emblems mounted on the rear end. Approach with caution. Because your lane may be sharply narrowed, do not attempt to pass unless you have a clear view of the road ahead. Do not speed past an SMV. From a distance you might not have accurately judged the "reach" of the equipment attached to the vehicle. Instead, pass slowly and carefully.

Watch for oncoming cars that might try to pass SMVs on the *other* side of the road. Drivers might not see you and may swerve into your lane as they pass. Even if they avoid colliding with you, they might hit the SMV and cause equipment on the vehicle to fall into your path. Move into the right side of the lane when passing oncoming SMVs to give yourself as large a space cushion as possible.

Off-Road Vehicles

In Illinois, vehicles designed exclusively for off-road use are classified as all-terrain vehicles, off-highway

Figure 11–11 The SMV emblem

Figure 11–12

Figure 11-13 Be aware of drivers of off-road vehicles who do not observe posted regulations.

factoid

Illinois has more than 300 miles (480 km) of snowmobile trails.

motorcycles, or snowmobiles. **All-terrain vehicles (ATVs)** are off-highway motorized vehicles less than 50 inches (125 cm) wide, weighing 600 pounds (270 kg) or less, ride on three or more low-pressure tires, and are equipped with a seat or saddle and handlebars or a steering wheel. ATVs include three-wheeled all-terrain cycles (ATCs), once popular but dangerously unstable and rarely found today, and "quads" (four-wheeled cycles).

Any two-wheeled motorized vehicle equipped with a seat or saddle and designed to travel off-road is considered to be an **off-highway motorcycle.** This type of vehicle includes enduro and dual-sport motorcycles, which are often used in competition, as well as a variety of so-called "dirt bikes." **Snowmobiles** are self-propelled vehicles designed for operation on snow or ice or natural terrain and are steered by skis or runners and supported in part by skis, belts, or cleats.

ATVs and off-highway motorcycles may only be operated on private property with the owner's consent. Some public lands, like the Hennepin Canal Parkway State Park and the Illinois and Michigan Canal State Trail, are open for use by snowmobiles when the snow depth is 4 inches (10 cm) or greater. Many equestrian trails double as snowmobile trails during the winter. All types of off-road vehicles are permitted to use designated portions of local public roads, usually those sections linking two off-road trails or riding areas.

There are a number of restrictions on the use of off-road vehicles in Illinois. ATV, off-highway motorcycle, and snowmobile operators may only cross a highway at a 90-degree angle after coming to a complete stop and yielding the right-of-way to highway traffic. Riders may only cross divided highways at intersections. All types of off-road operators are forbidden from crossing or using limited-access highways. Snowmobilers are permitted to ride on the shoulder of other roadways as long as they remain 10 feet (3 m) or more away from a traffic lane.

All off-road vehicles must be titled just like other motor vehicles. Registration is not required for ATVs and off-highway motorcycles, but snowmobiles must be registered every three years with the Illinois Department of Natural Resources. An Illinois driver's license is required to operate any off-road vehicle on or across a public roadway. Riders between the age of twelve and sixteen may operate a snowmobile only off public highways. They must also be supervised by a rider who is sixteen or older unless they have successfully completed a snowmobile training course and have obtained a Snowmobile Safety Certificate. Riders must be at least ten years old to operate a snowmobile on any of the trails or riding areas managed by the Illinois Department of Natural Resources and, until they turn twelve, be accompanied by an adult. Snowmobiles must be equipped with lights, brakes, and a muffler.

When driving in rural parks and recreational areas, watch for signs warning of off-road activity. Many off-road vehicles are noisy, so use your ears as well as your eyes to detect them. Be alert to "off-roaders" who ignore or do not know the rules. Some can get so caught up in the excitement of what they are doing that they drop their guard or behave irresponsibly when driving on or near public roadways. No matter what type of off-road vehicle you may encounter, keep in mind that all are generally less stable than passenger vehicles and provide little protection to operators in the event of a collision.

Livestock Crossing Areas

About 76,000 farms cover nearly 80% of the total land area of Illinois. Although it may seem dangerous, many farmers must cross their livestock across open highways. Sometimes these crossing points are marked with yellow warning signs showing a cow, but most are not. Always be on your guard when traveling through pasture or grazing land. If you suddenly encounter a herd of animals crossing the roadway, slow your speed to a crawl and stop far from the herd to avoid frightening them. Always obey the instructions of livestock handlers. Once the farmer has moved the herd across the highway and well beyond the shoulder, proceed forward with caution. If necessary, wait for any dust stirred up by the animals to

Figure 11–15 Most serious collisions with wild animals in Illinois involve deer.

Wild Wheels

Figure 11–14 What is the name of this early steam-powered version of an off-road vehicle that carried passengers up a steep, winding mountain road to a remote hotel in Colorado?

factoid

Car–deer collisions result in more than $1 bilion in insured losses each year in the United States.

DRIVING MYTHS — Do Not "Look for the Whites of Their Eyes"

The saying "look for the whites of their eyes" may have worked in the early days of motor vehicles when a deer would be easily surprised by an approaching vehicle and look in that direction. Today, however, many animals are so used to vehicles that they may not even turn around to look at you. When traveling at night, your best precaution when you are in an area populated by deer is to slow down and keep scanning for signs of their presence.

factoid

About 90% of collisions with deer occur between dusk and dawn.

Highway heroes

While responding to a call in the early hours of October 10, 1993, Deputy James W. Cook of the DeWitt County Sheriff's Department was killed when he struck a deer on County Highway 10 near the intersection with County Road 1650 and went off an embankment.

settle to make sure that you can see ahead clearly.

Horse-Drawn Vehicles and Horseback Riders

In areas where you are likely to encounter horses (such as farms, ranches, horse-friendly suburbs, and parks with specially marked equestrian trails), exercise extreme caution and watch for warning signs. Horseback riders and horse-drawn vehicles are restricted to the shoulder or far-right side of the road, but some may ignore the law or have difficulty controlling their mounts.

If you come across a horse-drawn vehicle or a person riding or leading a horse, reduce your speed and maintain a reasonable distance as you pass. If you are on a two-lane highway, you may have to wait until the road is clear of oncoming traffic to pass safely. Stop and allow a rider that is crossing your path or approaching you, either from the front or rear, to proceed. Never use your horn, which may "spook" the animal and endanger the life of the rider or those being pulled by a horse. Pull over and stop if the animal appears frightened or the rider asks you to do so.

Wild Animals

A deer is a large animal, and it can do substantial damage to your vehicle if you hit one. Collisions with wild animals, primarily deer, account for more than 18,000 crashes, about 700 of which result in injuries or fatalities, each year in Illinois. The deer population has swelled in the last decade, making crashes with deer a growing risk on rural roadways.

Smaller animals such as raccoons, skunks, opossums, foxes, coyotes, and rabbits can also cause deadly crashes by causing drivers to swerve or brake sharply to avoid a collision. It is instinctive not to want to hit an animal, but driving into another vehicle or tree or causing a car to rear-end you by slamming on your brakes is a far worse scenario.

The first thing you can do to avoid a crash with an animal is to watch for warning signs indicating that wildlife inhabits the area. These signs are usually posted at places where collisions have occurred in the past. When driving near wooded country or farmland, control your speed and stay alert. Be especially careful during twilight hours in the spring and in the months of November and December, when deer are most active. Deer rarely travel alone, so if you see one animal you should assume that others are nearby. They typically move across a road in single file.

If you see an animal by the side of the road, reduce your speed and drive very slowly past it. Remember that frightened animals can bolt in any direction, so keep your brake covered, watch for oncoming traffic, and have an escape route planned. If an animal is on the road directly in front of you, slow down and honk your horn. Do *not* flash your lights. This can cause the animal to "freeze" in its tracks.

If you think you are going to collide with an animal, even a large one, it is better to brake than to swerve out of the way. Swerving might confuse the animal as to which way you are going and cause it to run right into you. You might also get into a worse collision. Some insurance companies will hold *you* liable if you have a crash trying to avoid an animal, but not for damage resulting from hitting it. If it is a large animal, take your foot off the brake the moment before you actually make impact. This will cause the front of your car to rise and reduce the chance of the animal hitting your windshield.

If you hit a large wild animal, do not attempt to treat it or drag it off the road. Not only might the animal attack you, but you can put yourself at risk of a crash with another vehicle. Mark the scene with flares or emergency triangles. If your vehicle is disabled, get it as far off the road as you can and seek professional roadside assistance.

Contact the nearest law-enforcement officer immediately. If an officer does *not* appear at the scene, you must go to the nearest police station or sheriff's office and have an officer there fill out a crash report. You must also submit an "Illinois Motorist Report" (form SR-1) to the Illinois Department of Transportation within ten days of the collision if anyone was injured or killed or if it resulted in $500 or more worth of damage to your vehicle.

WHO'S AT FAULT?

1. Driver 2 was proceeding southbound on Illinois Highway 145 through Shawnee National Forest at night. His speed was about 55 miles per hour (90 km/h), just at the posted limit. Driver 1 was traveling northbound on the same highway at approximately 60 miles per hour (100 km/h) when a coyote appeared in his headlights on the road ahead. Surprised, Driver 1 swerved into the oncoming lane to his left. Driver 1 impacted Driver 2 head on. *Who's at fault?*

2. Driver 1 and Driver 2 were proceeding along Elizabeth Drive in Carroll County at about 50 miles per hour (80 km/h). Driver 1 suddenly realized that she was coming up behind a slow-moving tractor. She attempted to pass but discovered that she was in a no-passing zone. When she applied her brakes to retreat back to her position in her lane, she was hit from behind by Driver 2. *Who's at fault?*

YOUR TURN

11-1 The Rural Driving Environment
1. What makes rural roads and intersections dangerous?
2. What is the proper way to enter, cross, and exit a divided highway with a median strip?

11-2 Defensive Driving on Highways
3. What is the importance of scanning ahead and reducing speed in rural driving?
4. What are some of the dangers of passing on highways?

11-3 Other Dangers of Highway and Rural Driving
5. What are velocitation and highway hypnosis?
6. What should you do if you encounter a slow-moving vehicle or wild animal on the road?

SELF-TEST

Multiple Choice
1. Divided highways are highways that:
   a. are built on two levels.
   b. have a fixed barrier or space between opposite directions of travel.
   c. do not have a median strip.
   d. are maintained by both state/provincial and federal agencies.
2. Which of the following is *not* a common feature of a rural highway?
   a. unmarked driveways
   b. narrow lanes
   c. passing lanes
   d. short sight distances
3. The SMV emblem is:
   a. orange and red, and triangular.
   b. orange and red, and round.
   c. blue and red, and triangular.
   d. red and triangular.
4. On highways, how far ahead should oncoming vehicles be for you to safely attempt to pass?
   a. ⅓ mile (500 m)
   b. ½ length of a football field (45 m)
   c. ¾ length of a football field (70 m)
   d. ½ mile (1 km)
5. A crossover is:
   a. a rural railroad crossing.
   b. a place to turn around on a divided highway.
   c. a type of rural intersection.
   d. a pedestrian walkway over a highway.

Sentence Completion
1. A _____ is a main public roadway designed to carry traffic for long, uninterrupted distances at medium to high speeds.
2. The area of ground that separates some divided highways is called a _____ .
3. In rural areas, _____ are often the site of junctions of major and minor roads.
4. _____ occurs when you find yourself driving much faster than you intended.
5. _____ are required to cross highways at a 90-degree angle.

Matching
Match the concepts in Column A with examples of the concepts in Column B.

Column A
1. __ Off-road vehicle
2. __ Unmarked property entrance
3. __ Uneven road
4. __ Velocitation
5. __ Crossover
6. __ Highway hypnosis
7. __ Slow-moving vehicle
8. __ Livestock crossing area

Column B
a. "Washboard"
b. Mailbox
c. Unconsciously speeding
d. Reserved for emergency vehicles
e. Feeling like you just woke from a dream
f. WIDE LOAD
g. ATC
h. Large cloud of dust by roadway

Short Answer
1. How do you turn left off a divided highway?
2. What are the dangers of T-intersections?
3. Why can it be hazardous to enter or exit a highway?
4. How do traffic safety engineers compensate for reduced visibility on rural roadways?
5. What dangers do slow-moving vehicles pose to motorists?

Critical Thinking
1. You are driving on a level two-lane highway with heavy shrubbery on both sides. A pickup carrying a large load of hay is moving slowly in front of you in a passing zone. Scanning ahead, you see no oncoming vehicles for more than half a mile (1 km). You signal and pull into the oncoming traffic lane to make your pass. Just as you pass the pickup and are signaling your intent to return to the right-hand lane, you see a car preparing to pull out from a hidden property entrance 100 yards (90 m) ahead to your right, and headed in your direction. What should you do?
2. It is early evening in December, and you are driving to your grandmother's house out in the country. The road to her house is a narrow and winding two-lane highway with a steep embankment to your left and a shallow wide ditch running alongside to your right. As you go around a curve at 35 miles per hour (55 km/h), a skunk appears in your headlights directly in front of your car. What should you do?

PROJECTS

1. Find a quiet and level rural road with a speed limit of 50 miles per hour (80 km/h) or more. While you are driving, select two landmarks ahead of you that you estimate to be half a mile (1 km) away from each other. When you reach the first landmark, check your odometer. Make sure that you are maintaining a constant speed at or just below the road's speed limit. When you reach the second landmark, check your odometer again. Note how accurate your estimate of the distance between the landmarks was. Repeat the procedure several times to see if you get more accurate with practice.

2. Take a drive on a local rural road or highway. Have a friend write down every potential hazard you encounter—for example, sharp curves, debris on the road, unmarked driveways, livestock crossing areas, and short acceleration lanes. Share your observations with your class.

12 chapter

Freeway Driving

Unlike city streets and rural highways, access to freeways is controlled and limited. Because there are, except in extremely rare circumstances, no intersections, traffic flows smoothly with a minimum of interruptions and a lower risk of collisions. Freeways are also better maintained and equipped with more safety features than highways. Slow-moving vehicles, bicyclists, and pedestrians are normally not permitted. The high rates of speed, however, coupled with congestion, can still make freeways dangerous. Even though there are fewer crashes on freeways than on highways and fatality rates are far lower, freeway driving still demands special skills.

Chapter Objectives

Upon completion of this chapter, you should be able to:

12–1 The Freeway Driving Environment

1. Describe some of the hazards and safety features of freeways.
2. Describe the basic types of interchanges and where they are used.
3. Understand the importance of scanning, space cushioning, and lane positioning in freeway driving.

12–2 Entering Freeways

4. Describe what you should do when driving on an on-ramp.
5. Understand the purpose of acceleration lanes.
6. Describe the proper way to merge into freeway traffic.

12–3 Exiting Freeways

7. Describe the proper way to exit a freeway.
8. Describe some of the dangers of off-ramps.

12–4 Other Dangers of Freeway Driving

9. Identify the dangers of "weave" lanes and double-merge lanes.
10. Describe the proper way to approach a tollbooth.

KEY TERMS

freeways	deceleration lanes	tollways
interchanges	off-ramp	toll plaza
on-ramps	"weave" lanes	I-Pass
acceleration lane	double-merge lanes	
merging area	express lanes	

12–1 THE FREEWAY DRIVING ENVIRONMENT

Freeways are controlled-access divided roadways with at least two lanes going in each direction that are designed to carry heavy traffic efficiently and rapidly. All interstate highways, expressways, and tollways in Illinois are freeways. Portions of some state routes, such as Illinois Highway 255 west of Edwardsville, as well as some U.S. routes, including U.S. Highway 24 west of Galesburg, are also freeways.

Freeways are built to maximize safety at high speeds. Lanes and shoulders are wider than average, and fixed objects such as signs and bridge supports are kept well clear of the roadway. Grades are even and curves banked to improve visibility and vehicle control. However, you have less time to react to an emergency situation on a crowded freeway than on an open highway. Escape routes are limited. There are also more vehicles to worry about because freeways can have five, six, or even more travel lanes in addition to special lanes for cars entering and exiting the freeway.

Interchanges

To reduce confusion, congestion, and the dangers of high-speed collisions, most intersections between freeways and other roadways (including other freeways) are grade-separated—that is, one roadway passes over the other—and connected by one-way ramps. These types of intersections are called **interchanges**, and they come in many forms depending on the volume or flow of traffic on the intersecting routes and the amount of construction space available.

A *diamond* interchange is used when a freeway intersects a roadway with relatively light traffic. In a classic diamond, drivers enter and exit the freeway on four relatively short directional ramps. When one or more of the ramps is "flipped over" to make a semi-loop, the result is a *partial cloverleaf,* also called a "folded diamond." The advantage of the diamond design is that it provides drivers quick, easy access to both roadways. The disadvantage of this type of interchange is that there is little room available for drivers on the surface street to turn left onto the freeway. Also, if there are signals at the point where the ramps meet the surface street, traffic can build up on the

Figure 12–1 Despite high speeds, freeways are among the safest roads, featuring wide shoulders and median strips.

factoid

Nearly one-quarter of all collisions, but only 14% of all fatal collisions, occur on freeways in Illinois.

factoid

The I-290/I-88/I-294 interchange located just outside of Hillside is known as "The Hillside Strangler."

factoid

The I-80/I-74/I-280 interchange near Moline is nicknamed "The Big X."

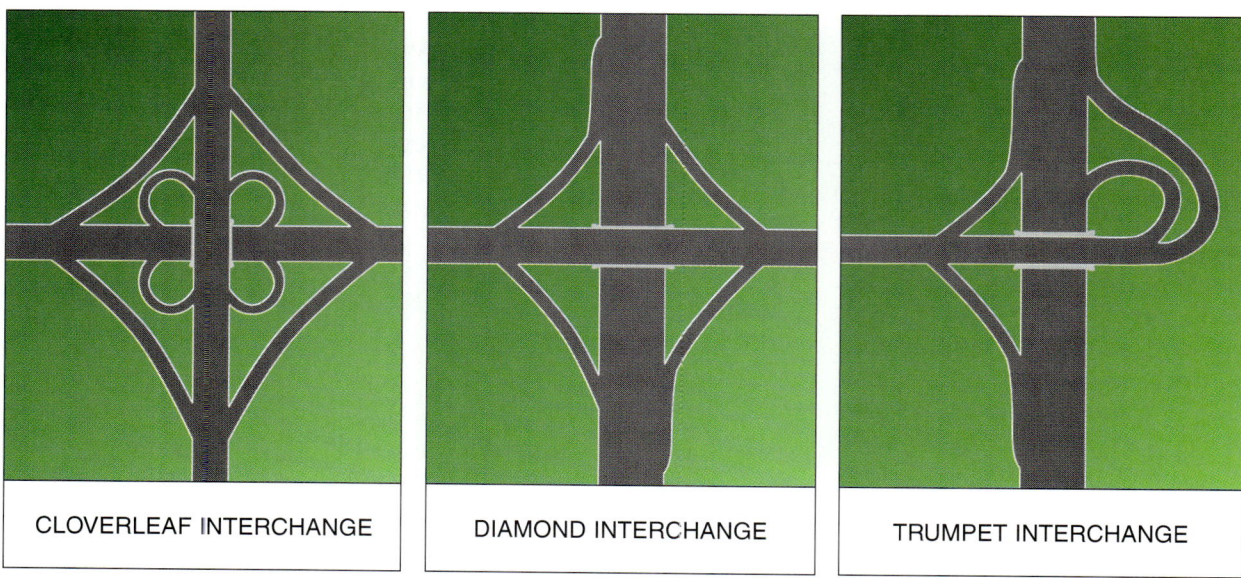

Figure 12–2 Interchanges are used on freeways instead of intersections.

ramps—even to the point of backing up onto the freeway.

A *full cloverleaf* interchange is used when a freeway intersects a roadway with relatively heavy traffic, including another freeway. Eight connecting ramps—four looping and four curving—allow drivers to move from one roadway to the other without having

Figure 12–3 What type of interchange is this?

to stop or turn left or right. The advantage of this design is that there are two ways to exit in the direction you wish to go, so you have more than one chance to exit. The disadvantage of this type of interchange is that drivers coming off one loop have to merge with and cross traffic entering another loop, increasing the risk of collisions and slowing down traffic.

A *direct connection* interchange is commonly used when two major freeways intersect. This type of interchange uses a complicated series of crossing elevated ramps that gives drivers a direct link to the intersecting freeway. The advantage of this design is that drivers do not have to loop around or weave, making a smoother and safer transition from one freeway to the next, and visibility is excellent. The disadvantage of this interchange is that it is expensive and requires a lot of space.

A *trumpet* interchange is used when one freeway ends at another. In this case, drivers leaving the stem of the "T" either follow a directional ramp (to go right) or loop back (to go left). Drivers transitioning from the top of the "T" to the stem either follow a directional ramp (if they are coming from the left) or veer right and then circle back (if they are coming from the right).

There are multiple variations of these basic designs to deal with specific problems or concerns at any given interchange. Loops can be "unrolled," ramps may be "flipped" or "stretched," left on-ramps and off-ramps may be used, and special channeling devices can be added to smooth traffic flow. In some cases, "collector/distributor roads" are used to transition traffic from one freeway to another.

Scanning for Danger

When driving on the freeway, you should *repeatedly* scan ahead, behind, and to your sides. Note where other vehicles are at all times and which lanes have a heavy volume of traffic. Pay close attention to signs, signals, and roadway markings on freeways to identify upcoming areas of merging traffic, exits, and lanes or shoulders closed for construction. Increase your following distance behind large vehicles that can block your view of the surrounding area.

famous collisions

On the morning of September 26, 1992, all five passengers of a car that jumped the median on Interstate 57 in Chicago and landed in oncoming traffic were ejected, resulting in three fatalities and three injuries.

DRIVING TIPS *Facing Your Fear of the Freeway*

Facing a high volume of fast-moving traffic on multiple lanes of traffic for the first time can be scary. If you are uncomfortable driving on the freeway, stick to "surface streets"—city streets that you can use as an alternative to the freeway—until you gain more confidence behind the wheel. Practice driving on the freeway at quiet times, such as early weekend mornings, in the company of an experienced driver.

KNOW YOUR NEIGHBOR — *Freeway Potholes*

Although freeways generally offer a smoother ride than rural roads or highways, many of the nation's freeways are in bad shape. Although only a quarter of Chicago's freeway miles are in good condition, you can expect an even bumpier ride in St. Louis or Indianapolis.

Try to identify possible hazards long before you reach them. Watch especially for traffic back-ups—caused by collisions, obstacles in the road, severe weather, and vehicles entering the freeway—to give you time to change lanes or slow down gradually. Long-distance sight on freeways is normally very good, so you can usually spot a mass of cars—or brake lights when it is dark or hard to see—long before you get there.

Figure 12–4 The speed and volume of traffic, the type of vehicle you are driving, and the distance from your intended exit all determine which lane you should use on the freeway.

Be on the lookout for speeders, constant lane-changers, tailgaters, and drivers suddenly stopping or having difficulty staying in their lanes.

Increasing Your Space Cushion

Another key to defensive driving on freeways is to leave as large a gap as possible between yourself and other drivers. Remember that higher driving speeds require larger space cushions. The faster you travel, the more time and room you need to stop or swerve to avoid a collision or roadway hazard. Because freeways are more crowded than highways and escape routes are limited, you must work harder to increase the space cushion around your vehicle. This requires *constantly* responding to the actions of drivers around you both in your lane and in adjoining lanes.

Choosing a Lane

To achieve a smooth flow of traffic, all freeway drivers have a responsibility to use the appropriate lane based on the number of available lanes, the volume and speed of traffic, and the desired exit. You should, as a general rule, drive in the left lane only when passing slower-moving traffic. During rush-hour times, however, the left lane is used as a through lane by faster-than-normal traffic. In that case, the left lane is often the smoothest lane because it is farthest from vehicles merging onto the freeway. On some occasions, however, travel in the left lane can be interrupted by vehicles merging from a left-side on-ramp or a connecting ramp from another freeway.

On four-lane freeways or on six-lane freeways with light traffic, you should use the right lane as the travel lane. On six-lane freeways with heavy traffic or on freeways with more than six lanes, however, you should generally avoid the right lane unless you are preparing to exit. Not only will you have merging conflicts with drivers entering and exiting the freeway, but you will also have to drive behind trucks, buses, recreational vehicles, and other large, slow-moving vehicles that can block your surrounding view of the roadway.

On multilane freeways, you should travel in the center lane(s). Because you do not have to deal with vehicles merging into or out of your lane on a regular basis and a "passing lane" is available to your left, center lanes are the ones that normally have the smoothest flow of traffic. The closer you get to the far-left lane, the faster the through traffic generally is. In reality, however, it does not always work out that way. It is up to you to assess traffic conditions and pick a through lane appropriate for your relative speed.

Changing Lanes

Changing lanes at high speeds in a crowded, multilane environment requires extreme caution and patience. When several lanes are moving in the same direction, drivers often compete for the same space. It is often difficult to guess other drivers' intentions, and sometimes you cannot see their signal—if they do signal. Openings in traffic can

factoid

Illinois has more than 2,000 miles (3,220 km) of interstate highway.

famous collisions

Early in the morning of August 20, 1999, a driver heading north on Interstate 290 just south of the Northwest Tollway in Schaumburg was killed when he veered onto the median and hit a concrete sign post, which split the car completely in half.

LATRELL SPREWELL

On March 2, 1997, Latrell Sprewell, a three-time All-Star guard with the Golden State Warriors basketball team, was changing lanes and cutting through traffic at high speeds on Interstate 680 near the town of Pleasant Hill in Northern California. After first moving into an EXIT ONLY lane, he lost control of his car while attempting to re-enter the freeway. He struck several sand barrels, hit a wall, and collided with another vehicle, flipping it over and injuring both the driver and the passenger. Sprewell pleaded no contest to a charge of reckless driving and was sentenced to three months of home detention, placed on two years probation, and fined $1,000. Weaving in and out of traffic to avoid slowing down or to pass other vehicles is both dangerous and illegal.

factoid

The interstate system was created in 1956, and was the largest peacetime public works program in the world up until that time.

factoid

At 3 miles (5 km), I-90 is the shortest freeway in Illinois; I-57 is the longest at 358 miles (576 km).

appear and disappear in seconds, and many motorists change lanes without signaling or thoroughly checking their blind spots because they do not want to "waste" an opportunity.

When changing lanes on a freeway, remember to *change only one lane at a time*. Even if two or three adjoining lanes appear to be clear, resist the urge to change more than one lane at a time. Most people simply cannot process the amount of visual information required to execute a multilane maneuver safely.

To safely change lanes on a freeway, follow these steps (see Figure 12–5):

1. Make sure that there is enough space between your vehicle and the car in front of you. If not, reduce your speed to increase your following distance. Watch for signs that the other driver intends to change into the same lane. Is the turn signal on? Is the driver looking over his or her shoulder? Is the car veering toward the other lane?

2. Assess possible merging conflicts in the lane you wish to enter. Look for vehicles ahead of you in the other lane, behind you in your lane, behind you in the other lane, and, if there is one, in the lane *next* to the lane you wish to enter. Use both your rearview and sideview mirrors, and look over your shoulder to check your blind spots in the direction you want to move with both eyes.

3. If the way is clear, signal your intention to change lanes by activating your turn signal. At freeway speeds, you should leave your turn signal on for at least 5 seconds before changing lanes. Thoroughly check traffic around you *again*. Make sure that the situation has not changed since you turned on your signal. Is a car ahead of you in the other lane suddenly slowing down? Is a motorcycle accelerating behind you in the other lane? Give other drivers around you plenty of time to see your signal. Delay your lane change if necessary until it is safe to move over.

4. If the way is clear, move into the other lane *with your head turned in the direction you are moving*. Check for any late-developing potential conflicts with quick, darting glances around you.

5. As soon as you have entered the lane, face forward again and establish your position in the new lane. Cancel your signal.

When you encounter other drivers about to enter the freeway near you, avoid potential conflicts by changing lanes to allow them to safely enter the freeway. Even if you are not in the far-right lane (or far-left lane if the on-ramp is on the left), you may need to allow other vehicles in that lane to move over to make room for cars entering the freeway.

Passing on Freeways

Just as in all other driving environments, the general rule is to pass on the left. However, freeway

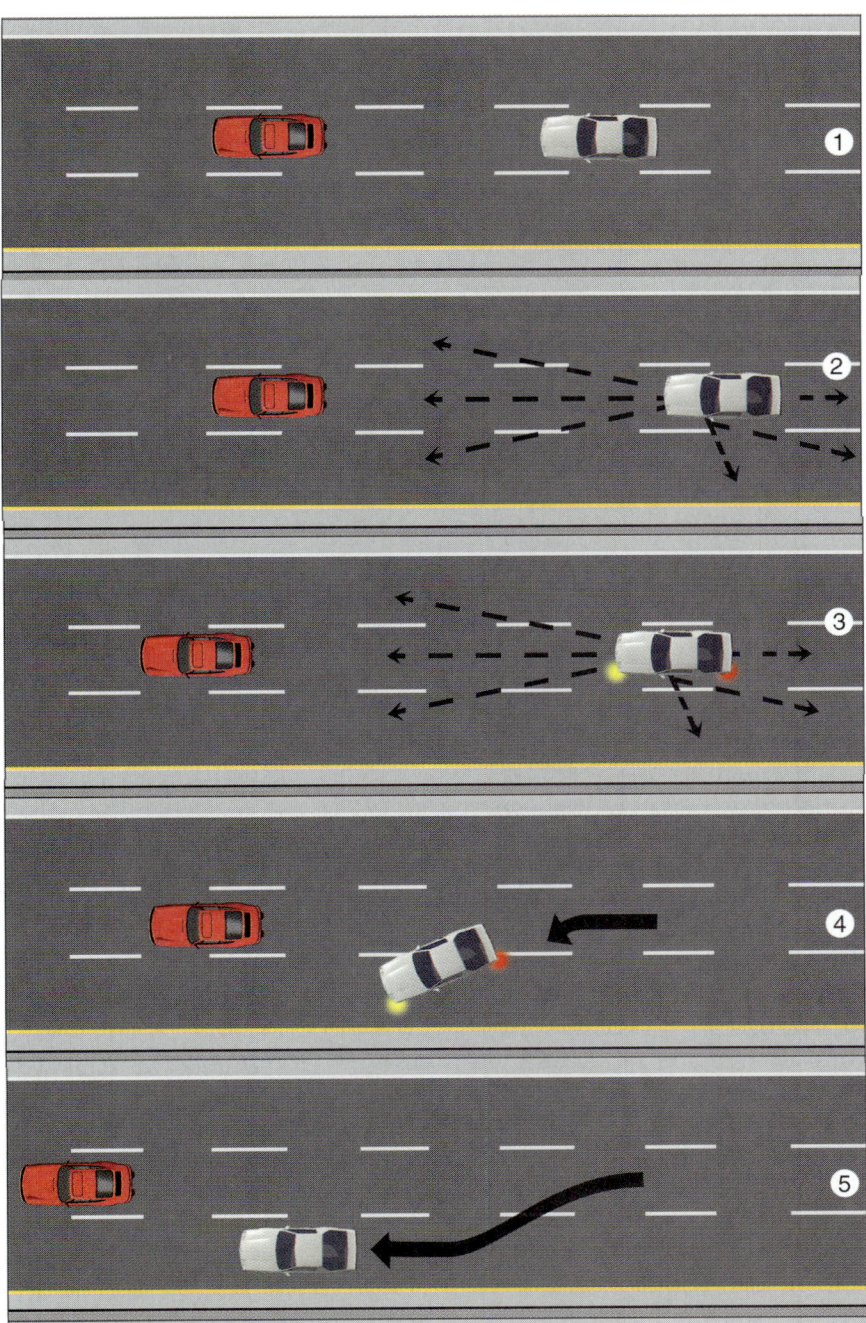

Figure 12–5 Changing lanes on a freeway

factoid

I-55 and I-270 are the two interstates that replaced the legendary Route 66 in Illinois.

traffic conditions sometimes prevent you from doing so. Many drivers drive too slowly or too fast for the lane they are in, causing traffic to back up where it should be flowing smoothly. For example, if you are behind a dangerously slow car in the middle lane of a normally flowing five-lane freeway, and there are no opportunities to use the two left lanes to pass but the right lanes are clear, you may have no choice but to pass on the right.

If you are constantly being passed on the right, you are probably driving too slowly for your lane. This is dangerous because it leaves you no escape path in case of an emergency. Work your way over to the right—one lane at a time—until other vehicles are passing you only on your left.

When passing other vehicles on a freeway, continually scan around you for potential conflicts. Use both your rearview and sideview mirrors and check your blind spots by looking over your shoulder. Watch for brake lights, turn signals, and people turning their heads in preparation for a lane change. As you change lanes, flash your lights or tap your horn if necessary to get the attention of other drivers who may be merging into the space you are about to occupy.

Always try to maintain as large a space cushion as possible around your vehicle when passing or being passed. If a car passing you is too close, move away from the other vehicle within your lane until the other driver has passed. When passing other cars, make sure that you have adequate room on both sides of your vehicle. If possible, avoid passing vehicles on both sides of you at the same time. If you have *any* doubts about the safety of your pass, wait for a better opportunity. Remember that conditions change quickly on a freeway, so even if you "lose" a chance to pass, another is likely to come along shortly.

12–2 ENTERING FREEWAYS

Entering freeways can be challenging, especially for inexperienced drivers. If you are entering a freeway from a surface street, you must quickly adjust both to a rapid increase in speed and a high volume of traffic. The first step in learning how to safely and comfortably enter freeways is to familiarize yourself with the three major parts of a freeway entrance: the on-ramp, the acceleration lane, and the merging area.

On-Ramps

Access to freeways from surface streets is limited to one-way on-ramps. On-ramps may go up or down to the freeway, depending on whether the surface street passes above or below it. Watch for guide signs and route markers directing you to the freeway. Many on-ramp signs provide the freeway name, route number, direction, and destination (usually the name of a city).

On-ramps for a particular direction on the freeway may be lo-

Wild Wheels

Figure 12–6 What is the name of this car, capable of violating the freeway speed limit even in FIRST gear?

Figure 12–7 Access to freeways is limited to on-ramps.

cated on either side of the surface street, so pay close attention to signs indicating which lane you should get into on the surface street as you approach the underpass or overpass. If the surface street has numerous lanes and traffic is heavy, it might be difficult changing from the far-left lane to the far-right lane or vice versa to reach the appropriate on-ramp. If you miss the on-ramp you want, continue on the surface street until you can safely turn around and try again from the opposite direction. Do not try to move across two or more lanes of traffic at the last second or try to back up on the shoulder to reach the ramp.

Figure 12–8 It is sometimes easy to confuse on-ramps with off-ramps, so pay careful attention to all signs.

Figure 12–9 Scan for a gap in freeway traffic from the on-ramp to give you more time to prepare to merge.

Once you have entered an on-ramp, make sure that you maintain a safe following distance behind any vehicle in front of you. Depending on freeway traffic conditions, the driver ahead of you may suddenly brake to look for a gap. This is especially true on on-ramps with YIELD signs posted at the end, which are common on some Chicago freeways. Because you will be scanning freeway traffic as you move along the on-ramp, your full attention will not be on the vehicle directly in front of you. A long following distance will give you extra time to react if the other driver suddenly slows or stops.

Remember to also keep an eye on drivers behind you on the ramp. Tailgaters on freeway on-ramps may rear-end you if *you* have to suddenly hit your brakes. Their aggressive behavior is also a sign that they may try to cut you off as you prepare to merge, accelerating past you onto the freeway before you can enter it.

As you reach the end of the ramp, adjust your speed to traffic conditions. Watch for backups caused by freeway traffic or ramp signals, and tap your brakes if necessary to warn drivers behind you to slow down and prepare to stop. If you mistakenly take the wrong on-ramp, continue onto the freeway and get off at the next available exit to turn around on the freeway or retrace your steps on surface streets to reach the correct on-ramp. *Never* back up on a freeway on-ramp under any circumstances.

On-ramps may be short, steep, sharply curved, or separated from freeway lanes by walls, landscaping, or other obstacles that block your view. It is important that you be ready to scan for a gap in traffic as soon as you have a clear line of sight. Observe traffic on all lanes of the freeway as well as ahead of you and behind you on the ramp. Look over your shoulder and use sideview and rearview mirrors to find a gap that you can safely

factoid

One mile in every five of the interstate system must be straight; the straight sections serve as airstrips during war or other emergencies.

Figure 12–10 Use extra caution when using an on-ramp with double-merge lanes.

enter. Activate your turn signal well before you reach the freeway. If you have to stop at the end of the on-ramp, keep in mind that you will have less roadway to use for accelerating into traffic.

Acceleration Lanes

Unlike most rural highways, freeways usually provide special "speed-change lanes" to get on and off the roadway. An **acceleration lane** is a temporary lane on the freeway that is an extension of the on-ramp and that allows entering vehicles to match freeway driving speeds before merging into traffic.

Never be tempted to "dive into" a stream of traffic, counting on freeway drivers to get out of your way, without properly searching for an opening. If you have not found an opening in traffic by the time the acceleration lane ends, be

---

USING SAFE — Entering Freeways

When entering a freeway from surface streets, you must adjust to a rapid change in speed. Use the SAFE process to reduce the risk of a collision. As you make your way up or down the on-ramp, *scan* toward the freeway lane you will be entering. Look over your shoulder and check your mirrors for vehicles that may be approaching from the rear. Watch for the brake lights of the car directly ahead of you in case the driver is forced to stop or abruptly slow down to merge. *Assess* the relative speed of traffic on the freeway. Consider the types of vehicles to determine where you are likely to end up when you reach the merging area. The far-right lane is often used by "big rigs" and other slower-moving vehicles, but you also have to be careful of fast-moving drivers who are preparing to exit the freeway or those who have just entered the freeway. This problem often occurs in metropolitan areas, where access points are spaced close together and cars are constantly entering and exiting the freeway. *Find* a gap in traffic and adjust your speed as necessary to *execute* your merge safely.

282 ◆ UNIT 4 *The Driving Environment*

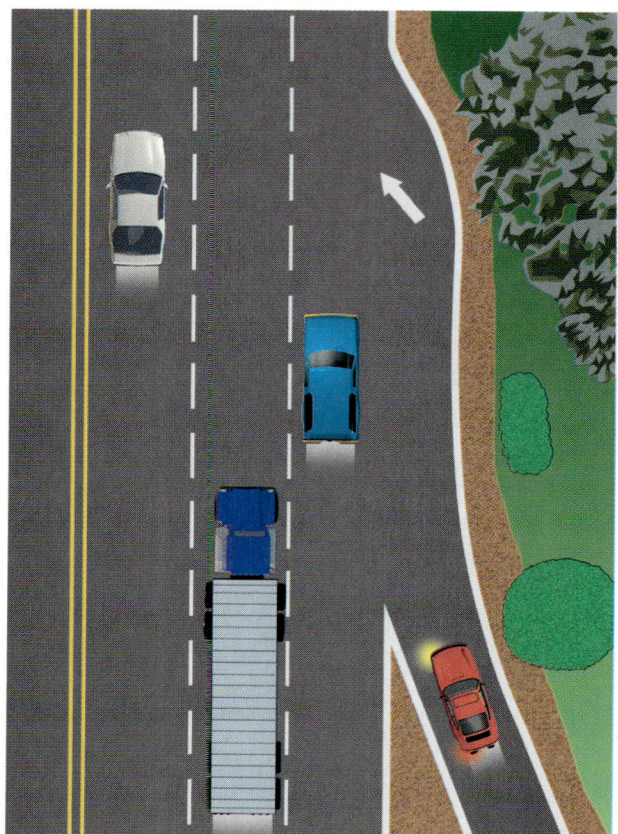

Figure 12–11 Acceleration lanes give you time to match the speed of freeway traffic before merging.

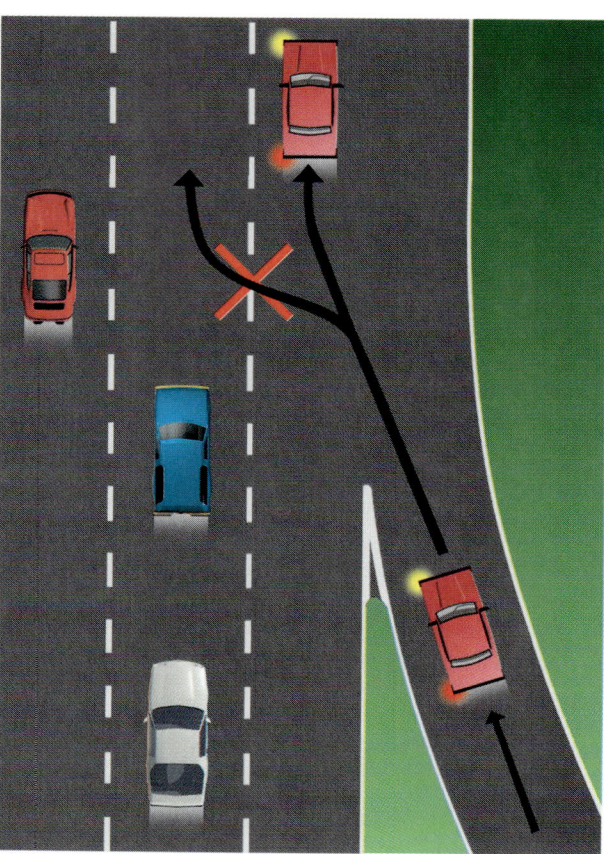

Figure 12–12 Always merge into the first through lane of freeway traffic.

factoid

Chicago's Dan Ryan Expressway, which opened in 1962, was named for the president of the Cook County Board of Commissioners.

patient and reduce your speed. Tap your brakes to warn drivers behind you on the on-ramp to slow down. Allow stopped or slowing drivers directly in front of you to find a gap first before stopping and merging onto the freeway yourself. Never try to rush past another vehicle in the acceleration lane. The driver of the other vehicle may not see you and could attempt to merge onto the freeway just as you are trying to pass, resulting in a crash.

Merging Area

The **merging area** is the space where the acceleration lane merges with the freeway. Before reaching the merging area, you should be traveling at about the same speed as other vehicles already in the traffic flow. You should also target which vehicle you are going to get behind after you merge.

As you enter the merging area, steer gradually into the through lane. Make periodic speed adjustments to blend into traffic smoothly. Continue to check traffic in all lanes, and watch for drivers that unexpectedly slow down, accelerate, or change lanes. Do not rely on your mirrors. Turn your head in the direction that you are merging to make sure that your blind spots are clear.

Position your car at a safe distance behind the vehicle you decide to follow. Once you are safely on the freeway, cancel your turn signal. Check for vehicles around you once again. Be especially careful of drivers still in the acceleration lane who may cut in front of you or directly behind you. If traffic is stop and go, common courtesy often comes into play, and merging cars and through-traffic cars take turns. Most drivers will cooperate with you, but be prepared for the occasional drivers who will not let anyone in ahead of them. Let these drivers pass and take the next spot.

Always merge into the first through lane of the freeway. Never try to cross one or more freeway lanes while merging. In those instances when you enter a freeway on the left side, keep in mind that you will be merging into the fastest lane of traffic and must accelerate more quickly than you would if entering the freeway on the right.

12–3 EXITING FREEWAYS

Exiting a freeway is normally smoother than entering one. Exits are easier to identify on a freeway than on-ramps are from a city street or highway. You also have more time to prepare to exit than you do to merge onto the freeway. Also, exit ramps have no signals to worry about. However, exiting freeways can still be dangerous. Ramp over-flows, short or sharply curved exit ramps,

Interstate 90 in downtown Chicago is the state's busiest freeway, carrying more than 400,000 people each day.

Figure 12–13 Get into the proper exit lane well ahead of the exit to avoid last-second merging conflicts.

famous collisions

On the night of January 12, 1996, three people were injured on Interstate 88 in Chicago when a driver illegally crossed the median wall through an emergency U-turn opening and was struck by opposing traffic, triggering a series of other collisions.

REALITY CHECK A Missed Exit Is Not the End of the World

If you have missed your exit, do not panic and stop, back up, or swerve across several lanes in a last-second desperate attempt to make it. Instead, continue to the next exit and use surface streets to drive back to your intended point of exit. If you have missed an exit for another freeway, take the next exit and follow signs to get back on the freeway going the opposite direction.

famous collisions

On the evening of June 24, 1999, a speeding van collided with a flatbed trailer pulled by a SUV, also speeding, and careened off Interstate 55 north of Springfield, killing one woman in the van and injuring ten others inside including her husband and eight of her children

and crossing acceleration and deceleration paths all can cause problems.

Most freeways give you plenty of time to prepare for an exit. For example, guide signs for an exit may appear 5 miles, 2½ miles, and ¼ mile before the exit as well as at the exit itself. Yellow EXIT or EXIT ONLY panels are often posted on, above, or below guide signs to give you plenty of time to prepare and to indicate whether the exit ahead will be located on the left or right. If the panel appears on the left side of an overhead guide sign, you should move to the leftmost lane to prepare for your exit; if the panel appears on the right, you should move to the rightmost lane.

Do not wait until the exit is just ahead of you to begin making your move toward the proper exit lane. Sudden, surprise exit moves could cause other drivers around you to make evasive maneuvers that may result in a crash. If possible, signal and move into the proper lane at least 1 mile (1½ km) before your exit. This is especially critical in heavy traffic, when it may be difficult to find an opening for lane changes.

If you see an EXIT ONLY sign for your exit, move into the lane indicated by the arrow as soon as it is safe to do so.

Figure 12–14 Deceleration lanes give you time to reduce your speed before getting on surface streets.

Deceleration Lanes

Just as acceleration lanes allow drivers entering a freeway to adjust to high-speed conditions without disrupting the flow of traffic, **deceleration lanes** allow drivers exiting a freeway to adjust to slower-speed conditions without blocking traffic in the through lanes. Just as with acceleration lanes, you will sometimes be able to recognize a deceleration lane by its distinctive lane-line markings. Make sure that you signal and enter a deceleration lane well in advance. Check your rearview and sideview mirrors for other vehicles, and look over your shoulder to be certain that your blind spots are clear. Watch for vehicles to your rear that may try to sneak past you to exit ahead of you.

Avoid slowing down to enter the deceleration lane. Instead, wait to reduce your speed until you are completely out of the freeway traffic flow. Once you have entered the deceleration lane, cancel your signal and make sure that you have a comfortable space cushion with vehicles ahead of you and behind you. If a driver behind you is too close, tap your brakes to warn him or her that you are slowing down. If you are too close to a vehicle in front of you, first take your foot off the accelerator and then use your brakes if necessary to increase your following distance. This is especially important if you are behind a large vehicle that can block your view of the off-ramp ahead.

Gradually reduce your speed as you drive along the deceleration lane. Watch for posted speed reduction signs or warning signs about sharply banked off-ramps. Keep in mind that some deceleration lanes are very short, requiring you to reduce your speed more quickly than normal.

Off-Ramps

Deceleration lanes lead directly to one-way **off-ramps.** Off-ramps direct vehicles exiting the freeway up or down to another roadway. Because off-ramps can be very short or have sharp curves, you must adjust to the off-ramp speed limit before actually getting onto the ramp. In some cases, this requires a rapid decrease in speed. Off-ramps with two lanes may require you to merge with other vehicles exiting the freeway into a single lane, a challenging maneuver if the ramp is curving sharply up or down. Warning signs are usually posted on dangerous off-ramps, but there is no substitute for properly scanning ahead to assess potentially hazardous conditions.

When off-ramps merge directly into street or highway traffic, watch for YIELD or MERGE signs and use caution. Off-ramps may also end in an intersection with a STOP sign or signals. After you exit the freeway, anticipate the dangers associated with city and highway driving—intersections, undivided two-way traffic, bicyclists, and pedestrians. It is often difficult to adjust to slower speeds after driving on the freeway, so keep an eye on your speedometer to make sure

Where am I?

Figure 12–15 Built in 1852, this gristmill provided access to a different kind of "freeway," the Underground Railroad.

Guess the vanity plate

Figure 12–16

286 ◆ UNIT 4 *The Driving Environment*

Figure 12–17 Drivers entering and exiting the freeway must sometimes share the same lane.

that you are driving at the proper speed.

12–4 OTHER DANGERS OF FREEWAY DRIVING

Although freeways are among the safest roadways in Illinois, a number of hazards are associated with freeway driving that can challenge even the most experienced drivers.

"Weave" Lanes

One of the most dangerous situations facing freeway drivers is that of shared speed-change lanes, often called **"weave" lanes** because vehicles exiting and entering the freeway have to weave in and out of the same lane. This situation occurs when an off-ramp is placed immediately after an on-ramp and both share at least one access lane. Vehicles entering and exiting the freeway at the same place *share* the right-of-way. The result is a nerve-racking jockeying for position between drivers who are trying to do opposite things.

When driving in one of these lanes, actively scan so that you know where vehicles around you are at all times. Increase your space cushion as much as possible, and communicate your intentions to other drivers using turn indicators, hand signals, a flash of your headlights, or a tap of your horn. Watch for brake lights. Find an escape route in case another driver makes an unexpected move.

Double-Merge Lanes

Some on-ramps have two lanes that merge together at the end of the ramp so that a single **double-merge lane** enters the freeway. This situation calls for drivers to merge twice—once with other vehicles into the acceleration lane, and then again into freeway traffic.

If you are entering the freeway from the lane closest to through traffic, you must find an appropriate merging space with the through traffic *and* watch for vehicles in the outer lane. Outer-lane vehicles often dart ahead of cars in the inner lane. Vehicles in the inner lane behind you may move over quickly and occupy the gap in through traffic which you had intended to use. Continue to scan on all sides of your vehicle when executing this tricky maneuver.

Express Lanes

Special reversible **express lanes** are used on a 7-mile (11-km) section of Chicago's Kennedy Expressway to accommodate the high volume of commuter traffic. From late evening until late morning, the center two lanes are open to vehicles entering the city, while at other times they are open to vehicles leaving the city. On each side of the freeway is a concrete barrier separating the express lanes from the other lanes. There are three access points, each controlled with special gates and overhead lane-use signals to prevent wrong-way traffic from entering. Electronically operated overhead signs indicate whether the lanes are "OPEN," "CLOSED," or "CONGESTED." During those times when traffic is reversed, variable message signs with flashing chevrons indicate "GATES CLOSING" and "STAY IN YOUR LANE."

A 5-mile (8-km) stretch of the Dan Ryan Expressway in Chicago uses a different type of express lane to separate through traffic from vehicles using local exits. These express lanes are also located on the left side of the freeway and separated from the other lanes by a concrete barrier, but they are not reversible. At any given point, there are at least three express lanes and two "LOCAL" lanes, with access points to the "LOCAL" lanes provided at regular intervals.

If you are certain that you will not need to exit the freeway at any point before the destination posted on the overhead "EXPRESS LANES" sign, taking the express lanes will allow you to bypass congestion caused by vehicles leaving the freeway at any prior exit. If you are unsure when your exit will appear, or if you know that your exit appears before the one posted on the sign, you should take the "LOCAL" lanes. Choose carefully, because once you commit to the express lanes you will not be able to exit the freeway until they rejoin the "LOCAL" freeway lanes.

Rush-Hour Traffic

Freeways are designed for high-speed driving, but at rush-hour times, traffic can be extremely congested. As city populations in Illinois increase, freeways are becoming increasingly crowded

Since 1975, travel congestion on the urban interstate system has increased by nearly 70%.

Figure 12–18 Access to the Kennedy Expressway's reversible lanes is controlled by special gates.

On average, a Chicago commuter wastes 44 hours and 65 gallons (245 l) of gasoline each year by being stuck in traffic on freeways and major surface streets.

factoid

There are approximately 275 miles (440 km) of tollways in Illinois.

places. Local jurisdictions have attempted to tackle this problem with more road construction, tollways, express lanes, ramp signals, and various forms of public transportation. Space and funding for road improvements are limited, however, and many drivers simply refuse to give up the convenience of using their own vehicle to get to and from work or school.

Rush-hour traffic can be frustrating. Drivers might wait an hour or more to cover a distance that might take half as long at nonpeak hours. The problem becomes worse when there is a crash, breakdown, or lane closing. Rush-hour traffic can also be hazardous. The most dangerous situation occurs when traffic backs up periodically and unpredictably, so drivers continually alternate between high and low speeds on different sections of the freeway. Traffic flow can vary greatly by lane as well. Vehicles can be completely stopped in one lane and traveling 65 miles per hour (105 km/h) in the next!

Under these kinds of conditions, many impatient drivers try to jump from one lane to the next to get into a "faster" lane. This is almost guaranteed to cause a crash. Vehicles in fast-moving lanes may be forced to suddenly stop, causing one or more rear-end collisions. On congested freeways, sometimes *no* escape paths are available in emergencies. One crash can easily lead to a chain of other collisions. If you get stuck in rush-hour traffic on the freeway, stay calm. Keep your radio tuned to traffic reports and look for variable message signs for updates on conditions ahead. If necessary, carefully consult a map for an alternative route when you are completely stopped. If you lose all patience, get off at the next exit and wait it out before putting yourself or others at risk.

Never attempt to use the freeway shoulder or median to get past a traffic backup on the freeway. This is both hazardous and strictly illegal. These spaces are often the only available routes for emergency vehicles to get past traffic. If you get in the way of a fire truck, ambulance, or police car, you could be putting other people's lives at risk.

Toll Facilities

You must pay a fee, or toll, to travel on several controlled-access highways known as **tollways** (also called turnpikes or toll roads in other states) in northeastern Illinois. Tolls are also required on some of the bridges that connect Illinois with neighboring Iowa, Missouri, and Indiana.

Tollways

The Tri-State Tollway is the main trunk of the state's tollway system. It comprises I-94 from the Wisconsin border to where I-94 and U.S. 41 join to form the Edens Expressway, and all of I-294, which loops around inner suburban Chicago and reconnects with I-94 just short of the Indiana border. The East-West Tollway branches off from the Tri-State Tollway, taking travelers along I-88 to the junction with U.S. 30 southeast of Sterling.

Nearly 30% of all tollway users in Illinois have I-Pass.

Figure 12–19 Before entering a tollbooth lane, make sure that you know whether you have to provide exact change.

The North-West Tollway follows I-90 from the Tri-State Tollway to Beloit, Wisconsin, skirting Rockford along the way. The Chicago Skyway traces I-90 from the Dan Ryan Expressway (I-90/I-94) in South Side Chicago to the Indiana border, where the East-West Tollway continues. Finally, the North-South Tollway, which comprises all of I-355, connects Army Trail Road in Du Page County with I-55 and will, in the future, extend to I-80.

At various points along their routes, the tollways are connected to other freeways. Signs are posted along these freeways indicating which lanes lead to the tollway ahead. At some point on a tollway, you must pass through a line of tollbooths at a **toll plaza.** These toll plazas are placed at intervals to collect a fixed fee for a designated stretch of roadway. The farther you travel, the more often you have to stop and pay a toll, which may be collected at an entrance, an exit, or at one or more barrier points.

As you approach the toll plaza, you are directed to reduce your speed and choose a lane based on how you will be paying the fare. Marked by light red signs, "MANUAL" lanes lead to staffed tollbooths, where you can get change. "AUTOMATIC" lanes have purple signs and go to unattended booths with machines utilizing plastic buckets to collect exact change. If you do not have exact change, you must pay the fare at a staffed tollbooth. Be aware that only cash is accepted. Multiple tollbooths are provided at toll plazas, so you generally have more than one lane from which to choose.

Open lanes at older toll plazas may be indicated by a green light above the open lane. Closed lanes have lowered gates across the roadway and, in some cases, a red "X" signal. Pick an appropriate toll lane as early as possible, and *stay in it.* Many crashes at toll

bumper sticker sightings

TRAFFIC TAKES ITS TOLL . . . PLEASE HAVE EXACT CHANGE READY

factoid

When it opened on July 12, 1940, the Rock Island Centennial Bridge became the first four-lane bridge built over the Mississippi River.

plazas are caused by drivers who change lanes at the last second to get to a booth with the shortest line. Also be alert to tollway workers who might cross your path on foot to get to or from a tollbooth.

Every "MANUAL" and "AUTOMATIC" lane has a red-and-green traffic signal, mounted at eye level on the other side of the tollbooth, indicating when you may proceed after you have paid the fare. Be aware that there are cameras mounted at toll plazas to record the license plate number of those who drive through without paying the correct fare. Violators will receive a citation in the mail. Since cameras also record the transaction itself, you may proceed free of anxiety if you do deposit the correct change into a plastic bucket and the light for some reason does *not* turn green. As you leave a toll plaza, accelerate gradually and watch for vehicles on both sides that merge into your lane.

I-Pass

Using I-Pass is simpler and much faster than the traditional stop-and-pay method. I-Pass is an electronic toll collection system used on all five tollways. With this system, you prepay a required minimum amount of money into an account. Each time that you use an I-Pass facility, the cost of the toll is deducted automatically from your account. When you sign up for I-Pass, you are issued a radio transponder or "tag" to be mounted on the inside of your windshield. As you pass through a toll facility, an antenna "reads" the tag for your account information so there is no need to roll down your window and handle money.

All tollbooths are equipped to detect tags. If you use a "MANUAL" or "AUTOMATIC" lane, you must come to a complete stop. If you take an I-Pass "CARS ONLY" lane, marked by a light blue sign, you are only required to slow to 30 miles per hour (50 km/h) and observe the signals on the other side of the booth. The signal will flash blue once your transponder has been read, indicating that you may proceed. It will flash yellow if it is time to replenish your account. Special "EXPRESS" lanes, also marked by light blue signs,

---

DRIVING TIPS — Ferries

Four toll ferries conduct motorists as well as foot passengers across the Mississippi River at various points to and from Missouri. Signs will direct you to the ferry landing, where you will be instructed to remain with your vehicle until boarding. When the ship is ready to be loaded, you will be signaled to proceed up and over a metal ramp onto the deck in single file. Pay close attention to the marine crew directing you, and keep your brake covered so that you can stop as soon as you are instructed to do so. When you arrive at the dock, wait for instructions from deckhands to drive off the boat, again in single file.

allow I-Pass users to bypass toll plazas completely and maintain a normal freeway speed as they pass below the antennas.

If you do not have an I-Pass account, or your account has expired, and you drive through a toll plaza or take an I-Pass "EXPRESS" lane, a camera will record your license plate number and you will receive a ticket in the mail.

Toll Bridges

Some bridges linking Illinois with its neighboring states collect tolls. The Fort Madison Bridge connects Niota with Fort Madison, Iowa, while the Rock Island Centennial Bridge joins the city of that name and Davenport, Iowa. The McKinley Bridge takes travelers between Venice and St. Louis, Missouri. The New Harmony Bridge and the Wabash Memorial Bridge, both in White County, take you to New Harmony and Mount Vernon, Indiana, respectively.

The toll plazas at these bridges generally function in the same way that they do on the tollways, but they have no form of electronic toll collection. Some will use prepaid tokens or coupons, which are made available to frequent-use customers, but otherwise they accept only cash.

Because bridges are subject to stronger weight restrictions than other roadways, be alert for signs directing oversize vehicles into the far-right or far-left lane at a toll plaza. Also, be prepared for occasional delays when crossing the Fort Madison Bridge, which is occasionally swung open to allow large watercraft to continue up or down the Mississippi River. Bridge tollbooths will either be located at one end or in the middle of the bridge. Some combination of gates, overhead lane-use signals, neon signs, cones, or barricades are used to indicate when a toll collection lane is closed. When the Fort Madison Bridge is open, lights and sirens are activated to warn motorists that the lanes are closing to traffic.

WHO'S AT FAULT?

1. Driver 1 was in a tractor-trailer truck proceeding westbound on the Dan Ryan Expressway in Chicago at about 5 miles per hour (3 km/h) in the right lane in heavy traffic. He spotted the 31st Street off-ramp ahead and began moving right to prepare to exit. Driver 2 had just come from the 35th Street on-ramp and was attempting to merge left into the same lane. Her speed was about 30 miles per hour (50 km/h). While merging right, Driver 1 collided with Driver 2. *Who's at fault?*

2. Drivers 1 and 2 were entering westbound Interstate 80 on a two-lane on-ramp from southbound Interstate 57. As the on-ramp lanes merged into one, both drivers "floored" their accelerators and raced ahead. Attempting to avoid hitting each other, both drivers swerved violently and hit a bridge abutment. *Who's at fault?*

CHAPTER 12 Freeway Driving ◆ 293

YOUR TURN

12–1 The Freeway Driving Environment
1. What are some of the hazards and safety features of freeways?
2. What are the basic types of interchanges and where are they used?
3. How are scanning, space cushioning, and lane positioning important in freeway driving?

12–2 Entering Freeways
4. What should you do when driving on an on-ramp?
5. What is the purpose of acceleration lanes?
6. What is the proper way to merge into freeway traffic?

12–3 Exiting Freeways
7. What is the proper way to exit a freeway?
8. What are some of the dangers of off-ramps?

12–4 Other Dangers of Freeway Driving
9. What are the dangers of "weave" lanes and double-merge lanes?
10. What is the proper way to approach a tollbooth?

SELF-TEST

Multiple Choice
1. When changing lanes on a freeway, you should always:
   a. look over your shoulder.
   b. turn on your headlights.
   c. signal at least 1 mile (1½ km) ahead of time.
   d. use a "weave" lane.
2. When driving on a freeway, signal and move into the proper exit lane at least _____ before your exit appears.
   a. 2 miles (3 km)
   b. ½ mile (1 km)
   c. 5 miles (5 km)
   d. 1 mile (1½ km)
3. On a freeway, you are likely to encounter trucks, buses, and recreational vehicles in:
   a. the far-left lane.
   b. the far-right lane.
   c. deceleration lanes.
   d. double-merge lanes.
4. A temporary lane on the freeway that allows entering vehicles to meet freeway driving speeds before merging into traffic is called a(n):
   a. acceleration lane.
   b. express lane.
   c. deceleration lane.
   d. on-ramp.
5. If you are consistently being passed on the right side on a freeway, you should:
   a. move to the left.
   b. speed up.
   c. move to the right.
   d. stay where you are.

Sentence Completion
1. Two major freeways intersect at a _____ interchange.
2. A _____ interchange is where a freeway intersects a highway with relatively light traffic.

3. A _____ interchange is where a freeway and highway form a "T."
4. The _____ is the space where the acceleration lane merges with the freeway.
5. You must pay a _____ to use many controlled-access highways and bridges in Illinois.

Matching
Match the concepts in Column A with examples of the concepts in Column B.

Column A
1. __ Shared right-of-way
2. __ Tollway
3. __ Used by trucks and buses
4. __ Interchange
5. __ Sign at entrance to on-ramp
6. __ Sign of lane change
7. __ Ramp danger
8. __ Sign at end of off-ramp

Column B
a. Carpool lane next to a regular lane
b. Far-right lane
c. I-Pass
d. "Weave" lane
e. INTERSTATE 55 NORTH
f. Driver turning head
g. WRONG WAY
h. Diamond

Short Answer
1. What should you do if you mistakenly take the wrong on-ramp?
2. What should you do if you have not found an opening in traffic and the acceleration lane ends?
3. How fast should you be driving in the merging area?
4. Who has the right-of-way in a "weave" lane?
5. What is the most dangerous situation you face in rush-hour traffic?

Critical Thinking
1. You are in the far-right lane on a freeway preparing to get off at the next exit. Merging traffic is causing a backup in your lane. Just as you commit yourself to enter the "weave" lane, you notice that a car entering the freeway is moving up fast behind you on your right. What should you do?
2. You are traveling in the center lane of a three-lane freeway behind a slow-moving tractor-trailer truck. The lanes to the left and to the right are clear, so you decide to pass the truck on the left. As you pull alongside, you observe that several other big rigs are in a line ahead of the truck you are passing. Although you are traveling at 65 miles per hour (100 km/h), the posted maximum limit, another driver in a sports car moves up close behind you in your lane and flashes his lights at you to speed up. What should you do?

PROJECTS
1. Look at a detailed road map of metropolitan Chicago. Find the freeway interchanges and classify each one as a diamond, cloverleaf, direct connection, or trumpet. How do the actual interchanges differ from the models described in the text?
2. Select a point at the edge of the same map and plot out the most direct route to get to downtown. Next, plot out one or two alternative routes that you think would bypass the heaviest traffic during rush-hour times. Drive each of your routes during both rush hour and at a time with light traffic. Were you able to follow your planned routes? Did you get lost, or did the trip go smoothly? Did the alternative routes save you time and hassle?

13 chapter

Sharing the Road

As a driver, you have a responsibility to share the road with many types of roadway users. Most people that you will encounter on streets and highways will be driving cars or light trucks. However, bicycles, motorcycles, large trucks, and buses are also permitted on public roadways, and pedestrians use or cross streets to get to their destinations. Understanding the dangers you face from other roadway users and the dangers they face from you is an important part of defensive driving.

CHAPTER OBJECTIVES

Upon completion of this chapter, you should be able to:

13–1 Pedestrians
1. Understand what responsibilities motorists have to pedestrians.
2. Understand what responsibilities pedestrians have to motorists.

13–2 Bicycles
3. Understand what responsibilities motorists have to bicyclists.
4. Understand what responsibilities bicyclists have to motorists.

13–3 Motorcycles
5. Understand what responsibilities motorists have to motorcyclists.
6. Understand what responsibilities motorcyclists have to motorists.

13–4 Trucks and Buses
7. Describe the "no zones" of a large truck.
8. Describe some of the dangers of passing trucks.
9. Understand what precautions you should take when driving near stopped buses.

KEY TERMS

pedestrian
jaywalking
bicycle
motorcycle
moped
motor scooters
"no zones"
"off-tracking"

Figure 13–1 City streets are often crowded with pedestrians.

13–1 PEDESTRIANS

The term **pedestrian** broadly refers to any person who uses or crosses a roadway on foot or by means of a self-propelled device other than a bicycle. People who walk, run, jog, skateboard, or rollerblade are pedestrians, as are disabled persons who use wheelchairs, walkers, or crutches. You can expect frequent encounters with pedestrians on city streets, especially near commercial districts, parks, shopping areas, and schools. On weekends and after school and work hours, residential areas can also be crowded with pedestrians.

Driver Responsibilities

In Illinois, you must give the right-of-way to any pedestrian legally crossing your side of a street at an intersection, using a crosswalk, or using the sidewalk when crossing an entrance to a driveway, private road, or alley. You must also give the right-of-way to any person working in a construction zone and to any visibly disable pedestrian wherever he or she is. In general, however, you should always give the right-of-way to *any* pedestrian you meet for the simple reason that while you are safely enclosed in a metallic shell, pedestrians only have as much protection as their clothes provide.

More than 200 pedestrians are killed and more than 1,000 are injured in traffic collisions in Illinois each year.

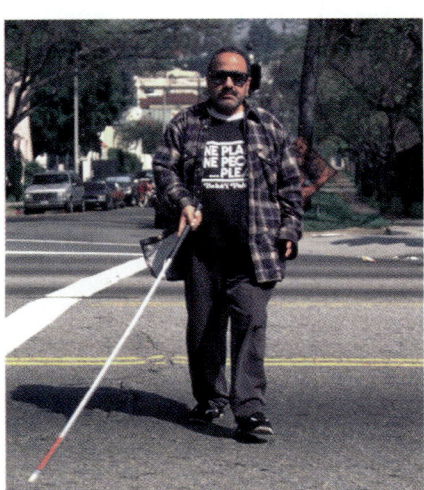

Figure 13–2 You have a responsibility to look after the safety of pedestrians.

REALITY CHECK — Help the Unseen Pedestrian

We have all seen it happen. A pedestrian is trying to cross a busy street and the vehicle in the lane nearest the pedestrian stops to allow him or her to cross. Unfortunately, approaching drivers in adjoining lanes do not see the pedestrian or understand why the vehicle ahead has stopped. They continue at full speed and reach the pedestrian's path just as the pedestrian is coming into view from behind the stopped vehicle. If you are the driver of that stopped vehicle, you should do everything you can to help protect the pedestrian. Warn the motorists behind you to stop by using the hand signal for slowing/stopping or by tapping your brakes. If necessary, honk your horn or flash your lights to alert the pedestrian that danger is approaching.

bumper sticker sightings

> IF YOU DON'T LIKE THE WAY I DRIVE, GET OFF THE SIDEWALK!

Nearly every vehicle–pedestrian collision results in injury or death for the pedestrian.

Most collisions between vehicles and pedestrians occur in or near intersections. Some of these crashes are the result of pedestrians who cross illegally against a red light or pedestrian signal. Others are caused by drivers who run a red light or fail to give the right-of-way. Still others are caused by carelessness on the part of both drivers and pedestrians. Drivers may hit a pedestrian hidden from view by a vehicle in another lane as they move to the corner to turn. Similarly, pedestrians may step off the curb right into the path of a car without looking because they think they have the right-of-way or they are looking only at the signal.

As you approach an intersection or mid-block crosswalk, watch for drivers ahead of you in adjoining lanes who have stopped to allow a pedestrian to cross. You are required by law to stop as well, even if the pedestrian is far away. Keep in mind that a truck or bus ahead of you in an adjoining lane may block your view of the crosswalk area. To avoid hitting any pedestrians who may be crossing near the vehicle blocking your view, make sure you stop *behind* the other vehicle at the stop line or nearer crosswalk line. If there is no limit line, stop before where the first crosswalk line would normally be. If you are preparing to make a right turn, carefully creep forward to the

Figure 13–3 Be alert to mid-block crosswalks, especially in large cities.

Figure 13–4 If another vehicle is blocking your view of pedestrian traffic at an intersection, make sure that you come to a stop behind the front of the other vehicle.

factoid

One out of three pedestrians killed in traffic collisions in Illinois is under the age of ten or older than sixty-five.

factoid

Nearly 65% of fatal pedestrian crashes in Illinois occur at night or in low-light conditions.

corner and check for pedestrians who may be hidden from view.

As a driver, be especially careful of children and the elderly. Children, especially those under the age of ten, have a difficult time judging speed, spatial relations, and distance. They also are too young to understand the dangers of roadway traffic, and can be unpredictable. Those over the age of sixty-five are also at greater risk, in part because their vision tends to be impaired and their reaction times are generally slower than those of other adults.

Pedestrian Responsibilities

Just as drivers have a responsibility to watch out for the safety of pedestrians, pedestrians have a responsibility to avoid creating dangerous situations such as suddenly walking or running from a curb or other safe place into the path of a vehicle. Pedestrians have a less obstructed view of the road than do drivers, have fewer distractions to worry about, and can stop or change directions much more quickly than a person in a vehicle. Although they have the right-of-way in many situations, pedestrians bear much of the burden for their own safety on the road.

Use common sense and remember how vulnerable you are. Always use a pedestrian tunnel or overhead crossing when one is provided. At night and during twilight, wear light-colored or reflective clothing and carry a flashlight to make yourself more

REALITY CHECK *Pedestrian "Don'ts"*

- Do not assume that you are safe in a crosswalk. Stay to the right, and continue to watch for cars and other vehicles approaching you.
- Do not cross an intersection that has a pedestrian "push to walk" button if you or another pedestrian has failed to depress the button. The pushbutton causes the signal controller to extend the length of both the green light and the pedestrian WALK signal, and if the button is not pressed, the green light may not be long enough to permit you to safely cross the roadway.
- Do not begin to enter an intersection if the pedestrian signal is already flashing DON'T WALK or WAIT, regardless of how fast you can walk or run. If there is no pedestrian signal, do not start to cross on a "stale" green or yellow light.
- Do not cross a divided highway unless a crossing place is designated.
- Do not cross an intersection diagonally unless there are traffic-control devices that authorize it.
- Do not cross or enter a street from between parked cars.
- Do not cross a street outside of an intersection except in rural areas where it would be impractical to walk for long distances to get to one. In these situations, always give the right-of-way to vehicles on the roadway.
- Do not assume that just because you see a driver that he or she sees you. Always try to establish eye contact with drivers as you cross in front of them.
- Do not chase or follow balls, toys, or pets into the street without first checking for traffic-this is the most common cause of vehicle-pedestrian collisions involving children.
- Do not walk in the street if you do not have to. If there are sidewalks, use them, even if you are jogging. If there are no sidewalks, use the shoulder or, if there is none, walk on the left side of the street facing traffic and try to stay as far away from oncoming vehicles as possible.
- Do not step into the street from the curb before you can do so legally and safely.
- Do not assume that just because you are wearing light-colored or reflective clothing at night that you are visible to all drivers.
- Do not stand on or next to a roadway to guard a parking space; solicit a ride, employment, or business; or distribute any materials, pamphlets, or products. Be aware that hitchhiking is illegal in Illinois as well as dangerous.
- Do not walk on freeways except in an emergency, especially at night.
- Do not walk or be on a roadway if you are under the influence of alcohol or other drugs.
- Do not walk on roadways or cross streets when your hearing is limited by a radio or a portable CD or cassette player. Turn your music or radio off or, better, take off your headphones.
- Do not walk through, around, over, or under any crossing gate at a railroad crossing or drawbridge. Never walk on railroad tracks.
- Do not throw or drop any object from a bridge or overpass onto a highway.

Figure 13-5 Pedestrians have a responsibility to stay off the roadway when they can.

visible to drivers. If you come across a work zone, and the sidewalk is blocked off, choose an alternative route rather than try to negotiate your way through. When crossing a divided highway, cross one-half of the highway at a time stopping at the median or island and checking for traffic again before continuing across the other half of the roadway. Like motorists, you must give the right-of-way to in-service emergency vehicles.

If you are crossing a street at a crosswalk where there may be a pedestrian crossing sign but no other traffic controls, wait for a gap in traffic to cross. This is especially true on roads with high-speed traffic. Even if vehicles are always required to stop for you, it is safer to attempt to cross when no vehicles are nearby. The more time a driver has to see you cross, the better prepared he or she will be to slow down. You will also have less distance to cover to reach the other side of the street in case the driver does not see you. Drivers appreciate the courtesy—and good sense—of pedestrians who avoid interrupting the flow of traffic.

Jaywalking

Jaywalking is crossing the street without regard for traffic rules or signals. The most hazardous form of jaywalking is crossing a street at a place other than an intersection or crosswalk. Because of different urban environments and customs,

The average pedestrian walking speed is 3.1 miles per hour (5 km/h) for men and 2.5 miles per hour (4 km/h) for women.

DRIVING TIPS **Roads That Are Not "Pedestrian Friendly"**

Always watch for pedestrians on the road in areas without sidewalks. If you approach a pedestrian whose back is to you, especially on a narrow road, reduce your speed and flash your lights or tap your horn as you pass to let him or her know of your presence.

Figure 13–6 Nearly 70% of pedestrians killed in traffic collisions in Illinois are males.

some localities rigorously enforce jaywalking laws, whereas others are very lax about it. Regardless of whether or not it is enforced, jaywalking is dangerous. As a pedestrian, you should cross streets only at intersections and crosswalks when it is legal and safe to do so.

Although it is difficult to predict when people will jaywalk, drivers who pay careful attention to their surroundings can anticipate jaywalking situations. For example, people are more inclined to take shortcuts across the street on long blocks or in bad weather. Jaywalkers might also dash across a street to reach a crowded restaurant or to get to a movie theater if they are worried about being late. Unsupervised children in residential areas may jaywalk to and from school and playgrounds. The more people present, the higher the risk that you will encounter a jaywalker; so if the sidewalks are crowded, be on your guard!

13–2 BICYCLES

A **bicycle** is any two-wheeled, human-powered vehicle propelled through a belt or chain. Bicyclists are commonly found in residential areas, in some business districts, near schools, on scenic highways, and by parks, lakes, rivers, and other recreational ar-

factoid
On average, about 800 bicyclists are injured or killed in traffic collisions each year in Illiniois.

REALITY CHECK *Right-Turning Bicyclists*

Bicyclists often fail to stop or even slow down when making right turns. If you see a bicyclist approaching from the right at an upcoming cross street and he or she shows no sign of stopping or yielding, expect the rider to go through the intersection or fail to turn sharply enough to safely execute the turn. Reduce your speed and prepare to stop until the danger is past.

eas. They are permitted on all public roadways in the state except freeways.

Because bicycles are used for both transportation and recreation by people of all ages and sizes, you should expect to find them almost anywhere. In addition, because bicyclists ride close to traffic and are vulnerable to injury in a collision, drivers have a special duty to pay attention to them and provide for their safety.

Driver Responsibilities

When sharing the road with bicyclists, expect sudden moves on their part at all times. Debris, minor oil slicks, a pothole, an opening door of a parked car, and other hazards can force a bicyclist to swerve suddenly into the lanes of traffic. Especially in residential areas, bicyclists will often ride in the middle of the street and disregard STOP signs and other traffic controls. Be alert for children on bikes and tricycles who are riding on the sidewalks, weaving between cars and pedestrians, riding against traffic, and darting out from behind parked cars and other obstructions.

When turning left at intersections, remember that you must yield to oncoming bicyclists just as you would to any other vehicles. Be careful not to underestimate the speed at which a bicyclist is traveling, which can be as fast as 25 miles per hour (40 km/h). Many collisions involving bicycles occur when a driver miscalculates the distance from an approaching cyclist and turns across the rider's path too early.

Figure 13–7 Bicycling is a popular form of transportation and recreation throughout the Prairie State.

Slow down and create a wide space—at least 4 feet (120 cm)—when passing a bicyclist, especially on high-speed roadways where your pass may create a wind burst that could interfere with the safe handling of the bike. If possible, use an adjacent lane. Never overtake a bicyclist if the street is too narrow for you to pass safely. If you are unable to pass, do not follow the bicyclist too closely. Remain behind at a safe interval, and warn the rider that you want to pass with a "toot" of your horn. Do not "blast" your horn, which may frighten the bicyclist and cause him or her to lose control of the bike.

Check your blind spots for bicyclists when you execute a turn or open your door to exit your vehicle after parking. When preparing to turn right at intersections, slow down and stop, if necessary, to allow nearby riders traveling in your direction to safely pass you.

Because bicyclists are smaller and harder to see than other vehicles, be especially careful in low-light conditions and when driving in traffic. Use extra caution at intersections; when entering or leaving alleys, driveways, or buildings; and when crossing sidewalks. Dim your "brights" at night when approaching a bicyclist just as you would with other motorists. On rainy days, remember that bicyclists may be unable to stop or slow down as quickly because many brakes on bicycles operate less effectively when wet.

Do not enter a bicycle lane except to enter or exit an alley or driveway, prepare for a right turn (where the bicycle lane line is dotted), avoid a conflict with other traffic, access a parking space to the right of the bicycle lane (a situation sometimes found on very wide streets), or respond to an emergency situation or instructions from a law-enforcement officer. When legally crossing or entering a bicycle lane, remember that you must yield the right-of-way to any bicyclist using it.

REALITY CHECK Bicyclist "Don'ts"

- Do not turn quickly or erratically on a bike. This can surprise motorists and cause them to instinctively swerve out of their lane.
- Do not carry any package, bundle, or article while riding a bicycle if it prevents you from keeping at least one hand on the handlebars.
- Do not carry more persons on a bicycle than it was designed for. If it is not a tandem bike (a bicycle built for two riders), then this means only *you* should be riding!
- Do not ride on the handlebars, on the center frame bar, or over the rear tire of someone else's bicycle.
- Do not hitch a ride by holding on to or attaching your bicycle to a moving motor vehicle.
- Do not install any type of siren or whistle on your bike.
- Do not modify a bicycle in such a way that it prevents you from stopping it, supporting it in an upright position with at least one foot on the ground, and restarting it in a safe manner.
- Do not perform any "trick" riding maneuvers such as weaving or riding with no hands on roadways.
- Do not ride your bike if you have a leak in a tire, which can lead to a blowout and loss of control of the bike.
- Do not use your bike on slippery roadways or in the rain or snow.
- Do not use portable music devices with headphones when riding a bike.
- Do not operate a bicycle under the influence of alcohol or other drugs.

Bicyclist Responsibilities

Bicyclists often ignore, forget, or are unaware of the fact that they have an obligation to obey the rules of the road just like any motorist or motorcyclist. This means riding *with* traffic, obeying all traffic controls, yielding the right-of-way to emergency vehicles, stopping for school buses loading and unloading children, and signaling all turns with proper hand signals. Signal your intentions early so that motorists or other bicyclists behind you will have time to slow down.

You should always try to ride in a designated bicycle lane or on a special hard-surfaced bike or multiuse path separated from motorized vehicular traffic by an open space or barrier. If neither is available, ride as close as practicable to the right curb or edge of the road, or on the right shoulder, except when preparing to turn left. Give yourself a space cushion of at least 3 feet (1 m) on the right to avoid the risk of colliding with opening doors of parked cars. If the street is extremely narrow, claim the entire lane for yourself rather than invite passing vehicles to skim your elbow. Do not, however, delay traffic; pull over and let motor vehicles pass if you are causing a backup.

Always give the right-of-way to pedestrians in a crosswalk, multiuse path, or sidewalk, and to workers in a construction zone. Give an audible signal with your voice, a horn, or a bell when overtaking a pedestrian or another rider. Even if local ordinances permit riding on sidewalks, avoid doing so if they are busy. Never lay your bike down on its side or otherwise park it in such a way as to interfere with other traffic. You may park your bicycle on the road at any angle to the curb or edge of the roadway where parking is allowed.

Use both hands to control a bicycle, except when signaling, and follow well behind moving cars in case they stop quickly. If you are with a group of riders, do not ride more than two abreast on roadways. Ride single file when passing or being passed by other vehicles. Watch your speed as you ride down hills, and do not pass other traffic. Establish eye contact with motorists when turning, and yell if necessary to get their attention. Always turn your head and use both eyes to make sure the way is clear before moving into a traffic lane to prepare to turn or get past an obstacle.

Although it is legal to turn left from a turn lane with roadway traffic at an intersection, it is safer to stay to the right and cross the streets one at a time. Even better, dismount and walk your bicycle across with pedestrians, especially in heavy traffic. If you do use a left-turn lane, always choose the one furthest to the right. If you are proceeding straight through an intersection, try to keep to the left of right-turning traffic; if there are right-turn lanes, stay out of them. To avoid being boxed in by right-turning vehicles at a STOP sign or red light, stop several feet from the intersection rather than pull all the way up to the curb. When riding through an intersection, avoid exceeding the flow of traffic.

A modern bicycle can have up to nine gears in twenty-seven different configurations.

factoid

Bicycling is up to twenty times more dangerous at night than it is during the day.

No matter how old you are or where you ride, you should *always* wear a helmet. Helmets reduce the risk of head injuries, which pose the greatest danger to bicyclists, by as much as 85%. You can also attach a rearview mirror to your helmet to improve your visibility in heavy traffic.

Make certain that your bicycle is in proper working condition and includes safety devices as needed for the type of riding you are doing. Each rider and passenger must be on a separate, securely attached seat. Illinois law allows an adult rider to carry a child firmly attached in a backpack or sling, but no child under the age of one should be carried on a bicycle at any time and those between one and four years old should ride in special bicycle child carriers.

If you are riding at night, you must have a white light on the front of the bicycle and a red taillight or reflector on the rear. You should also have reflectors on the pedals and on the tires or spokes. Special flashing red or yellow warning lights that can be attached to the bike or your clothing are also available. All bicycles must be equipped with effective brakes. If you are wearing long pants, wear clips on your ankles to prevent the pant legs from getting caught in the gears. Toe clips or straps on the pedals also increase control for experienced riders.

Nearly 40% of bicyclists injured and 14% of bicyclists killed in Illinois are between the ages of five and fourteen. If you are a parent, keep in mind that it is up to you to be sure that your child can operate a bicycle safely, and knows and obeys the traffic laws. If he or she commits a violation or causes a collision, you may be held responsible. If your child is injured or killed while riding a bicycle, you will be second-guessing yourself for the rest of your life.

13–3 MOTORCYCLES

A **motorcycle** is any two- or three-wheeled motor vehicle having a seat or saddle for riders, excluding tractors. More powerful than many automobiles, motorcycles are capable of incredible speed and acceleration. They are also more maneuverable and easier to stop than cars. On the other hand, cars provide protection from weather and collisions, are more stable, and are easier to see because of their size and wide array of lights.

Motorcyclists are directly exposed to rain, snow, gusting wind, and other types of bad weather. Noise from their own engine and wind reduces their hearing ability. Rocks and dirt kicked up from vehicles ahead can cause them to lose control of their vehicle, and slippery patches or obstacles on the road can instantly result in a crash. Like bicyclists, motorcycle riders have virtually no protection in a collision with another motor vehicle or fixed object. It is no surprise that motorcycles are one of the most dangerous forms of transportation on the road.

Driver Responsibilities

Failure of drivers to spot motorcyclists in traffic is the primary

More than 100 motorcyclists are killed and about 1,200 are injured each year in traffic collisions in Illinois.

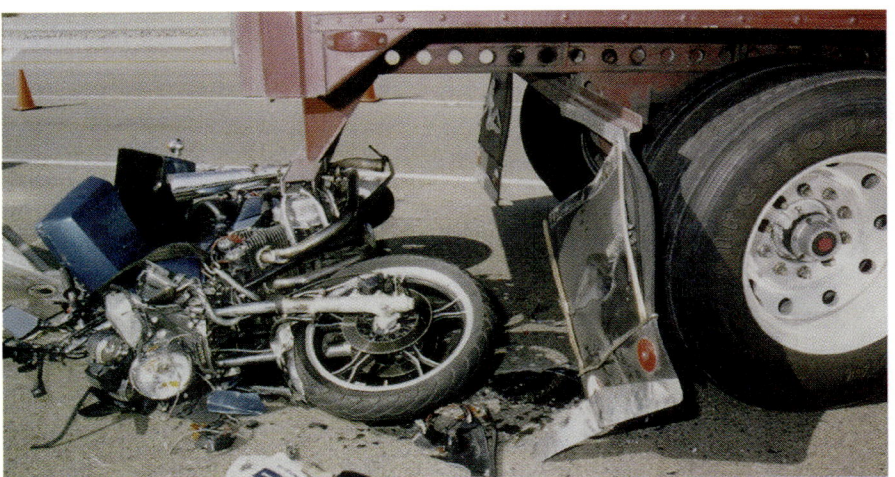

Figure 13–8 The motorcycle fatality rate is sixteen times that of passenger vehicles.

factoid

Motorcycles are more likely than other vehicles to be involved in a fatal or injury collision with a fixed object.

cause of crashes involving motorcycles. Because motorcycles are smaller than motor vehicles, drivers who are not consciously scanning for motorcycles can miss seeing them in other lanes or at intersections. Train yourself to check for motorcycles at all times while you drive, especially before passing, changing lanes, turning, exiting alleys, or backing out of driveways. Motorcycles are typically louder than cars, so use your ears as well as your eyes.

Before executing any maneuvers, double-check your estimates of a motorcyclist's speed and distance, which are easy to misjudge because of the bike's small size. Be especially careful turning left in front of a motorcycle. Also, keep in mind that the relatively small sideview mirrors on motorcycles offer a limited field of vision and often vibrate when the bike is in motion, so if you are behind a motorcyclist, you should assume that he or she cannot see you.

Motorcycles are legally entitled to the full use of a traffic lane, and they need all the space they have to make evasive maneuvers. Never drive in the same lane alongside a motorcyclist, even if the lane is wide or the cyclist is riding to one side of the lane. Be prepared for riders to make sudden moves, including lane changes, to avoid a pothole, pavement seams, debris, or other roadway hazards, or when being passed by large vehicles.

Increase your following distance to at least five or six seconds on city roads, and eight or nine seconds on high-speed roadways. Remember that rain, wind, smog, and dust limit a cyclist's vision even when he or she is wearing a face shield or goggles, or has a windshield. You should always increase the space cushion between yourself and a motorcyclist in bad weather, on slippery roads, in gusting wind, on grooved or rutted roadways, and while crossing

factoid

Three out of four motorcycles involved in fatal collsions with other vehicles are struck in the front.

factoid

Motorcyclists with less than six months of experience on their bikes account for more than half of all collisions involving motorcycles.

Figure 13-9 Always check for motorcycles in your blind spots before changing lanes.

railroad tracks or metal-grated bridges.

When being passed by a motorcyclist, maintain your lane position and speed and let the rider pass. Be aware that motorcycle turn signals usually are not self-canceling and that a motorcyclist's signal may be "left over" from an earlier maneuver. Do not assume that a signaling motorcyclist is preparing to turn so that you can suddenly pass on the rider's left or right. Also, be very cautious when passing a motorcycle with a passenger. Any wrong movement of the passenger can throw off the motorcycle and result in either erratic recovering movements or a crash.

Motorcyclist Responsibilities

As a motorcyclist, you can do a lot to reduce the risk of being involved in a collision or to minimize the severity of a crash should one occur. The most important

Figure 13-10 You must use your headlights at *all* times while riding a motorcycle in Illinois.

GARY BUSEY

Gary Busey is a popular TV and movie star with more than fifty film credits. After his breakout performance in *The Buddy Holly Story*, a role that earned him an Academy Award nomination for Best Actor in 1978, Busey went on to appear in a number of movies, including *Big Wednesday*, *Lethal Weapon*, and *Under Siege*. As a member of a group of celebrity motorcycle enthusiasts, Busey campaigned against mandatory helmet laws in California. On December 4, 1988, he lost control of his Harley-Davidson, flew off the cycle, and hit the curb of the roadway head first. Busey was not wearing a helmet and suffered extensive head trauma requiring brain surgery. He eventually recovered, but after this near-death experience, he became an outspoken advocate of helmet use. If you ride a motorcycle, remember that head injuries are the leading cause of death in motorcycle crashes. Protect yourself by always wearing a helmet that fully complies with safety standards.

precaution you can take is to understand your motorcycle well before taking to the roadways.

Keep your motorcycle in proper working condition, and make sure it complies with all state laws. In Illinois, motorcycles must be equipped with one or two white headlights, a red taillight, a stop light, and a license plate light. You *must* keep your headlights on at all times while riding to be more visible to other roadway users. Many newer models of motorcycles are designed so that the lights come on automatically when the ignition is activated.

All motorcycles must also have at least one sideview mirror and a seat and footrests for the driver. If the cycle is designed to carry more than one person, there must be a separate, securely fastened seat and footrests for the passenger. It is unlawful to carry passengers if the motorcycle is not equipped to do so. All motorcycles must also have a horn, brakes, and a properly functioning muffler. Do not add dangerous accessories to your bike or modify it so that it becomes a menace to other drivers—for example, by extending the front forks or by putting on "sissy bars." The handlebars should be adjusted in such a way that the handlebar grips are even with or below your shoulders.

In the event of a crash, proper safety equipment often means the difference between life and death. Although not required by law, *all* riders and passengers regardless of age, lifestyle, or riding experience should wear an approved helmet. Eyeglasses, goggles, or a face shield must be worn unless the cycle is equipped with a windshield or screen high enough to protect your eyes and face.

Other important safety gear for motorcycle riding includes leather boots or over-the-ankle shoes for protection from hot metal or moving parts, full-fingered leather gloves, and abrasion-resistant clothing that covers both arms and legs in case of a collision. A rainsuit will make riding in severe

factoid

Motorcycle riders who do not wear helmets are 40% more likely to sustain a fatal head injury in a motorcycle crash than riders who do wear helmets.

factoid

Most motorcycle collisions occur on trips less than 5 miles (8 km) long, shortly after starting out.

Figure 13–11 Despite the dangers, only one out of three motorcyclists in Illinois wears a helmet.

Figure 13–12 You should ride a motorcycle on the side, rather than in the center, of a lane for better visibility.

weather less uncomfortable. Finally, because remaining visible helps avoid collisions with motorists, wear bright colors with reflective stickers or striping to make yourself easier to see.

Ride to the *side* of a lane, not in the middle of a lane, to get a better picture of what the driver in front of you sees. The left side of a lane is safest to use because it gives you the best view of traffic ahead, makes you more visible to other drivers, and reduces the risk of encountering fluid drippings on the roadway. Do not ride too closely to the rear of motor vehicles. The closer you get, the less time you have to react to quick stops by those vehicles. Avoid riding in drivers' blind spots or weaving in traffic. Riding between lanes of slow-moving or stopped traffic, known as "lane-splitting," is illegal as well as extremely dangerous. An unsuspecting motorist who cannot see you could easily hit you while changing lanes or passing another vehicle.

At intersections, driveways, parking spaces, and other dangerous spots for motorcycles, slow down and anticipate trouble. Reduce your speed when approaching areas of the road you suspect are slick, oily, or wet. Exercise caution when crossing railroad tracks, bridge gratings, and potholes. In wet weather, ride in the wheel tracks of vehicles ahead of you for maximum traction and remember to use both front and rear brakes in braking situations.

Always communicate your position and intention, and assume that you are invisible to other motorists. In addition to your turn signals, which are hard to spot

guess the vanity plate

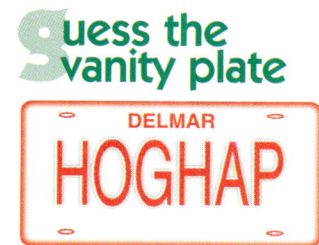

Figure 13–13

from a distance, use hand signals when possible. When you use hand signals, keep a firm grip on the handlebars with your right hand. At all other times, keep both hands firmly on the handlebars.

Reduce speed before entering turns. Although this is one of the first things new motorcycle riders are taught, they seem to forget it or ignore it after riding for some time. Once you bank for a turn, it is difficult, if not impossible, to apply brakes firmly without losing control of your motorcycle. Also, avoid the temptation to "pop wheelies," stand on your seat, "do donuts," or otherwise show off on a motorcycle, especially on a public roadway. This reckless behavior is sure to cause a crash.

Never overload your motorcycle or try to carry cargo with one hand. If anything obstructs your vision or your access to the controls, it drastically increases the chances of a crash. Remember how vulnerable you are and leave something at home.

Do not try to take advantage of your motorcycle's speed and maneuverability when passing other motorists. Before passing another vehicle, position yourself on the left side of the lane behind it. Always use an entire lane to pass, and avoid crowding a car or truck as you overtake it. Never rely on your sideview mirrors when passing or making a lane change. Most bike mirrors are convex, which means that vehicles appear farther behind you than they actually are. Always look over your shoulder as well as use your mirrors.

When riding in groups, break up into four or fewer motorcycles. Large groups of motorcycles can become confusing both for riders and other roadway users. Never ride more than two abreast in a single lane. Use single-file formation on all curves, turns, entrances, exits, and narrow roadways, and when approaching the crest of a hill. When passing others, always do so individually, not in pairs or groups.

Ride with passengers only once you are a good enough rider to be able to do so safely. Passengers should get on a motorcycle only after it has been started. Have them sit as far forward as possible with both feet on footrests at all times, keeping their legs and feet away from the muffler. Instruct them to hold on firmly to your waist, hips, or belt and to keep their movement and talking to a minimum. If you need to communicate with a passenger while riding a motorcycle, use intercom helmets. Have your passenger lean into turns when you lean. Remember that this extra weight affects the handling of your motorcycle, especially during turns and at slow speeds. Your braking distance is also increased.

Mopeds and Motor Scooters

A **moped** is as any two- or three-wheeled motor vehicle that is incapable of exceeding 30 miles per hour (50 km/h) and which has an engine displacement of 50 cubic centimeters or less. A cross between a bicycle and a motorcycle,

Where am I?

Figure 13–14 This 1910 Harley-Davidson, one of the most valuable motorcycles in the country, is one of the many vintage vehicles featured at this museum.

factoid

There are more than 10,000 crashes involving tractor-trailer trucks each year in Illinois.

Figure 13–15 Even though they have motors, mopeds and motor scooters should be ridden on the right side of the roadway like bicycles instead of in the lanes of traffic.

a moped can be propelled by pedals as well as by a low-power helper motor. **Motor scooters** are more powerful than mopeds. They have a platform for the rider's feet instead of pedals, an enclosed engine, and are capable of reaching speeds of 40 miles per hour (65 km/h) or more.

Mopeds and motor scooters are generally subject to the same rules of the road and equipment requirements as motorcycles. All mopeds and most motor scooters are classified as "Class L" motorcycles (under 150 cubic centimeters), and as such are entitled to full use of a lane as long as they do not impede the flow of traffic. Like bicycles, mopeds and Class L motor scooters should be ridden on the far-right side of the road or shoulder when it is practicable and safe to do so. Neither may be ridden on freeways.

Both mopeds and motor scooters must be registered like other

Figure 13–16 Trucks have several "no zones" that cars do not have.

motor vehicles. You must have a valid driver's license (of any class) to operate a moped; a Class L or Class M motorcycle license is required to ride a motor scooter. Although not required by law,

factoid

About 36% of crashes with big rigs occur in "no-zones."

> **KNOW YOUR NEIGHBOR** **Motorcycle Helmets**
>
> All motorcyclists in Missouri are required to wear helmets. Riders under the age of twenty-one must be helmeted in Kentucky. In Wisconsin and Indiana, motorcycle helmet use is mandatory for all riders under the age of eighteen.

moped and motor scooter riders should always wear a helmet.

13–4 TRUCKS AND BUSES

Commercial tractor-trailer trucks, often called "semis" or "big rigs," are easy to see, but their enormous size makes them a far greater hazard to motorists than other vehicles. In general, the bigger a vehicle, the slower its acceleration, the larger its blind spots, the more room it needs to maneuver, and the longer it takes to stop or be passed.

A typical tractor-trailer combination can weigh up to 80,000 pounds (36,300 kg) and, depending on the length of the trailer, can exceed 70 feet (21 m). Trucks with multiple trailers are especially challenging to operate, and are prone to overturning or "jack-knifing" (folding up like a pocket knife) in emergency braking situations. Because trucks are so much larger and heavier than cars, the driver of the car is almost always the one killed in a fatal crash involving both types of vehicles.

"No Zones"

Many people think that because truck drivers ride much higher than other drivers that they can see better. Because of their large size, however, truck drivers have larger blind spots, called **"no zones,"** than do passenger-car drivers. In addition to blind spots on either side of the cab, there is a deep blind spot up to 200 feet (60 m) long directly behind large trucks in which the driver cannot see you and in which your own view of traffic is severely limited. Drivers in truck cabs with long hoods cannot see up to 20 feet (6 m) in front of their bumper. This is enough room for a car to slip into a position of danger and be completely unnoticed by the driver. Even truck cabs with no hood, called "cab-overs," can have a front blind spot up to 10 feet (3 m) long.

Driving Behind Trucks

Always increase your following distance when driving behind trucks. This is the only way that you will be able to see down the roadway beyond the truck. If you are directly behind a big rig, stay far enough away that you can see the truck's side mirrors; if you are in an adjoining lane, make sure that you can see the driver's face in the side mirror. On inclines, compensate for the truck's loss of speed by slowing down yourself.

When you are behind a truck at a stop, especially on an incline, allow extra room for the truck to

factoid

More than 80% of crashes involving tractor-trailer trucks in Illinois occur in urban areas.

famous collisions

Four people were killed and nine injured on the night of December 29, 1988, when a "bobtail" heading north on the Edens Expressway in Chicago lost its brakes while approaching the Clavey Road intersection and rolled over on top of several cars, setting off a fuel-tank explosion.

USING SAFE — Passing Big Rigs

Passing large tractor-trailer trucks requires respect for the limitations and special performance characteristics of these vehicles. Use the SAFE method to reduce the risk of a collision when passing a big rig on a two-lane highway. Before you make the decision to pass, maintain a long following distance for as long as it takes to properly *scan* far ahead. If you are too close to the rear of the truck, you will not be able to see potential hazards down the road that could force the truck driver to take evasive action and collide with your vehicle or cause an oncoming car to crash into you. You must also be able to see whether there is enough space ahead of the truck for you to easily fit after passing. *Assess* the road and weather conditions to determine whether it is safe to pass. Do you and the truck driver have good visibility? Is it raining, dark, or foggy? Is the sun in your eyes? Is the road icy or curving? Is the road uneven or full of potholes or debris? Is it narrow? Are you in a no-passing zone? Are there soft shoulders or drop-offs? Are you moving uphill or downhill? Is there a lot of oncoming traffic? Note the length of the truck, the nature of its cargo, and the condition of its tires. After observing the truck driver's behavior, is he or she likely to help you pass by slowing down? If it is safe to pass, *find* the proper place to *execute* your pass with a minimum of interference to the truck driver. Make sure that throughout the pass, you have an escape route in case of an unexpected danger.

maneuver. Because of gravity, gearing, and the type of load it is carrying, the truck may first rock backwards toward you before going forward. If you are too close and you are unable to back up, you may get into an easily avoidable crash.

When following or being passed by trucks in rain or on wet or muddy roads, watch out for "splash and spray" from rear wheels that can temporarily block your view.

Passing Trucks

In general, it is not safe to drive beside a truck. Not only will you be in the driver's blind spots, but in an emergency the driver may need to swerve into your lane. You also want to reduce the risk of having recapping material from a truck tire, which is more prone to blowouts than a passenger-car tire, impact your vehicle. Before you attempt to pass a semi, make sure you have a good feel for its size. If you commit to a pass without correctly assessing how long it will take you to get by the truck, you may find yourself stuck facing oncoming traffic with no escape route.

You should also take into account the type of terrain you are on when passing a truck. On an upgrade, trucks often lose speed, making them easier to pass than on a level roadway. On a downgrade, a truck's momentum will cause it to go faster. Truck drivers also use downgrades to get up enough speed to make it over the next hill. In general, never try to pass a truck going downhill.

Do not cut right back in front of a truck after passing it. You may end up directly ahead of the truck and force the driver, who may be unable to see you, to abruptly slow down, causing a crash. Instead, wait until you can see the entire truck in your rearview mirror before pulling back into the lane.

Remember to compensate for the change in air pressure, or "wind effect," that occurs when you pass a truck or are being passed by one. The truck itself can generate air currents that may rock your car, or it can temporarily block crosswinds. The return of strong gusts once your vehicles have separated may surprise you and cause you to lose control of your vehicle.

Trucks sometimes drive in clusters so that they can "draft," or ride behind the wind break of vehicles directly ahead, to reduce gas consumption. Never try to pass more than one truck at a time. If you find yourself stuck behind a slow truck or a line of drafting trucks, be patient and wait for a safe opportunity to pass. Do not assume that you will be able to squeeze in between two trucks.

Do not attempt to pass closely behind or in front of a truck backing into a loading dock or other parking area. The driver might be fully concentrating on maneuvering into a tight space and may not see you as you try to sneak past.

Merging Trucks

When entering highways and freeways, truck drivers often need extra time to adjust to high-speed traffic. Considering the reduced visibility a truck driver has, as well as the limitations of having to shift up to eighteen gears, it is both courteous and good defensive driving to slow down and move to another lane to allow the truck to enter the roadway smoothly and safely.

Oncoming Trucks

Many motorists make the mistake of judging an oncoming truck to be traveling slower than it really is. This optical illusion is caused by the truck's large size, which allows it to be seen from far away. One of the main causes of collisions between cars and trucks at intersections is the inability of motorists to accurately determine the speed of an approaching truck before making a left turn. When in doubt about the speed of an oncoming truck, do not turn left in its path or drive toward it in an attempt to pass another vehicle. The truck may be going faster than you think.

Large trucks use air brakes rather than the hydraulic brakes found in cars and light trucks. These brakes take longer to engage, adding many feet/meters to a big rig's stopping distance.

Most fatal crashes involving large trucks and autos are caused by the drivers of the cars.

Figure 13–17 Adjust for the "wind effect" when passing or being passed by a truck.

driving tips

If a truck behind you is picking up speed and its lights are flashing and/or its horn is sounding, it may have lost its brakes, so get out of the way.

factoid

There are about 5,000 commercial buses serving the needs of more than 300 million riders in Illinois.

Figure 13–18 Be careful not to underestimate the speed of oncoming trucks before turning or passing.

It takes more than 100 yards (90 m)—the length of a football field—for an average fully loaded tractor-trailer truck traveling at 55 miles per hour (90 km/h) to stop. It takes even longer if the truck's brakes are hot. Contrary to what you might think, tractor units without trailers, known as "bobtails," take almost as long to stop as regular tractor-trailer combinations. If you guess wrong when passing another vehicle or turning in front of an oncoming truck, the driver will not be able to stop to avoid a collision.

"Off-Tracking"

Due to their large size and limited turning radius, big rigs must often swing wide on turns to avoid riding up over the curb, hitting parked vehicles or the median, or veering into oncoming traffic on the cross street. As a result, the rear wheels of the truck do not follow the path of the front wheels, what is known as "off-tracking." Because right turns are sharper than left ones, off-tracking is greater on right turns, especially if the cross street has only one or two lanes or the lanes are narrow. As if often indicated on warning signs on the back of trucks, drivers will first swing wide to the left before turning right. The sharper the turn, the narrower the intersection, and the longer the truck—the wider the driver must go.

Off-tracking can confuse drivers, who misinterpret the preliminary move to the left as preparation for a left turn or lane change. If a truck is stopped at or approaching an intersection, never attempt to "cut in" along the right side of the roadway as the driver first maneuvers left, or you will find yourself sandwiched between the turning truck and the curb. Trying to pass a right-turning truck driver on the left can also present dangers. If the truck swings wide enough, it can force you to stray into oncoming traffic or the median. Pay attention to a truck's turn signals before attempting to

Wild Wheels

Figure 13–19 What is the name of this 200-ton mining truck, which can share the road with a car by driving directly *above* it?

REALITY CHECK "Freeze" in Your Tracks!

Trucks and vans selling frozen desserts are usually found in residential areas near schools, parks, and public recreational facilities like municipal pools. Because children, especially very excited children, are inevitably present, always exercise extreme caution when approaching a stopped ice cream truck or van. Also, keep an eye out for children racing down the street on foot or on bicycles to catch up to such a vehicle cruising for business.

Figure 13–20 "Off-tracking" presents a danger to vehicles on both sides of a right-turning truck driver.

pass, and always give the driver sufficient clearance and time to complete a turn safely.

Commercial Buses

Buses present many of the same problems as large commercial trucks for the motorist. They are up to 60 feet (18 m) long, heavy, difficult to maneuver and stop, and have large blind spots. Like big rigs, they can be deceptively fast when approaching from a distance, take a long time to pass, tend to go slower uphill and faster downhill, swing wide on turns, and produce a wind effect at high speeds.

The unique danger presented by buses, especially those used for public transportation, is that they make frequent stops to drop off and pick up passengers. Buses pull into and out of traffic at designated curbside stops, and on major urban streets, they can tie up traffic in the right lanes. Therefore, some cities in Illinois have special bus pullouts to separate buses from the through-traffic lanes.

When driving behind a bus, increase your following distance to improve your visibility and avoid a rear-end collision. Expect the bus to suddenly stop or change lanes. When approaching a stopped bus, exercise extreme caution. Buses often stop at corners, blocking your view of cross traffic. Pedestrians may suddenly rush across the street to catch a

There are more than 1,200 commercial bus stops in metropolitan Chicago.

Figure 13–21 Watch for pedestrians around stopped buses.

factoid

More than 500 school bus crashes occur each year in Illinois, killing or injuring between 200 and 400 people.

Figure 13-22 Exercise extra caution around school buses.

factoid

There are more than 17,000 yellow school buses in service in Illinois.

bus. Never try to pass a bus on the right unless it is in the left lane of a multilane roadway. You may pass a bus on the left only if you use a separate lane going in the same direction or if the bus pulls out of through traffic into a special bus-stop lane.

School Buses

School buses are normally yellow and are generally found in residential areas early in the morning and in mid-to-late afternoon. Some school buses have a flashing white strobe light mounted on the roof that remains in continuous operation whenever any children are on board and the bus is stopped, preparing to stop, or is moving very slowly. Because children are excitable and less self-conscious than adults about traffic safety, especially when traveling in groups, you must be especially careful when driving near school buses and special vans designated to transport pupils.

If you see a school bus with flashing yellow lights, slow down and prepare for it to stop to load or unload passengers. When approaching from the front or passing from the rear a school bus that is stopped to load or unload passengers, you must come to a complete stop *unless* it is on the other side of a highway where traffic is moving in two or more lanes in each direction. The bus should have its red rights flashing and/or its mechanical STOP sign extended, but you must stop even if these devices are not present or activated when passengers are getting on or off. As a rule of thumb, you should stop at least 30 feet (9 m) away from the bus to effectively scan around and under the vehicle for children.

Watch for children getting on or off the bus, especially small children, whom the bus driver may not see or who dart across the street at the last second without checking for cross traffic. Glance periodically in your rearview and sideview mirrors. If you see a vehicle approaching from behind in another lane that shows no indication of stopping, tap your brakes and/or use the hand signal for slowing/stopping to get the driver's attention. If necessary, honk your horn to alert the bus driver and children.

Resume driving only once the lights of the school bus have stopped flashing and the STOP sign arm has been retracted, or if the bus driver has indicated that you should continue. Proceed slowly and make sure that the street is clear of children. Always give a moving school bus extra space cushioning, especially while passing. Bus drivers may become distracted by rowdy children and be unable to give their full attention to the road.

Passing a stopped school bus is regarded as a very serious offense in Illinois, and if you are convicted you will be fined $150 and have your driver's license suspended for three months. For a second or subsequent offense within five years, the license suspension will be increased to one year and the fine raised to $500.

WHO'S AT FAULT?

1. Driver 1 was stopped behind a bus in the right lane of Court Street in Pekin when a passenger exited the bus and crossed the road between the car and the back of the bus. When the pedestrian walked into the left lane, she was hit by Driver 2. **Who's at fault?**

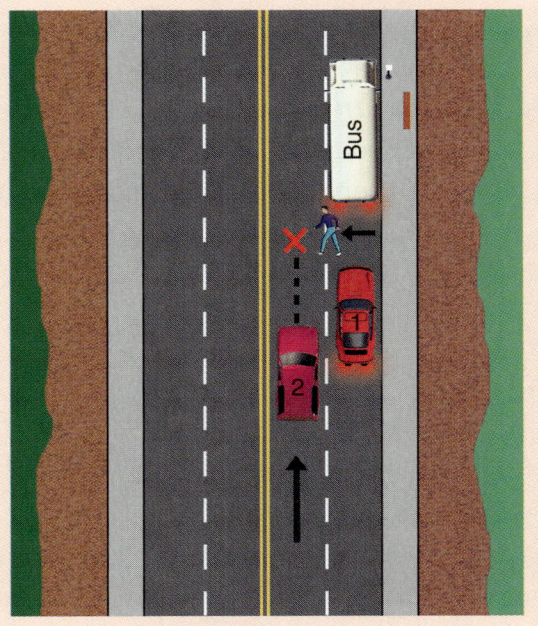

2. A motorcyclist traveling 25 miles per hour (40 km/h) was splitting lanes in bumper-to-bumper traffic on University Avenue in Urbana. Driver 1 was traveling in front of the motorcyclist and changed lanes without looking over his shoulder, striking the motorcyclist. ***Who's at fault?***

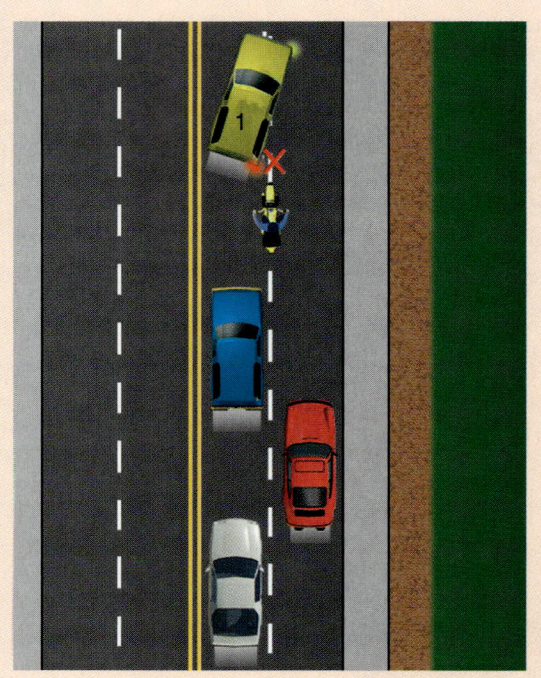

YOUR TURN

13–1 *Pedestrians*
1. What responsibilities do motorists have to pedestrians?
2. What responsibilities do pedestrians have to motorists?

13–2 *Bicycles*
3. What responsibilities do motorists have to bicyclists?
4. What responsibilities do bicyclists have to motorists?

13–3 *Motorcycles*
5. What responsibilities do motorists have to motorcyclists?
6. What responsibilities do motorcyclists have to motorists?

13–4 *Trucks and Buses*
7. What are the "no zones" of a large truck?
8. What are some of the dangers of passing trucks?
9. What precautions should you take when driving near stopped buses?

SELF-TEST

Multiple Choice

1. Most collisions between vehicles and pedestrians occur in or near:
   - a. driveways.
   - b. intersections.
   - c. parking lots.
   - d. bus stops.
2. In Illinois, motor scooters are classified as:
   - a. motorcycles.
   - b. mopeds.
   - c. off-road vehicles.
   - d. bicycles.
3. Bicyclists ride on all types of roads except:
   - a. freeways.
   - b. scenic roads.
   - c. service roads.
   - d. divided highways.
4. The bigger the vehicle, the _____ its blind spots.
   - a. smaller
   - b. larger
   - c. higher
   - d. lower
5. The blind spot directly behind a commercial truck can be up to:
   - a. 50 feet (15 m).
   - b. 100 feet (30 m).
   - c. 150 feet (45 m).
   - d. 200 feet (60 m).

Sentence Completion

1. Your following distance behind a motorcycle on city roads should be at least _____ seconds.
2. When you are passing a large truck, wait until you see the _____ of the truck in your rearview mirror before pulling back into the lane.
3. _____ can be propelled by either pedals or the motor.
4. The _____ occurs when you pass or are being passed by a large vehicle such as a truck or bus.
5. Many motorists make the mistake of judging an oncoming truck to be traveling _____ than it really is.

Matching

Match the concepts in Column A with examples of the concepts in Column B.

Column A
1. __ Swinging wide on turn
2. __ Dangerous motorcyclist practice
3. __ "No zone"
4. __ Dangerous pedestrian practice
5. __ Trucker maneuver to save fuel
6. __ School bus
7. __ "Big rig"
8. __ Commercial bus

Column B
a. Can be up to 60 feet (18 m) long
b. Retractable STOP sign arm
c. Drafting
d. Jaywalking
e. "Lane-splitting"
f. 20 feet (6 m) in front of bumper
g. "Off-tracking"
h. Can be up to 105 feet (32 m) long

Short Answer

1. What is the major cause of collisions between cars and motorcycles?
2. What should you do if you approach a pedestrian from behind on a road with no sidewalks?
3. If you are stopped behind a school bus loading or unloading passengers, when can you resume driving?
4. What precautions should you take when carrying a passenger on a motorcycle?
5. When is it legal for a pedestrian to cross a roadway outside of a crosswalk?

Critical Thinking

1. You are traveling uphill on a two-lane highway behind a slow-moving big rig. As you round a curve, you notice that there is another large truck in front of it. After the road straightens out, you enter a passing zone about 3 miles (5 km) from the crest of the hill. What should you do?
2. You are traveling in the center lane of a three-lane freeway. A fast-moving motorcyclist in the left lane pulls up and changes lanes in front of you. You notice that a short distance ahead traffic is backed up in the center and right lanes. The motorcyclist still has his right-turn signal on, and the left lane is clear for you to pass if you act quickly. What should you do?

PROJECTS

1. To get a feeling for what it is like to share the road from a few different perspectives, take a walk that extends around at least two square blocks near your home. Note which way you look when you cross the street, whether there is a sidewalk for you to walk on, whether drivers respect your right-of-way, and how much of the roadway and oncoming traffic you can see from the side of the road. Take the same route on a bicycle and in your car. Describe how your perspective and concerns are different in each of your roles as pedestrian, bicyclist, and motorist.

2. Observe the activity of pedestrians and motor vehicles at a busy intersection. Are drivers being considerate of pedestrians and bicyclists? Do pedestrians take unsafe actions such as step off the curb when the WALK sign goes on without looking each way first or jaywalk across the street? Do bicyclists making right turns stop at red lights or ride right through them? Do they turn left from the left lane or walk their bike across? How would you drive defensively through this intersection?

Challenging Driving
unit 5

Chapter 14 Challenging Driving Conditions
Chapter 15 Driving in Bad Weather
Chapter 16 Collisions
Chapter 17 Emergencies

chapter 14

Challenging Driving Conditions

Before going out on the road, it is important to realistically evaluate the driving conditions that you will encounter. Your driving environment will rarely be perfect. An important part of defensive driving is learning how to respond to challenging driving conditions. If your visibility is reduced or if you are negotiating bad roads or challenging terrain, you must take extra precautions.

Chapter Objectives

Upon completion of this chapter, you should be able to:

14–1 Reduced Visibility

1. Describe some of the dangers of driving with reduced visibility.
2. Understand what it means to "overdrive" your headlights.
3. Understand how you can prevent being blinded by the headlights of other vehicles.
4. Describe ways to minimize the risks of sunshine glare.

14–2 Challenging Road Conditions

5. Understand what precautions you should take when driving on poor road surfaces.
6. Understand how to "rock out" your vehicle if you get stuck.
7. Describe how to safely approach narrow roads, bridges, and tunnels.

14–3 Hills and Mountain Roads

8. Describe how to drive uphill and downhill using either an automatic or a manual transmission.
9. Describe some of the special dangers of mountain driving.

KEY TERMS

overdriving your headlights
"rock out"
shoulder
soft shoulders
drop-offs
switchbacks
pullouts

CHAPTER 14 *Challenging Driving Conditions* ◆ **327**

Figure 14–1 Darkness severely limits your ability to scan the roadway and identify possible dangers.

14–1 REDUCED VISIBILITY

Although there are far fewer vehicles on the road at night, 40% of all fatal crashes in Illinois occur between the hours of 8 PM and 4 AM. The most obvious reason why night driving is so dangerous is darkness. Ninety percent of a driver's ability to react depends on vision. Darkness makes it both harder for you to see and for others to see you. It is more difficult to determine the size, speed, color, and distance of objects ahead of you. This is especially true for older drivers, who may need twice as much light to see as well as younger drivers.

Looking Beyond Your Headlights

Headlights can give you a false sense of security on the road. At

factoid

A nighttime crash is twice as likely as a daytime crash to be fatal.

SAM KINISON

Stand-up comedian Sam Kinison made his reputation as a loudmouthed wild man. His brand of humor was not for everyone, but he had a loyal following of fans. On the evening of April 10, 1992, Kinison was driving from California to Laughlin, Nevada, on U.S. Highway 95 to perform at a sold-out show. At about 7:30 PM, as he neared the California–Nevada border, an oncoming pickup swerved into Kinison's lane as the driver attempted to pass traffic. The truck hit Kinison's Pontiac Trans Am head-on. Kinison was able to pull himself from his wrecked car but died within minutes of internal injuries at the scene of the collision. His wife, with whom he had just returned from a honeymoon in Hawaii, was severely injured. When driving at night, you need to be extremely cautious when executing a difficult maneuver such as passing. If you cannot see well enough to know it is safe, be patient and stay where you are. Taking driving risks in challenging driving conditions is far more likely to cause collisions and cost lives.

factoid

Headlights give off only $1/2{,}250$ as much light as the sun on a clear day.

night, a person's 20/20 daylight vision can be reduced to 20/50, so do not assume that just because you have "perfect" eyesight you do not need to pay more attention to the road. In addition, the narrow width of headlight beams can limit your view of the surrounding area to merely outlines and forms, and off-road areas may not be visible at all.

Always try to look beyond your headlights to get a better picture of possible dangers ahead. Scan beyond the center of the lane in which you are driving, not just to the edge of the area illuminated by your headlights. Remember that speed limits posted on roadways are considered reasonable for daylight hours but too fast for nighttime. Reduce your speed and give yourself an extra space cushion to react to unexpected hazards, especially on unfamiliar or narrow, winding roads. Look for flashes of light in the distance that may signal the presence of another vehicle or traffic signals. Watch for reflections off of signs, roadway markings, "delineators" (reflectors on posts or flexible rods that are placed along the edge of the roadway on curves), and pedestrian or bicyclist clothing.

Avoid using a light *inside* your car while driving at night. This will greatly reduce your night vision. If you need to check your map or read the song titles on a new CD, pull over to a safe area and do it there, not while you are driving.

Overdriving Your Headlights

On a pitch-black night, large objects can only be seen about 500 feet (150 m) away with high beams and 150 feet (45 m) away with low beams. People, animals, and small objects may be visible only at much shorter distances. If pedestrians are wearing dark clothes, you may not see them until they are right in front of you.

driving tips

Because it is difficult to judge relative speed and distance at night or any other time visibility is limited, allow extra time and space to pass others on the highway.

Figure 14–2 Headlights allow others to see you at night long before you meet them on the road.

CHAPTER 14 Challenging Driving Conditions ◆ **329**

DRIVING TIPS **Dirty Windows and Lights**

Over time your vehicle will build up a film of dirt, especially if you have been driving on wet or dusty streets. Make sure that you keep your windows clean. Dirt on the glass will reflect rays of light, either from the sun or from headlights, and add to glare. Do not forget your headlights. Dirt on headlights can reduce their effectiveness by as much as 75%. When you stop to get gas, it is a good idea to clean your windshield *and* wipe your headlights.

Figure 14–3 Avoid overdriving your headlights.

Do not "overdrive" your headlights. That is, do not travel at speeds that prevent you from stopping within the distance lighted by your headlights. To prevent **overdriving your headlights,** follow these steps:

1. Select a fixed object ahead of your vehicle, such as a sign or telephone pole, the moment your high-beam headlights pick it up.

2. Count off 6 seconds. If your high beams are on, or if there is enough background light—from roadside light fixtures, the lights of other vehicles, or a bright moon—for your low-beam headlights to see well into the distance, you should be able to see that far ahead even at highway speeds.

3. Is the object still ahead of you or is it behind you? If it is ahead of or even with you, you are driving at a safe speed. If you have passed it, you are driving too fast! Slow down and repeat the process again until the next object you choose is within range.

Give your eyes at least 5 minutes to adjust to the dark before driving at night.

guess the vanity plate

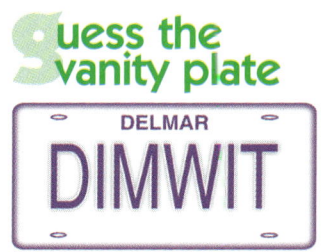

Figure 14–4

Figure 14–5 If you replace your headlights when they burn out, do not forget to realign them.

Make sure that your headlights are aimed properly. Misaimed headlights make it harder for you to see and can blind other drivers. Also, if one of your headlights burns out, replace it immediately. Not only is it both dangerous and illegal to drive with only one headlight, but if your second light goes out, you will be in serious trouble.

High and Low Beams

When traveling on dark or poorly lit roads with little oncoming traffic, use your high beams to get a bigger picture of the road. Switch to low beams as soon as you see the headlights of an oncoming car to avoid temporarily blinding the other driver. You should also switch to low beams when you see the taillights of a vehicle in front of you. High beams will reflect off the rearview and sideview mirrors and into the eyes of the other driver. In Illinois, you must dim your headlights within 500 feet (150 m) of an approaching vehicle and within 300 feet (90 m) of a vehicle ahead of you, even if that vehicle is in a different lane.

Blinded by the Light

Staring into the headlights of oncoming cars can blind you, especially if they are high beams. Reduce your speed and look off to the right-hand side of the road as other cars with their high beam lights on approach you. Make brief, frequent glances ahead of you to keep your bearings, always keeping oncoming cars in the cor-

REALITY CHECK The Blind Leading the Blind

If an approaching driver in the oncoming lanes of traffic forgets to turn off his or her high beams and they are interfering with your vision, briefly flick on your high beams as a reminder to turn them off. If the other driver does not switch to low beams after you signal, do not play tit-for-tat by keeping your high beams on or by wildly flashing your lights. Driving is not child's play, and you may cause a collision by trying to "teach the person a lesson." The offending driver may have an emergency situation or some other valid reason for using high beams that you do not know about.

ner of your vision. If you do get blinded by oncoming traffic, slow down or pull off the road until your eyes recover.

Unless you are in dense traffic, adjust your rearview mirror to the "night" setting for night driving. This will reduce the glare caused by the headlights of cars behind you, especially if they have their high beams on. Never wear sunglasses at night to reduce the glare caused by oncoming cars. This will darken your *entire* view of the road, which is even more dangerous than a momentary flash of blinding light.

You should also avoid tinting your windows, which can severely limit your ability to scan while driving at night. Be aware that it is illegal in Illinois to add tinting to your front-side windows. Windshield tinting is restricted to a 6-inch (15-cm) strip applied along the top-most portion (a standard factory feature on many newer cars). Tinting can be added to the rear windows of your vehicle only if it is equipped with two external sideview mirrors that enable you to see at least 200 feet (60 m) behind you.

Dawn and Dusk

Pay very close attention to your driving during "twilight" hours, the periods of dawn and dusk. Your eyes must constantly adapt to changing light conditions. Many drivers do not have their headlights on, yet the low level of available light can still make it hard to see other vehicles. Morning and evening commuter traffic is often heavy during these times, and many drivers are tired and not at peak alertness. Traffic controls, roadway markings, bicyclists, and pedestrians are often difficult to spot.

In Illinois, you are required by law to drive with your headlights on from sunset until sunrise, or at anytime you cannot see 1,000 feet (300 m) ahead of you. Even if the headlights do not help *you* see better, you will be more visible to other roadway users.

When driving during twilight hours, use your low-beam head-

driving tips

If you use your headlights while driving during daylight hours, do not forget to turn them off after you park your vehicle.

factoid

Wearing sunglasses during daylight hours will help your eyes adapt better to the dark by protecting a chemical known as "visual purple."

AUTO ACCESSORIES *Glare Guards*

If sunshine glare through your side windows is causing you problems, you can buy adjustable glare guards that attach to any side window with clips and suction cups. These inexpensive devices can make life much easier for children in safety seats who are unable to shield their eyes from the harsh glare of the sun. Use glare guards, however, with caution. Because they can interfere with your vision while looking over your shoulder to change lanes, pass, and merge with traffic, you must take extra time to make sure that the way is clear. Always remember to remove glare guards at night and when you do not need them.

Figure 14–6 Sunshine glare can limit your vision just as much as darkness.

lights, *not* your parking lights. Parking lights are designed to let other drivers see you when you are temporarily parked. Driving only with your parking lights on is both dangerous and illegal. When turning, changing lanes, or passing, use your turn signals earlier than normal to give other drivers more time to see them.

Sunshine Glare

Sometimes sunshine can be dangerous, especially at daybreak or in late afternoon when the sun's rays shine directly into your windshield or rearview mirror. Glare can make it hard to see and can contribute to driver fatigue. If you are driving into the sun, you may not see traffic signals or the brake lights or turn signals of cars in front of you even if your visor is down. If the sun is behind you, you may be unable to see in that direction at all.

When sunshine glare is a problem, assume that other drivers cannot see you or your turn signals as you execute a maneuver. Maintain an extra space cushion around your car. Take more time to prepare for lane changes. If you are at an intersection, look in each direction one extra time before proceeding. Activate your turn signal earlier, and if neces-

DRIVING TIPS   *The Blinding Commute*

Many of us have to face driving with the sun shining at a low angle directly into our eyes on a near-daily basis in the morning or late afternoon when we commute to and from work or school. In case of glare, keep a pair of sunglasses handy in the car. If possible, rearrange your schedule so that you will not have to commute when the sun is blinding you.

sary, use hand signals. Tap your brake pedal lightly several times as you slow or come to a stop. Because glare can prevent other drivers in front of you from seeing your own vehicle, turn on your low-beam headlights to make yourself more visible.

14–2 CHALLENGING ROAD CONDITIONS

Some roadways have dangerous stretches that can limit your view ahead. You must take extra precautions to maintain control of your vehicle and avoid a collision.

Unpaved and Gravel Roads

Dirt and loose gravel can reduce your traction, causing you to slide or skid. At higher speeds, your car can swing from side to side, or "fishtail." Stopping distance is also increased. When driving on dirt or gravel roads, always increase your following distance behind other vehicles. Reduce your speed, especially on curves, and use lower gears to avoid spinning your wheels. Firmly grip the steering wheel to maintain control. Use the tire tracks left by other vehicles to keep your tires closer to the road surface. On dirt roads, keep your headlights on so that oncoming drivers can better see you through the clouds of dust kicked up by your vehicle.

Deep Sand and Mud

Unpaved roads often contain stretches of deep sand that are nearly impassable. Rain can also turn unpaved roads into virtually undrivable muddy messes. Even a small amount of mud can make an unpaved road slippery, and if the mud is thick, tires of other vehicles can form deep ruts that fill up with water. Always approach muddy roads with caution. Avoid sudden stops or sharp turns, and maintain a fast enough speed to let the vehicle keep its momentum without driving so fast as to lose your traction.

If you get stuck in mud or deep sand, first try to back out using the tracks you just made to help you steer. If this does not work, you can **"rock out"** as follows:

1. Start the car slowly in a low gear and keep the front wheels straight. *Gently* step on the gas pedal. Do not spin the wheels, as this will literally get you deeper into trouble.

Figure 14–7 What is the commercial name of this ultimate high-traction machine, known to the United States armed forces as the "High Mobility Multipurpose Wheeled Vehicle"?

DRIVING TIPS **Check Your Clearance**

If you are traveling on rough terrain and notice an object on the road, remember that you cannot always clear it even though it may look as though you can. The chassis, differential, tailpipe, and muffler all may have a lower clearance point than you think. Damaging these parts can lead to serious and expensive repairs.

factoid

According to a recent study, an estimated 90% of freeway road surfaces in Illinois are not in good condition, second only to Oregon.

2. Go forward as far as you can. Firmly press the brake pedal to hold the car in place, and shift into REVERSE.

3. Back up slowly as far as you can. Again, do not let your wheels start to spin. Step on the brake and hold it. Shift back to a low gear and go smoothly forward again, expanding the path you are making.

4. Repeat steps 1 to 3 in rapid succession, rocking the vehicle backward and forward until it springs free. To avoid doing damage to your transmission, always use the brake while switching gears. After a few moments of rocking, shift into NEUTRAL and gently press on the accelerator to help the transmission cool down. If someone is helping you rock out, make sure that they are clear of the tires before you press the gas pedal to avoid being struck by flying debris.

If the tire is still stuck, put floor mats, small tree branches, cardboard, flat rocks, or some rough material such as stiff canvas or burlap under the stuck tire. You might also try letting a little air out of the tire to get better traction.

Potholes

A pothole is a hole in the roadway surface caused by weather, overuse, or a combination of both. Cracks in the roadway can often develop into dangerous potholes. If you hit a pothole at a high speed, you can lose control of your car or do serious damage to the tires, wheels, or underside of your vehicle. Try to avoid potholes by carefully driving around them if you can, but do not swerve into another lane of traffic to do so.

If you must drive over a pothole, slow your vehicle down to maintain better control of your car and to minimize any damage. If you hit a deep pothole, pull over into a safe spot and examine your vehicle. If you notice any damage or loss of tire pressure, have a mechanic check it as soon as possible. If the damage is serious, have the vehicle towed to a repair facility rather than try to drive to one on your own.

Figure 14–8 Large potholes can cause you to lose control of your vehicle and damage it as well.

Dips

Watch for areas of the road that sharply dip down. These areas are usually marked with warning signs and speed reduction signs. Some

Figure 14–9 Soft shoulders are common on rural roads and highways.

dips are a natural part of the roadway; others are intentionally created to channel water runoff. Always reduce your speed when approaching a dip. If you drive through a dip too fast you can seriously damage the front and rear ends of your vehicle, especially if it is long or loaded down with cargo or passengers.

Shoulders and Drop-offs

Normally a roadway has a **shoulder**, a continuation of pavement or other stable surface that extends beyond the road boundary line. Shoulders are designed to provide space for disabled vehicles, work crews, and evasive maneuvers in emergencies. They reduce hydroplaning by helping channel water away from the travel lanes. They also increase the distance between motorists and bicyclists or pedestrians who might also be using the roadway.

Soft shoulders are shoulders that either slope downward or do not provide effective traction. These are typically found on old highways and rural roads. You should avoid driving on soft shoulders unless absolutely necessary or in an emergency situation.

If you ever drift onto a soft shoulder, resist the temptation to jerk the wheel to get back on course. If one or two of your tires are on dirt or gravel, this might cause you to skid completely off the road into a tree or ditch. Instead, ease off the gas pedal, firmly grip the wheel, and gently come to a stop. Allow any dust you may have stirred up to settle. Look over your shoulder to check for traffic behind you with both eyes, signal, and slowly accelerate back onto the road once the way is clear.

Drop-offs are areas where the terrain literally drops off from the edge of the roadway without any shoulder. These are most common on mountain roads and on elevated roadways. Drop-offs are extremely dangerous, especially at night on unlit rural roads when it is almost impossible to see be-

Highway heroes

While sealing cracks along the closed right lane of Illinois 1 about 2 miles (3 km) north of Georgetown, Vermillion County, IDOT worker Thomas Petersen was struck and killed by a pickup truck pulling a load of piglets whose driver failed to merge into the left lane at the last second on July 12, 1999.

yond the road boundary line. If you have to pull off the roadway, make sure that you have a shoulder to drive onto!

Narrow Roads

Narrow roads can be very dangerous because there is little room for driver error. Be prepared to slow down when reaching a narrow stretch of roadway to better gauge whether there is enough clearance for two cars to pass each other. If there is, proceed with caution, hugging as close as you can to the right-hand edge of the roadway. If there is not enough room, wait on the side of the road and signal the other driver through first with a hand gesture or some other form of communication. After the other driver has gone by, proceed through the narrow stretch safely.

Figure 14–10 Crossing the double-yellow lines to pass on a narrow road is illegal and *extremely* dangerous.

Sometimes you will encounter "temporary" narrow roads in construction zones. It is common for two-way traffic to be reduced to one lane, with workers posted at opposite ends of a section of roadway letting traffic go in one direc-

USING SAFE *Construction Zones*

Using the SAFE method will help you negotiate construction zones. *Scan* ahead for orange signs, cones, barricades (sometimes mounted with flashing yellow lights), safety barrels, truck- or trailer-mounted arrow boards, and other traffic controls. Watch for flaggers in fluorescent vests and hard hats directing traffic. Keep an eye out for slow-moving vehicles, heavy equipment, road hazards such as cargo in the street, sand, mud, tar, oil, and work-related debris. After identifying a construction zone, *assess* potential dangers. Keep your speed low and your attention level high. Is construction going on now? Are any lanes closed ahead? Is traffic being rerouted? Are vehicles moving in your direction being stopped at intervals to let oncoming cars go by or to let a large truck in or out of the work site? Could delays cause frustrated or confused drivers around you to make sudden or unexpected movements? *Find* the best lane to get in as early as possible to avoid last-second lane changes and *execute* your decision safely. Increase your following distance, and look for escape routes on the shoulder or on the other side of cones or barriers. At night, especially in rural areas, it is hard to see what dangers are concealed in construction zones. Make sure that you can see what is behind any barrier set up in adjoining lanes before you have to take evasive action, if necessary, to avoid a collision.

CHAPTER 14 *Challenging Driving Conditions* ◆ **337**

Figure 14–11 The number of available lanes is reduced in construction zones.

tion while vehicles going in the other direction wait their turn. Obey construction or maintenance workers just as you would police or emergency personnel when they are directing traffic.

Bridges

Bridges can be noisy, making it hard to hear what is happening around you. They can also distract you by tempting you to look over the edge; if you have a fear of heights, bridges can cause feelings of anxiety that make it difficult to concentrate on driving. Do not let yourself gaze at a spectacular view while driving over a bridge. Keep your eyes on the road. If you must see the view, cross the bridge again later as a passenger!

Another challenge of bridges, especially in rural areas, is that

factoid

Illinois has more than 26,000 highway bridges.

Figure 14–12 Always use caution when approaching a bridge.

Where am I?

Figure 14–13 This is one of the 33 highway bridges that span the Mississippi River.

factoid

Each year, nearly 5,000 collisions occur on bridges and in underpasses in Illinois.

they often have narrow lanes. On older roads, bridges may even be reduced to one lane. All or part of the roadway surface on a bridge may be made up of metal grating on which your tires are more likely to slip or skid. Guardrails and bridge support structures reduce your options in the event of an emergency. If you are stranded on a bridge, just getting out of your vehicle can be dangerous.

Before you cross a bridge, reduce your speed and turn on your low-beam headlights to make yourself more visible to other drivers. On narrow bridges, expect oncoming vehicles to crowd the yellow center lines. Carefully study any signs or signals ahead for special instructions or warnings. Watch for narrowing lanes and disappearing shoulders, and note where other vehicles are in relation to you. Be on the lookout for pedestrians and bicyclists crossing the bridge. Pick a lane and stay in it until you are off the bridge. Be aware that it is illegal as well as extremely dangerous to stop or park on a bridge.

Tunnels and Underpasses

Like bridges, tunnels and underpasses provide few escape routes. They can also alter your perception of the driving environment. Sounds can be distorted, and light can reflect off the walls. Many tunnels and underpasses are not lit or are poorly lit, and the sudden transition from driving outside on a sunny day to plunging into darkness can be momentarily disorienting. The effect is even worse if you face an oncoming driver with his or her high beams on. If you have claustrophobia, it can be difficult driving through relatively long tunnels like those below the runways of O'Hare International Airport or so-called "Hubbard's Cave" under the railroad yards in Chicago.

When driving in tunnels, reduce your speed, activate your low-beam headlights, and observe any special signs or signals. Before entering a tunnel, remember to remove sunglasses if you are wearing them. Even if other drivers honk their horn to be funny, do not give in to the temptation to do the same thing. As with bridges, never stop or park in a tunnel or underpass.

14–3 HILLS AND MOUNTAIN ROADS

Even though Illinois is known as the "Prairie State," much of the landscape is in fact covered with gentle, rolling hills. There are more rugged hills in the extreme northwestern and southern portions of Illinois. Even if you never drive out of the state, you must be prepared to deal with the special challenges posed by steep, curving roads.

Hills

When driving on hills, you must consider the force of gravity on your vehicle. The steepness of the grade, the speed you are traveling, and the weight of your vehicle can all impact how you

I'D RATHER BE OVER THE HILL THAN UNDER IT

drive. Whether you go uphill or downhill, you must consider the proper gear to use, how much power you need, and how to apply your brakes. Whether you use a vehicle with a manual or automatic transmission, you need to exercise more caution when driving on hills.

When you drive uphill, you are driving *against* the force of gravity. This means that you need more power to keep moving at the same speed. If you have an automatic transmission, you must apply more pressure on the gas pedal to avoid losing speed. When you get to a comfortable speed, keep your foot at that point until the road starts to level off. On very steep hills, switch to a lower gear to help the engine work more efficiently. If you have a manual transmission, downshift to a lower gear to avoid losing power and speed. The grade of the hill will determine which gear to use. The steeper the grade, the lower the gear you should select.

When you drive downhill, you are driving *with* the force of gravity. This means that you need less power to keep moving at the same speed. If you have an automatic transmission, ease off the gas pedal as you drive downhill. If you start going too fast, gradually apply the brakes to slow down until you reach a comfortable speed. On long or steep hills, switch to a lower gear to maintain better control of your vehicle and reduce wear on your brakes. Avoid riding your brake.

If you have a manual transmission, switch to a lower gear to maintain better control of your vehicle and reduce the drag on your engine *before* you start driving downhill. If you switch gears after you begin to move downhill, you may be forced to apply your brakes while shifting. This is especially true on steep hills, where if you wait too long to downshift you may not have enough braking power to switch to a lower gear without braking continuously for a long time.

Highway heroes

While attempting to pass another vehicle on a curve in a hilly area on Illinois 171, Officer Daniel Jarema of the Lockport Police Department lost control of his vehicle and collided with an oncoming truck on April 1, 1989, dying from his injuries twelve days later.

Figure 14–14 You often have to reduce your speed and downshift when driving on hills.

factoid

Charles Mound is named for Elijah Charles, a farmer who built a house at the summit, which is still on private property, in 1828.

Mountain Driving

The highest point in the state, Charles Mound, is only 1,235 feet (377 m), so you will not have to deal with the unique dangers of mountain roads on a daily basis. However, many popular national parks have mountains, and some of the nation's greatest scenic highways cross mountainous areas.

Mountain roads are typically narrower than average roads. They also have sharp curves and steep grades that make it difficult to see the road ahead and put more stress on your vehicle's engine, transmission, and brakes. Some "hairpin" turns, called **switchbacks,** are so sharp that they reverse direction. If you are behind a truck, bus, or any other large vehicle, your view of the road is even more limited. The absence of walls, barricades, or guardrails on many stretches of mountain roads leaves little margin for error. If you stray into oncoming traffic on a mountain road, you risk not only colliding with another vehicle but of driving off the edge of a cliff or embankment.

Always pay careful attention to signs and roadway markings on mountain roads. Keep your speed low and stay as far to the right as you can, especially on curves. Expect problems and delays caused by fog, snow, and ice. Watch for washed-out areas and road damage or debris from landslides. Give yourself an extra space cushion when driving behind other vehicles. Use your low-beam headlights to make yourself more visible. When you approach any area in which it is difficult to see ahead of you such as a blind curve, lightly tap on your horn or flash your lights to warn any drivers that might be coming your way.

Driving down a mountain is riskier than going up. As you move downhill, you can accelerate quickly unless you resist the

Figure 14–15 Switchbacks are a common hazard of mountain driving.

Figure 14–16 Mountain roads require patience and extra caution.

force of gravity. *Never* coast down a mountain road in NEUTRAL or with the clutch disengaged. This is both dangerous and illegal.

Some mountain roads have off-ramps designed to stop vehicles, primarily tractor-trailer trucks, that experience braking problems on long, steep downgrades. Yellow rectangular warning signs alert motorists to their presence. These "ramps" are actually short, uphill roads that help the driver of a runaway vehicle stop by quickly reversing the effects of gravity. They typically have long pits filled with gravel and plastic barrels filled with water or sand at the end to stop the vehicle if it cannot do so on its own. Do not mistake a runaway vehicle ramp for an off-ramp, especially at night, or you may find yourself in serious trouble!

Passing on Mountain Roads

It is common to get stuck behind large trucks, recreational vehicles, and cars with trailers when driving uphill in the mountains, especially during holidays. As more and more cars back up behind you, you may feel more pressure to pass so you do not "hold them up." Be patient! Wait for an appropriate place to pass with a clear view of the road ahead. Many mountain roads have special passing lanes at periodic intervals for this very reason.

If *you* are the one holding up traffic, you are required by law to pull aside at the first safe opportunity to allow the other vehicles get by you. Sometimes paved, usually gravel or dirt, **pullouts** are widened off-road areas or extra-wide shoulders that appear on rural roads and highways, especially on steep, narrow ones. Because U-turns are illegal as well as dangerous on winding mountain highways where you cannot see far ahead of you, pullouts generally provide enough room for you to turn your vehicle around and head the other way. Upgrades of major mountain roads often have special paved areas, identified by rectangular TURNOUT signs and roadway markings, at periodic

factoid

For every 1,000 feet (300 m) of elevation, your car's engine loses 5% of its power.

intervals for slow vehicles to get out of the way of faster traffic.

What do you do if you meet a car going in the opposite direction on a one-lane or very narrow mountain road? The general rule in Illinois is to give the right-of-way to the driver having the wider shoulder or access to a side road, driveway, or pullout that provides maneuvering space. Each situation, however, may call for a different solution. If you get in this kind of jam, be courteous.

Effects of High Altitude

The lower oxygen available at high altitudes can make you sleepy, especially if you are not used to it. High altitudes can also increase your heart rate and cause shortness of breath and headaches. The effects are worse if you are already tired or ill. If you have any of these symptoms while driving on a mountain road, stop at a pullout, scenic overlook, or some other safe place clear of traffic to rest. If you are traveling with others, try changing drivers to give yourself a break. Also, consider selecting an alternative route at a lower altitude.

Because the air is thinner, cars tuned up at lower altitudes do not run as efficiently at higher altitudes. Engines heat faster, and the risk of stalling is greater. Keep in mind that you will have less power going uphill, a time when you need it the most. While driving on mountain roads, periodically check your temperature gauge. If the car is close to overheating, stop at a safe place and allow your engine to cool down. If you cannot pull over right away, turn on your heater to draw off some of the heat built up in the engine coolant. Always carry a bottle of water in the car for your radiator when you are driving in the mountains, especially if it is hot outside.

WHO'S AT FAULT?

1. Driver 1 was traveling uphill on the Eddyville Blacktop in Pope County behind several slow-moving cars. Losing patience, she attempted to pass three cars at once. Just as she reached the top of the hill, she saw Driver 3 moving down the hill in the oncoming lane at a high rate of speed. Driver 1 tried to get over into the right-hand lane, but Driver 2 refused to let her in. To avoid a head-on collision, Driver 1 moved over to the right and hit Driver 2, forcing him off the road. **Who's at fault?**

2. Driver 1 was traveling on the single downhill lane of a three-lane mountain road and approached a line of cars stopped by a flagger at a construction zone. Despite the presence of several orange signs, he failed to realize that the cars were stopped until the last second and slammed on his brakes, skidding into the left oncoming lane of traffic. Driver 1 hit Driver 2, who was driving at the posted speed limit in the left lane and did not have time to get into the free right lane. **Who's at fault?**

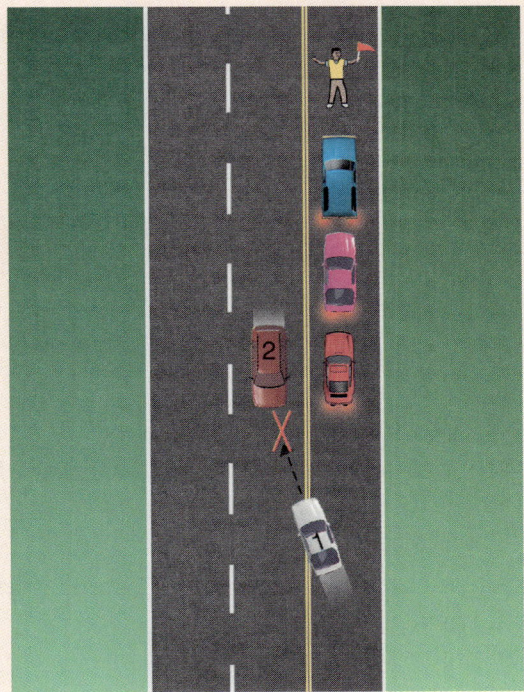

YOUR TURN

14–1 Reduced Visibility

1. What are some of the dangers of driving with reduced visibility?
2. What does it mean to "overdrive" your headlights?
3. How can you prevent being blinded by the headlights of other vehicles?
4. What can you do to minimize the risks of sunshine glare?

14–2 Challenging Road Conditions

5. What precautions should you take when driving on poor road surfaces?
6. How do you "rock out" your vehicle if you get stuck?
7. How do you safely approach narrow roads, bridges, and tunnels?

14–3 Hills and Mountain Roads

8. How do you drive uphill and downhill using an automatic transmission? a manual transmission?
9. What are some of the special dangers of mountain driving?

SELF-TEST

Multiple Choice

1. Large objects can be seen at about _____ away with low beams.
   a. 50 feet (15 m)
   b. 100 (30 m)
   c. 150 (45 m)
   d. 200 (60 m)
2. You must dim your headlights within at least _____ of an approaching vehicle.
   a. 200 feet (60 m)
   b. 300 feet (90 m)
   c. 400 feet (120 m)
   d. 500 feet (150 m)
3. When entering a tunnel, you should turn on your:
   a. low beams.
   b. high beams.
   c. parking lights.
   d. emergency flashers.
4. "Overdriving" your headlights means:
   a. driving too fast to stop within the distance illuminated by your headlights.
   b. using your headlights on a sunny day.
   c. leaving your high beams on too long.
   d. driving with headlights aimed too high.

5. You must put your headlights on from sunset until sunrise or at anytime you cannot see _____ ahead of you.
   a. 700 feet (215 m)
   b. 2,000 feet (610 m)
   c. 500 feet (150 m)
   d. 1,000 feet (300 m)

Sentence Completion
1. Dirt and gravel roads can reduce your _____, causing you to slide or skid.
2. To avoid damaging your transmission while "rocking out," always use the _____ while switching gears.
3. _____ either slope downward or do not provide effective traction.
4. The _____ on your rearview mirror reduces the glare from headlights behind you.
5. Some mountain roads have _____ in case vehicles experience brake failure on long, steep grades.

Matching
Match the concepts in Column A with examples of the concepts in Column B.

Column A
1. __ Bridge hazard
2. __ Construction zone hazard
3. __ Switchback
4. __ Shoulder
5. __ Effect of high altitude
6. __ Pullout
7. __ Dip
8. __ Pothole

Column B
a. Used to channel water runoff
b. Sleepiness
c. Two-way traffic reduced to one lane
d. "Hairpin" turn
e. Widened off-road area
f. Noise
g. Space for broken-down vehicles
h. Can damage wheels

Short Answer
1. What is dangerous about switchbacks?
2. How can you test whether you are overdriving your headlights?
3. How do you communicate to an oncoming driver that his or her high beams are blinding you?
4. When driving downhill, what can you do to maintain better control of your vehicle?
5. Who generally has the right-of-way if two drivers going in opposite directions meet on a narrow mountain road?

Critical Thinking

1. You are driving up a two-lane mountain road approaching a sharp curve with a steep drop-off. Suddenly you notice that a small pile of rocks is in your lane ahead. You cannot see oncoming traffic from where you are, and no shoulder is on either side of the road. What should you do?

2. You are on a two-way, four-lane road approaching an intersection with a green traffic signal. An oncoming driver in the left lane has his left-turn signal on, preparing to turn onto the cross street when the way is clear. A car is to your rear, but because the sun is directly behind you, it is difficult to determine how far back it is. The left-turning driver is not slowing down, and you suspect that the oncoming driver cannot see you because of sunshine glare. What should you do?

PROJECTS

1. Use a newspaper or the Internet to determine what time sunset is on a given day. On that day, select a safe location where you and a friend can view traffic on a major roadway. Every 5 minutes, note on a piece of paper whether some, most, or all vehicles have their headlights on from sunset until it is completely dark outside. Describe the weather conditions. Is it cloudy? raining? clear? What percentage of drivers comply with the law about when they must turn on their headlights? Share the results with your class.

2. As a passenger, go for a drive and find five different challenging road conditions where you live, including gravel roads, unpaved roads, roads with dips or potholes, roads with soft shoulders or drop-offs, narrow roads, bridges, tunnels, and construction zones. Identify their location and what makes them dangerous. Share the results with your class.

15 chapter

Driving in Bad Weather

It is always best to avoid driving in bad weather. Sometimes, however, you cannot delay a trip because of rain, snow, ice, fog, gusting wind, extreme cold or heat, or other hazardous conditions that may test the limits of both you and your vehicle. When driving in severe weather, which is common in Illinois, you must exercise common sense and good judgment. Because you face both reduced visibility *and* challenging road conditions, you must be extra cautious when executing all driving maneuvers.

Chapter Objectives

Upon completion of this chapter, you should be able to:

15–1 Rain

1. Describe some of the dangers of driving in rain.
2. Understand what hydroplaning is and how you can prevent it.
3. Understand what you should do if you encounter deep water on the road.

15–2 Snow and Ice

4. Understand how to prepare for driving in winter conditions.
5. Describe some of the dangers of snow-covered and icy roads.

15–3 Other Severe Weather Conditions

6. Understand what precautions you should take when driving in fog.
7. Understand what to do if you are caught in a severe storm.

15–4 Cold and Hot Weather

8. Describe some of the problems caused by extremely cold temperatures.
9. Understand what to do if your engine overheats.

KEY TERMS

hydroplaning
vapor lock

15-1 RAIN

Illinois receives moderate precipitation, averaging about 37 inches (86 cm) each year. Southern Illinois generally receives more rain, though spring and summer thunderstorms are common throughout the state. Occasionally, storms can unleash torrential downpours like the one that dropped nearly 17 inches (43 cm) of rain on East St. Louis on June 24, 1957, the greatest 24-hour rainstorm in state history. Wherever you live, you must be prepared to handle the special driving challenges posed by rain.

Rain makes road surfaces slippery. It also makes it harder for you to see others and for others to see you. Water pounding on your windshield, the motion of wiper blades, and darker skies significantly reduce your visibility. "Splash and spray" from oncoming or passing traffic can temporarily blind you. Painted roadway markings can virtually disappear, making it difficult to stay within your lane. In heavy rainfall, both hearing and seeing in the driving environment can be difficult.

Driving in Rain

When driving in the rain, always use your low-beam headlights to increase visibility. In Illinois, you are required by law to turn on your lights whenever your windshield wipers are on or rain or other atmospheric conditions reduce visibility to less than 1,000 feet (300 m). Watch for other vehicles that may not have their lights on. Activate your windshield defroster and rear window defogger to keep your windows clear of condensation. Remember that other roadway users may have trouble controlling their vehicles in the rain, especially bicyclists and motorcyclists. Also, pedestrians may unexpectedly dart across the street to get out of the bad weather.

factoid

Rain contributes to more than 20,000 traffic collisions in Illinois each year.

Figure 15-1 Visibility is severely limited in heavy rain.

factoid

The first windshield wipers were offered by Trico in 1916.

DRIVING TIPS   *A Slick Combination*

The first half-hour of any rainfall can be the most dangerous. Oil and grease that have built up on the road combine with the water to create a very slippery surface. If possible, wait until the road oils have been washed off the roadway by more rain before driving.

When driving in the rain, increase your space cushion in case you or other drivers begin to skid. Increase your normal dry weather following distance by at least 5 seconds. Signal your turns, stops, and lane changes earlier than usual. To increase contact with the road and improve traction on wet road surfaces, drive in the "tracks" left by the vehicle in front of you.

Never drive at normal, fair-weather speeds when it is raining. A good rule of thumb is to reduce your speed by 25% below the posted limit on straight roads and 50% on curves. If you have a manual transmission, avoid quick downshifts because this can cause

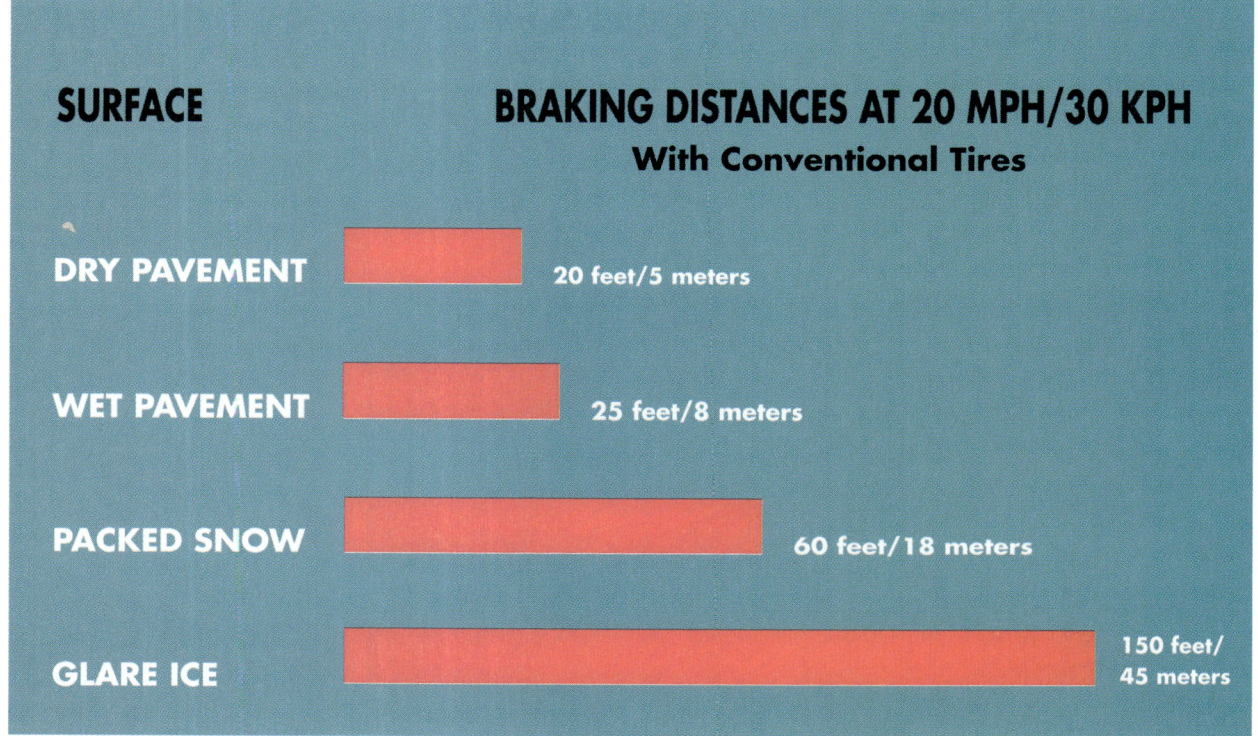

Figure 15–2 Braking distance varies dramatically in different weather conditions.

Figure 15–3 Driving through standing water increases the risk of hydroplaning.

a skid. Do not make sudden movements of the steering wheel or turn too sharply. Use lower gears on hills for added control.

Avoid applying your brakes too suddenly or too hard on wet road surfaces. Because wet brakes may "pull" or stop more slowly than dry brakes, plan ahead and brake earlier than normal. Let the car coast to reduce speed when possible.

If the rainfall becomes so heavy that you cannot see with the wiper blades at maximum speed, pull over to a safe area and wait it out. Make sure that you park far away from traffic. Vehicles parked on the side of the road are often struck by drivers who cannot see well in bad weather. Do *not* set the parking brake. In the event of a rear-end collision, the less resistance your car offers to the impacting vehicle the better.

Hydroplaning

Hydroplaning occurs when a thin sheet of water gets between the road surface and a vehicle's tires, causing them to lose contact with the road. The vehicle then begins a skatinglike movement across the road. Hydroplaning can start at speeds as low as 30 miles per hour (50 km/h) and in water little more than $\frac{1}{8}$ inch (3 mm) deep.

factoid

Raindrops are not teardrop-shaped but either spherical or in the shape of a hamburger—round at the top and flat on the bottom.

REALITY CHECK Wet Leaves

Especially in the fall, roads may be covered with leaves that become very slippery when they get wet from rain or fog. Even if the leaves are dry on top, they may still be wet underneath. Wet leaves can reduce the braking distance of your vehicle and even cause you to skid. When you approach a road surface covered with leaves, slow down and try to stay on clear pavement.

factoid

On September 11, 1954, Chicago was deluged with almost 7 inches (18 cm) of rain in 48 hours, causing the Chicago River to flood; a three-day rainstorm in December 1895 left water 12 feet deep (366 cm) in some places in the city.

Highway heroes

While responding to a report that a vehicle had gone off the Interstate 80 overpass into the Rock River, Auxiliary Officer Ty E. Massey of the Colona Police Department was swept away by flood waters and drowned on the evening of May 11, 1993.

When your speed increases, so do the chances of hydroplaning. Your ability to slow, stop, or even steer your vehicle is greatly reduced. Once you begin to hydroplane, any sudden jerking of the wheel or even a strong gust of wind can send you into an uncontrollable skid that can result in a crash. If you do begin to hydroplane, take your foot off the gas pedal. *Resist your instinct to use the brakes.* Keep your steering wheel straight and let the car's momentum ease down until the tires grip the road again and you regain control. If you begin to skid while hydroplaning, turn the steering wheel in the direction that the rear of the car is moving to regain control.

Before going out in rainy weather, check your tires. Tires with deep, open treads allow water to escape and help prevent hydroplaning at moderate speeds. Tires with worn tread and those that are underinflated have less grip on the road surface, increasing your chance of hydroplaning. To properly handle skids, it is better to have tire pressure on the high side, rather than the low side, of the manufacturer's specifications.

Be alert to warning signs of standing water on the roadway, which can lead to hydroplaning. These include visible reflections on the surface of the water, "dimples" created by rain drops as they hit the water, a "slushing" sound made by your tires, and a "loose" feeling in your steering wheel. If you are driving after a rainstorm has just ended, continue to be cautious. Rain leaves puddles in the road for several hours, sometimes even days, after the showers stop. These can cause your car to hydroplane as well as conceal dangerous potholes.

Deep-Water Driving

If you come to a point on the road with deep water, it is best to use an alternative route, even if it

Figure 15–4 Think twice about driving through deep water.

is far out of your way. If water on the road ahead is running, which is typical of areas prone to "flash floods," the worst possible thing you can do is try to cross. Moving water as shallow as 18 inches (45 cm) can carry away an average car. If there are no alternative routes, find a warm, dry place to wait it out, and try again later when the water level drops.

If you have to go through a stretch of standing deep water, make sure that the water level does not reach the bottom of your car. Consider the difference in ground clearance between other vehicles trying to cross the water and your own vehicle. Use signposts, fire hydrants, and other fixed objects to help gauge how deep the water is. If the water is too deep, you risk getting water into your engine through the carburetor, air filter, or exhaust pipe that can cause it to stall. Water can also short out your electrical system. Do not assume that just because you are in a jeep, truck, or sport utility vehicle you can make it safely across deep water. There may be underwater dips or trenches that you cannot see at all.

To cross safely, wait until other cars ahead of you go through the water and clear it. If they stall, you do not want to be behind them with nowhere to go. Shift to a lower gear to keep the engine turning faster, which reduces your chance of stalling. Proceed slowly. Crossing quickly will splash water and drown your engine. Stay close to the higher center of the road and away from the soft shoulder, where you may have reduced traction. Gently "ride" your brakes—apply pressure to the brake pedal with your left foot and the accelerator with your right foot at the same time—as you go through the water to keep your brake linings in contact with the brake drums. The heat from the brake pressure will prevent your linings from getting soaked and keep them operating more efficiently. After passing through the water, continue to ride your brakes to dry off any water that may have accumulated on your brake pads.

15–2 SNOW AND ICE

Northern Illinois is subject to heavy snowfall in the winter, averaging 40 inches (102 cm) each year. More moderate conditions prevail in the south, which gets around 14 inches (36 cm) of snow annually. Ice storms are more common in the northern and central parts of the state, but they can occur anywhere. Powerful winter storms, like the one in 1967 that belted Chicago with nearly 23 inches (58 cm) of snow and caused $150 million in damage, can create havoc on the roadways and leave motorists stranded.

Snow, especially in combination with gusting winds, can greatly limit your vision. Even a few inches (8 cm) of snow combined with wind can cause snowdrifts, especially in rural areas, that make roads virtually impassable. Road markings are often covered up with accumulations of snow, and traffic signs and signals

Snow and sleet contribute to more than 10,000 traffic collisions in Illinois each year.

Figure 15–5 Make sure that your windows, mirrors, and lights are completely free of snow and ice before going out in winter conditions.

may also be hard to see. A blanket of snow can hide familiar landmarks, making it easier to become lost. If the sun is shining, sunlight can reflect off snow and ice on the ground and add to glare.

"Winterizing" Your Vehicle

Make sure that your vehicle is properly prepared for winter driving by having a mechanic thoroughly check all of the car's systems. Add a deicing solvent to your windshield washer solution—*not* antifreeze—to prevent icing. Carry tire chains properly fitted to your tires and chain repair links, especially if you will be driving in the mountains. Also carry along a tarp to lay on to install the chains, if you want to keep your clothes dry. Keep a bucket or bag of sand, cat litter, or rock salt, or pieces of carpet, in the trunk to help you "rock out" of deep snow.

Check your spare and make sure that you have a flashlight, ice scraper, snow brush, ice pick (for hardened ice under the fender), snow shovel, gloves, flares, jumper cables, dry matches or a lighter, rags or paper towels, a first aid kit, and a large coffee can and candles (to melt snow into drinking water). If you will be driving in remote areas, make sure that you have the ability to call for help, either with a cellular phone or citizen's band (CB) radio. Take along extra clothes and warm blankets, snacks, water, and any needed medication in case you get stuck.

DRIVING TIPS Winter Driving

- Check the current road conditions and weather forecast before starting your trip. Stay off the road when traveler's advisories are issued. Once you are on the road, periodically check the radio for updated changes in the weather.
- Before going on a long trip, let a family member, friend, or co-worker know where you are going, when you plan to get there, and which routes you plan to take. When you arrive at your destination, let that person know you have made it safely.
- Leave early. Expect your trip to take up to twice as long as normal because of delays and roadway closures.
- Keep your gas tank at least half full at all times. Bad storms can cause traffic to be delayed or rerouted away from roadside services that are usually available. Also, the more gas you have in your tank, the longer you will be able to run the heater if you get stranded.
- Use low-beam headlights to increase visibility in winter driving conditions.
- On a long drive through snow, maximize visibility by making sure that you periodically check for and remove ice and snow from your windows, mirrors, roof, and hood, and from under the fenders.
- Keep your window cracked open while driving to prevent any carbon monoxide from seeping into the passenger compartment of your car from a leak in your exhaust system.
- Put an extra car key in your pocket. If you accidentally lock yourself out of your vehicle, you do not want to wait outside in the cold weather for help to arrive.
- If you get trapped in a blizzard, stay in your car unless help is visible within 100 yards (90 m). Hang a distress flag from your radio antenna or, if stranded in a remote area, spread a large, brightly colored cloth over the snow to attract the attention of rescue personnel. Run the engine and heater about 10 minutes each hour to keep warm. If your battery is strong, turn on your emergency flashers and an interior light, especially at night. Because the cold weather reduces your battery's reserve capacity, turn them on only when you hear another vehicle or individual approaching if your battery is low.
- If you ever get stuck in snow, make sure that your exhaust pipe is not blocked. This can cause your engine to stall and carbon monoxide to build up inside your vehicle.
- Never drink alcohol—this increases the *loss* of body heat.

Starting on Snow and Ice

When starting your car on a snow-covered road, first clear a path in front of the wheels by gently driving back and forth. Use a shovel if the snow is deep. With your wheels pointed straight, shift to DRIVE if you have an automatic transmission or SECOND gear if you have a manual transmission. Gently press on the accelerator. Ease out without spinning

DRIVING TIPS — While You Were Sleeping

During winter, take extra precautions in the morning when you first drive your vehicle. You may not know what the weather conditions were like overnight. Even though the roads may appear to be dry, moisture from light rainfall or even fog could have frozen in certain areas.

factoid

It takes about 10 inches (25 cm) of wet snow and up to 50 inches (125 cm) of dry, powdery snow to melt down to 1 inch (2½ cm) of water.

the wheels of your car, which will only get you deeper in the snow. If you are starting on ice, try spreading sand, kitty litter, or other abrasives about 15 feet (4½ m) around the tires, front and back, for better traction. You can also let some air out of the tires to increase contact with the road surface to get more grip.

Driving on Snow and Ice

In extremely cold temperatures, such as around 0°F (−18°C) or lower, fresh, powdery snow offers decent traction. At warmer temperatures, however, snow becomes a watery slush as it melts, creating a very slippery surface. The greatest danger occurs when temperatures approach the freezing point (32°F or 0°C) and slush turns into a slick sheet of ice. If it is raining at the freezing point, or sleeting, ice can form instantly on the road. Ice is one of the most hazardous conditions you can face as a driver. Not only does ice reduce your traction to almost nothing, it is virtually impossible to see. Roads that look wet may in fact be glazed with what is called "black ice."

Because cold air circulates below bridges and overpasses, they tend to freeze before other roadway surfaces. Even if the roads are completely dry, bridges and over-

Figure 15–6 After a snowfall be especially careful at intersections, where packed snow and slush can freeze and form an invisible layer of ice.

passes may still have ice on them. If it is cold enough, water from melting snow can even freeze in the shaded areas of an otherwise dry, sunlit roadway, creating dangerous patches. Intersections and other areas of the roadway with heavy traffic also tend to be icy because the weight of vehicles continually packs down the snow and slush into ice.

During snowstorms, maintenance crews in Illinois often lay down sand and/or salt to melt ice and improve traction on the roadway, especially on bridges, at intersections, and in other dangerous areas. Do not assume, however, that a treated area is free of ice. Remember that a road with *any* ice is extremely dangerous.

To improve your traction in snow, try to stay within the tire tracks created by other vehicles and off the snow and ice between the tracks, on the road shoulders, and in the central part of the road as much as possible. Reduce your speed to no more than half the posted maximum on snow-covered or icy roads, and execute all maneuvers slowly and smoothly to avoid losing traction. If you ever feel your drive wheels start to slip, immediately ease off the gas pedal.

Increase your following distance by at least 10 seconds on winter-slick roads. Stay far enough behind other vehicles that you only have to stop when absolutely necessary because it may be difficult for you to get enough traction to start again. Always give yourself extra room to stop when approaching signs and signals. You could easily get into a crash if you slide into oncoming traffic at an intersection. Use lower gears on hills for added

DERRICK THOMAS

Derrick Thomas was a linebacker for the Kansas City Chiefs, regarded as one of the National Football League's best-ever pass rushers. A first-round draft pick in 1989, he spent his entire eleven-year career in the Red and Gold, racking up $126\frac{1}{2}$ sacks, numerous team career records, and nine Pro Bowl appearances. Off the field, Thomas was active in community charity work. On the afternoon of January 23, 2000, during an icy snowstorm, Thomas was driving on Interstate 435 with two companions to the Kansas City International Airport to fly to a playoff game. Witnesses later reported that he was going more than 70 miles per hour (115 km/h) and weaving in traffic. At around 1:30 PM, near Woodland Avenue, the thirty-three-year-old father of three hit a patch of ice and lost control of his sport utility vehicle, which rolled over several times on the median. Thomas and his close friend, Michael Tellis, were ejected. Tellis was killed while Thomas suffered broken vertebrae in his neck and back that left him paralyzed from the chest down. The third passenger, the only one in the vehicle wearing a safety belt, had only minor injuries. Thomas was transferred to a hospital in Miami, where he grew up, and died on February 8. "A light has gone out," reflected Kansas City Chief Coach Gunther Cunningham. "I hope people do one thing as a result of this accident—buckle up."

factoid

Eighteen storms struck Illinois in the winter of 1981–1982, the second worst of the century, including one in late January that dumped 17 inches (43 cm) of snow on Paris.

Highway heroes

While directing traffic around a crash scene on an Interstate 57 overpass in Dixmoor (a suburb of Chicago), State Trooper Anthony Millison slipped on a patch of ice as he tried to avoid a vehicle that had spun out of control during a severe snow storm and fell over a guardrail nearly 50 feet (15 m) to his death on October 27, 1997.

control. Be careful not to downshift or brake abruptly as you are going downhill or the car may skid. Never use cruise control on snow- or ice-covered roadways.

On a snow-covered or icy road, brake as you approach a curve but coast around it smoothly without touching either the brake pedal or accelerator. Accelerate only *after* straightening your wheels. Braking or accelerating during a turn on snow may cause you to oversteer or understeer, resulting in a skid and possible loss of control of your vehicle.

Snow Tires and Tire Chains

Road and weather conditions can change rapidly during winter. It is therefore important that your car be equipped with proper snow tires or chains. Snow tires have a large, open tread that allows the snow to escape from under the tire to better grip the roadway. Tire

Figure 15–7 Tire chains can greatly improve traction on snow-covered roadways.

chains provide even better traction on snow-covered roads and, with practice, take only about 5 minutes to install per wheel.

When installing chains, pull well off the roadway at a flat, open space, turn off your engine, and make sure that your emergency brake is engaged. Chains should be placed on the drive wheels. When leaving an area when chains are necessary, wait until you can get off the road safely before removing them. Always clean and properly lubricate tire chains after use before packing them away. Never use chains on dry pavement; this will severely *reduce* your traction.

Parking on Snow and Ice

Avoid parking on snow-covered or icy roads. It may be extremely difficult to later get out of the space. If you have to park downhill, always make sure that there is enough room to pull out of the parking space without backing up, which can be very tough with reduced traction. Never set your parking brake after driving in snow unless absolutely necessary. It can freeze and leave you unable to move your vehicle. Instead, leave your car in PARK or, if you have a manual transmission, REVERSE.

Snowplows

After a heavy snowfall, maintenance crews will often clear the road with snowplows, graders, and trucks equipped with special blades. Be patient if you get stuck

Figure 15–8 Watch for oncoming traffic crowding the center line(s) on roads cleared by snowplows.

The Illinois Department of Transportation has more than 1,600 snowplows in its inventory.

behind a snowplow, which can travel 30 to 50 miles per hour (50–80 km/h) slower than other vehicles. Blowing snow from a snowplow can reduce visibility, so increase your following distance—as a rule of thumb, stay two car lengths behind a snowplow for every 10 miles per hour (15 km/h) of your vehicle speed. Do not blindly follow a snowplow. If you are not careful, you might end up on the median, which the snowplow may cross to reverse direction.

Avoid passing snowplows if you can. The width of the blade is often difficult to judge, and, depending on when the other side of the roadway was plowed, available passing room may offer poor traction. Also, snowplows often move sideways to avoid hitting drifts or to cut through a snow pack. If you do have a safe opportunity to pass a snowplow, make sure that you are on a straight section of roadway, you can see clearly ahead, and there is no oncoming traffic. Do not cut back right in front of the snowplow after passing or brake suddenly ahead of one. Remember that snowplows are heavy and need more braking distance than lighter vehicles like your own.

Snowplows often open up only one central lane in each direction of a roadway. Because it is often difficult to see exactly where the road markings are, watch for oncoming traffic that may be traveling closer to the center line(s) than normal. If a snowplow is approaching you from the opposite direction, move as far to the right as possible. If you are uncertain how much room you have, or are near the edge of a cliff or embankment, tap your brakes if other vehicles are behind you, pull over, and activate your emergency flashers until the snowplow has passed.

15–3 OTHER SEVERE WEATHER CONDITIONS

In addition to rain, snow, and ice, you may encounter other severe weather conditions on the roadway.

driving tips

Practice installing chains in your driveway before you have to do it "for real" to make sure that they are the proper size for your tires and in good working order.

Fog and Smoke

Fog is most common early in the morning, late at night, at high altitudes, and near bodies of water. You may not think of fog as severe weather, but fog can be one of the most dangerous driving environments you will encounter. Some of the worst crashes in history have occurred in fog. Fog can reduce your visibility drastically, sometimes to within only a few feet (1 m) of your car. If you cannot see other vehicles or pedestrians, remember that they cannot see you either.

Greatly reduce your speed when approaching a patch of fog. Maintain an extra space cushion around your vehicle. Fog can alter your depth perception, and vehicles may be closer to you than you realize. Use your low-beam headlights to increase visibility. Fog reflects light, so using your high-beam headlights can actually *reduce* your view of the surrounding area. If the fog becomes so thick that you cannot see at all, pull over as far away from the roadway as you can and wait for it to clear. Make sure that your headlights are turned off. Drivers approaching from behind you might use your taillights as a steering guide and mistakenly rear-end your vehicle.

Forest fires, derailed freight trains with flammable liquids or gases, and industrial accidents can produce blinding levels of smoke just as dangerous as heavy fog. If you have to drive through smoke, exercise the same preventive steps you would in fog. As an added precaution, make sure that your windows and vents are closed to avoid taking in dangerous fumes or thick smoke, especially if you have asthma or other breathing problems. Check the radio for news of a serious health risk, road closures, or traffic problems associated with the fire or accident. If you get trapped in an area with dangerous levels of smoke, pull off the road and seek shelter indoors. If you start to feel sick, or think you may have been exposed to toxic smoke or gas, drive to a hospital or emergency

Figure 15–9 Always use your low-beam headlights when driving in fog.

DRIVING TIPS — When "Hazard Lights" Are Hazardous

If you have to pull over on the side of a roadway in heavy rain, thick fog, smoke, or a severe storm, do *not* put on your emergency flashers. This can mislead other motorists into running off the highway and into your vehicle. As vehicles approach you from the rear, drivers might mistake your emergency lights for fleeting glimpses of your regular brake lights. In conditions of low visibility, drivers often follow the rear taillights of vehicles ahead of them to gauge their own lane position. If you are stopped on the side of the road, these approaching drivers could easily drift off the roadway and rear-end your vehicle.

clinic immediately and get yourself checked.

Dust Storms

Dust storms are common in agricultural areas. Do not attempt to drive through such a storm, even at a reduced speed. Other drivers behind you may be driving much faster than you and could rear-end you. Also, your visibility can be reduced to zero. While the storm may not be so bad where you are, dust storms often change very quickly in intensity over an area and you could get in real trouble down the road. If you see a thick cloud of dust ahead, gradually slow down and activate

With an average wind speed of just over 10 miles per hour (1.6 km/h), Chicago ranks only sixteenth among the nation's windiest cities.

USING SAFE — Driving in the Wind

Severe wind can come from any direction, making steering difficult. If you are not alert, crosswinds can literally force you into an adjoining lane, off the road, or into the path of oncoming traffic. Use the SAFE process. *Scan* ahead for warning signs in areas prone to strong, gusting winds. Observe how the wind is affecting other vehicles. Look at trees and foliage to try to gauge the direction and intensity of the wind. *Assess* how the wind may affect your own vehicle. Will tailwinds cause your vehicle to build up speed, increasing your stopping distance? Are you facing a head wind, making it necessary to accelerate harder to maintain a steady speed or pass? Are you in a high-profile vehicle such as a van, recreational vehicle, or pickup with a camper shell, making handling more difficult? Are the roads slippery, making gusting winds even more dangerous? Are you passing over or under a bridge, emerging from a tunnel, or driving between hills or tall buildings that can create a funneling effect? *Find* an "out" in case buffeting winds push you out of your lane and *execute* the appropriate defensive driving techniques to maximize control of your vehicle. Increase your space cushion, especially around vehicles that may have trouble handling the wind, such as motorcycles and cars pulling trailers. Slow down if you feel the wind "pulling" your vehicle, and firmly grip the wheel. If you are on a multilane road, stay in the right lane. If you are on a two-lane highway, move to the right side of your lane when facing oncoming traffic or when being passed by trucks, buses, or other large vehicles that can magnify the effects of wind.

Figure 15–10

Hailstones can fall at speeds up to 110 miles per hour (175 km/h).

factoid

Illinois lightning strike victims include championship golfer Lee Trevino, who survived, and Norma Jean, a 6,500-pound (2,950-kg) circus elephant, who did not.

your windshield wipers and low-beam headlights. If visibility is severely reduced, pull off the road as far as you can and stop the car. Roll up the windows, turn off your headlights, and stay where you are until the storm passes.

Hail

Hail usually accompanies spring and summer thunderstorms. Heavy downpours of pea-size hail can dramatically reduce visibility and layer the road with slippery balls of ice. Larger golf ball–size hail can crack windshields and dent car bodies beyond repair. If the weather forecast calls for a severe storm, and you value your safety and the appearance of your vehicle, find a covered place to park your car and keep it there until the storm passes.

Lightning Storms

During a lightning storm, get indoors or stay in your car. Stay away from open areas and high objects such as tall poles or trees that can attract lightning. Avoid any contact with water, metal, or electrical devices such as a radio or cellular phone. If you happen to be in the open far away from your vehicle and cannot get to shelter, head for the lowest area possible and get on the ground.

Tornadoes and Hurricanes

Although the Prairie State falls outside of "Tornado Alley," it ranks seventh among all states in the number of reported twisters. On average, about thirty tornadoes touch down each year in Illinois, sometimes with devastating results. On March 18, 1925, the Great Tri-State Tornado, the most destructive in United States history, killed 488 people as it cut a swath across the southern portion of the state. In 1974, more than 100 tornadoes were reported in Illinois, the busiest on record. Spring is tornado season, but twisters can occur at anytime during the year. Most are reported in the afternoon and early evening, with nearly half touching down between 3 and 7 PM.

What makes a tornado so deadly is the speed, unexpectedness, and deadly force with which it strikes. Should you ever be driving and see a tornado, pull over

AUTO ACCESSORIES *Weather Radio*

To get the most complete and up-to-date weather information while on the road, consider purchasing a weather radio. Available at most electronics stores for a moderate price, weather radios are designed to pick up high-frequency 24-hour broadcasts over National Oceanic and Atmospheric Weather Administration (NOAA) Weather Radio from the National Weather Service. Forecasts, current weather conditions, radar reports, and other information important to travelers are repeated every 3 to 6 minutes.

immediately, get out of your vehicle, and find shelter in a basement or low-lying area such as a bridge underpass. Never try to "chase" a tornado to take photographs or to videotape it. Tornadoes can quickly change directions, and before you know it, you are trapped.

15–4 COLD AND HOT WEATHER

Illinois has a varied climate. The state has cold winters, with average daily temperatures at or below the freezing point in December and January while February temperatures remain below 40°F (4°C). Summers are humid and warm, with normal temperatures in the seventies and eighties. During "heat waves," however, the thermometer can climb over the century mark, straining the limits of both you and your car. Vehicles are designed to operate in both cold and hot weather, but extreme temperatures create special problems.

Cold Weather

Air is denser in cold weather, resulting in a leaner mixture of fuel and air in your engine. This means that in colder temperatures your engine has to work harder to vaporize fuel. If your car has any vacuum leaks, such as loose or cracked vacuum hoses or a loose carburetor, sticky valves, faulty seals around an injector, or a faulty thermostat, the air–fuel mixture can become even leaner. If your engine is performing poorly in cold weather but normally in warm weather, have a mechanic check your engine.

If your engine does not have enough antifreeze or has the wrong kind of antifreeze, the radiator can freeze in cold weather. This will block the coolant from getting to your engine, causing it to overheat. Make sure that you follow the directions in your owner's manual regarding the proper maintenance of your cooling system.

Starting in Cold Weather

Cold weather puts added strain on your engine, especially when you first start it. The battery has less power, and oil, transmission fluid, and other lubricants tend to thicken. In winter, keep your battery charged on a regular basis. You do not want your battery

factoid

On February 3, 1996, the mercury sank to −35°C (−37°C) in the northwestern Illinois town of Elizabeth, the coldest temperature ever recorded in the state.

DRIVING TIPS   **"Locked" Out by the Cold**

Water that splashes onto your door or trunk locks can freeze over and prevent you from getting into your vehicle. If the ice is not too thick, pouring hot water on the lock may do the trick. If you have a pair of fire-retardant gloves to protect your hands, you can also try heating your car key over a flame and then inserting it into the lock. You can also spray or squirt a chemical deicer or isopropyl alcohol into the lock. To prevent locks from freezing during winter, treat them in the fall with liquid graphite.

factoid

Tall tales of chickens being frozen in their tracks and men being frozen to their saddles were inspired by the famous "Sudden Change" of 1836, when temperatures dropped in central Illinois from 40°F (4°C) to 0°F (−18°C) in mere minutes.

to fail when you are out in severe weather and far from a service station. To avoid unnecessary strain on the electrical system, make sure that the radio and any other electrical accessories are turned off before attempting to start the engine.

Depending on what type of battery you have and how old it is, your car can only sit in cold weather for a limited time. If possible, keep your vehicle parked in a closed garage. If you are not going to use your car for a long time, have someone else periodically start it and run the engine for about 10 minutes. If the car is going to be parked in extremely cold weather, attach an electric engine-block heater to the engine to keep it warm.

Frosted-up Windows

Cold temperatures can cause your windows to ice up, reducing your visibility. Scrape the outside of the windows thoroughly before driving. Once the engine is warmed up, use the defroster to completely clear the frost from the inside of the windows. Trying to drive with limited visibility because you are too impatient to wait for the windows to defrost is just asking for a collision. Keep an eye out for other drivers who may have a hard time seeing you through *their* frosted-up windows.

During winter, moisture from your breath, perspiration, snow on your shoes, or even a leak in your heater core can cause your windows to continue fogging up even with the defroster on. The more people there are in the vehicle, the worse the problem can get. Keep a window open and the ventilation system set on "fresh air" to help eject moist air from your car. You can also use a quick blast from your air conditioner, which can actually clear your view faster than air from your defroster. Although defroster air is warm, air from your air conditioner is very dry and evaporates the condensation much more effectively.

Figure 15–11 Wait until your front and rear windows are completely defrosted before you start driving.

AUTO ACCESSORIES *Heat Shields*

To prevent heat from building up inside your vehicle, and to protect vinyl and leather finishes from the sun's damaging rays, consider purchasing a heat shield. These accessories, which range from inexpensive cardboard panels to custom-fit designs finished in felt and laminated with silver reflective fabric, fold up like an accordion for easy storage. When parking your car on a hot day, simply place the heat shield on the dashboard, reflective side up, and fold down your sun visors to hold it in place.

Hot Weather

Just as the cold temperatures of winter put added stress on your vehicle, so do the hot temperatures of summer. In general, heat causes liquids to evaporate. Motor oil, automatic transmission fluid, and power-steering fluid are all crucial to the safe operation of your vehicle and should be checked regularly during hot weather conditions. Heat can also shorten the life of a car's battery and cause belts and hoses to crack and tear. Before summer driving season, make sure that you take proper preventive maintenance measures as described in your vehicle owner's manual.

Vapor Lock

In very hot weather, gasoline can boil and turn to vapor. This may cause some engines to "lock," or stop running. If this happens to you, turn off your engine and let it cool down. Once the vaporized fuel in your fuel line returns to its liquid form, you should be able to restart your engine.

Vapor lock is almost a thing of the past. Today's cars have fuel pumps in the fuel tank rather than on the engine, which "pushes" rather than "sucks" gasoline into the engine. This results in a more efficient flow of pressurized, cool fuel in the fuel lines, making vapor lock almost impossible. Find out what kind of fuel system your vehicle has. If you have an older car, be aware of the risk of vapor lock in hot weather.

Overheating

Driving for long periods in hot weather, especially in heavy traffic, going up steep inclines, or

Figure 15–12 Wait for your radiator to cool down before adding water.

factoid

Driving with the air conditioner on reduces gas mileage by approximately 2.5 miles per gallon (6 km/l).

factoid

Although many portions of Illinois have experienced temperatures above 110°F (43°C), East St. Louis holds the state record for the hottest, 117°F (47°C) on July 14, 1954.

Where am I?

Figure 15–13 There was no air conditioning in this building when Abraham Lincoln launched his political career here in the 1830s.

using your air conditioner at full blast on a very hot day can all cause your engine to overheat. The risk of damage to your engine from overheating is even greater if you have not maintained the proper coolant level or used the wrong type of coolant.

If your engine starts to overheat, move to a safe place off the roadway and stop your vehicle. Turn off your air conditioner if it is on. Open your windows. Put your heater on the maximum setting to draw heat away from the engine. Shift the transmission into NEUTRAL, and give the engine a little gas by pressing on the accelerator. If this fails to work, turn off the engine and get out of your vehicle. Stand away from the hot engine until steam has stopped escaping. Cover your hand with a thick towel or a heat-resistant glove and open the hood to allow more heat to escape.

In an emergency situation, you can add water to the radiator to serve as a temporary coolant. With a thick towel or heat-resistant glove, give the radiator cap a one-quarter turn. Keep your face turned away to prevent accidental scalding. Retreat several steps from the vehicle while the pressure under the cap escapes. Once the pressure is released, turn the radiator cap and remove it. Add a measure of water into the radiator after it is *completely* cooled down. This can take as long as an hour, so be patient. Putting cold water in your ra-

DRIVING TIPS *Summer Driving*

- If you are going on a long trip or are driving for long periods of time, check the radiator fluid level each time you stop for gas. If you do not have a sealed battery, check the battery fluid at least once a day.
- Check tire pressure daily. Do not "bleed" air from hot tires. Instead, prevent an excess buildup of pressure in your tires by driving your vehicle only for moderate time periods. If you think the car is riding hard on the tires, pull over and let the tires cool off.
- Do not make your engine work too hard by driving fast.
- Avoid driving during the hottest period of the day.
- Carry at least two extra bottles of water in your car—one for you, to avoid dehydration, and one for the radiator.
- Keep good-quality sunglasses in the car to help reduce sun glare. It is also a good idea to keep a hat in your car. If you have to walk a long distance to get help, you will need protection from the sun.
- Never enclose a child or pet in a parked vehicle when it is hot outside for even a few minutes. Temperatures can rise rapidly and be fatal!
- Do not drive through or park in areas with dry grass, which can be ignited by the intense heat generated by a vehicle's catalytic converter.

diator right after the engine has overheated can cause the radiator to crack.

Start the engine again and return to the radiator. As water circulates through the engine, the fluid level in the radiator will drop. Add water as necessary. If it appears that the engine is running again without overheating, replace the radiator cap tightly in place and resume your travel. If your engine starts to overheat again after you add water, do not drive the vehicle. Call a tow truck or your automobile club for roadside assistance.

Get your car checked out at as soon as possible in case you have a leak in your coolant system or another engine problem. Have the radiator pressure-tested and flushed, if necessary. Replenish the coolant in the container if the level is low. Replace the thermostat in case it was damaged.

Wild Wheels

Figure 15–14 Which auto manufacturer made this "dune buggy" for the moon, where temperatures range from −184°F (−120°C) to 320°F (160°C).

WHO'S AT FAULT?

1. Driver 1 was proceeding northbound in the far-left lane of the Tri-State Tollway in Chicago between Irving Park Road and 95th Street about 65 miles per hour (100 km/h) in heavy rain. Driver 2, worried about driving on a wet surface, was traveling about 25 miles per hour (40 km/h). The posted speed limit was 55 miles per hour (90 km/h). Because of the great difference in their speeds, Driver 1 came upon Driver 2 very rapidly and attempted a last-second pass that caused his car to begin hydroplaning. A few seconds later, Driver 1 spun 180 degrees around and struck a barrier on the center divider. Driver 2 was not involved in the crash. **Who's at fault?**

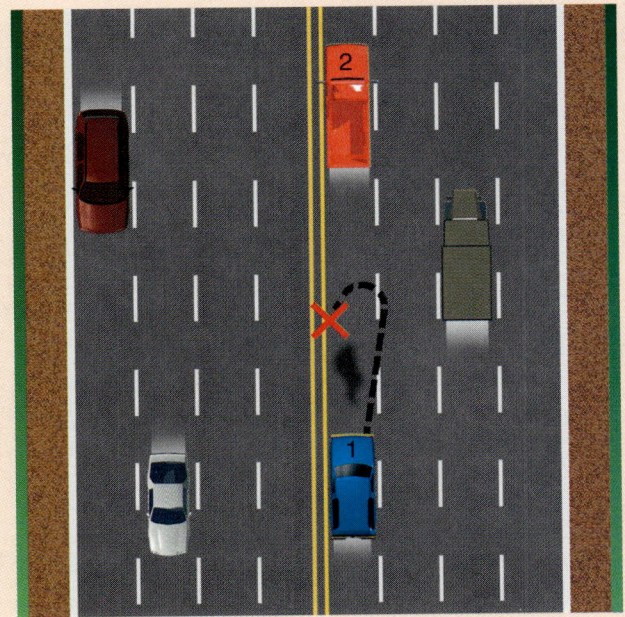

2. Driver 1 was proceeding eastbound about 45 miles per hour (70 km/h) as she rounded a curve on Illinois 16 near Windsor. In the middle of the curve she came upon Driver 2, who was proceeding about 35 miles per hour (55 km/h). The posted maximum speed limit for the turn was 55 miles per hour (90 km/h). As Driver 1 attempted to apply her brakes to avoid hitting Driver 2, she encountered a patch of ice on the road. The brakes locked and Driver 1 slid toward the edge of the eastbound lane until her vehicle left the road and overturned in a ditch. Driver 2 was not involved in the crash. **Who's at fault?**

YOUR TURN

15–1 Rain

1. What are some of the dangers of driving in rain?
2. What is hydroplaning and how can you prevent it?
3. What should you do if you encounter deep water on the road?

15–2 Snow and Ice

4. How should you prepare for driving in winter conditions?
5. What are some of the dangers of snow-covered and icy roads?

15–3 Other Severe Weather Conditions

6. What precautions should you take when driving in fog?
7. What should you do if you are caught in a severe storm?

15–4 Cold and Hot Weather

8. What are some of the problems caused by extremely cold temperatures?
9. What should you do if your engine overheats?

SELF-TEST

Multiple Choice

1. Hydroplaning can occur at speeds as low as:
    a. 45 mph (70 km/h).
    b. 30 mph (50 km/h).
    c. 15 mph (25 km/h).
    d. 35 mph (55 km/h).
2. The fastest way to clear moisture from your windshield is by using:
    a. the air conditioner.
    b. the defroster.
    c. the heater.
    d. washer fluid.
3. On a winter-slick road, increase your following distance by at least _____ seconds.
    a. 3
    b. 4
    c. 5
    d. 10
4. If your car engine starts to overheat, the first thing you should do is:
    a. drive to the nearest mechanic.
    b. turn off the air conditioner and turn on the heater.
    c. pop open the hood slightly to let some heat escape.
    d. pour cold water into the radiator.
5. A car traveling at 20 miles per hour (30 km/h) will take _____ to stop on ice with conventional tires.
    a. 75 feet (25 m)
    b. 120 feet (35 m)
    c. 150 feet (45 m)
    d. 175 feet (55 m)

Sentence Completion

1. Using your high beams can actually reduce your view of the surrounding area when driving in _____.
2. Use lower gears on _____ to maintain better control on wet or icy roads.
3. Do not set your _____ after driving in snow because it can freeze and leave you unable to move your vehicle.
4. The greatest danger of reduced traction on winter-slick roads occurs when temperatures approach the _____.
5. _____ occurs when a thin layer of water gets between the road surface and your vehicle's tires, causing them to lose contact with the road.

Matching

Match the concepts in Column A with examples of the concepts in Column B.

Column A
1. __ Way to improve traction on wet roads
2. __ Safe speed for driving in rain
3. __ Indication of hydroplaning
4. __ Crossing deep water
5. __ Response to gusting winds
6. __ Winter driving tip
7. __ Summer driving tip
8. __ Desert driving tip

Column B
a. Ride your brakes
b. 25% below posted limit
c. Slow down and keep firm grip on wheel
d. Carry extra water for radiator
e. Drive in tracks of vehicle ahead of you
f. Check tire pressure daily
g. Allow extra traveling time
h. "Loose" feeling in the steering wheel

Short Answer

1. How would you approach a curve on a snow-covered or icy roadway?
2. Why is it dangerous to use hazard lights if you are parked by the roadway in bad weather?
3. What are the advantages of increasing your space cushion on slick roads?
4. Why should you avoid parking on snow-covered or icy roads on hills?
5. Why is vapor lock rare in modern cars?

Critical Thinking

1. You are in the right lane of a four-lane highway behind a large tractor-trailer truck driving at the speed limit. Although it is no longer raining, there are large puddles all over the roadway. "Splash and spray" from the truck is reducing your visibility. What should you do?
2. You parked on an uphill stretch of road, only to discover in the morning that it snowed overnight. What should you do?

PROJECTS

1. Go to a local auto supply store and see what products they offer for driving in bad weather. Write down a list of items you would include in an emergency preparedness kit for winter driving and desert driving.
2. Contact the Illinois Department of Transportation's Division of Traffic Safety to determine the number of collisions and traffic-related fatalities that occurred each month of the past year. Which month had the most crashes? The fewest? Which month had the highest percentage of traffic-related deaths? Discuss the possible reasons for these statistics in class.

16 chapter

Collisions

Most collisions are easily preventable. If drivers paid more attention to the road and made better choices before an impending crash, nearly 1,500 fatalities and more than 65,000 injuries would be prevented each year in Illinois. By understanding what causes collisions and how to avoid them, you will be a safer driver. If you are ever involved in a crash, it is important for you to know what to do and what your responsibilities are at the scene. Your own life, and the lives of others, may depend on it.

Chapter Objectives

Upon completion of this chapter, you should be able to:

16–1 Collisions
1. Understand why a vehicle crash is a "collision" rather than an "accident."
2. Describe the different types of collisions.

16–2 What to Do at the Scene of a Crash
3. Understand what responsibilities you have at the scene of a collision in which you are involved.
4. List the steps you should take when giving aid at a crash scene.

16–3 Preventing Collisions
5. Describe the four choices you have to avoid an impending crash.
6. Understand what you should do if a collision is unavoidable.

16–4 Vehicle Restraint Systems
7. Describe the purpose of safety belts and air bags.
8. Understand the proper way to wear safety belts.

KEY TERMS

solo collision	rollover	safety belts
head-on collision	"hit and run"	passive restraint
side-impact collision	"brake and hold"	air bags
rear-end collision	head restraints	child safety seat
chain reaction		

16–1 COLLISIONS

On average, about 168,000 traffic collisions occur each year in Illinois. When a collision occurs, *everyone* pays. Besides the direct costs to those involved in a crash of medical care, legal fees, funeral costs, and property damage, there are indirect costs to society in the form of higher auto and medical insurance premiums, traffic delays, crowded court dockets, and loss of wages and worker productivity. Being involved in a crash or having a friend or loved one hurt or killed in a collision has emotional costs that are impossible to calculate.

Causes of Collisions

Collision is a more accurate description of a vehicle crash than *accident*. The word *accident* implies that nothing could be done to prevent the crash. In fact, almost all collisions are caused by driver error. The most common causes of crashes include speeding, failure to give the right-of-way, driving under the influence of alcohol or other drugs, failure to obey traffic controls, improper turning, and following another vehicle too closely. If you obey the law, manage your speed, and remain aware of the driving environment around you at all times, you should be able to avoid getting into a crash in almost all circumstances.

Types of Collisions

All crashes are really *two* separate collisions. The first is the collision of the vehicle with some obstacle. The second collision is that of the human occupants with the interior of the vehicle. It is the second collision that often causes injury

The economic cost of one traffic fatality is nearly $800,000.

Figure 16–1 What are the direct and indirect costs of this collision?

About 85% of the population will be involved in a serious crash at some point in their lives.

factoid

A 150-pound (68-kg) person involved in a crash at 30 miles per hour (50 km/h) exerts a force of more than 2 tons (1,800 kg).

and death. Injuries may be sustained because of the jarring effect of a collision—for example, whiplash in a rear-end crash—or the collapse of a vehicle's protective shell, especially in high-speed collisions or crashes involving a physical mismatch between the vehicle and the obstacle. However, most people are hurt when *they* impact the wheel, dashboard, or windshield of their own vehicle after it has been stopped by the force of the impact.

There are many types of collisions. A **solo collision** involves only one vehicle, such as when a driver drives into a ditch because he or she fell asleep at the wheel or failed to take a curve properly. These collisions can be deadly if the crash involves a hard obstacle such as a concrete barrier, bridge support, tree, utility pole, or large animal or if the vehicle falls off a cliff or into a deep ditch.

A **head-on collision** occurs when two vehicles collide front to front. This type of collision results in the most serious injuries and the highest rate of death because the force of impact combines the kinetic energy of both vehicles. A **side-impact collision** occurs when two vehicles collide side to side, side to front, or side to rear. The worst injuries in this type of collision result when one vehicle directly impacts the passenger compartment of another. A **rear-end collision** occurs when one vehicle hits another from behind. This type of crash results in the most damage when the vehicle impacted is not moving.

A **chain reaction** is a series of collisions involving vehicles that impact one after another. In most cases, chain reactions involve rear-end collisions. These are most common on highways and freeways where cars are driving close together at high speeds or in conditions of low visibility.

Some collisions can result in a **rollover,** in which a vehicle is

EDWARD GIVENS

Edward Givens' lifelong dream was to be an astronaut. As an Air Force pilot and instructor, he had logged more than 3,500 hours of flight time. After completing the USAF Aerospace Research Pilot Program, the thirty-six-year-old Texas native was selected by NASA in 1966 to be one of nineteen American astronauts to train for manned space flight. Unfortunately, Givens never got his chance to walk on the moon. On June 6, 1967, a fatigued Givens was returning from a training mission at the Manned Spacecraft Center near Houston, Texas, shortly after midnight. He fell asleep at the wheel, plunging off the road into a ditch, killing himself and injuring two passengers. As fellow Apollo crewmen carried his flag-draped casket to its final resting place, Givens was mourned by his wife, his three children, his parents, and several hundred people whose lives he had touched. Givens' untimely death was all the more tragic because his "accident" was entirely preventable.

Figure 16–2 Rear-end collisions often lead to chain reactions on crowded roadways.

flipped upside down or literally rolls over one or more times. Rollovers can be very serious because the occupants may impact with all sides of the vehicle's interior, often resulting in head and neck injuries, or be ejected. If you are thinking about buying a sport utility vehicle or other vehicle with a high center of gravity, make sure that it has some form of rollover protection.

16–2 WHAT TO DO AT THE SCENE OF A CRASH

If you are involved in a crash, you have certain legal and moral responsibilities. You must stop, regardless of the extent of damage. You are also obligated to exchange information with others involved in a crash, to notify

Figure 16–3 When rollovers occur, there is a risk of the vehicle's roof collapsing or being crushed on impact with the ground.

REALITY CHECK — "Set-up" Collisions

Some collisions are staged by criminals who seek to defraud insurance companies for fake or exaggerated injuries and property damage. In the "swoop and squat" scheme, for example, the driver of a car you are following (the "squat" car) suddenly slams on the brakes when another vehicle (the "swoop" car) makes a sudden lane change directly ahead, giving the driver of the "swoop" car no time to stop before rear-ending the "squat" car. Pay close attention to drivers who drive slowly for no apparent reason, act rashly by trying to cut in quickly ahead of you, or otherwise behave out of the ordinary. The best defense against these criminals is to maintain a large space cushion around your vehicle at all times. Do not fall into the trap of tailgating someone who *wants* to be tailgated!

Where am I?

Figure 16–4 You would not want to collide with this vehicle, one of many tanks at this museum dedicated to the history of the "Big Red One."

owners of property damage you may cause with your vehicle, and to report the incident to the proper authorities. As a citizen, you have a duty to give aid to any person injured in a crash, or make arrangements for the person to receive medical treatment, as well as to warn other drivers by marking the scene.

Stopping at the Scene

The first duty you have if you are involved in a collision is to stop immediately, and as close to the site of the crash as safety will allow. This means getting your vehicle out of the traffic lanes and away from any debris or wreckage left by the collision that puts you or others at the scene at risk of another crash in the wake of your own. You need not have suffered injury or damage to your own vehicle to be "involved" in a crash. You must stop if you directly *or* indirectly cause a collision to occur.

If you fail to stop at the scene of a crash in which you are involved, you could be prosecuted for the crime of **"hit and run."** This is an extremely serious offense that carries severe penalties if you are convicted. Failure to stop at the scene of a property-damage-only collision in which you are involved is a misdemeanor punishable by up to one year in jail and up to $2,500 in fines. If the damage is more than $1,000, your driver's license will also be suspended for a year.

If there are injuries or deaths and law enforcement does not arrive, you are permitted to leave the scene after stopping provided that you report the incident to a local police or sheriff's station *within 1 hour*. If you are injured and taken to a medical facility before law enforcement arrives, you must report the collision within 1 hour of your release if an officer does not appear at the clinic or hospital in person to take a report.

Failure to either stop at the scene of a collision involving injuries or fatalities or to properly notify the authorities of such a collision is a felony. If someone is hurt, you face up to three years in prison and up to $25,000 in fines as well as the revocation of your license for a year. If someone is killed, your time in prison could be extended to up to fourteen years.

As soon as you have pulled off the road to a safe spot and come to a stop, turn on your car's hazard lights. Watch for traffic from both directions before you exit your vehicle. Locate any crash victims by looking under and through all wreckage and in nearby grass or bushes. Account for any persons who may have been thrown from their vehicle by talking to those who are uninjured or only slightly injured. If victims cannot turn off their ignition, do it yourself. Be alert to fire, downed power or telephone lines, flammable materials that may be leaking, and other hazards. Make sure any lighted cigarettes or cigars are extinguished. Use extreme caution if you choose to approach a burning vehicle to rescue occupants. If you choose to do so, put out the fire using a fire extinguisher, dirt, or heavy fire-retardant blanket.

If you are not involved in a crash, you are not legally required to stop at a crash scene. You should consider stopping, if it is practical and safe to do so, if you are a good witness to a crash. Never stop, however, just to look, because by doing so you may block the route of law enforcement, fire and paramedic personnel, or tow trucks. You will also contribute to traffic congestion.

Giving Aid

Many people who die from injuries sustained in a collision might otherwise live if they were to get immediate care from others. It may take anywhere from a few minutes to an hour or longer for emergency medical help to arrive, depending on the site of the crash. If you are one of the first to arrive at the scene of a crash or are yourself involved in a crash and are uninjured, you can do a lot to prevent further injuries and save lives. However, provide care to the injured only if you feel confident that you know what you are doing. If you are nervous, uncertain, or scared, do not attempt to render first aid. Ask others at the scene as to their knowledge of first aid. If someone else is more qualified or capable, let him or her take charge.

On the other hand, do not hesitate to help someone in need if you are the only person available or no one else steps forward. In Illinois, you are required by law to provide "reasonable" assistance to any person injured in a crash in which you are involved, including making arrangements for the person to be taken to a physician or hospital for medical treatment. Failure to do so is a misdemeanor with penalties similar to those for "hit and run" involving only property damage.

Do not move crash victims or allow them to move unless their location puts them in immediate life or death danger. Hazards that would justify moving an injured person include being in the path

Three out of four drivers killed in traffic collisions in Illinois are male.

highway heroes

While sweeping debris away from a northbound lane on the Kennedy Expressway near the Madison Street overpass in Chicago in the early morning hours of May 24, 1996, construction worker Tonya Greer was killed after being struck by two cars and a semitrailer truck, all of which fled the scene.

> **REALITY CHECK** **Protect Yourself While Helping Others**
>
> Because of the potential of transmitting disease via the mouth, especially when other injuries may increase the chances of blood being in the mouth, it is prudent to wear some form of mouth-to-mouth resuscitation barrier that prevents you from making actual contact with the mouth of the injured. Have several of these barriers in your car at all times. To reduce the risk of exposure to acquired immunodeficiency syndrome (AIDS), hepatitis, and other blood-borne diseases, use latex gloves, waterproof material such as a plastic bag, or clean clothing to stop bleeding in other areas of the body. You can also try applying pressure with the victim's own hand.

of oncoming traffic, being submerged in water, being nearby downed power lines, or being near fire. Moving an injured person could make the injury much worse, especially in the case of head or spinal trauma. These types of injuries are often difficult to diagnose because there may be no blood or other evidence of a wound. If you have to move an injured person, make sure that his or her head and neck are in alignment, then delicately drag the person backward by holding on to the person's clothes or armpits. Do not drag the person sideways. Do not attempt to remove an injured motorcyclist's helmet.

Assess who needs help the most. Check for any crash victims who may be unconscious. If a victim is neither moving nor talking, ask the person if he or she is okay. If the victim does not answer, his or her airway may be blocked. Open the injured person's airway by slightly lifting the chin. Do *not* tilt the head backward. Place your ear next to the person's mouth and nose to feel and listen for any sign of breath. Look at the person's chest to see if it rises and falls, indicating breathing. If an injured person is not breathing, begin rescue breathing as soon as possible. Only a few minutes of nonbreathing can result in permanent damage to the brain or death.

Once you have made sure that the injured are conscious and breathing, check for bleeding. Apply a sterile bandage, if available, using additional dressings over the existing bandage as needed. If blood is moving rapidly from a specific wound, apply pressure with your fingertips over the bleeding area. Carefully elevate the wound as necessary to reduce bleeding, moving the injured person's body as little as possible. Encourage walking injured persons to sit or lie down; if their lower face or jaw is bleeding, have them lie on their side.

Talk to the victim and make him or her as comfortable as you can. Cover the injured with blankets or clothing to minimize shock, and loosen tight clothing. Do not give any food or liquid to the injured. Once you have begun to administer first aid, do not

One out of four traffic fatalities in Illinois is a passenger.

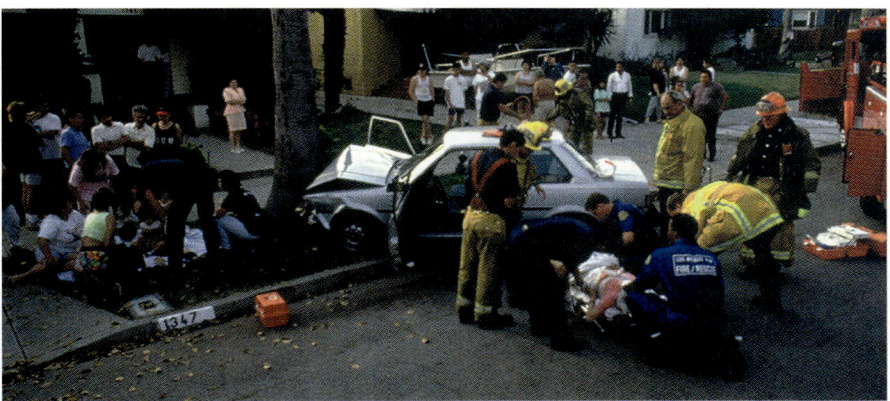

Figure 16-5 Stay with victims of a collision until professional help arrives.

Highway heroes

While standing on the shoulder of I-55 near Harlem Avenue in Chicago clearing wreckage from a fatal crash that had occurred an hour earlier, IDOT "Minuteman" Santo "Sonny" Danno was struck and killed by a motorist, who also injured a state trooper and tow-truck operator in the crash.

abandon a victim unless it is to attend another injured person in more need of assistance. Never take any injured to the hospital by yourself or in your own vehicle. Wait for professionals who know how best to transport people with different types of injuries.

Warning Others

After attending to the injured, mark the crash scene with flares, reflective triangles, or other warning devices. Depending on the speed of traffic, they should be placed immediately behind and up to about 150 feet (45 m), half the length of a football field, behind the vehicles involved in the collision. Generally, the higher the speed of traffic, the farther back the warning devices should go. If the crash occurred near the crest of a hill or in a sharp curve, place warning devices where approaching drivers who do not have a good view of the crash site will be able to see them well in advance.

Figure 16-6 Place flares behind disabled vehicles to warn other drivers.

Never light a flare anywhere near gasoline, diesel fuel, motor oil, or other flammable materials that might have been discharged on the road following a crash. If you have no flares or marking devices, raise the hood of your vehicle to draw more attention to the scene. Tie a piece of cloth or an article of clothing to the radio antenna, a door handle on the side of the vehicle facing traffic, or the corners of the trunk or hood. If

factoid

Motorola, a leading manufacturer of personal communication devices, is headquartered in Schaumberg.

factoid

Three out of four traffic collisions in Illinois involve only property damage.

you have help to spare, have a person warn drivers approaching from the rear with brightly colored clothing during the day or a flashlight at night. Make sure he or she is well to the side of the road and can be seen easily by other drivers.

Calling for Help

If you have a cellular phone, dial 911 or, if you are on a freeway in the Chicago area, *999, to obtain help for any injured parties, even if the injury appears to be slight. If you have a citizen's band (CB) radio, you can also dial channel 9 for emergency assistance. If you do not have your own phone or radio, try to find a public pay phone or ask to use the phone of a nearby business or residence. Some freeways have emergency roadside phones that you can use.

Be prepared to tell the dispatcher your name and telephone number, what happened, the number of victims, and the exact location of the crash. If you are on a stretch of roadway with no specific address, be able to provide the names of nearby landmarks, mileposts, cross streets, off-ramps, or any distinguishing geographic features. Ask the dispatcher for advice on how to care for the victims until help arrives. Make sure that you stay on the line until he or she hangs up.

If your vehicle is undrivable after a collision, contact your automobile club or insurance company, or a private towing service, to have it towed to a garage or your residence.

Exchanging Information

Once first aid has been rendered to any victims, you have a responsibility to gather and exchange information with other parties to the crash. You are required by law to provide your name, current address, and vehicle license plate number. If requested, you must also exhibit your driver's license. If you are not the vehicle owner, you are obligated to present the vehicle's registration certificate as well. If law enforcement arrives at the scene, you must be prepared to provide the reporting officer with your telephone number in addition to your insurance company name, policy number, and policy expiration date. Be aware that failure to exchange required information in a collision involving damage to a vehicle, no matter how minor, is a misdemeanor punishable by up to one year in jail and a fine up to $2,500.

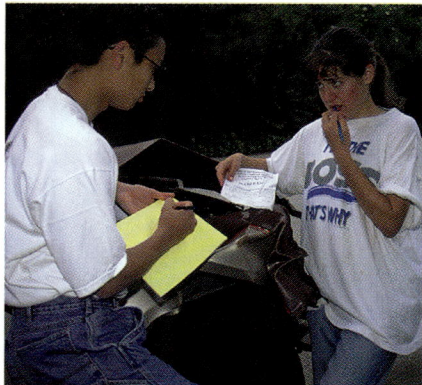

Figure 16–7 Gather and exchange as much information as you can after a collision.

Keep calm and do not argue at the crash site with other drivers and passengers about who was "at fault." Instead, gather any evidence you can that will support what you believe to be the true version of events. Do not trust your memory. Write down any relevant information and file it for future reference. Get the seat locations of different passengers, and note the extent and nature of everyone's injuries. If you have a camera, take photos of the crash scene. Make a sketch of the collision indicating each vehicle's speed, lane, and direction of travel, and record any relevant circumstances such as the weather, time of day, and location of the crash. Do not sign any papers given to you.

If the police do not appear at the scene, note the make, model, and year of the other vehicle(s) involved. Talk to any witnesses and write down their names, addresses, telephone numbers, and what they saw. If the other party or parties to the crash flee or fail to produce the required information, remain at the scene and

driving tips

In the event of a collision, always keep pencils, pens, a pad of paper, and a recyclable camera in your glove compartment.

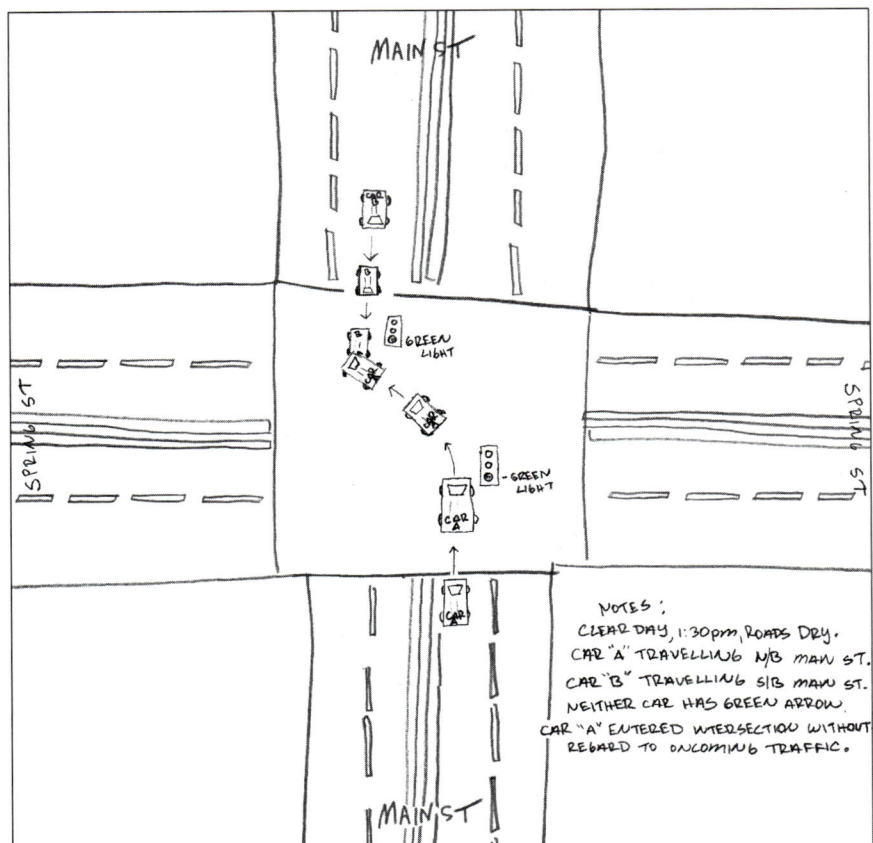

Figure 16–8 Make a sketch of the collision to help protect yourself in the event of a dispute with other parties involved in the crash.

factoid

More than half of all insurance claims for collision-related injuries are for neck sprains.

attempt to contact the police, even if the damage appears to be minor. If you do not a cellular phone or CB radio or access to a nearby phone, have a passenger, witness, or other third party make the call. If another driver does leave the scene, give the police a physical description of him or her along with the other vehicle information you recorded.

Property Damage

If you hit an unattended vehicle or cause damage to someone else's property, such as a mailbox or fence, you have a legal as well as moral obligation to locate and notify the owner. If you are unable to do so, report the incident to the nearest police station or sheriff's office and leave a note in a conspicuous place securely attached to the damaged vehicle (such as beneath a windshield wiper blade) or property. The note should include your name and address, the name and address of the owner of the vehicle (if you are not the owner), your telephone number, your vehicle license plate number, and a brief statement of what happened. Failure to comply with this requirement carries the same penalty as not exchanging information in a collision with another driver.

If you kill or injure a domestic animal such as cat, dog, or cow, pull over at a safe place and try to find the owner or custodian. If you are unable to do so, take steps to arrange for the animal's treatment by contacting the local police or humane society—do not simply leave an injured animal to die. As in the case of wild ani-

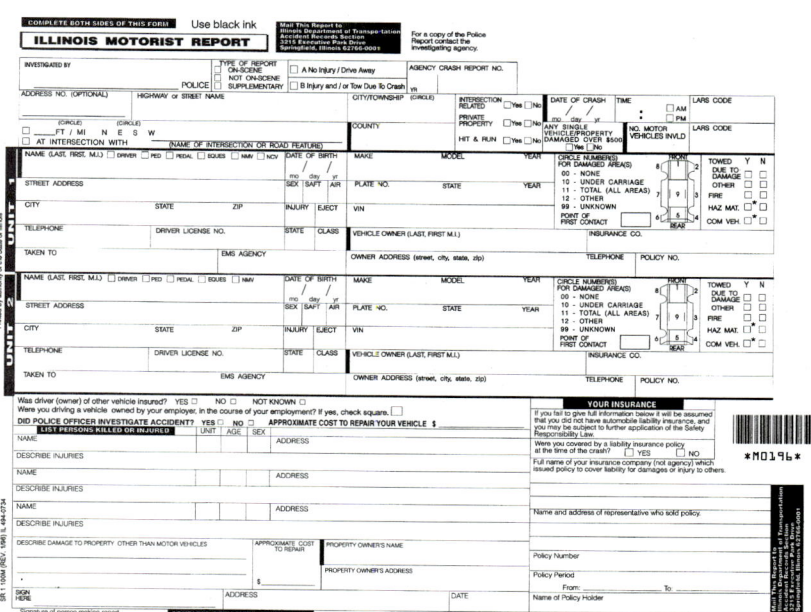

Figure 16–9 If you are involved in a collision, you may be required to file a formal report.

mals, however, never attempt to treat or move an injured domestic animal yourself.

Reporting the Crash

The final responsibility you have after a collision is to submit a written report of the incident to the Illinois Department of Transportation (IDOT) if there was death or injury involved, or more than $500 in property damage occurred. The law-enforcement officer who appears at the scene or to whom you report the crash will provide you with a blank "Illinois Motorist Report" (SR-1) form. You must complete and submit this form to IDOT's Accident Records Section within ten days of the crash. Failure to do so may result in the suspension of your driver's license.

Telephone your car insurance company as soon as possible after the collision and give them all the details you can. Provide them with copies of any reports, photos, or any other information that they require. Contact a physician if you suspect that you may have been hurt. It can take hours or even days for symptoms of a neck sprain or other injuries to appear.

16–3 PREVENTING COLLISIONS

In most cases, you can avoid an imminent collision by executing an evasive maneuver. By properly scanning and remaining aware of your driving environment at all times, you will give yourself more time and space when the unexpected occurs.

"Brake and Hold"

Most drivers have a reflex action to **"brake and hold"** when faced with an impending collision, no matter where it is coming from or what type it is. Although it may not be the best thing to do in a given situation, sometimes it is the only option you have. If you find yourself in congested traffic and there is no escape route to avoid the collision, you may have no other choice but to stop as quickly as possible. In lower-speed situations, such as driving less than 30 miles per hour (50 km/h), it may be more effective for you to brake and hold than to swerve.

When braking, hold the upper portion of the steering wheel with both of your hands. If you do not have an air bag, keep your arms

---

USING SAFE — Preventing Collisions

The SAFE process is the key to responding to specific circumstances that may lead to a collision. Always *scan* the road well ahead of you, *assess* potential hazards and other drivers' actions, *find* an escape route to avoid a collision, and *execute* the best alternative smoothly and safely. Many crashes occur because drivers fail to respond, or respond improperly, to an emergency situation. Whenever and wherever you drive, constantly be thinking of an escape route if a collision were to be imminent.

between your body and the steering wheel, not to the sides, to help cushion your head and upper body in the event of a crash. Press your body up against the back of your seat and place your head directly against the headrest. Follow procedures for emergency braking according to whether you have an antilock braking system or non-ABS system.

Speeding Up

You can often avoid an impending crash by speeding up. This is especially true when you are about to be hit from the side or rear. Situations in which speeding up can help you avoid a crash include when you are in an intersection, when a driver traveling next to you in the same direction moves toward you without seeing you, or when you are about to be rear-ended and you have clear space ahead to accelerate and escape danger. Even if you cannot avoid a crash by speeding up, you may be able to shift the impact toward the rear of your vehicle, thus minimizing injuries to yourself and passengers. Keep in mind that although it is legal to exceed the speed limit in an emergency situation, you must resume a normal driving speed once the immediate danger is past.

Steering Left or Right

When there is not enough time to brake and you have some room to maneuver on one or both sides of an impending obstacle, you may be able to avoid a crash by steering to the left or to the right.

If you are traveling more than 30 miles per hour (50 km/h), you can generally react more quickly by turning the steering wheel than by applying your foot to the brake pedal and trying to stop the car in time.

Remember that to your left you can usually expect either oncoming traffic or traffic traveling in your own direction but at higher speeds. To your right you will most often find slower lanes of traffic, the shoulder of the road, the curb, or parked cars. Avoid making too sharp a turn at higher speeds to maintain control, especially if you are driving a vehicle with a high center of gravity.

Driving off the Roadway

Driving off the roadway can be a good solution to avoiding crashes. First, however, consider your environment. Certain terrain off the roadway can lead to more problems than a collision. It is better to sideswipe a car than it is to drive off a cliff or run into a concrete barricade. On the other hand, driving off the road onto farmland, front lawns, level highway medians, or shallow ditches may be a safe option. You must also consider your vehicle. A high-clearance 4×4, for example, can manage harsher terrain than a low-to-the-ground sports car.

When driving off the roadway, you may have to deal with several types of surfaces, making traction unpredictable. Slamming on your brakes can cause a skid, an uncontrolled re-entry across other lanes of traffic, or a rollover. Instead, ease

Figure 16–10 What was the name of this "little death trap" manufactured by BMW in the 1950s?

your foot off the accelerator to slow your vehicle down naturally.

Avoiding Head-on Collisions

The possibility of serious injury and death is more likely with a front-impact crash than any other type. Be aware of situations in which a head-on collision is more likely, such as on two-lane highways, on narrow lanes around roadway construction, and near obstacles on the road.

If a car is coming at you head-on and is likely to collide with you, flash your headlights and/or honk your horn—loudly and continuously if necessary—to get the other driver's attention so that he or she will slow down or veer away. Move your vehicle to the right. A head-on driver will be more likely to move back onto his or her side of the roadway to your left than toward your right. By turning to your right, you have a better chance of being sideswiped by the driver than being struck head-on.

Avoiding Side-Impact Collisions

Despite recent safety advances in side protection such as reinforced steel beams in doors and side-mounted air bags, most vehicles are less well-equipped to withstand a side impact than they are a head-on impact. If you are at risk of colliding with the side of another vehicle, honk your horn and flash your lights to warn the other driver. Swerve right rather than left when there is no time to look first.

If you are the one about to be impacted, your best option is to accelerate rather than brake if the way is clear. Accelerating will get you past the danger more quickly. Braking may actually contribute to a side-impact collision, especially if the other driver has judged that your speed is sufficient to avoid a crash. If the way ahead is not clear, another alternative is to turn in the direction that the other vehicle is moving to force the impact behind you to the rear of your vehicle. If you turn in the direction of the approaching car, you risk colliding head-on with it.

Avoiding Rear-End Collisions

One of the most common types of multiple-vehicle collisions, rear-end crashes are most often caused when a driver follows another vehicle too closely. If you practice defensive driving and maintain a large space cushion and following distance, you should be able to avoid rear-ending a vehicle in front of you in *all* situations.

Drivers who risk hitting you from behind are another matter. To prevent an imminent rear-end crash with a vehicle behind you, tap your brakes rapidly and continuously to flash a warning to the approaching driver. Release the parking brake if it is on. Position your steering wheel straight, especially if you are stopped waiting to make a left turn. If your wheels are already pointed in the direction you want to turn, the impact may send you into lanes of oncoming traffic. This might result in a more serious head-on or side-impact

Highway heroes

On the night of November 20, 1997, Sergeant Wilford R. Lewis was struck and killed by a passing car while assisting state police in the investigation of a crash scene on Interstate 57.

driving tips

If your vehicle is rear-ended, prepare to begin braking immediately after you have recovered from the force of the impact to minimize the chances of your vehicle moving into any other cars or obstacles in front of you.

collision, as well as the initial rear-end crash.

If the way ahead is clear, move forward as rapidly as possible to approach the speed of the oncoming car. This will reduce the force of the impact and give the approaching driver more time and space to stop if he or she recognizes the imminent crash. If moving forward means crossing an intersection, do so only if it is clear of traffic. If it is not clear, you risk a side-impact collision that may be more dangerous than being rear-ended. If you cannot move forward, look for another avenue of escape to the right or left if those areas are clear.

When a Crash Is Unavoidable

Recognizing that a crash is imminent does not mean that the world is coming to an end and that you have absolutely no control over the situation. You may have only seconds to do something about it, but if you do the right thing, you can greatly reduce the severity of the impact.

The first thing to remember is to *hit something soft if you can*, such as a crash barrier, clump of shrubs, or chain-link fence. The second is to *choose where your vehicle will be hit*. Finally, remember that, in general, it is better to *collide with another vehicle moving your way than with a fixed object, and with a stationary object rather than with an oncoming vehicle*. All vehicles are designed with crash-resistant features that work to protect the passengers by spreading out the force of the impact throughout the car. Familiarize yourself with these safety design features in your vehicle. Knowing your car's strengths and weaknesses may save your own life and others in a crash situation.

16–4 VEHICLE RESTRAINT SYSTEMS

Every vehicle is equipped with devices to protect the driver and passengers in the event of a collision or an evasive driving maneuver. These restraints significantly help to reduce injuries and deaths.

---

factoid

Since safety belts became mandatory for front-seat occupants in Illinois in 1985, use has increased from 16% to 64%.

DRIVING MYTHS — Safety Belts

Some people argue against wearing safety belts. One common myth about safety belts is that in a crash they can trap you inside your vehicle. In fact, it takes less than a second to undo a safety belt, and if you are wearing one, you will be in a much better condition to undo it and get out of the car than if you did not wear it. Another myth is that safety belts are unnecessary on short trips or when driving on local streets. The reality is that more than half of traffic-related deaths occur within 25 miles (40 km) of home and on roads with speed limits under 45 miles per hour (70 km/h). A third myth is that if you are not wearing safety belts you may be thrown clear of the vehicle in a crash and walk away without a scratch. In fact, you are 25 times more likely to be killed in a crash if you are ejected from your vehicle.

Head Restraints

Head restraints, also known as head rests, are designed to reduce the risk of serious neck injury in a rear-end collision. Most vehicles have poorly designed head restraints, meaning that they are too low, and too far away from front-seat occupants' heads. If your vehicle has adjustable head restraints, make sure that they are positioned so that the center is even with your ear, not at the base or curve of your neck. For most drivers, this means raising the restraint to the full "up" position.

Safety Belts

Safety belts are among the most important safety features in your vehicle. They provide protection against most kinds of collisions, they keep you behind the wheel so you can steer to avoid a crash, and they prevent you from being thrown from your car. Most vehicles have a lap belt and a shoulder belt, although much older vehicles only have lap belts. On most vehicles, when you turn on the ignition a light appears on the instrument panel and a buzzer or chime sounds to remind the occupants to attach their safety belts.

The lap belt is designed to restrain your pelvic area, and should be adjusted to fit snugly across your hips and below your stomach. If the lap belt is too loose, it may slide up over your pelvis and injure your abdomen in a crash. The shoulder belt goes over the shoulder and across the sternum at the center of the rib cage. It should be loose enough to allow you to fit your fist between the belt and your chest, but not so loose that the belt hangs over your arm. If the shoulder belt is

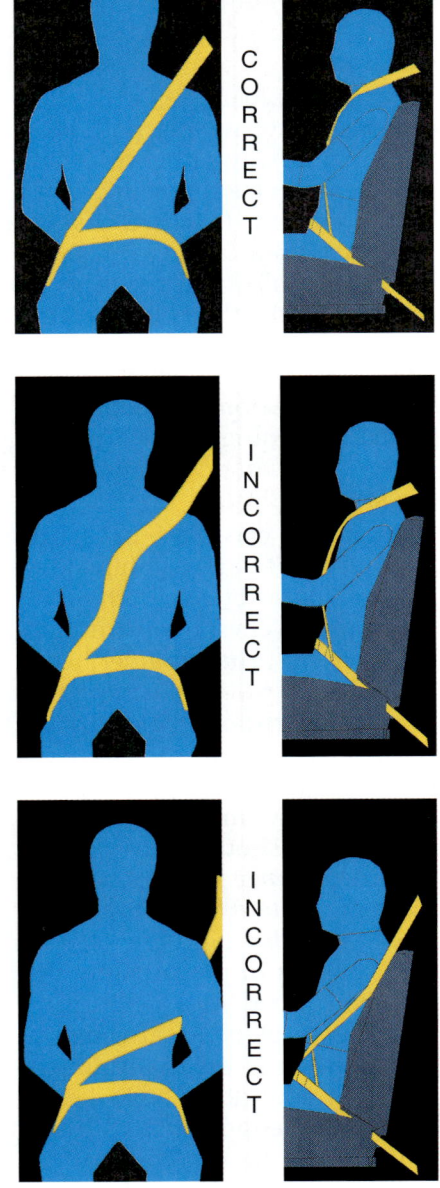

Figure 16–11 Safety belts include lap and shoulder belts.

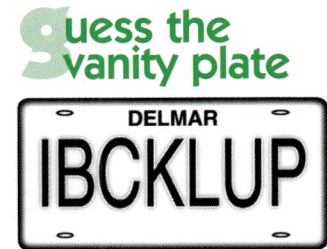

Figure 16–12

know your neighbor

Wisconsin and Kentucky require rear-seat as well as front-seat occupants to wear safety belts.

factoid

It costs about $350 to replace an air bag after it has been deployed, a cost covered by most insurance companies.

too loose, there is a greater chance that your upper body will strike the vehicle interior. Never leave the shoulder belt *under* your arm. This will not only result in unrestricted forward motion of your upper body, but also can cause severe internal and spinal injuries in a crash.

Most shoulder belts will adjust automatically as you pull them across your body. Other belts have a tension device that works like a window blind. If it is too short, you pull it out farther. If it is too long, you pull it tight gently and let it roll back into its housing until it reaches the length you want. Always take the time to untwist safety belts before buckling them. When fastening a safety belt, make sure that you hear the buckle click shut.

In some cars, lap belts and shoulder belts are combined, allowing you to buckle both across your lap and chest in one motion. In other cars, the shoulder belt is a **passive restraint**, meaning that it operates without action by the occupants. When you sit in the car and shut the door, the belt automatically moves into place across your shoulder. *If you have separate lap and shoulder belts, remember to buckle your lap belt across your hips;* otherwise, you can slide out of the shoulder belt and be injured in a collision. If your safety belts are attached to the doors, it is especially important that your doors be locked. If your door flies open in a collision, which locks are designed to help prevent from happening, you will lose the protection offered by your belt. Safety belts differ from vehicle to vehicle, so read your owner's manual for instructions on proper belt use.

As a driver, it is *your* responsibility to make sure that your vehicle's safety belts are in good working order and that you and all passengers are properly wearing them whenever the vehicle is moving. Statistics indicate that *wearing properly fitted safety belts doubles your chance of surviving a serious crash.* Illinois law requires all front-seat occupants aged six or older to be buckled up. If the driver is under the age of eighteen, *all* occupants must be restrained. If a law-enforcement officer stops you for a suspected motor vehicle violation and you are not wearing a safety belt, you will be cited and fined or required to perform community service. You will also be ticketed if any front-seat passenger under the age of sixteen is not in compliance with the law.

Air Bags

A major cause of serious chest and facial injuries in head-on collisions is driver contact with the steering wheel rim or hub. **Air bags** are a type of passive restraint system that cushion the force of impact and distribute it over a wider surface of the torso. Air bags are concealed in the steering wheel and dashboard and are designed to instantly inflate in front-end collisions that have a force of impact equivalent to hitting a

DRIVING MYTHS *Air Bags*

Many people believe that air bags can seriously injure occupants if they deploy at the wrong time or too forcefully. In fact, the chance of an air bag deploying when it is not needed is almost zero. They deflate almost as quickly as they inflate, so there is no possibility of suffocation. Most air bag–related injuries are minor cuts and scrapes. In rare instances, air bags can cause major injuries to drivers who are very short or sit too close to the steering wheel. Infants in rear-facing carriers placed in the front seat are also at risk for serious injury by a passenger-side air bag. For these reasons, you should get pedal extenders if you are unable to sit at least 10 inches (25 cm) from the steering wheel, and you should *never* put a rear-facing infant carrier in the front passenger seat.

solid barrier at speeds of 15 to 18 miles per hour (25–30 km/h) or higher. Some air bags also deploy from a vehicle's doors, side pillars, or seats in side-impact collisions. When used with safety belts, air bags provide greater protection against contact between a vehicle's occupants and the interior.

It is important to remember that air bags are supplemental restraint systems. *Even if your car has air bags, you must still wear your safety belts!* Air bags will protect you from hitting the steering wheel, dashboard, or windshield only once on the first impact. They will not secure you behind

"Vince" and "Larry" are the famous crash-test dummies used by the United States Department of Transportation.

Figure 16–13 Air bags prevent contact with the interior of the car in a collision.

As many as 85% of parents use child safety seats incorrectly.

the wheel of your vehicle. Only a safety belt can do this.

Child Safety Seats

The factory-installed restraint systems in cars are designed for adults. To protect infants and young children, you must install a separate **child safety seat.** Child safety seats are placed directly on top of the car seat and are usually held in place by the vehicle's adult safety belts. In some cases, a special buckle or clip is needed to hold the safety seat firmly in place. Some newer vehicles have built-in attachments. The child seat itself often has its own restraints, and these must be correctly fitted to the child to be safe and effective. Parents using child safety seats should always carefully follow both the instructions that come with the seat and those included in the vehicle owner's manual.

There are three basic types of child safety seats. Rear-facing seats are designed for infants from the time they are born until they weigh 20 pounds (9 kg) or are one year old. Most have a detachable base so that the seat can double as a carrier. Convertible seats are for toddlers who can sit up alone and who weigh between 20 and 40 pounds (9 to 18 kg) or are one to four years old. These seats convert from rear-facing to forward-facing as the child gets older. Booster seats are used for older children not quite ready for regular safety belts.

Illinois law requires that all children under the age of four be restrained in a federally approved safety seat. Children between the ages of four and six must either ride in a safety seat or wear a safety belt. It is generally recommended that children ride in a safety seat until they are eight years old or weigh about 80 pounds (36 kg). Be aware that police officers can also stop and fine you solely for violating the child safety seat law.

REALITY CHECK   Child Safety Seats

Child safety seats can create anxiety for parents who cannot see their child while driving. Constantly looking over your shoulder to check up on your child, however, is a major distraction that can easily lead to a crash. Instead, as part of your normal scanning technique you can occasionally—and very briefly—flick on the "night" setting on your rearview mirror. During daylight hours, this will reflect the back seat rather than the rear window, and can let you see what your little one is up to. If you have an infant in a rear-facing seat, you can buy a small mirror that mounts on the rear window that will reflect the child's face in your own rearview mirror. Another danger of many child seats is that they partially block the driver's view through the rear window. The only way to deal with this problem is to take even more time than normal scanning for danger. When you are not using the seat, it is safer to remove it and stow it in the trunk. If this is impractical, at least make sure that the bar is fastened down so that it is less of an obstruction.

Child safety seats should always be placed in the back seat to separate children as far as possible from the force of a head-on collision. This is especially true of rear-facing seats. If possible, they should be centered in the back seat to reduce the risk of injury from a side-impact collision as well. It is generally recommended that all children under the age of twelve, even if they have outgrown a child safety seat, sit in the back, especially in vehicles with passenger-side air bags.

WHO'S AT FAULT?

1. Driver 1 was going north on Coventry Lane in Crystal Lake and stopped at the intersection with Berkshire Drive. Driver 2 was stopped directly behind Driver 1. Driver 3 was speeding and failed to stop in time to avoid colliding with Driver 2. Seeing that she was about to be rear-ended, Driver 2 hit her brakes and gripped the steering wheel with both hands. The force of impact pushed Driver 2 forward into the rear of Driver 1. ***Who's at fault?***

2. Drivers 1 and 2 were stopped in the right-hand lane of Freemont Street in Galesburg after a fender bender. As they stood outside their vehicles arguing about who was responsible, Drivers 3 and 4 approached the site while driving side by side in adjoining lanes. Blocked from entering the left-hand lane by Driver 3 and going too fast to stop, Driver 4 went off the road into a bush to avoid hitting either Driver 1 or 2, damaging the front of her car. ***Who's at fault?***

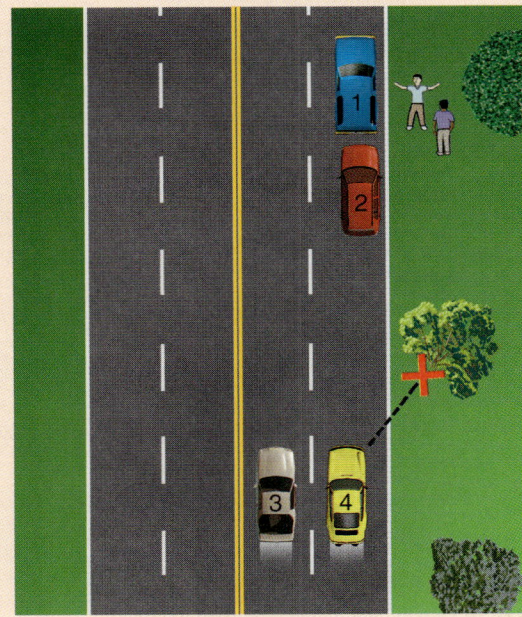

YOUR TURN

16–1 Collisions
1. Why is a vehicle crash a "collision" rather than an "accident"?
2. What are the different types of collisions?

16–2 What to Do at the Scene of a Crash
3. What responsibilities do you have at the scene of a collision in which you are involved?
4. What steps should you take when giving aid at a crash scene?

16–3 Preventing Collisions
5. What are the four choices you have to avoid an impending crash?
6. What should you do if a collision is unavoidable?

16–4 Vehicle Restraint Systems
7. What is the purpose of safety belts and air bags?
8. What is the proper way to wear safety belts?

SELF-TEST

Multiple Choice

1. Almost all collisions are caused by:
   a. accidents.
   b. miscommunication.
   c. driver error.
   d. defective vehicles.
2. You can legally exceed the speed limit if you are:
   a. driving off the road.
   b. avoiding a collision.
   c. traveling on rural roads.
   d. going with the flow of traffic.
3. Serious injury and death are more likely in which type of collision?
   a. rear-end
   b. head-on
   c. side-impact
   d. chain reaction
4. If you are facing an impending head-on collision, the first thing you should do is:
   a. swerve to the left.
   b. swerve to the right.
   c. flash your lights and/or honk your horn.
   d. "brake and hold."
5. Front air bags are designed to prevent:
   a. collisions with other vehicles.
   b. serious chest and facial injuries.
   c. windshield breakage.
   d. damage to the steering wheel.

Sentence Completion
1. Most drivers have a reflex action to _____ when faced with an impending collision.
2. A _____ is a series of collisions involving vehicles that impact one after another.
3. If you fail to stop at the scene of a crash in which you are involved, you could be prosecuted for _____.
4. A shoulder belt that operates without involvement by the occupants of a vehicle is an example of a _____.
5. In a _____ collision, the occupants may impact with all sides of the vehicle's interior.

Matching
Match the concepts in Column A with examples of the concepts in Column B.

Column A
1. __ "Set-up" collision
2. __ Crash scene responsibility
3. __ Option when crash is unavoidable
4. __ Evasive maneuver
5. __ Solo collision
6. __ What not to do at a crash scene
7. __ Rear-end collision
8. __ Rollover

Column B
a. Common on crowded roads
b. Choose where you will be hit
c. "Swoop and squat"
d. Discuss who is "at fault"
e. Colliding with a tree
f. Setting out flares
g. Flipped upside down
h. Driving off the roadway

Short Answer
1. In which situations might speeding up prevent a collision from occurring?
2. If you are involved in a crash, what information should you be prepared to exchange?
3. Why can all crashes be considered two separate collisions?
4. Why is it important to report a collision?
5. What are the dangers of a vehicle with passive restraints?

Critical Thinking
1. You are stopped at a red light at the intersection of two four-lane roads in the right lane. Looking in your rearview mirror, you see a car approaching from behind traveling too fast to avoid rear-ending you. At that moment, you notice another car approaching the intersection from your left in the left lane of the cross street. What should you do?
2. As you are traveling on a lonely rural road, you come across a collision between two vehicles in the middle of an intersection. The driver of one vehicle is slumped forward over the wheel, and the driver of the other vehicle is lying on the ground bleeding. What should you do?

PROJECTS

1. Clip articles from your local newspaper describing automobile collisions. Select three crashes, and for each one, write down the most likely cause and what could have been done to avoid it. Share the results with your class.

2. Because people have been reportedly killed by air bags, they remain a controversial passenger restraint system. Call your or your parent's insurance company to find out what their policy on air bags is. Do they offer a reduction in your premium if you have them installed on your vehicle? Call a local automobile dealer to find out how much it costs to have air bags installed, or if they are already installed, how much it costs to replace them once they deploy. Contact the National Highway Traffic Safety Administration to find out what the current federal guidelines are for the safe use of air bags. How do you apply to get a turn-off switch installed? Are all new vehicles required to have air bags? What about side-impact air bags?

17 chapter

Emergencies

Mechanical failures can lead to emergencies if your vehicle is in motion when the malfunction occurs. Blowouts can cause you to lose control of your car. If your brakes or steering fail, you are at great risk for a collision. Loss of traction caused by weather or road conditions can cause you to skid. Dead batteries and other system failures can leave you stranded. No matter how alert you are or how well you take care of your vehicle, you must be prepared to deal with unexpected and unusual emergencies. Knowing what to do in these situations can mean the difference between life and death.

CHAPTER OBJECTIVES

Upon completion of this chapter, you should be able to:

17–1 Blowouts and Flat Tires
1. Understand what to do if your vehicle has a blowout.
2. Understand how to replace a flat tire.

17–2 Mechanical Failures
3. Understand what to do if your vehicle breaks down on a freeway.
4. Understand what to do if your brakes fail.
5. Understand the proper way to jump-start a dead battery.

17–3 Skids
6. Describe the four types of skids.
7. Understand the proper way to respond to a skid.

17–4 Other Emergencies
8. Understand what to do if you get trapped in deep water.
9. Understand what to do if your car is on fire.
10. Understand what precautions you can take to avoid being the victim of a car jacking.

KEY TERMS

blowout	skid	cornering skid
brake fade	power skid	downshifting skid
jump-start	braking skid	car jacking

17–1 BLOWOUTS AND FLAT TIRES

A **blowout** is a sudden loss of air pressure in a tire. Hitting a deep pothole or a sharp object on the road (such as a nail or screw) can sometimes cause a blowout, even of a brand-new tire. However, properly maintained steel-belted tires can withstand most damage from road debris. In most cases, if you get a nail stuck in your tire, it will not cause a blowout and you can have it removed and the tire repaired for a small cost at a service station.

Most blowouts are caused by low tire pressure, which causes the tire to separate from the wheel rim. Excessive tire wear also contributes to blowouts. The average automobile tire has a tread life of about 60,000 miles (96,000 km), but you can shorten a tire's life by hard cornering and braking, driving in hot temperatures, driving at high speeds, "burning rubber" on quick starts, driving on rough road surfaces, and overinflating your tires. Do not wait until your tire is "bald," the rubber is cracked, or there are bulges on the sidewalls to replace them. By practicing simple preventive maintenance—regularly checking your tire pressure and tread depth, rotating and balancing your tires, and having your wheels aligned—you can prevent a blowout from ever occurring.

Responding to a Blowout

Blowouts can be scary. They usually occur suddenly and unexpectedly and can cause you to lose control of your vehicle. The first sign of a blowout is often a thumping sound, followed by the vehicle pulling to one side. If you experience a blowout while driving, your natural instinct will be to slam on the brakes. However, this can cause you to completely lose control of your vehicle. Instead, hold on tightly to the steering wheel, which may be vibrating badly. Gradually remove pressure on the accelerator, and allow the car to slow down by itself.

Concentrate on staying inside your lane. If a front tire blows, the vehicle will pull hard in the direction of the blowout and the steering will vibrate. Steer away from the direction of the blowout to regain control of your vehicle. If the rear tire blows, the back of the car will weave from side to side just like a skid. In that case, steer in the direction of the skid. Do not use the brakes until the vehicle is once again under control.

After checking that the way is clear, slowly and gradually pull completely out of the traffic lanes, even if this means driving for a while on your wheel rim. Try to stop the car on a paved shoulder or other level surface.

Flat Tires

If you have had a blowout or if a tire has gone flat during the course of your drive, you must either call for roadside assistance or replace the flat tire yourself. You must first determine whether it is safe to change the tire in the location where your car is stopped. If the

Figure 17–1 Inspect your tires often *before* problems occur.

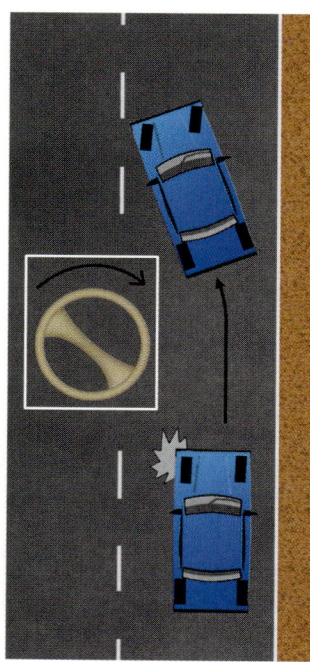

Figure 17–2 If you have a blowout, concentrate on staying within your lane.

Wild Wheels

Figure 17–3 What is the name of this 1911 vehicle, which could potentially experience *eight* flat tires?

ground is uneven or you are too close to passing traffic and there is no way to move your vehicle to a safer location, call a tow truck or law enforcement rather than try to change the tire yourself.

Fixing a flat tire is fairly easy given level ground, plenty of room to operate, and the proper equipment, but many people either are not prepared or do not know how to do it. Newer vehicles usually include a repair kit with instructions and a spare tire in the trunk, but before you get on the road, you should check to make sure that your vehicle has a spare in good condition and the necessary equipment to replace a tire, including a jack, "chocks" (special blocks for the wheels), penetrating oil, and a lug wrench.

To replace a flat tire, follow this procedure:

1. If your vehicle has an automatic transmission, switch gears to PARK. If you have a manual transmission, put the stick in REVERSE. Turn on your emergency flashers and apply the parking brake. Make sure that everyone is out of the vehicle, and set out flares or warning triangles immediately behind and up to 150 feet (45 km) behind the rear of your vehicle.

2. To make sure that the vehicle does not roll or fall off the jack, put two blocks—for example, bricks, pieces of wood, stones, or heavy metal bars—in front of and behind the tire *diagonally opposite* the flat tire.

3. Put the jack under the vehicle as indicated in your owner's manual. Slowly crank the jack to raise the car. Make sure that the jack is securely in place when you first make contact with the vehicle's undercarriage. Continue cranking until the flat tire is partially up but still making contact with the ground and is incapable of being turned. Be careful not to expose any part of your body underneath the car in case the jack fails and the vehicle falls down.

4. If there is a wheel cover, remove it to get access to the lug nuts, the bolts that hold the wheel in place on the car. Loosen, but do not remove, each of the lug nuts using a lug wrench. If your wheel has not been removed for a long time, the lug nuts may be difficult to take off. If you have penetrating oil, apply it to any rusted nuts and wait a few minutes. If necessary, position your body sideways to the car and use your foot to press downward on the handle of the wrench to loosen the nuts. Work your way around the wheel, loosening the bolts in a criss-cross pattern.

5. Finish jacking up the car until the flat tire is just off the ground. Remove the lug nuts with the lug wrench. Take off the entire wheel with the flat tire and set it down under the car as a safeguard in case the jack fails. Line up the holes on the spare tire with the lugs, and slide it onto the wheel mount. Replace the lug nuts and securely tighten

Figure 17–4 Make sure that your vehicle is equipped with a complete tire repair kit.

them with the lug wrench, again in a criss-cross pattern.
6. Set aside the flat tire and lower the car down by cranking the jack in the opposite direction from which you cranked it up. Stay clear of the vehicle until the car is all the way down. Put the wheel cover back onto the spare. Remember to recover the blocks you used to keep the car in place.

On many cars, the spare is not a full-sized tire. Emergency spares are not designed to be used for normal driving but to get you home or to an auto repair shop. Full-sized spares that come with cars are often of inferior quality to your regular tires. Whatever type of spare you have, check your owner's manual or with the manufacturer to see how long you can safely drive on it and what its limitations are. As a rule of thumb, you should replace your spare with your second set of tires. Dry rotting and heat can cause a spare to deteriorate over time. It will also lose air pressure. If you have a car with a spare that has been sitting in the trunk for years, odds are the spare tire is not safe to use or needs air before it can be used.

17–2 MECHANICAL FAILURES

Tire failures are one of the most common causes of vehicle breakdowns, but other systems in your car can malfunction and cause an emergency while you are driving.

Brake Failure

One cause of vehicle breakdown is brake failure. In most cases, brakes do not literally "fail." Most cars on the road today have some type of power-assisted braking system. Engine failure or interruption of the flow of brake fluid can cause the power-assist or full-power feature of the brakes to fail. The brakes still work, but you have to apply much more pressure on the pedal.

driving tips

Practice changing a tire in a driveway under the supervision of an experienced driver to familiarize yourself with the procedure as well as any special instructions in your vehicle's operator's manual.

Figure 17-5 Emergency spares should never be used for normal everyday driving.

reality check

Most emergency spares are designed for speeds up to 50 miles per hour (80 km/h), and anything faster is dangerous and not recommended.

Brakes can also "go out" if your brake pads become totally worn out. However, it takes time for pads to wear out, and you will notice a gradual reduction in the effectiveness of your brakes before they reach this point. The brake pedal will start to feel spongy, and will, with increased use, press closer and closer to the floor. If one day you cannot stop your vehicle because you ignored grinding noises coming from your brakes, *you* failed, not your brakes!

In most vehicle braking systems, the front and rear wheels are controlled by a separate mechanism. If one part of the system ever fails, you can still use the foot brake to stop the car. In the unlikely event that both parts of your regular braking system fail, you must respond quickly. If you do *not* have an ABS braking system, pump your brakes rapidly to engage any remaining brake fluid pressure to the brakes. If you do have ABS brakes, just brake and hold.

If this does not work, use the power of the engine to reduce your speed by downshifting. If you are driving a vehicle with an automatic transmission, do not be afraid to downshift to SECOND gear and then to FIRST gear in quick succession to slow the car down. Do not switch to PARK. If you do *not* have power steering or power brakes, you can also turn off the engine while keeping the vehicle in gear to slow it down. Make sure, though, that you do not turn the ignition key to the "lock" position as this will cause the steering to lock.

Gradually apply the parking brake. Because this brake controls the rear two wheels of the vehicle, you must use the brake gently so that the brakes do not lock and throw your vehicle into a skid. If you have a hand-cranked parking brake, slowly apply pressure on the release button of the parking brake as you pull the lever up. If you have a foot-pedal parking brake, hold the parking brake release handle while you are applying pressure on the brake with your foot.

As you are carrying out this procedure, find an escape route. Head for a safe place to stop the vehicle on the side of the road and steer safely out of traffic. If there is no safe place to go or if your speed is still too high, use whatever means are available to produce enough friction to stop, such as brush vegetation on the side of the road or an uphill slope.

Overheated Brakes

When you use your brakes continuously over time—for example, when descending a long, steep mountain road—your brakes may overheat. To prevent this overheating, called **brake fade,** try to let your engine do most of your braking for you by changing the transmission to a lower gear. However, if the road is extremely steep or a lot of traffic is going downhill, staying in a low gear may not slow you down enough. In this case, pull off to the side of the road and give your brakes a rest. If this does not resolve the problem, wait for roadside assistance. Do *not* continue to drive the vehicle.

Steering Failure

Most cars today have some type of power-assisted steering. A power-steering failure is most common when the engine stalls. Without the power provided by the engine, your steering wheel moves, but it is much more difficult to turn. In the event of a power-steering failure, grip the steering wheel firmly and exert as much force as you can to get it to turn in the direction you need to go to get off the roadway and out of traffic. Also apply the brakes, which may take a lot of force. Activate your emergency flashers and use your horn, if necessary, to warn other drivers.

If the wheel will not turn at all or the vehicle does not respond as the wheel turns, your steering system has failed completely. Any attempt to steer the vehicle could be very dangerous, so shift the transmission into NEUTRAL and apply the brakes immediately until you stop.

Stalled Engine

Engine failure usually happens as a result of a mechanical failure, but it can also occur as a result of a common driver error—forgetting to put gas in the car! An engine can also stall if it gets wet or is exposed to excessively cold or hot weather.

If the engine dies or stalls while you are driving, first try to re-engage the engine. Shift the transmission to NEUTRAL and try turning the ignition several times. If the engine comes back to life, shift out of NEUTRAL and back into a driving gear to proceed on your way. If shifting to NEUTRAL does not work, take a strong grip of the steering wheel, pull off the roadway to a safe place, and stop your vehicle. Keep in mind that it will be extremely difficult to turn the wheel and apply the brakes with a loss of engine power.

Stuck Accelerator

A stuck accelerator is rare and is often caused by an obstruction on the floor in front of the driver's seat, like hardened mud or ice, or an object wedged against the accelerator pedal such as a floor mat. However, it may also be caused by a mechanical problem with the vehicle.

Where am I?

Figure 17–6 This museum features a unique collection of telephones by a former Illinois Bell employee, who bought his first phone in 1949 for $10 at a gas station while waiting for a mechanic to arrive and repair his broken-down truck.

DRIVING TIPS — *Freeway Breakdowns*

A freeway is the most dangerous place that you can have a breakdown. There is not much room to pull over, and other vehicles *constantly* pass you at high speeds. If you have to stop on the freeway, signal and drive completely off the freeway onto the shoulder. Activate your emergency flashers. Set out warning devices, raise the hood of your car, or attach a white cloth or handkerchief to your antenna or roadway-facing door handle to get the attention of law-enforcement officers, tow-truck operators, or "Minutemen" that regularly patrol the area. If you have a cellular phone or citizen's band (CB) radio, use it to call for help. Stay in your vehicle if it is completely off the freeway, and lock your doors until help arrives. If other drivers stop to offer their assistance, roll your window down slightly and ask them to call the police. Avoid walking on a freeway after a breakdown, especially at night or in bad weather. If no one is stopping and you are close to an exit with a source of help, such as a service station or place with a telephone, walking may be the best option if you can do so without risking your safety. Always walk on the right side of the roadway facing traffic. Never attempt to cross a high-speed, multilane freeway on foot.

To try to free or "unstick" your accelerator, first try lifting the accelerator pad up with the tip of your shoe. If this does not work, shift into NEUTRAL immediately. Pull off the roadway, using your brakes to slow the vehicle down. Do not turn off the engine until you have come to a complete stop.

Headlight Failure

If your headlights go out, try flipping the "on" and "off" switch several times. If this does not work, try your high beams. These usually use different bulbs and may still work even if the low-beam bulbs will not go on. If neither your high beams nor low beams work, activate your emergency flashers. These lights will help you see the road ahead, and more important, they will make you visible to other drivers. Your parking lights or turn signals may also provide enough light to see ahead.

If you are on a well-lit highway, you should be able to navigate your way to a safe spot off the roadway. If you are on a dark, deserted highway, it is best to take your foot off the accelerator and keep your vehicle on the pavement until you can slow down enough to pull off onto the shoulder.

Dead Battery

If you attempt to start your vehicle and you do not hear the ignition turn, your battery is probably dead. It may have been old, weak, or drained by a defective alternator or extremely cold weather. The battery cables may be corroded. Perhaps you left the lights on by mistake after parking your car.

> **REALITY CHECK** *"Stuck in the Middle with You"*
>
> If your vehicle becomes disabled in the middle of traffic and you cannot pull over, which sometimes happens in stop-and-go traffic, keep your wheels pointed straight and turn on your emergency flashers. When it is safe to do so, get everyone out of the vehicle and onto the shoulder or median far from traffic. If possible, use the shoulder rather than the median. The shoulder is generally wider than the median, traffic is slower in the right lanes, and emergency vehicles can reach you more easily. If the shoulder is dangerously narrow, carefully walk to a wider point and face traffic so that you can see oncoming vehicles.

You can **jump-start** your car to recharge your battery. This involves using a set of jumper cables to start your vehicle by drawing on the charge from another vehicle's battery. If you do not carry your own set of cables, ask other drivers or check with a nearby parking garage or service station to see if they will loan you a set.

To jump-start your vehicle's battery, first make sure that the other vehicle's battery is in good working order and that it is the same voltage as your battery. Make sure that any fluid in either battery is not frozen and that any vent caps are tightly closed and level.

The front ends of both vehicles should be face to face, or at such an angle that the cables can easily reach both batteries. *Do not let the vehicles touch each other.* The other vehicle's engine, lights, and accessories should all be off. The transmissions of both vehicles should be set to PARK, with the emergency brakes activated. To avoid being pinned between the vehicles in case one is rear-ended in a crash, stand safely to the side and away from any passing traffic.

Now you are ready to attach the jumper cables and jump-start your vehicle. Note that there are two separate cables with hand-operated clamps at each end. The "positive" jumper cable is usually marked with a "+" sign and is red. The "negative" jumper cable is usually marked with a "−" sign and is black. To avoid accidentally burning yourself, *make sure that your hand remains on the plastic grip of the clips at all times.*

1. Attach one of the clips of the positive cable to the positive terminal of the working battery.
2. Attach the clip at the other end of the positive cable to the positive terminal of the dead battery.
3. Attach one of the clips of the negative cable to the negative terminal of the working battery.
4. "Ground" the other end of the negative cable by attaching the clip at the other end to your vehicle's frame, to the metal engine block, or to an engine bolt. Be aware that there may be a spark when the last cable is connected. For this reason,

factoid

A typical car battery, which carries 12.68 volts when fully charged, "goes dead" when the voltage drops below 11.7.

driving tips

If your battery is older than its projected life span (usually three to five years), replace it rather than wait for it to die.

Figure 17–7 Keep a pair of jumper cables in your vehicle in case you need to jump-start your battery or someone else's battery.

you should attach it as far away as possible from the battery and any moving engine parts. *Never attach the other end of the negative cable to the negative terminal of the dead battery.* This can cause damage to the other vehicle's battery or electrical system or, worse, cause the dead battery to explode. Make sure that none of the cables are near fans or belts.

Try starting the vehicle with the dead battery. It should start on the first try. If not, try it again. Once your vehicle has started, let it run for a few minutes, then remove the cables one at a time while the engines on both vehicles are still running. Remove the negative cables first, then the positive cables. Be aware that the first cable you disconnect may cause a small spark. *Do not allow the clamps at the ends of the cables to touch each other or to touch car parts.* Close the hoods of both vehicles.

To recharge your battery, you must leave your engine running for a while. If you can, drive around for at least half an hour. If you turn your engine off too soon, you will have to repeat the jump-start process. If your battery dies again, even when you have

AUTO ACCESSORIES *Jump-Start Without Jumping Back!*

While generally safe, using jumper cables to recharge a dead battery can put you at risk of chemical burns, eye damage from flying sparks, and other injuries. There are now products available to eliminate this risk. One allows you to jump-start your battery by connecting your vehicle to another via the cigarette lighters or accessory outlets rather than the batteries. A similar device conveniently enables you to recharge your battery without the use of another vehicle's power source at all. By plugging this rechargeable apparatus into your own vehicle's cigarette lighter/accessory outlet, you will be able to start your engine within 15 to 30 minutes. Yet another alternative now available is a lightweight, compact battery that you can attach with relative ease to your car's dead battery with two short cables. This product, which runs on AA batteries and retains its charge for more than a year, provides a burst of power sufficient to restart most vehicles.

allowed plenty of time to recharge it, have a mechanic check the condition of the battery and look for problems in your vehicle's electrical system.

Windshield Wiper Failure

Like every other system on your vehicle, windshield wipers must be well maintained to work. Over time, excessive heat and cold causes wipers to rot or crack, reducing their effectiveness. If the rubber on the blades is very worn or has fallen off, the metal bracket that holds them will have direct contact with your windshield and will scratch the glass when the wipers are turned on. If you do not replace your blades regularly, you will have an unpleasant surprise when you find out that they are useless when you need them most.

Windshield wiper failure can be dangerous if it occurs while you are driving in the rain. Without your wipers to clear away the water beating on your front windshield, your visibility may instantly be reduced to zero. If your windshield wipers fail, first flip the wiper switch several times. If this does not work, roll down a side window, look ahead, and maneuver your way to a safe place off the roadway as soon as possible. Call a tow truck or wait out the rain before starting back on the road again.

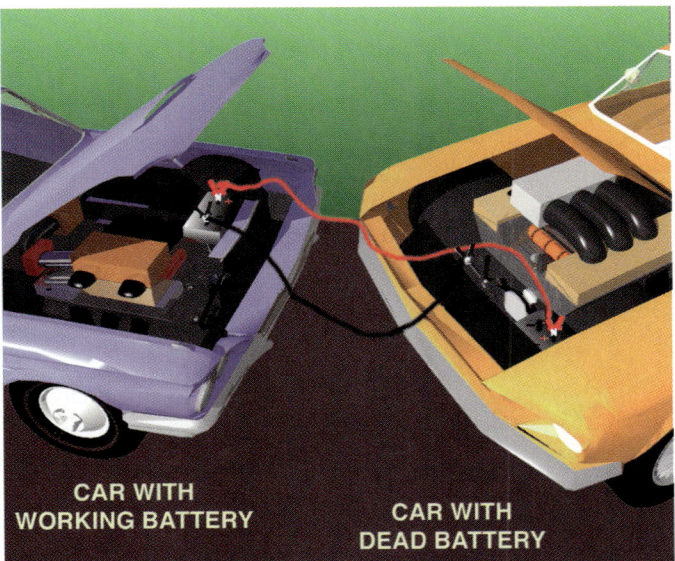

Figure 17–8 Make sure that you know the proper connections before trying to jump-start a dead battery.

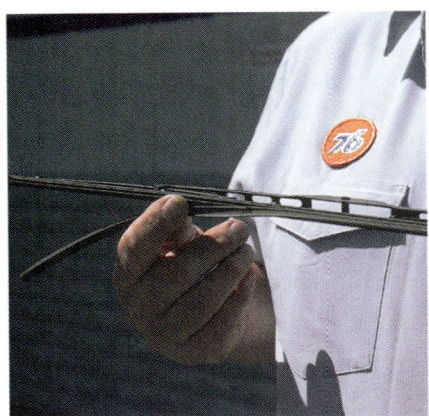

Figure 17–9 Replace the windshield wiper blades if they start to wear, streak, or smear.

Hood Latch Failure

If your hood latch fails, your hood can pop open all the way and completely block your view of the road. Although it may be your first instinct, the worst reaction you can have in this situation is to slam on your brakes. Stay calm. Switch on your emergency flashers and look through the opening between the hood and the dashboard. On most cars, this space is large enough for you to see traffic and get your vehicle safely off the roadway. If you cannot peek through the bottom of the hood, lean out of the side window to see. Ask any passengers to help by directing you and checking whether you have clear passage to the side of the road.

Once you are off the roadway and stopped, check to see whether the latch was not closed properly or if the latch mechanism is actually broken. If the latch seems to be functioning, shut the hood properly. When you begin to drive again, do so slowly and remain in the far-right lane so that you can be ready to move off the roadway quickly if the hood opens up again. If the mechanism is broken, you may be able to secure the hood shut with some wire, cable, or rope. Drive the vehicle to your mechanic as soon as possible to have the latch mechanism replaced or fixed. If you cannot secure the hood in such a way that it does not block your vision, call a tow truck.

17–3 SKIDS

When a vehicle experiences a loss of traction, the result may be a **skid.** When a skid occurs, all or some of the tires slide over the roadway, the engine loses its pulling effect, and the brakes and steering are less effective in controlling the vehicle. Skids are most likely to happen when you are driving at high speeds in bad weather, on dirt or gravel roads, and on wet or icy pavement. However, they can also occur on dry surfaces and even at low speeds, given certain conditions.

Types of Skids

Although roadway conditions contribute to skids, in most cases they are caused by drivers going too fast for the current road or weather conditions, steering the car incorrectly, changing the car's direction too quickly, braking improperly, or applying too much pressure on the gas pedal at the wrong time.

A **power skid** is caused by accelerating too quickly, especially on a slick road surface. If you have rear-wheel drive, suddenly pressing on the gas pedal too hard can cause the back end of your car to skid to the side. The best way to avoid a power skid is too ease off the accelerator when

Figure 17–10 A skid can cause you to lose directional control of your vehicle.

DRIVING TIPS — Emergency Traffic Patrols

The Illinois Department of Transportation (IDOT) offers free help to stranded motorists on nearly 80 miles (130 km) of selected freeways in Chicago and East St. Louis. A fleet of more than 50 "Emergency Traffic Patrol" tow trucks operate 24 hours a day, seven days a week, to get disabled vehicles off to the shoulder or median out of harm's way and to get freeway traffic moving smoothly again. Drivers of these roving trucks, known familiarly as "Minutemen," can provide a jump-start, gasoline, assistance in changing a flat tire, emergency telephone service, and minor repairs. If you get stuck on the freeway in either of these areas, dial *999 on your cellular phone or look for the yellow trucks with the distinctive logos and flashing yellow lights. The response time to most incidents is within 20 minutes.

you feel the tires start to spin or lose contact with the road. On slick roads or roads with poor traction, remember to always accelerate and decelerate very gradually.

If you apply the brakes while the vehicle is going too fast, one or more of the wheels may lock up and result in a **braking skid.** This type of skid most commonly occurs in vehicles without an antilock-braking system and happens more frequently on slippery or uneven roads. If the front wheels lock, the vehicle will plow straight ahead, building up a wedge of water or debris beneath that will lift the tire off the pavement. If the rear wheels lock, your car may start to spin sideways. If you sense a braking skid, you should immediately take your foot off the brake pedal until the wheels start turning again and drop to the road surface.

A **cornering skid** occurs when your tires lose traction on a curve. This type of skid is usually caused by driving too fast. If the road is slippery or you have worn tires, the chances of a cornering skid are higher. If you feel your rear wheels start to skid away from the turn, take your foot off the gas and make steering adjustments to straighten out.

If you downshift from a high gear to a low gear too quickly, you may cause a **downshifting skid.** If you are driving a vehicle with a manual transmission, make a

Figure 17–11

Figure 17–12 The "Illinois Minutemen" respond to nearly 100,000 incidents in the Chicago area and more than 14,000 in East St. Louis each year.

habit of engaging NEUTRAL each time that you downshift instead of moving from one driving gear directly to the next.

Responding to a Skid

If you experience a skid, your natural instinct will be to hit the brakes. However, slamming on your brakes will only cause you to skid more. The appropriate response to a skid depends on the road conditions. If you are skidding on a dry surface at a high speed, you must turn the steering wheel sharply in the appropriate direction to correct the skid. On a slippery road, however, you should only turn the steering wheel slightly to bring your vehicle back under control.

With any skid, the objective is to keep the rear of the vehicle from outrunning the front by either slowing the rear wheels or speeding up the front wheels. On rear-

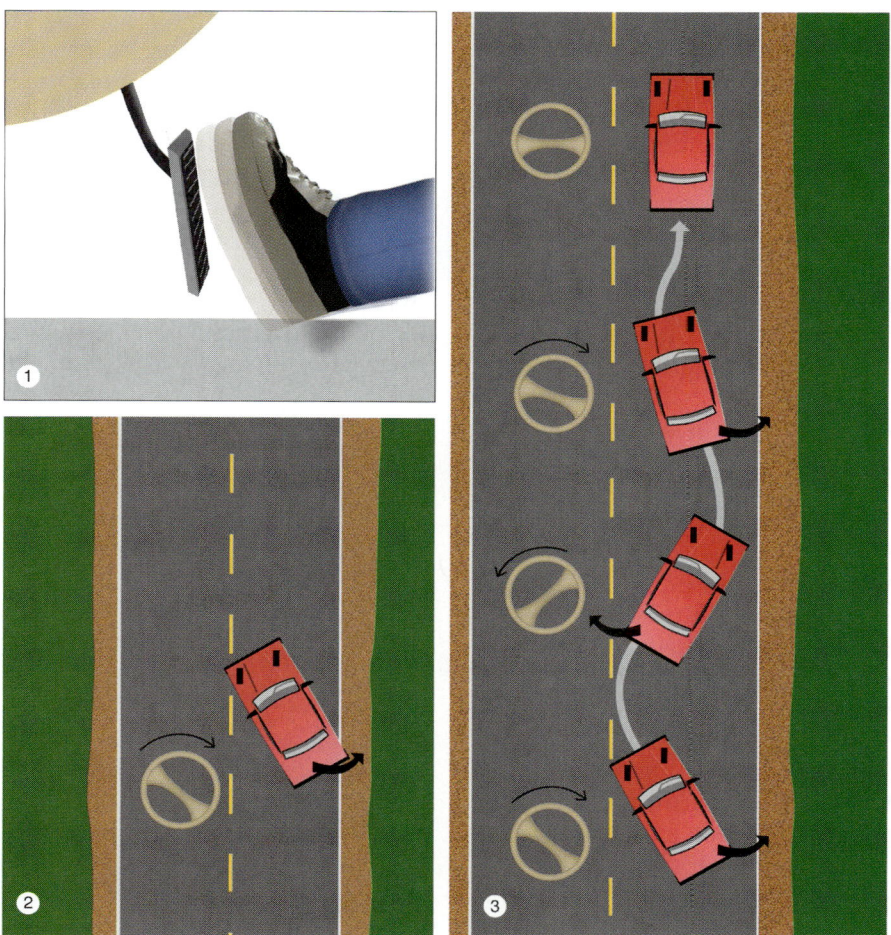

Figure 17–13 Always steer in the direction of a skid.

wheel-drive vehicles, follow these steps to slow the rear wheels to get out of a skid (see Figure 17–13):

1. Immediately take your foot off the gas to decrease your speed. *Do not use the brakes.*
2. Steer the car gently in the same direction as the skid. This will get the front end of the car to head in the direction that you are skidding. At higher speeds on a dry surface, you should turn the wheel more sharply. If you overcompensate by steering too sharply or quickly, you may start to skid in the other direction and "fishtail" from side to side.
3. When the vehicle begins to straighten out, turn the steering wheel back the other way, correcting the car's path of travel. You may need to correct your steering several times, turning slightly left and right until you get out of the skid. Once the vehicle is back under your control, gently apply your brakes to reduce speed.

On front-wheel-drive vehicles, stepping lightly on the accelerator will increase the front-wheel speed and allow the front of the vehicle to match the rear skid. If you do not have enough room to accelerate, disengage the transmission by shifting to NEUTRAL or press and hold down the clutch pedal if you have a manual transmission.

"Spinning Out"

If your vehicle starts to spin in circles during a skid, corrective steering will not help bring the vehicle back under control. In this situation, you have no choice but to hit the brakes as hard as you can, holding them down until the vehicle ultimately stops. If you have standard brakes, this will cause the brakes to lock up. Although your vehicle will continue to spin, it will travel along a straight line in its original general direction of travel. If you have ABS brakes, holding down the brake pedal will engage the brakes of all four wheels.

17–4 OTHER EMERGENCIES

Many emergencies are not caused by a vehicle malfunction or driver error. Debris on the roadway, deep water, and downed power lines are all potentially hazardous obstacles. Car fires are rare but require an immediate response. Car jacking is increasingly common in today's world. A defensive driver should be prepared to deal with each of these situations.

Roadway Debris

Occasionally you will encounter auto parts, pieces of tire, boxes, rocks, tree branches, furniture, and other unusual objects on the roadway. Avoid these when possible. Watch for other drivers ahead of you maneuvering around or stopping for an unexpected obstacle. Never swerve into another lane unless you are first sure that the path is clear. It is better to hit an object and suffer some damage to your car than to uncontrollably swerve into another lane of traffic

DRIVING TIPS: When an Emergency Becomes Another Emergency

When your vehicle breaks down or you are experiencing another kind of emergency, do not let your emergency get worse by stopping your vehicle where visibility is limited, such as around a curve or just over the crest of a hill. If you are unable to avoid stopping in one of these vulnerable positions, place flares and emergency warning triangles well behind your vehicle to give other drivers sufficient warning. This means that you need to place your first warning device *before* the curve begins or on the *other side* of the crest of the hill. Do not forget to use flares with extreme caution, keeping them away from flammable liquids such as leaking gasoline and oil.

and cause a collision. If you do not know whether any vehicles are occupying the lanes to your sides, apply your brakes and slow down for the obstacle, stopping in front of it if necessary. Use your mirrors and look over your shoulder to make sure that the way is clear, and then drive around the obstacle at a slow speed.

Although extremely rare, your windshield may be blocked by a sheet of cardboard, plastic, canvas, or other similar material blowing around the road or that has come off a vehicle ahead of you. Loose materials such as hay, grass, or leaves may also get trapped on your windshield and block your view. If this happens, follow the same steps you would in the case of hood latch failure and steer your vehicle carefully off the roadway to remove the debris.

Deep Water Escape

Cars do not float long in water. If your vehicle gets trapped in deep water, you need to get out fast.

ALAN PAKULA

Born in New York in 1928, filmmaker Alan Pakula rose from an assistant in the cartoon department of Warner Brothers to a critically acclaimed producer, director, and/or writer of more than two dozen motion pictures. He received many awards, and Oscar nominations for his work on *To Kill a Mockingbird, All the President's Men,* and *Sophie's Choice.* On November 19, 1998, Pakula was driving on the Long Island Expressway about 30 miles (50 km) east of Manhattan when a metal pipe lying on the road was sent careening into his windshield by the vehicle ahead of him. Severely injured in the head, he lost control of his Volvo and crashed into a fence. Pakula was later pronounced dead at a hospital. Even when driving on a relatively safe modern freeway, you must be constantly alert to the potential for encountering obstacles on the roadway. Increasing your following distance behind other vehicles will give you a better view of the road and more time to respond to an emergency situation.

Make sure that all other occupants' safety belts are undone. Do not, however, remove your safety belt *before* entering deep water. If you fall from a bridge or other elevated point into deep water, safety belts will help protect you from the impact.

Resist the instinct to panic. Your best chance of escape will be through the windows, because the pressure of the water on the doors may make them impossible to open. If you have power windows, you should activate them immediately. If you have hand-cranked windows, it may be easier to wait until the vehicle fills up with water and then open them rather than have to contend with a powerful inrush of water. Keep in mind that the weight of the engine will force the front of the vehicle to sink first, creating an air pocket in the back of the vehicle where you can move. While you are waiting, breathe normally and make sure that the door is unlocked. Once the pressure has equalized, open the window or, failing that, the door. If neither will open, attempt to kick out a side window or the rear window to escape.

Downed Power Lines

If you come across downed power lines on the road, avoid driving over them. Wait for a safe opportunity and maneuver around them slowly. If the power lines completely cover the road, especially if it is wet or standing water is nearby, turn around and choose an alternative route.

If power lines or other electrical wires fall onto your vehicle while you are inside, do not panic. Remember that your tires, which are your only points of contact with the ground, are made of rubber and will insulate you from any electrical shock. Stay in your vehicle until emergency personnel arrive. Do not try to exit the vehicle and hopscotch your way around very dangerous electrical wires. Switch on your emergency flashers so that other drivers will see you and know that you are in trouble. If necessary, use body movements to warn others to stay away from the scene and call for help.

Car Fires

The smell or sight of smoke, either coming from under the hood of the engine or in the passenger compartment, is a sign of a vehicle fire. If your vehicle is moving and you suspect that a fire is under the hood or inside the car, immediately pull off the roadway into a safe, open area without people. Turn off the ignition and make sure that all the occupants, especially infants, small children, and those with limited mobility, are out of the vehicle. Move as far away as possible in case of a fuel-tank explosion, and call the fire department.

If the fire is far from the fuel tank and is not widespread or if a passenger is trapped inside the vehicle, you may choose to try to control the fire. Aim to put out the fire at its source or base. *Never use water to fight a car fire.* If the fire is being fed by gasoline or oil,

Highway heroes

Just after midnight on July 1, 1968, Deputy Clarence J. Bridwell of the McHenry County Sheriff's Department was struck and killed by an oncoming vehicle while removing a tree that had fallen after heavy rains on Crystal Lake Blacktop near the intersection with Edgewood Road.

Figure 17–14 Unless passengers are trapped inside, it is best not to try to put out a car fire by yourself.

the fire will literally float on the water or oil and may spread to other areas. If you do not have a dry-chemical fire extinguisher in your car, try to smother the fire by throwing dirt, sand, snow, large blankets, or clothing onto it.

Never attempt to fight an uncontrollable fire that should be handled by professionals, especially if no one is trapped inside. It is not worth risking serious burns, or worse, when you can just walk away. A car can always be replaced, but you cannot.

Car Jacking

In recent years, there has been an alarming increase in assaults on drivers to steal their vehicles or rob them of money or personal belongings. One of the most effective ways to prevent being a victim of a **car jacking** is to plan a route that does not go through dangerous areas. Try not to drive alone if possible, especially if you must go through dangerous or unfamiliar areas. Make sure that your doors are always locked and your windows are shut. Never,

AUTO ACCESSORIES *Preparing for Emergencies*

You can purchase emergency kits for your car to cope with unexpected events like a car fire or mechanical breakdown. These kits typically contain such items as flares, booster cables, safety gloves, a flashlight with extra batteries, a utility knife, tape, assorted fuses, a safety vest, an emergency blanket, a rain poncho, a first-aid kit, an aerosol flat-tire sealer, and a bright fluorescent banner to attract the attention of the police. You might also want to purchase battery-operated illuminating emergency triangles, which are easier to see than those which are only reflective. Finally, you might also want to obtain a small, portable A-B-C fire extinguisher, making sure to secure it tightly to the inside of the vehicle with clasps or bolts so that it will not roll around.

> **DRIVING TIPS** — **Unwelcome Passengers**
>
> A good way to check whether anyone has been hiding on the back floorboard of your car while you have been away from it is to tip forward the front seats when you leave your vehicle. Upon your return, check whether the seats have been returned to their original position. If so, someone may have broken into your car, and may still be hiding there!

under any circumstances, pick up a hitchhiker.

Constantly scan while you are driving, and remain aware of what is happening around you. Because car jackers can most easily approach if you are moving slowly or stopped, always leave enough room in front of you and to the sides to give yourself an escape route whenever you are stopped at an intersection or in heavy traffic. If you are stopped at a deserted intersection and someone threatening approaches you, it is safer to run a red light rather than risk an assault.

Be wary of anyone who approaches you with fliers or requests for change or directions while you are stopped. Do not leave valuables exposed on the seat, dashboard, or console that might attract thieves. At gas stations, turn off your engine and lock your doors when going to pay the attendant.

If you have a vehicle breakdown, stay in your car and keep all your windows and doors locked tight until the police arrive or professional roadside assistance arrives. If anyone approaches to help you, write a note, motion with your hands, or "mouth" to them to call the police or a tow truck.

Be aware that car jackers often work in teams, using two vehicles to corner a victim's car. Do not let yourself be trapped between two other suspicious-looking vehicles. Carjackers also use deception to get you to pull over—typically in a dark, deserted place—by gesturing to you that there is something wrong with your vehicle.

When parking your vehicle, make sure that you park in well-lit areas with other people around. If parking at home and you have a garage, use it. If you are parked in a lot, avoid returning to your vehicle late when the lot is vacant, especially if you are alone. Have your door key ready in your hand when returning to your vehicle. Inspect your car from the outside first to see if anyone is on the other side, underneath, or hiding inside. If you do spot someone in or near your vehicle, move away from the area as fast as you can and call the police.

Some car jackers stage collisions by intentionally rear-ending your vehicle or causing a similar type of "set-up" crash. When you get out of the car to inspect the damage, they use the opportunity

factoid

About 25,000 car jackings occur annually in the United States.

to take off with your vehicle. If you feel suspicious about the circumstances of a collision in which you are involved, and you are alone or feel scared, communicate to the other driver (without rolling down your windows) to follow you to the nearest police station. If you do not know where a nearby police station is, drive to a well-lit place close by where there are a lot of people. Leaving a crash scene may cause you trouble with law enforcement if your fears prove unjustified, but you have no other choice if you truly believe that your life is at stake.

The most important thing to remember if you are the victim of a car jacking, especially if you are threatened with a weapon, is to not resist. Hand over your keys, money, or whatever the car jacker wants. A car or wallet is not worth your life!

Police Activity

It is possible that at some time you may find yourself in the path of a police chase or of police officers responding to an emergency. Officers may take great driving risks, disregarding traffic controls, speed limits, and laws regulating the movement and direction of traffic, and fleeing suspects often drive erratically and recklessly. If you find yourself in the middle of high-speed police activity, do not do anything sudden or rash. Stay calm, keep your lane position and speed constant, and follow officers' instructions.

WHO'S AT FAULT?

1. Driver 1 was traveling west in the right-hand lane of Morton Avenue in Jacksonville. Driver 2 was traveling in the same direction in the left lane behind Driver 1. Driver 3 was pulled over on the shoulder and changing her left-front tire. When Driver 1 saw the disabled car, he veered into the left lane to give it more room. Driver 2 reacted by swerving left, hitting the concrete barrier and ricocheting into Driver 1. Driver 1 slammed on the brakes, spun across the road, and flipped over. **Who's at fault?**

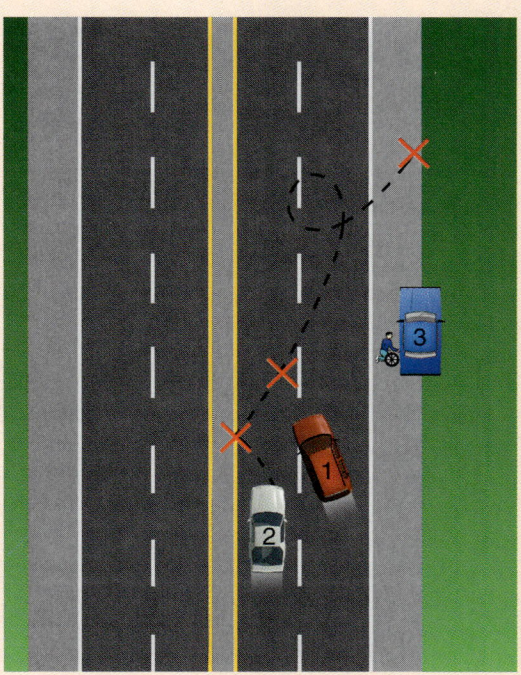

2. Driver 1 was rounding a curve on an icy portion of North Bloomington Street in Streator at the speed limit. When the tires lost traction, she hit the brakes and skidded from the right lane all the way into oncoming traffic. Driver 2 was traveling in the opposite direction in the left lane, also at the speed limit, and had just gone around the curve when he saw Driver 1 spinning into his path. Both cars collided in the middle of the highway. **Who's at fault?**

YOUR TURN

17–1 Blowouts and Flat Tires
1. What should you do if your vehicle has a blowout?
2. How do you replace a flat tire?

17–2 Mechanical Failures
3. What should you do if your vehicle breaks down on a freeway?
4. What should you do if your brakes fail?
5. What is the proper way to jump-start a dead battery?

17–3 Skids
6. What are the four types of skids?
7. What is the proper way to respond to a skid?

17–4 Other Emergencies
8. What should you do if you get trapped in deep water?
9. What should you do if your car is on fire?
10. What precautions can you take to avoid being the victim of a car jacking?

SELF-TEST

Multiple Choice

1. Braking skids only occur in vehicles that:
   a. have antilock brakes.
   b. are traveling too slowly.
   c. are traveling on dry pavement.
   d. do not have antilock brakes.
2. Most tire blowouts are caused by:
   a. driving too fast.
   b. too little air pressure.
   c. too much air pressure.
   d. excessive tire wear.
3. Power-steering failures are more likely to occur when the:
   a. engine stalls.
   b. coolant is low.
   c. brakes overheat.
   d. car skids.
4. If you are changing a left front tire, put two blocks in front of and behind the:
   a. left front tire.
   b. right front tire.
   c. left rear tire.
   d. right rear tire.

5. Brake fade occurs when:
   a. brakes overheat.
   b. the parking brake fails.
   c. the brakes lock up.
   d. you fail to "pump" the brakes.

Sentence Completion

1. The best way to avoid a power skid is to take your foot off the _____ when you feel the tires lose traction.
2. The "positive" jumper cable is _____ in color.
3. _____ hold the wheels in place on a car.
4. On many vehicles, the emergency _____ is not full sized.
5. Your _____ will insulate you from electrical shock if power lines fall onto your vehicle.

Matching

Match the concepts in Column A with examples of the concepts in Column B.

Column A
1. __ Skid in circles
2. __ Cause of power skid
3. __ Response to brake failure
4. __ Cause of cornering skid
5. __ Response to headlight failure
6. __ Cause of braking skid
7. __ Response to engine stall
8. __ Skid from side to side

Column B
a. Downshift
b. "Spin out"
c. Activate turn signals
d. Loss of traction
e. Wheels lock up
f. Accelerate too quickly
g. "Fishtail"
h. Shift to NEUTRAL

Short Answer

1. Why should you never use water to put out a vehicle fire?
2. What other emergencies might a stalled engine cause?
3. What direction do you steer in a skid?
4. What should you do if you are trapped in deep water and the windows will not open?
5. How do you "ground" jumper cables?

Critical Thinking

1. You are driving in the center lane of a rural interstate highway at night when your front left tire goes flat. You have an emergency spare and flashlight in the trunk. There is a narrow shoulder and no median. You remember seeing a sign for an exit about 10 miles (15 km) ahead. What should you do?

2. You are waiting at a traffic light in an unfamiliar area when a car behind you rear-ends your vehicle. Two young men get out of the car and approach you, one on either side. They seem less interested in the damage than in confronting you. There are no other pedestrians or vehicles nearby. What should you do?

PROJECTS

1. Practice jacking up your car and changing a front and rear tire so that you will be prepared in case you have a flat. Make sure that you have all the necessary equipment handy in your trunk.

2. If you do not already have a set, purchase a pair of jumper cables. Read the instructions carefully. Determine which end of the cables is positive and negative. Look at your battery to see where the voltage is indicated and where the terminals are located. Go to a service station or garage and practice attaching the cables to another vehicle under the supervision of a certified auto mechanic.

Driving Responsibly at Home and Away from Home

unit 6

Chapter 18   Driving Under the Influence
Chapter 19   Licensing and Vehicle Ownership
Chapter 20   Vehicle Maintenance, Equipment, and Security
Chapter 21   Driving Away from Home

18 chapter

Driving Under the Influence

Driving under the influence means driving under the influence of any substance that affects your ability to drive safely, whether that substance is alcohol, an over-the-counter (OTC) medication, a prescription drug, or an illegal drug. Driving under the influence is one of the dumbest things you can do. Even if you avoid a collision driving home from a party or bar, you stand a good chance of ending up in jail if you are pulled over by the police. Being arrested for drunk driving is just the start of a long, costly, and humiliating process that could result in the loss of your driver's license, the confiscation of your vehicle, extremely heavy fines, and time behind bars. You will be punished, and rightly so, if you choose to endanger lives by drinking and driving.

CHAPTER OBJECTIVES

Upon completion of this chapter, you should be able to:

18–1 Alcohol and Other Drugs
1. Understand what determines a person's BAC.
2. Describe the physiological effects of alcohol.
3. Understand how alcohol affects judgment.
4. Understand what happens when drugs are combined.

18–2 Preventing Drunk Driving
5. Describe the role of peer pressure in driving under the influence.
6. Understand how to avoid drunk drivers on the road.

18–3 The Law and Driving Under the Influence
7. Understand what it means to be charged with DUI.
8. Understand what the implied consent law is.
9. Understand what the statutory summary suspension law is.
10. Describe the possible penalties for a DUI conviction.

KEY TERMS

blood alcohol concentration (BAC)
intoxication
oxidation
double vision
tunnel vision
inhibitions
synergistic effect
peer pressure
designated driver
Driving Under the Influence (DUI)
field sobriety tests
preliminary breath test (PBT)
implied consent law
breath test
blood test
urine test
statutory summary suspension law
alcohol/drug counseling
victim impact panels
zero tolerance law
judicial driving permit (JDP)
ignition interlock device (IID)
open container law

18–1 ALCOHOL AND OTHER DRUGS

If all drivers followed a simple rule—do not drink and drive—the roads of Illinois would be safer for everyone. Approximately 500 people are killed and nearly 5,000 people are injured each year in impaired-driving collisions. More than 40% of all traffic fatalities in the state involve alcohol. Tragically, many of the victims of drunk driving are sober drivers or passengers riding with a sober driver.

Alcohol and other drugs affect your judgment and physical abilities in ways that you might not expect. If you get behind the wheel while under the influence, you may not have the opportunity to second-guess your decision and realize that you are putting yourself, your passengers, and other roadway users at risk. When it comes to life and death, there are no second chances.

Types of Alcohol

Alcohol comes in a variety of forms, including beer, wine, and

Figure 18–1 Alcohol abuse by teenagers is a major contributor to collisions in Illinois.

spirits. The alcoholic content of beers ranges from about 3% for a "light" beer to 12% for some "microbrews." Wines have between 12% and 14% alcohol content. The strength of "hard liquor," or spirits, which include whiskey, gin, vodka, rum, brandy, and tequila, ranges from 20% to over 60% alcohol.

Blood Alcohol Concentration

The amount of alcohol in your body at a given time can be measured by determining your

factoid

Nearly two out of five people will be involved in an alcohol-related collision in their lifetime.

Type of Alcoholic Beverage	Percentage of Alcohol
Beer	
Domestic	3–5%
Imports	3–11%
Microbrews	4–12%
Wine	12–14%
Wine coolers	3–5%
Hard liquor	
Whiskey, vodka, rum, tequila	40%
Single malt scotch and some bourbon	40–63%
Fruit liqueur, blended liqueur,	
fortified wine (brandy, port)	20%

Figure 18–2 Alcohol content depends on the type of liquor that you are drinking.

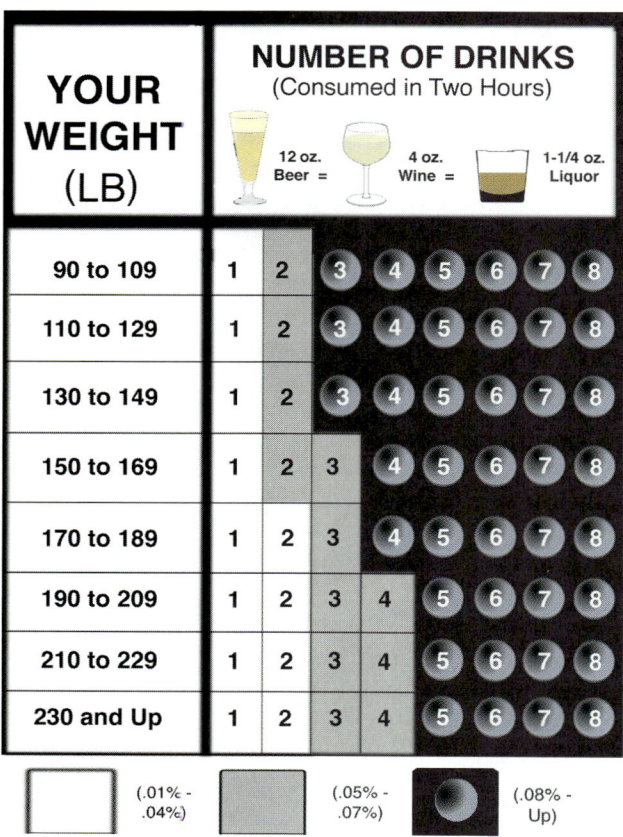

Figure 18–3 Your BAC determines your level of intoxication.

Alcohol can affect driving at a BAC as low as 0.02%, when reactions begin to slow and the first signs of poor judgment appear. At a BAC of 0.05%, driving skills begin to noticeably deteriorate. Reasoning is less reliable, muscles are too relaxed, and coordination is decreased. After reaching a BAC of 0.08%, your probability of a crash climbs rapidly. Judgment and inhibition are affected, as well as coordination, balance, vision, hearing, and speech.

If you have a BAC of 0.10% or higher, the chances that you will be involved in a collision dramatically increase. All mental and physical skills are extremely affected. Performing any task with your hands or feet, including walking without stumbling, is difficult. BAC levels above 0.30% can cause a driver to pass out and even go into a coma or die.

What Determines a Person's BAC

Two individuals having the same number of drinks may be affected quite differently by alcohol. A person's BAC depends on a number of variables:

- *The amount of alcohol consumed.* Alcohol is alcohol. Whether you drink a "shot" of scotch whiskey, a glass of wine, or a bottle of beer, your BAC will rise. Switching from beer to wine, or from hard liquor to beer, will not by itself affect your level of intoxication.

- *The amount of time over which the alcohol was con-*

blood alcohol concentration (BAC). BAC is a percentage measurement of the level of alcohol in your bloodstream. For example, if you have a BAC of 0.10%, it means that your bloodstream contains one-tenth of 1% of alcohol, or one part alcohol per 1,000 parts blood. BAC can be measured using chemical tests that analyze your breath, blood, or urine. Your level of **intoxication** is determined by your BAC reading.

factoid

Every 0.02% rise in BAC doubles the risk of highway death.

sumed. The faster you drink, the faster your BAC will rise. For example, your BAC rises faster if you have three drinks in 1 hour than if you have three drinks in 3 hours.

- *The amount of food in your stomach.* Some people falsely claim that "you cannot get drunk on a full stomach." In fact, food in your stomach will only slightly slow the absorption of alcohol into your bloodstream.
- *Your body weight.* A small person will be more affected by the same amount of alcohol than a larger person, who has more fluids in his or her body to dilute the alcohol.
- *Whether the alcohol is mixed with a carbonated beverage.* If alcohol is mixed with a carbonated beverage, it will enter your bloodstream faster and your BAC will therefore rise faster.
- *Whether you are a man or a woman.* Men can absorb alcohol better than women because their body contains a higher percentage of water, which helps dilute the alcohol.

How the Body Gets Rid of Alcohol

The body rids itself of alcohol by means of the liver, which turns alcohol into oxygen and carbon dioxide through a process called **oxidation.** Any alcohol that the liver cannot break down remains in the body. Your BAC increases when your alcohol intake and absorption rates are higher than the rate of the body's ability to oxidize the alcohol. You should never drive until you are absolutely sure that enough time has passed for your liver to oxidize the alcohol. This means considering how many drinks you have had in a given amount of time. For example, after three 1-ounce (30-g) drinks in the space of an hour, a person would need to allow over 4 hours to pass before being rid of the alcohol in his or her system.

Some people may tell you that you can "sober up" by taking a cold shower, drinking coffee, exercising, or other tricks. Even if they make you feel more alert temporarily, they will *not* decrease the amount of alcohol in your blood. Nothing can sober you up but time. It is also a myth that you can "build up a tolerance" to alcohol or that by driving slowly

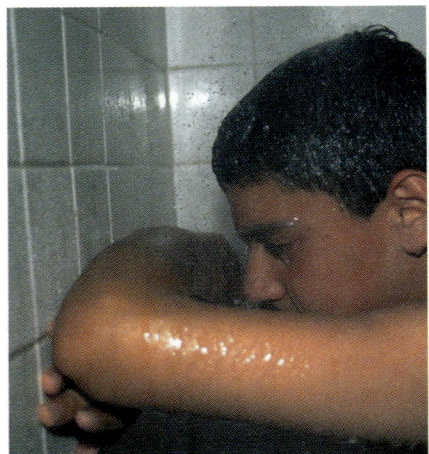

Figure 18–4 Only time, not cold showers or coffee, can sober you up.

famous collisions

On the night of December 26, 1986, a driver with a BAC of 0.199 traveling twice the legal speed limit struck a utility pole at the corner of Edgewood Road and Lewis Avenue in Waukegan, killing himself and two passengers when the roof was ripped off the car and all three were ejected.

you can reduce the effects of alcohol impairment.

Once you let some time pass after having a few drinks, it is easy to convince yourself that you no longer feel the effects of the alcohol and are sober. This state of mind is a delusion because you are comparing your peak feeling of impairment with the return of only some of your physical and mental abilities as the body eliminates alcohol from the blood. Do not assume that sleep can magically remove all the alcohol from your system. If you have been drinking heavily late at night, the next day you may still have a dangerously high BAC which is in excess of the legal limits for driving.

The Physiological Effects of Alcohol

You must exercise mental and physical skills in tandem to safely control your vehicle. Alcohol slows the central nervous system and dulls these skills to a great extent. Your vision, reflexes, and coordination, as well as your breathing and heartbeat, are all affected by alcohol. This makes performing multiple tasks at once, such as braking and steering, extremely difficult.

Vision

Good visual skills are crucial to carrying out the SAFE process. Alcohol relaxes the fine, delicate muscles that move the eyes and allow them to focus. This reduces your ability to scan effectively and blurs your vision. In some people, alcohol causes **double vision,** an uncontrollable rapid vibration of the eye that makes it virtually impossible to see at all. Alcohol further impairs your awareness of the driving environment by distorting your depth perception, narrowing the scope of your peripheral vision, and impairing your night vision. Colors and shapes can be distorted, and you will be less able to interpret what you see. Your vision need only be mildly impaired by alcohol to have trouble identifying hazards.

You need your depth perception to judge the speed of oncoming vehicles, your stopping distance, and your distance from other vehicles, objects, and people around you. If you have been drinking, you may perceive a red light to be farther away than it actually is and have to slam on your brakes at the last second to avoid entering the intersection. You will not have a sense of where other vehicles are in relation to your own to make decisions about how to position yourself safely on the road. Drivers under the influence of alcohol also tend to stare at one spot and forget to scan constantly with their eyes. The combination of blurred peripheral vision and fixation on a narrow field ahead can result in **tunnel vision,** a 70% reduction in your field of vision.

You need more light to see dimly lit objects. Alcohol impairs your night vision as much as 25% by reducing the time it takes for the pupils of your eyes to respond to changes in light levels. Because your eyes take longer to adjust, you may be blinded more easily by oncoming headlights

---

factoid

Although sixteen- to twenty-year-olds make up only 15% of the licensed drivers in Illinois, they are involved in nearly 30% of all fatal alcohol-related crashes.

and continue to have blurred vision long after the oncoming car has passed.

Physical Reflexes and Coordination

The more alcohol that enters the bloodstream, the more the portions of the brain that control physical reflexes and coordination become depressed. This obviously has severe effects on your ability to execute driving maneuvers, which depends on both skills.

Your reaction time gets longer the more you drink. The brain works less efficiently, and instructions to the muscles are delayed. After two or more drinks, a typical driver becomes physically slower and less alert. The effect of alcohol on your muscles is to decrease coordination. Your muscles react more slowly to commands from your brain and function with less control. This can make your driving sloppy. You will tend to oversteer, understeer, brake late, overbrake, or not brake at all. You could even lose control of the car. This impairment of coordination affects beginning drivers most because they have fewer reliable automatic reactions and ingrained skills.

Drinking can have many other unpleasant side effects. Consuming large quantities of alcohol can cause a driver to pass out behind the wheel. It can also produce "hangovers" hours later, after the immediate effects have worn off, that consist of feeling headachy, nauseous, dizzy, and tired. Heavy or "binge" drinking can cause death.

How Alcohol Affects Judgment

When driving, your mental condition is just as important as your physical condition. The greatest number of alcohol-related collisions actually result from faulty thinking on the drinking driver's part, not the poor execution of physical skills such as steering. When you drink, your ability to reason and make sound judgments is decreased. Your ability to concentrate and to remember things is drastically reduced, and it takes longer to process information. Alcohol-impaired drivers may be more inclined to run a yellow light or attempt to pass without sufficient room. Even worse, they lose the ability to know that they are making poor decisions. This loss of self-judgment leads people who should not be behind the wheel into thinking that they are quite capable of driving.

Often a person drinking alcohol will get a sense of well-being and will experience an increase in confidence. In this state, sometimes described as euphoria, a person may believe he or she can do anything, or is more skilled and entertaining than he or she actually is. **Inhibitions** are the elements of your personality that stop you from behaving without regard to possible consequences. The loss of inhibitions is one of the reasons people drink. This can lead an individual to take chances that he or she normally would not take if sober. In addition, alcohol can magnify the feelings of an emotionally upset person.

factoid

Drivers involved in fatal crashes who have been drinking use safety belts at a far lower rate than sober drivers.

Other Drugs

Alcohol is a drug, but what people normally think of as "drugs" can also greatly affect the mind and body. Many drugs remain in the body much longer than alcohol. The effects of some drugs can continue well after the drug itself can no longer be detected in the bloodstream. Other drugs can be detected long after the effects wear off. Most of the perceptual and motor abilities that are necessary for driving are affected when the driver is under the influence of either legal or illegal drugs.

OTC drugs such as aspirin and other pain relievers, cold and allergy remedies, and medicines for back pain and arthritis can cause dizziness, drowsiness, slower reaction times, reduced coordination, and other side effects that reduce your ability to drive safely. Prescription drugs, which usually contain either higher dosages of the same active ingredients found in OTC drugs or other, more potent drugs, can have even more powerful effects on your body and ability to drive. When you take any OTC or prescription drug, check the label for warnings and ask your doctor or pharmacist if you will be able to drive safely.

It is a good idea not to drive the first time that you take a prescription or OTC drug that is new to you. Note how it impacts you—how it makes you feel—and decide whether it is advisable for you *ever* to drive when taking that particular medication.

Illegal as well as legal drugs can put you and others at great risk for serious injury or death in a collision if you drive under their influence.

- Depressants produce side effects similar to alcohol, including irritability, confusion, drowsiness, dizziness, and poor eye–hand coordination, that are disruptive to basic driving skills.
- Narcotics can slow a driver's reaction time, impair motor skills, and cause dimness of vision. Stimulants adversely affect people's ability to drive by making the users aggressive and overconfident. Two common stimulants that people use every day are caffeine, which is found in coffee, tea, and cola soft drinks, and nicotine, which is found in cigarettes.
- A person under the influence of hallucinogens can become confused, unable to think clearly or concentrate, and have an altered sense of direction, space, and time.

Figure 18–5 Legal drugs can impair driving skills just like illegal drugs.

- It takes only a small amount of the main active chemical in marijuana or hashish to impair your ability to see, steer, brake, and make correct driving decisions. Using these drugs results in drowsiness, fragmented thought patterns, and problems evaluating spatial relationships and the passage of time.

The effects of mixing any one drug with another drug or with alcohol are complex and can lead to serious impairment of driving skills. Alcohol in the bloodstream can trigger what is called a **synergistic effect** that enhances the side effects of certain drugs, including some legal drugs. When this happens, your body focuses on the elimination of alcohol as the primary drug in the body and ignores the other drug that you have taken. Because a "normal" dosage assumes that your body will eliminate part of the drug, the effects of the drug are multiplied.

18–2 PREVENTING DRUNK DRIVING

The best way to combat the problem of driving under the influence is to prevent people impaired by alcohol or other drugs from getting onto the road in the first place. *Everyone* at a place or event where alcohol is consumed is obligated to take an active role in keeping drinkers from driving.

Resisting Peer Pressure

The desire to "fit in with the crowd" can lead people to engage in dangerous and illegal behavior such as drinking and driving. The influence that others of your own age have on you is called **peer pressure.** People do things they would never do on their own to impress friends, classmates, co-workers, or members of the opposite sex. If others decide to act irresponsibly, do not participate. Honestly explain your choice and ask them to respect it. Be aware that if you give a party or organize some other activity at which alcohol is provided, and a person attending the function causes an injury or fatality while driving under the influence, you and your family may be sued in civil court by the victim or victim's family.

Setting Limits on Yourself

Ideally, you should not have *any* alcohol in your system when you get behind the wheel. If you do drink, consider your weight and other factors that influence BAC. Most of us have some idea when

Where am I?

Figure 18–6 This steak house was once a "speakeasy" where Al Capone and other legendary gangsters once rubbed elbows.

Figure 18–7 If you decide to drink, set a drinking limit for yourself and stick to it.

Guess the vanity plate

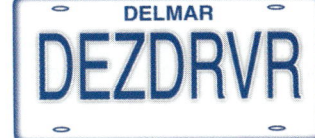

Figure 18–8

Figure 18–9 If you are not sober enough to drive, arrange for alternative transportation to get home.

Wild Wheels

Figure 18–10 Who was the "designated driver" of this famous TV car?

driving tips

As one alternative to drinking and driving, sign a "contract" with your parents or guardians in which they agree to give you a ride home with no questions asked at the time.

we have had too much to drink. Our speech slurs, we become overly friendly, our voices get louder, and we get a little dizzy. Recognize the signs that your body is giving you and stop drinking. Trust your friends when they express concern that you have had too much to drink or if they ask for your car keys. If your behavior tells others that you have been drinking, you definitely do not belong on the road.

Do not try to match the drinking of other friends or party-goers. "Nurse" a drink with small, occasional sips, skip a round, or have a nonalcoholic beverage. If you are afraid to admit that you do not want to drink, invent a creative excuse for not participating. Anything is better than drinking too much and later causing an injury or death on the road. Keep active and circulate among the crowd. If you are busy talking, dancing, or walking around, you will not be inclined to drink as much. *Never* accept a mixed drink, which may contain excessive amounts of alcohol that you will not be able to detect.

Alternatives to Drinking and Driving

If you know that you will be drinking, the best safeguard against getting behind the wheel is to have a nondrinking friend be your **designated driver.** In this way, you will not have to make a decision about whether to drive yourself home or not. Before going to a bar or party, choose someone who will volunteer not to drink during the event and who is willing to drive others who are drinking. You and your friends can take turns being the designated driver on different occasions. Some establishments will provide free nonalcoholic beverages to designated drivers to encourage responsible drinking.

If you have not appointed a designated driver, call a friend or relative for a ride. Do not worry about your car until the next day, even if you parked in a bad area or if you risk a ticket. If you are concerned about your vehicle being towed, vandalized, or stolen, ask the person who comes to get you to bring a friend who can drive your car home. If you cannot get a friend or relative to help you, call a taxi or use public transportation to get home. Some communities offer special programs on weekends and/or during special events that will provide you with a free cab or bus ride home if you are too impaired to drive and have no money to pay a fare. If none of these options is available at a party, ask the person giving the party if you can stay for the night.

Refuse to be a passenger in a car if you know or suspect that the driver has been drinking. You have the right to expect a driver to get you to your destination as safely as possible. If you are sober, offer to drive. If the driver refuses, find another way to get home.

Avoiding Drunk Drivers on the Road

The danger of encountering drunk drivers on the roadway varies depending on when you drive. The alcohol involvement rate in crashes is far higher at night than during the day. The risk of colliding with a drunk driver is also higher on weekends, holidays, and vacation periods. Always remember, however, that drunk drivers are on the road at all times.

No matter when you drive, be on the alert for the following signs of drivers who may be under the influence:

- Driving at inconsistent speeds, too fast, or too slow
- "Riding" the lane markers on the road or straddling the center line
- Weaving back and forth across the roadway
- Drifting into other lanes or opposing traffic
- Driving at the edge of the roadway
- Driving on or next to the shoulder
- Traveling in a turn lane
- Braking erratically or stopping without cause
- Stopping and starting suddenly
- Responding slowly to traffic signals
- Tailgating
- Driving at night without headlights or leaving high beams on
- Passing other vehicles recklessly
- Almost hitting another vehicle or object in the road
- Leaving turn signals blinking or signaling turns inconsistently
- Making wide or abrupt turns

Drivers may also exhibit indications of being drunk, including tightly gripping the steering wheel, slouching in the seat, making strange or obscene gestures, driving with windows open in cold weather, sticking their heads outside the vehicle, driving with their face too close to the windshield,

factoid

The proportion of fatal collisons in Illinois that are alcohol-related is about four times higher at night than during the day.

More than half of all alcohol-related crashes in Illinois occur during the weekend.

and not turning their head to scan ahead.

Once you have identified a driver you suspect to be under the influence of alcohol or other drugs, keep your distance as much as possible. If the other driver is ahead of you, do not pass him or her from behind. If the other driver is behind you, pull off the road or onto a side street when it is safe to do so. If you believe that the driver is jeopardizing the safety of the roadway and is likely to get involved in a collision, note the color, make, and model of the vehicle and its direction of travel, as well as the license number if you can see it without endangering yourself. As soon as possible, report the driver to the police. If you have a cellular phone, you can report a suspected drunk driver to the Illinois State Police by dialing *ISP (477). As in any driving situation, remember that the best way to protect yourself from injury or death in a collision with a drunk driver is to wear your safety belt.

18–3 THE LAW AND DRIVING UNDER THE INFLUENCE

Because of the high numbers of fatalities resulting from drunk driving, Illinois has enacted extremely strict laws against driving under the influence. Drunk driving is considered a serious *criminal* offense, and the public supports the need for harsh penalties. However, most people are not aware of just how much power the law has to stop drunk driving. Consider the following:

- You can be legally "impaired" without ever having a drink if you have taken other substances—including OTC and prescribed drugs—that affect your ability to drive safely.

- Thanks to the sophisticated tools of modern technology, you can be proven legally drunk even if all your physical and mental faculties appear in perfect working order.

- Police can and do arrest people whom they believe are under the influence of alcohol or other drugs before they start their cars, while they are sitting still in a parked or wrecked vehicle with the motor off, and even when they have pulled off the road to "sleep it off."

- You can be arrested for being under the influence while operating a golf cart or riding a horse.

DUI

It is illegal to operate a motor vehicle while under the influence of alcohol or other drugs. If you are arrested for violating this law, the police will charge you with **Driving Under the Influence (DUI).** You will be required to take a chemical test to determine your BAC. If you are of legal drinking age (twenty-one or older) and the test shows that you have a BAC of 0.08% or higher, you will automatically be presumed to be under the influence of alcohol. In other words, having a BAC of 0.08% is sufficient evidence to be convicted of DUI.

At any given time on the road, 1 of every 100 drivers has a BAC of 0.10 or more.

If you have a BAC below 0.08%, be aware that you can still be arrested and prosecuted for DUI if there is other circumstantial evidence, such as field sobriety test results and police testimony, that you were driving under the influence.

Illinois has a hard-line "Use it and lose it" policy regarding underage offenders convicted of drinking and driving. If you are under the legal drinking age, you can be arrested and prosecuted for DUI if *any* alcohol is detected in your system.

Field Sobriety Tests

If you are pulled over by a law-enforcement officer and he or she suspects that you are under the influence of alcohol or other drugs, you will likely be asked to get out of your vehicle and perform several **field sobriety tests** at a safe spot nearby. Field sobriety tests are scientifically designed to allow trained officers to detect physical or mental impairment from alcohol and/or other drugs. One such test requires the driver to follow a pen, small flashlight, or finger with his or her eyes as the officer moves it from side to side. Most people impaired by alcohol and certain illegal drugs have an exaggerated case of nystagmus, a natural condition in which the eyes involuntarily jerk as they change the direction of their gaze.

Other field tests check for divided attention impairment, the inability to perform more than one task at a time. Because both physical and mental tasks must be performed simultaneously when operating a vehicle, these tests are particularly useful in indicating whether you can drive

factoid

Nearly 50,000 people are arrested for driving under the influence in Illinois each year.

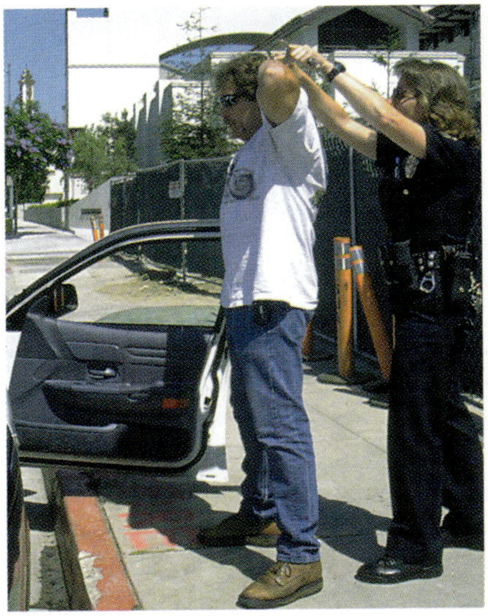

Figure 18–11 Illinois enforces severe penalties for driving under the influence.

Figure 18–12 Just one drunk-driving conviction can lead to jail time.

HOWARD E. ROLLINS, JR.

After a long career on the stage, actor Howard E. Rollins, Jr. (1950–1996), appeared in a number of critically acclaimed TV and feature film roles, including *A Soldier's Story* and *Ragtime,* for which he received an Oscar nomination for Best Supporting Actor in 1981. He was best known for his role as Detective Virgil Tibbs in the long-running series *In the Heat of the Night.* Rollins' promising career began to fall apart in 1988 when he was arrested near Baton Rouge, Louisiana, in March 1988 for DWI (Driving While Intoxicated), speeding, and cocaine possession. Rollins was arrested three times in 1992 in Newton and Rockdale counties in Georgia, where the series was filmed, for DUI, speeding, and reckless driving. In 1993, he was arrested for speeding and driving with a suspended license. Despite numerous fines and time in jail, including one 5-week sentence for violating his probation, Rollins continued to get into trouble. In November 1993, three days before his license was to be reinstated, he was arrested yet again for DUI and other related traffic charges. This time he was fined $3,000, sentenced to six months in the Rockdale County Jail, and banned from the area for two years. Rollins, whose numerous run-ins with local authorities had brought unwanted publicity to the set of the police drama and caused awkward changes in the script, was released from the show. If you have a serious alcohol or drug problem, get help. Being a repeat DUI offender will ruin you financially and make your recovery that much harder.

factoid

Nearly 40% of drunk drivers involved in fatal crashes in Illinois have a BAC over 0.08%.

safely. They measure not only your ability to follow simple instructions but also your balance and cognitive skills. For example, an officer might ask you to walk heel to toe along a straight line while counting off your steps aloud, turn on one foot, and return in the same way in the opposite direction. You might also be asked to stand on one leg for 30 seconds, recite the alphabet, or close your eyes and touch your nose with your finger.

Upon the completion of these tests, the officer may administer a **preliminary breath test (PBT).** This test calls for you to blow into a tube attached to a handheld device that registers an accurate estimate of your BAC. The results of a preliminary breath test cannot be used as evidence of insobriety against you in court, but a high reading would be considered as probable cause for the officer to arrest you.

Chemical Tests

Like most jurisdictions, Illinois has an **implied consent law** for drivers. According to this law, any person operating a motor vehicle is "deemed to have given consent" to have his or her breath, blood, or urine tested for the presence of alcohol or other drugs. This means that whenever or wherever you are in actual physical control of a vehicle, you are legally required to submit to a chemical test, if requested by any law-enforcement officer who has

AUTO ACCESSORIES — Portable Breathalyzer

To help prevent you or your friends from drinking and driving, consider purchasing your own breathalyzer. Portable models that run on batteries and can be stored in your car or kept on hand for parties are now available at auto supply stores. These models are designed for personal use and give fast and reasonably accurate BAC readings.

"probable cause" to believe that you are under the influence.

If you fail or refuse to perform the field sobriety tests, or exhibit other signs of intoxication such as a staggering walk, thick-tongued speech, flushed face, red or puffy eyes, or the smell of liquor, most likely you will be placed under arrest, read your rights, and taken into custody. Unless there is a sober passenger in your vehicle, it will be impounded. The arresting officer will administer or arrange to have administered one or more chemical tests to determine the concentration of alcohol or drugs in your breath, blood, or urine.

In most cases, you will be asked to take a **breath test**. This test requires you to blow into a tube attached to a breath-analyzing device, or "breathalyzer," that checks for the presence of alcohol using infrared light and/or "fuel cell" technology. The results of this test, unlike the preliminary breath test, are considered scientifically accurate and can be used as evidence against you in court.

A **blood test** is even more accurate in determining BAC. A nurse or other medical professional is required to take a blood sample. Blood tests are typically given if you are taken to a medical facility following a crash in which alcohol is suspected as being a factor.

If the arresting officer has cause to believe that you are under the influence of drugs, you will be asked to take either a blood test or urine test. A **urine test** is also typically performed by medical personnel at a medical facility. Because it is the least effective means of detecting BAC, the urine test is rarely used when alcohol is suspected.

After you take the chemical test(s) offered by the arresting officer, you have the right to re-

factoid

Drivers under the age of twenty-one are twice as likely as older drivers to cause a collision in which alcohol is a factor.

REALITY CHECK — Do Not Test "Fresh" Breath

If you think that using a breath freshener will help erase the evidence of drinking, think again. Because many breath fresheners have alcohol in them, rinsing your mouth before taking an alcohol breath test can actually increase the measurement of alcohol in your breath!

Figure 18–13 If a police officer suspects you of DUI, he or she will ask you to perform several field sobriety tests.

factoid

Nine out of ten people arrested for DUI in Illinois lose their driving privileges.

quest that another test of any type be administered by a medical professional of your choosing at your own expense.

If you refuse to take a chemical test, the arresting officer will issue you a temporary driving permit valid for forty-five days. During this time, you have the right to file a written request for a hearing before the court. If the court determines that the arresting officer had reasonable grounds to believe that you were driving under the influence, your license will be suspended for six months for a first refusal and three years for a second or subsequent refusal within five years. *These penalties apply whether or not you are subsequently convicted of DUI in criminal court.* For second and subsequent refusals, you may not apply for restricted driving privileges like going to work or

DRIVING TIPS   If You Are Pulled Over by the Police

If you see a police car's flashing lights in your rearview mirror or hear a siren, slow down until you are sure that the officer is signaling to you. Pull over to the right side of the road or where the officer instructs you to park and stop as soon as it is safe to do so. Stay in your vehicle and follow the instructions that the officer gives you. Sit still, keep your hands on the wheel, and avoid sudden movements. Be prepared to show your driver's license, vehicle registration, and proof of insurance. Answer all questions honestly and courteously. If you need to reach into the glove compartment, under your seat, or into another area of the vehicle not in full view of the officer, tell the officer what you are about to do first. If he or she shines a flashlight in your face or otherwise acts in a way that you find intimidating, keep in mind that many police officers are killed in the line of duty during routine stops. Therefore, they are extremely cautious when approaching a vehicle, especially at night or if they cannot see well inside.

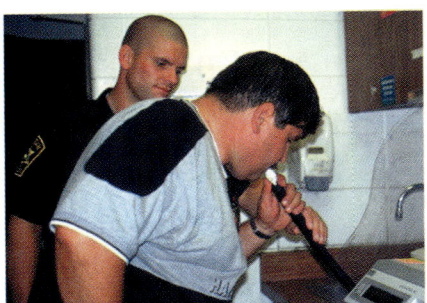

Figure 18–14 The breath test is the most common chemical test used to measure BAC.

school until the last year of your three-year suspension.

Administrative Penalties for DUI

Illinois has what is called a **statutory summary suspension law,** which mandates the immediate suspension of your driver's license for at least ninety days if you have a BAC of 0.08% or higher on a first offense. A second or subsequent offense within five years of the first will result in a one-year license suspension.

If you are under the age of twenty-one, keep in mind that the same penalties apply if you test positive for any alcohol. Many states set a maximum BAC of 0.01% or 0.02% for juvenile offenders, but in Illinois, it is 0.00%. The message is clear for teenagers who think they can avoid trouble by having "only" one drink before they get behind the wheel: *Use it and lose it.*

If you fail the chemical test, the arresting officer will seize your license on the spot and forward it to court, which will maintain possession of it until the end of your summary suspension.

Be aware that an administrative suspension is independent of any license suspension or revocation ordered by the court in your criminal trial. You have the right to appeal the suspension with the court within ninety days of the time you receive your notice of the suspension. The hearing must occur within thirty days of your requesting it or at your first appearance in court if responding to a traffic ticket received at the time you were arrested for DUI. As in the case of chemical test refusals, the arresting officer will issue you a temporary driving permit valid for forty-five days to give you the opportunity to get your affairs in order.

Criminal Penalties for DUI

If you are convicted of DUI, the criminal penalties you face generally depend on whether it is a first, second, or subsequent offense. For a first DUI offense, which is a misdemeanor, you face up to one year in jail and a fine up to $2,500. Driver Services will also revoke your driver's license for a minimum of one year. In addition, your vehicle's registration may be suspended and the license plates impounded.

First-time DUI offenders who plead guilty *may* be granted court supervision. In exchange for fully complying with all of the court's requirements for a certain period of time, individuals can avoid the

REALITY CHECK: Saturation Patrols and Roadside Safety Checkpoints

To help crack down on drunk drivers, law-enforcement agencies in Illinois conduct "saturation patrols" during times when impaired drivers are most likely to be on the road and in areas with a high concentration of alcohol-related crashes. "Roadside safety checkpoints" are also used throughout the state. These are roadblocks in which officers randomly stop drivers to determine whether any are impaired by alcohol or other drugs. They are also used to determine whether drivers are insured and to spot-check for motor-vehicle safety violations. Checkpoint boundaries are marked with cones, flares, portable lights, the emergency lights of police cars, or other devices. As vehicles enter the area, they are randomly stopped for observation or signaled to proceed through by an officer. Those drivers who are stopped and who exhibit signs of impairment are usually taken to a secured area away from the traffic lanes for field sobriety tests. Both saturation patrols and roadside safety checkpoints have proven extremely effective in deterring motorists from driving under the influence.

Highway heroes

While writing a ticket for a motorcyclist, Patrolman Barry A. Shalin of the Morton Grove Police Department was struck and killed by a drunk driver on the evening of October 9, 1989, at the intersection of Austin Avenue and Main Street.

criminal penalties and social stigma of a conviction. Be aware, however, that you can receive court supervision for a DUI offense only once in your life. You also are still subject to the statutory summary suspension of your license by Driver Services. If you fail to comply with all of the judge's requirements, your offense will be entered into the record as a conviction, with all the penalties of a conviction.

If you are found guilty of committing a second DUI offense within five years of the first, you will face a mandatory minimum of 48 consecutive hours in jail or ten days of community service. If you are convicted of a second DUI offense within twenty years of the first, your license will be revoked for a minimum of five years. A third or subsequent DUI conviction within a five-year period is a felony that can result in a sentence of one to three years in a state penitentiary, up to $25,000 in fines, and the revocation of your driver's license for a minimum of ten years. If you are convicted of a DUI for a fourth or subsequent time within five years, you will lose your license for life.

The criminal penalties for DUI increase significantly if there are injuries or death involved. If you seriously hurt, disfigure or kill others while driving under the influence of alcohol or other drugs, you could be sentenced to up to twenty-eight years in prison. You will be subject to a mandatory minimum sentence of either 48 consecutive hours in jail or thirty days of community service. These penalties apply even if you have no previous DUI offenses on your driving record.

Also, be aware that if you were carrying a passenger under the age of sixteen at the time of your arrest for DUI, the court will automatically increase the severity of your penalties even if no injuries or deaths occurred. These penalties include a minimum fine of $500 and five days of community service in a program benefiting children.

Other Costs of DUI

The costs of a DUI conviction do not end there. Because you will be considered a high-risk driver, your auto insurance will be *much* more expensive for years to come—if you can even find a company that will insure you! In fact, your policy will cost so much that, to make sure that you are indeed covered, the law requires you to maintain formal proof of financial responsibility with Driver Services. Your driver's license, vehicle registration, and license plates will not be reinstated until such proof is on file. You must maintain so-called "SR-22 insurance" (named for the form submitted to Driver Services, usually by your carrier, verifying proof of insurance) for three years. If at any time you cancel your policy or it lapses during that time, your driving privileges will be immediately suspended until you are in compliance.

Speaking of insurance, keep in mind that a growing number of health insurance companies are excluding drunk-driving injuries from their plans, meaning that if you sustain an injury in a collision in which you are convicted of DUI, you may have to pick up the tab for your own medical expenses even though you are insured.

In addition to higher insurance premiums, you must also pay for professional **alcohol/drug counseling,** which is mandatory for all DUI offenders in Illinois. Depending on the recommendations of your counselor and whether you have a history of alcohol or drug abuse, the court may impose a more extensive treatment or rehabilitation program, including inpatient therapy, as part of your sentence or as a requirement to reinstate your license. Such programs can cost anywhere from $150 to several thousand dollars, not to mention the time you will lose.

Many persons convicted of DUI in Illinois are required to attend, at their own expense, a victim impact panel one evening or weekend. **Victim impact panels** consist of volunteers who relate how their lives have been changed by an alcohol-impaired driver. They typically meet at least once each month, usually at a local courthouse, library, city hall, or traffic school.

Finally, you will have to pay for attorney's fees, court costs, a hefty license reinstatement charge, and depending on the circumstances, additional penalties. For example, if you caused a collision as a result of your impairment you will be charged for emergency medical services if they responded to the incident. If you damaged any public property, such as a utility pole or landscaping on a median strip, you will likewise have to pick up the bill.

DUI victims have the right to appear at your criminal trial and

famous collisions

In the early morning hours of April 3, 1998, a seventeen-year-old driver with a BAC almost twice the legal limit crashed into a tree on Ashbury Avenue near Trafalgar Court in Bolingbrook (a suburb of Chicago), killing her two teenage passengers, one of whom was sitting on the lap of the other.

factoid

Nearly half of all fatally injured drivers in Illinois between the ages of sixteen and twenty-four test positive for alcohol.

are permitted to make a statement concerning the impact on their lives of your impaired driving. Such statements will be entered as evidence in your trial and can result in more severe penalties. If you are subsequently sued in civil court and found liable for causing injury, death, or property damage, you as well as your family could face drastic economic consequences. If you are a minor (under the age of eighteen), your parents could be sued for negligently monitoring your behavior.

"Use It and Lose It"

Under the state's **zero tolerance law,** special criminal penalties apply to DUI offenders under the age of twenty-one. A first-time conviction will result in a two-year, rather than a one-year, license revocation. Those who commit a second offense will lose their driving privileges for a minimum of five years or until they turn twenty-one, whichever is longer.

Underage drinkers convicted of DUI may be required to attend a special program known as the Youthful Intoxicated Driver's Visitation Program as a requirement to get their license reinstated. This program requires youthful offenders to experience firsthand the devastating results of the irresponsible use of alcohol and drugs by attending one of the following: a rehabilitation facility treating victims recovering from injuries received in a DUI-related collision, a treatment facility caring for those in the terminal stages of alcoholism, or the local coroner's office.

Restricted Licenses

First-time DUI offenders who are eighteen or older can apply to the court for a restricted license known as a **judicial driving permit (JDP)** any time after they are notified of their statutory summary suspension by Driver Services. Applicants must demonstrate that they need to drive to or from work or school, or to obtain regularly required medical care or drug treatment for themselves or a family member. Those who qualify are granted restricted driving privileges after serving thirty days of their statutory summary suspension. The permit is valid until the end of the suspension period, at which time the grantee may apply to have his or her license reinstated.

Those who have been convicted of DUI for a second or subsequent time who meet similar criteria regarding the need to drive must apply directly to Driver Services for a standard restricted driving permit (RDP). Those who qualify are granted an RDP only after serving at least ninety days of their statutory summary suspension. RDPs are valid for only one year, so those who have had their license revoked for longer than that must reapply.

After serving one year of license revocation, adult DUI offenders under the age of twenty-one may also request an RDP. In addition to the restrictions mentioned earlier, you may be permitted to drive only between the hours of 5 AM and 9 PM. After driving for one year with an RDP, you may reapply for reinstatement of your license. Driver Services may either reinstate your driving privileges or extend the RDP, in one-year increments, until you turn twenty-one.

Minors convicted of DUI are not eligible for *any* type of restricted driving privileges.

Ignition Interlock Devices

If you are convicted of a second or subsequent DUI offense within a five-year period, Driver Services or the court may require you to equip your vehicle with an **ignition interlock device (IID)** when you apply for a restricted license or to reinstate your license. This handheld device attached to the vehicle requires you to breathe into a tube before the vehicle's ignition can be activated. The car will not start if you have a BAC above 0.025%.

You must pay to install an IID and have it checked after the first month and every two months thereafter. If you are caught driving without the device during the time that the court requires you to use it, your driving privileges will be withdrawn for at least an additional year. Once your license is reinstated, the length of time that you would be required to use an IID will also be increased.

The Open Container Law

To further discourage drinking and driving, Illinois has what is known as an **open container law,** making it illegal for any driver or passenger to drink while driving or to have an open container of alcohol in a motor vehicle. Any alcoholic beverage transported in

Figure 18–15 Interlock ignition devices are one of the high-tech weapons used in the war against drunk driving in Illinois.

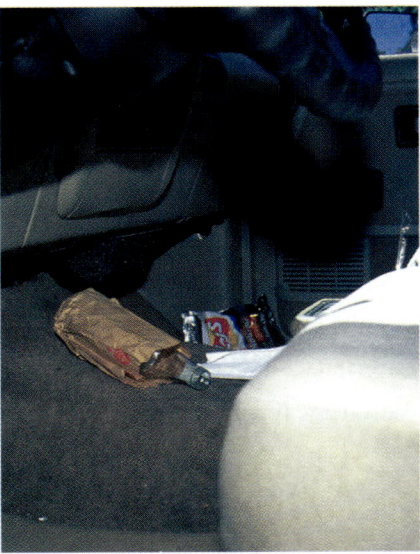

Figure 18–16 Having an open container of alcohol in your vehicle, even if it is empty, is illegal in Illinois.

the passenger compartment of a motor vehicle, including the glove compartment, must be full, sealed, and unopened.

Violators of this law face up to $500 in fines. Anyone under the age of twenty-one, or anyone convicted of violating the open container law a second time in five years, will also have his or her driver's license suspended for one year. If you are a minor and violate this law a second time within five years, your driver's license will be revoked.

Be aware that if you or any passengers in your vehicle are caught drinking, the offense will be recorded only on *your* driving record, but everyone in the vehicle can be cited, even if only one person is in violation of the law.

YOUR TURN

18–1 Alcohol

1. What determines a person's BAC?
2. What are the physiological effects of alcohol?
3. How does alcohol affect judgment?
4. What happens when alcohol and drugs are combined?

18–2 Preventing Drunk Driving

5. What is the role of peer pressure in driving under the influence?
6. How can you avoid drunk drivers on the road?

18–3 The Law and Driving Under the Influence

7. What does it mean to be charged with DUI?
8. What is the implied consent law?
9. What is the statutory summary suspension law?
10. What are the possible penalties for a DUI conviction?

CHAPTER 18 Driving Under the Influence ♦ **443**

SELF-TEST

Multiple Choice

1. A synergistic effect may be triggered by alcohol if you:
   - **a.** mix beer and hard liquor.
   - **b.** eat before drinking.
   - **c.** take drugs.
   - **d.** take a cold shower.
2. If you experience tunnel vision, your field of vision is reduced by:
   - **a.** 30%.
   - **b.** 50%.
   - **c.** 70%.
   - **d.** 100%.
3. A person who refuses to take a chemical test:
   - **a.** is free to continue driving.
   - **b.** is exempt from the implied consent law.
   - **c.** will have his or her driver's license suspended.
   - **d.** cannot be charged with driving under the influence.
4. According to the implied consent law, any law-enforcement officer can request a chemical test of your BAC if he or she:
   - **a.** has "probable cause" to believe that you are driving under the influence.
   - **b.** has a PBT.
   - **c.** has your permission.
   - **d.** sees an open container in your vehicle.
5. Penalties for a first-time offender convicted of DUI usually include:
   - **a.** a one-year license revocation.
   - **b.** up to $25,000 in fines.
   - **c.** community service.
   - **d.** installation of an interlock ignition device.

Sentence Completion

1. Your level of _____ is determined by your BAC.
2. Alcohol can cause _____, an uncontrollable rapid vibration of the eye.
3. A _____ is an "on the spot" roadside evaluation for intoxication.
4. The influence that others your own age have on you is called _____.
5. The best safeguard against drunk driving is to have a nondrinking friend volunteer to be a _____.

Matching

Match the concepts in Column A with examples of the concepts in Column B.

Column A
1. __ "Hard liquor"
2. __ OTC
3. __ PBT
4. __ Most accurate measure of BAC
5. __ Zero tolerance
6. __ Field sobriety test
7. __ Statutory summary suspension
8. __ Least accurate measure of BAC

Column B
a. Blood test
b. Aspirin
c. Estimates BAC
d. 0.08%
e. Urine test
f. Horizontal gaze nystagmus
g. Vodka
h. 0.00%

Short Answer

1. Why might an alcohol-impaired driver be willing to take more risks?
2. How does the body get rid of alcohol?
3. What is the purpose of roadside safety check points?
4. When can a law-enforcement officer request that you take a chemical test for the presence of alcohol or other drugs?
5. What are some of the signs that a driver may be intoxicated?

Critical Thinking

1. You are at a party, and the friend you came with has had several drinks. He appears to be sober and shows no symptoms of being even slightly impaired. He tells you that he has had a couple of cups of coffee and feels fine. You only live about 5 miles (8 km) away. What should you do?
2. You are driving a group of friends to a concert. One of your buddies has a case of beer he sneaked into the car with him. Even though you are all under the legal drinking age and there is an open container law where you live, he starts to open the beers and pass them around. What should you do?

PROJECTS

1. Visit your local drug store or supermarket and look at common OTC remedies. Read the labels and see which ones have side effects that might impair driving when taken with alcohol. Write down your results and share them with your class.
2. Contact the equivalent of Driver Services in Indiana, Iowa, Kentucky, Missouri, or Wisconsin to find out what the laws on driving under the influence are. What is the illegal BAC level? Is it lower for youth? Is a BAC test required if fatalities or serious injuries are involved? What is the mandatory length of license suspension or revocation for first, second, and third offenses? What are the penalties for refusing to take a chemical test? Is there an open container law?

chapter 19

Licensing and Vehicle Ownership

Many people think that owning a car is *the* ticket to freedom. Vehicle ownership, however, often involves obligations that are anything but liberating. You must be able to purchase, maintain, and insure a vehicle in addition to getting properly licensed to operate it. You must also budget for gasoline, registration, emissions inspections, repairs, and other operating costs. Both buying and selling a car take time and effort, as well as money.

Chapter Objectives

Upon completion of this chapter, you should be able to:

19–1 The Licensing Process
1. Understand the purpose of a driver's license.
2. Understand the purpose of an instruction permit.
3. Understand what the requirements are for obtaining an Illinois driver's license.

19–2 Registering and Insuring Your Vehicle
4. Understand what financial responsibility is.
5. Understand what a deductible is.
6. Describe different types of insurance coverage.

19–3 Your Driving Record
7. Understand why it is important to maintain a clean driving record.
8. Explain what can lead to the loss of your driving privileges.

19–4 Buying and Selling a Vehicle
9. Describe the relative advantages of buying a new or used car.
10. Understand what precautions you should take before purchasing a used car.
11. Understand what factors you should consider when getting an auto loan.
12. Understand how you should prepare a vehicle for sale.

KEY TERMS

Class D license
instruction permit
Driver Services Department
certificate of completion
25-hour certification
vision screening
written test
driving test
cooperative driver testing certificate
graduated licensing law
restrictions
registration
Vehicle Services Department
certificate of title
five-day permit
temporary registration permit
Air Team
financial responsibility
premium
policy
claim
deductible
liability insurance
uninsured motorist bodily injury (UM) coverage
underinsured motorist bodily injury (UIM) coverage
collision coverage
uninsured motorist property damage (UMPD) coverage
comprehensive coverage
medical payments coverage
Illinois Automobile Insurance Plan
questionnaire sampling program
driving record
court supervision
defensive driving course
license suspension
license revocation
restricted driving permit (RDP)
probationary license
warranty
depreciation
list price
"sticker price"
rebate
invoice cost
wholesale price
purchase price
mileage allowance
trade-ins

19-1 THE LICENSING PROCESS

The privilege of driving does not come automatically, and it does not come easily. You will receive your Illinois driver's license only after considerable training and testing. The state will check up on your knowledge of the rules of the road, your driving skills, and any physical limitations that may impact your driving before granting you the long-awaited credential that allows you to be alone behind the wheel.

Many teenagers consider getting a driver's license to be one of the first steps to becoming a mature adult. Permission to operate a motor vehicle on public roadways allows more independence than most teens have ever had before. With freedom, however, comes responsibility. If you fail to fulfill your obligations to the community with whom you share the road, you can be denied a license or have it taken away. Always remember that driving is a privilege, not a right. When the State of Illinois grants you a license, it is saying that it trusts you, not only with your own life, but with the lives of others. Never abuse that trust.

A driver's license has many functions. It demonstrates proof of legal ability to drive a vehicle in Illinois and, for limited periods of time, in other jurisdictions (including Canada as well as other countries). It also serves as proof of age, identity, and signature for check cashing, credit card purchases, and other financial transactions. You must have a license to operate a motor vehicle on a public highway or any private road or parking facility that is open to the public.

The Class D License

Most likely, you will be applying for a **Class D license,** which allows you to operate any motor vehicle weighing up to 16,000 pounds (7,265 kg). This category includes passenger cars, station wagons, pickups, panel trucks, vans, taxis, limousines, sport utility vehicles, and most recreational vehicles. A Class D license also allows you to tow any single vehicle, boat trailer, travel trailer, or trailer coach not in excess of 10,000 pounds (4,540 kg). Licenses of other classes allow you to operate motorcycles, mopeds/motor scooters, and various types of buses and commercial vehicles. Except in certain hardship cases, you must be at least sixteen years old to possess a Class D license.

The Instruction Permit

Unless you wait until you turn eighteen years old, you cannot apply for a Class D license without first obtaining an **instruction**

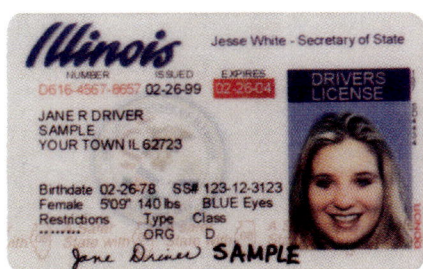

Figure 19-1 Your driver's license has many uses, including proof of identity.

Seven and a half million residents of Illinois, more than 60% of the state's population, are licensed drivers.

The first Illinois driver's license was issued in 1939.

ZSA ZSA GABOR

Zsa Zsa Gabor, actress and former Miss Hungary, has appeared in numerous films and television shows. On June 14, 1989, Gabor was pulled over by a Beverly Hills motorcycle police officer because the registration tags on her Rolls-Royce had expired. As the officer soon discovered, her driver's license had also expired. When he intercepted Gabor as she tried to drive away from the scene, she jumped out of the car and slapped him. Gabor was sentenced to three days in jail, nearly $13,000 in fines, a psychiatric evaluation, and 120 days of community service for neglecting to obtain a valid driver's license and for striking an officer. If you want to drive a motor vehicle, you must comply with all licensing and registration procedures no matter how time-consuming or inconvenient you might find them. If you are "busted" for being negligent, do not attempt to flee or commit other violations that will add to your problems.

permit. This permit, which is valid for two years, is available to applicants who are at least fifteen years old. Applicants between the ages of fifteen and seventeen years, nine months cannot receive an instruction permit unless enrolled in some form of driver education. You cannot take the behind-the-wheel portion of your driver education course without this permit.

Outside of your driver education course, an instruction permit allows you to practice driving with any parent, guardian, or responsible licensed driver approved by your parent or guardian in the front seat with you who is at least twenty-one years old and has been driving for at least one year. To be granted a license as a minor (under the age of eighteen), you must have had your permit for at least three months and received at least 25 hours of behind-the-wheel practice in addition to your 6 hours of driver education behind-the-wheel time.

Driver Services

Driver's licenses and instruction permits are issued by the Illinois Secretary of State's **Driver Services Department.** There are Driver Services facilities located throughout the state. Most are open from 8 AM to 6 PM on Tuesday and from 8 AM to 5 PM Wednesday through Friday, excluding legal holidays. Some are open on Saturday from 8 AM to noon. Most facilities offer a full range of services, including license and registration transactions and driver testing. There is at least

Figure 19–2 An instruction permit permits you to practice driving with supervision in Illinois as young as fifteen years old.

one Driver Services facility in each county. Some are mobile units that provide service to rural locations once or twice a month. Metropolitan Chicago has several "express" offices that exclusively handle license and registration renewals and duplicates.

Applying for a Permit/License

The first step in getting an Illinois instruction permit or driver's license is to apply in person at a local Driver Services facility. An examiner will ask you a series of questions and complete an application for you. Along with a required fee, you must submit original or certified copies of documents that prove your name, date of birth, residency, and Social Security Number (SSN). The easiest way to satisfy this requirement is to present your certified birth certificate (hospital-issued birth certificates are not accepted), a recent (within sixty days) utility bill, and your Social Security card. At least one of these documents must have your signature on it.

If you are a minor applying for an instruction permit, you must also provide proof that you are enrolled in a state-approved driver education course. Once you complete the 6-hour behind-the-wheel portion of your course, a **certificate of completion,** commonly called a "blue slip," will be issued to you. You must present this document when later applying for a regular license. To qualify for a certificate of completion, you must have maintained a passing grade in at least eight classes for the previous two semesters of school.

In addition to the blue slip, you must submit a **25-hour certification** verifying that you have had the required number of hours of behind-the-wheel practice before Driver Services will issue you a license. This form will be mailed to you after you are issued an instruction permit. If you lose it or do not receive it in the mail, you can obtain one from most Driver Services facilities and school driver education offices. Your parent or guardian must sign this form. He or she will also be required to cosign your license/permit application.

If you are applying for an instruction permit, you must successfully pass a vision screening and written test. If you are applying directly for a regular license, you must pass a driving test as well. If you are converting an instruction permit to a regular license, Driver Services may,

Figure 19–3 Driver's licenses are issued by a department of the Secretary of State's office.

reality check

Parents can request Driver Services to cancel the driving privileges of those children who are minors.

depending on the type of course in which you are enrolled, waive the driving test. In many cases, your driver education instructor can conveniently arrange to handle all the application paperwork and testing for both your permit and your license right at your school.

The Vision Screening

The **vision screening** checks your visual acuity and peripheral vision. The minimum standard of visual acuity for an Illinois driver's license is 20/40, with or without glasses or contact lenses. You must also have a peripheral field of vision of 140 degrees or greater. You can have the vision screening waived by submitting a "Vision Specialist Report" form, which you can obtain from any Driver Services examiner, that is completed and signed by your eye-care practitioner.

If you fail the vision exam, you will be required to consult a specialist to have your vision corrected to meet state requirements. Make sure that you obtain a Vision Specialist Report form, and have your eye-care provider fill it out. You will not have to retake the vision screening if you return the completed form to Driver Services while wearing your new glasses or contact lenses.

Figure 19–4 All permit and license applicants must pass a vision screening.

The Written Test

In addition to the vision screening, you must pass a **written test** at the time you apply for your license or instruction permit. This test evaluates your understanding of Illinois's traffic laws; signs, signals, and roadway markings; and vehicle equipment requirements. You

DRIVING TIPS Driving with an Instruction Permit

While driving with an instruction permit, ask the person supervising you to point out areas that need improvement. Spend extra time on difficult maneuvers. If possible, practice in the vehicle in which you will later take the driving test so that you are comfortable with its special characteristics. The more you practice, the more confident you will be on test day. If you do not think that you will be ready for your driving test before your permit expires, obtain a one-year extension from Driver Services. Be aware that if your instruction permit expires before you take the driving test, you will be required to retake both the vision screening and the written test.

Figure 19–5 The written test examines your understanding of basic traffic laws in Illinois.

will also be asked questions regarding collision prevention and safety. The test consists of thirty-five multiple-choice and true/false questions, and you must get twenty-eight of the questions correct to pass. There is no time limit. If you fail the test, you should wait until the next working day to retake it. You are permitted to retake the test twice on the same day, but if you fail a third time, you will have to reapply and pay another fee before you can retake it.

The Driving Test

In most cases, you must pass a **driving test** to obtain a Class D license. This test evaluates your ability to safely operate a motor vehicle and is offered at all full-service Driver Services facilities. You must be sixteen or older and have practiced at least three months with your instruction permit before you can take the driving test. If you are a minor, make sure that you bring your 25-hour certification and your blue slip along with your permit.

Driving tests are available on a walk-in basis only. If the weather is bad on test day, call ahead to make sure that the Driver Services facility is offering the test. You must provide the vehicle for the driving test. It must be the same class as the license for which you are applying. The person who accompanies you in the vehicle should remain at the Driver Services facility in case you fail the test.

The Driver Services examiner will first inspect the vehicle to make sure that it is safe to drive and in compliance with state regulations. He or she will also examine the registration and require you to show proof of insurance. Before the test begins, make sure that you put on your safety belt, check your mirror settings, and adjust your seat position so that you can drive safely. Know your car; the examiner will observe whether you are familiar with the location and function of important controls such as the headlights, windshield wipers, turn signals, brakes, gas pedal, emergency brake, hazard lights, and so on. You may also be asked to demonstrate hand signals.

Only you and the examiner will be allowed to remain in the vehicle during the test. Once you are out on the road, the examiner will check your ability to execute basic driving techniques such as changing lanes, maintaining a proper speed, yielding the right-of-way, entering and exiting intersections safely, turning, reversing direction by pulling into an

factoid

The written test is offered in English, Spanish, Polish, and Korean as well as orally with or without a translator present.

alley, backing up about 50 feet (15 m) at a slow speed, parking (uphill and downhill as well as level), stopping smoothly (also uphill and downhill as well as level), and passing and being passed by other vehicles. You must be extremely attentive to all conditions on the road and obey all signs and signals. The examiner will also pay attention to your posture, your judgment, your attitude toward pedestrians, and the overall smoothness with which you drive.

During the driving test, the examiner will give you clear instructions and tell you in advance what to do so that you have time to plan your actions. Listen carefully, and do not be tempted to steal glances at the notations made by the examiner. At the end of the test, he or she will give you an evaluation and discuss how you did. The examiner can immediately fail you if you get into a collision, cause a vehicle or pedestrian to take evasive action to avoid getting hit by you, do not follow instructions, violate a traffic law, or simply do not exhibit control and confidence behind the wheel.

Do not be shy or embarrassed if you fail the driving test. Use the opportunity to ask questions about how to improve your driving skills. Consider taking another driver education course. You must wait at least 24 hours to retake the test. As in the case of the written test, you will be required to reapply and pay another fee if you fail the driving test three times.

The Cooperative Driver Testing Program

Some school districts in Illinois are partners with Driver Services in what is called the Cooperative Driver Testing Program. Driver education instructors participating in this program give students their driving test as the final in their course. If you successfully complete the driving test as well as receive a final grade of A or B in the course, you will be issued a **cooperative driver testing certificate.**

Presentation of this certificate, which must be signed by your driving instructor and your parent or guardian, at the time of your application for a regular license will usually result in your driving test being waived. Be aware, however, that Driver Services monitors the effectiveness of this program by giving some applicants a second driving test even if they have obtained the certificate.

Getting Your Permit/License

If you are applying for an instruction permit, it will be issued on the spot at the Driver Services facility or your driver education classroom upon successful completion of the vision screening and written test. If you are applying for a Class D license, your photograph will be taken after successful completion of the driving test and then your license will be issued to you. If you do not wish to have your photo on your license for medical or religious

factoid

Driver Services administers nearly five million vision, written, and driving tests each year.

reasons, you may request a non-photo license. If you have an instruction permit and are applying for a Class D license, you will be required to surrender the permit before your license will be issued.

Current Illinois licenses contain a number of security features to prevent tampering and fraud, including a durable plastic construction, special lamination, a retroreflective hologram, and two bar codes on the back that contain the same information that appears on the front of the license. One of the bar codes is encrypted and can be "read" only by Driver Services and law enforcement. The digitized photo and signature make it possible to issue driver's licenses and duplicates quickly.

If you are under the age of twenty-one, a special notation will appear below "DRIVERS LICENSE," which along with the birth date field will be printed within a red block. In addition, the colored bar at the top of your photo, as well as the rim of the state seal, will be red instead of the standard blue. These features are designed to make it easier for law-enforcement officers, restaurant servers, bartenders, and bouncers to quickly determine whether you are of legal drinking age.

Take good care of your permit or license. Do not place adhesive material on it or mark on it. Most importantly, *remember to keep it with you at all times while driving.* You must present it to any law-enforcement officer who asks to see it and, if you get into a collision, any other driver involved in the crash.

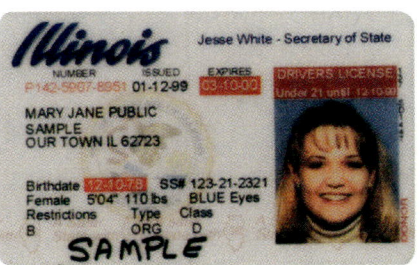

Figure 19–6 If you are under the age of twenty-one, this is the first driver's license that you will get.

Graduated Licensing

On January 1, 1998, a new **graduated licensing law** went into effect in Illinois that imposes special driving restrictions and criminal penalties on young drivers. Graduated licensing is a process designed to expose young drivers, who are statistically more likely to be involved in a collision than older drivers, to increasingly challenging driving experiences. Before being entrusted with full driving privileges, they must pass through several stages, demonstrating at each stage that they have the skill and confidence to move to the next level.

In Illinois, there are three graduated licensing stages: the permit stage, the initial licensing stage, and the full licensing stage. All instruction permit holders are prohibited from driving during "curfew" hours. Curfew hours can vary by locality, but they generally extend from 11 PM through 6 AM Sunday through Thursday and midnight through 6 AM on Friday and Saturday. (Exemptions are granted for employment, religious activities, and emergencies.)

factoid

Nearly half of all new drivers under the age of nineteen are convicted of a traffic violation in their first year of driving.

Digitized photo licenses have been issued in Illinois since February 1998.

Illinois teens make up only 6% of the driving population but account for 16% of all fatal collisions.

Nighttime driving restrictions have been shown to reduce crashes by up to 60% during curfew hours.

In addition, only one passenger may ride in the front seat and the number of passengers in the rear seat is limited to the number of safety belts installed. All vehicle occupants, including those in the rear seat, under the age of eighteen must be restrained. Permit holders are subject to a maximum legal blood alcohol content (BAC) of 0.00%.

Permit holders are limited to one court supervision for serious offenses. Those convicted of a serious moving violation must continue driving with a permit for six additional months before they can apply for a regular license. They are not eligible for either a restricted driving permit or a judicial driving permit. Those fifteen and under caught driving without a permit are not eligible to obtain a license until they turn eighteen.

Licensed individuals under the age of eighteen are generally subject to the same restrictions as permit holders. Curfew hours, however, apply only to drivers under the age of seventeen. Like permit holders, newly licensed minors are limited to one court supervision for serious offenses. A conviction for any moving violation will result in a warning letter from Driver Services both to the teen and his or her parent or guardian. A second conviction of a moving violation will result in a one-month license suspension, with longer suspensions for subsequent violations. Minors whose license is suspended must attend a defensive driving course and retake the tests as well as pay a license reinstatement fee.

Most of the restrictions that apply to minors are removed for licensed drivers between the ages of eighteen and twenty-one. The BAC limit, of course, remains 0.00%. The penalties that apply to drivers in this category are also similar to those for adults who are twenty-one or older. They are still limited to one court supervision (rather than two) for serious offenses. In addition, licensed drivers under twenty-one will have their license suspended for a minimum of one month if they are convicted of only two moving violations in a two-year period (rather than three within a one-year period).

DRIVING TIPS — On Your Own

When you earn the privilege to drive alone, you will be presented with many challenges and nervous moments. Ease your way into the world of driving. Continue driving with a licensed adult; it will be a good chance to increase your confidence behind the wheel and benefit from the tips and comments of an experienced driver. Do not jump right into advanced driving challenges like long trips, left turns across multiple lanes of traffic, nighttime driving, parallel parking, dense city traffic, freeway driving, and driving in bad weather. Master the basics, and then take on the big challenges one at a time. If you feel uncomfortable with a particular maneuver, practice it in a quiet area or empty parking lot until you can perform it smoothly and easily.

Restrictions

At the time you apply for your permit or license, you must indicate any medical condition that may impair your driving in any way. You can obtain a special form from a Driver Services examiner, to be filled out by your doctor, certifying that you can safely operate a motor vehicle.

Your permit/license may specify that you are permitted to drive only with special **restrictions.** For example, if you are disabled, you may be limited to driving a vehicle fitted with mechanical or prosthetic aids. If you have vision problems, you may be restricted to driving with prescription glasses or contact lenses or during daylight hours. If you are deaf or hard of hearing and require a form of communication other than speech such as sign language, a restriction can be added to your license at your request. Other restrictions include driving a vehicle equipped with outside mirrors or with an automatic transmission.

Restrictions are identified on the front of your permit/license by letter(s) only or the letter J combined with a number, keyed to a special code printed on the back. In most cases, you must take a driving test to remove a restriction. If a law-enforcement officer catches you violating a restriction, you could be taken into custody and have your car towed. Depending on the circumstances, you may face a stiff fine, the impoundment of your vehicle, and even jail time.

If you have any allergic reactions to medications or suffer from any ailment or condition that emergency medical personnel should be aware of, such as diabetes or hemophilia, a special restriction will be placed on your permit/license. You will be provided with a medical information card to carry with it. If you have a rare blood type, you may request a special sticker identifying your blood type/RH factor, to be affixed on the back of your permit/license.

Be aware that if you develop any medical condition that may effect your ability to drive safely, Illinois law requires you to report this to Driver Services within ten days of learning of the condition. Failure to do so may result in the cancellation of your driving privileges.

Organ Donation

At the time that you apply for your license, you will be asked if you are willing to donate your organs or other parts of your body in the event of your death to save the lives of others. If you wish to do so, a red "DONOR" designation will be printed on along the outside edge of your license. You will also be required to fill out the top half of the back of your license and sign it, along with two witnesses, preferably family members.

If you are a minor, you must have the written consent of your parent or guardian to have the donor designation placed on your license/permit. Be aware that even if you have indicated that you wish to be a donor, your next-of-kin must give his or her consent before your organs can be donated. For this reason, it is

Nearly half of all jurisdictions in the United States now have some form of graduated licensing.

There are twenty-four types of "J" license restrictions in Illinois.

important to discuss organ and tissue donation with your family.

Nonresidents

Visitors and out-of-state residents who work or attend school in Illinois may drive in the state as long as they hold a valid license from their home state. Nonresidents who have the equivalent of an instruction permit are subject to the restrictions of the Illinois instruction permit. Visitors from foreign countries should carry either an international driving permit (IDP) or an English translation of their national driver's license.

New Residents

Drivers with a current license from another jurisdiction who move to Illinois must apply for a new license at a Driver Services facility within ninety days of becoming a resident. They must provide proof of their Illinois residency and SSN and surrender their old license. Those with an expired out-of-state license are required to produce some other document as proof of name and date of birth. All new residents must pass the vision screening and written test. Most are exempt from the driving test.

REALITY CHECK   *License Refusal*

When you apply for a license, a background check is conducted to verify your name, birth date, and SSN. If you are converting an out-of-state license, Driver Services will check the number with a database called the Problem Driver Pointer System (PDPS) to determine whether the license has been suspended, revoked, cancelled, or denied. If it has, your license can be denied until the issue has been resolved. The Secretary of State can refuse to issue a driver's license for a number of other reasons, including the following:

- You are not yet of legal age to drive.
- You are not a resident of Illinois.
- You have not fulfilled the necessary steps for obtaining a license.
- You have lied on your application or submitted fraudulent documentation.
- You have refused to sign your application.
- You have attempted to alter or otherwise falsify a driver's license.
- You have impersonated someone else to qualify for a license.
- You do not understand traffic laws or signs.
- You do not have the skill to drive.
- You have a health problem that renders you unable to drive safely.
- You have a history of abusing alcohol or other drugs.
- You have a court order issued against you that prevents you from receiving a license.
- You have been recently convicted of a DUI-related or sex-related offense that includes denial of driving penalties as a penalty.
- You are more than ninety days late with your child support or any court payments.

Renewing Your License

Most Illinois driver's licenses are valid for four years and expire on your birthday. If you are under twenty-one, your license will expire three months after your twenty-first birthday. Seniors between the ages of eighty-one and eighty-six are required to renew their license every two years, and those eighty-seven and older must renew annually.

Driver Services will send a renewal notice to your last known address about one month before the expiration date, which is one of the reasons it is important to promptly inform them of any change of address. Even if you do not receive a renewal notice, it is *your* responsibility to renew your license before it expires.

If you are between the ages of twenty-two and seventy-four, you may be eligible to renew by mail or phone every other renewal period. To qualify, you must have had no moving violations or collisions noted on your driving record within the last four years. You also must be free of any medical problems requiring frequent examination. If you are eligible to renew by mail, you will receive a renewal form that must be returned to Driver Services with a required fee. Review it carefully to make sure that all the information is correct. If there are any changes or corrections, you should indicate them on the form. You will have the option of renewing by phone if you are willing to pay your renewal fee by credit card.

Three to four weeks after your renewal application is received by Driver Services, you will receive a "safe driver" renewal sticker, preprinted with updated information, in the mail that must be attached to the back of your driver's license. Remember to send your renewal application early enough that you receive the sticker before your license expires. Otherwise, you may be cited for driving without a license if you are involved in a collision or are stopped by a law-enforcement officer for a suspected moving violation. The sticker is valid for four years.

Figure 19–7 The back of your driver's license contains organ donor information and has a place to affix a "safe driver" renewal sticker.

Illinois law requires that all drivers renew their license in person and have their vision retested every eight years. Unless you have zero traffic convictions on your record, you also must take a written test once every eight years. Drivers who are seventy-four or older must take a driving test each time they renew their license. Drivers who have a poor driving record or a medical problem may be required to renew in person and be tested more frequently.

Your renewal letter will inform you whether or not you must renew your license in person at a Driver Services facility and what tests you will be required to take. When you renew in person, you will have your photograph and

REALITY CHECK *License Violations*

Be aware that if you are convicted of driving without a valid license in Illinois, you face up to one year in jail and up to $2,500 in fines. You face the same penalty if you lend your license to another, use someone else's license as your own, or take any driver's license test for someone else or have them take the test for you. You will face up to $25,000 in fines and up to six years in prison if you possess an altered, counterfeit, or stolen driver's license or if you are caught lying on your license application.

signature taken again and will receive a new license on the spot.

Correcting and Replacing Your License

If you move within the state of Illinois, you must notify Driver Services of your new address within ten days in person at any Driver Services facility or by mail. You can send in a completed "Notice of Address Change" (form VSD-165.13), which is available over the Internet at the Secretary of State's website. You may also mail a letter providing your driver's license number, full name, date of birth, and new and old address. For a fee, you may apply for a duplicate license with your new address.

If you legally change your name, you must go to a Driver Services facility in person and surrender your old license. After providing proof of your name change (such as a marriage certificate), a new photo and signature will be taken, and you will be required to pay a fee to get a new license.

If you lose your driver's license, or if it is stolen or damaged, you should immediately request a duplicate license in person at a Driver Services facility. You will need to present the same type of identification required to obtain an original license. After paying a fee, you will be issued a new license on the spot.

Motorcycle/Moped Licensing

To operate a motorcycle in Illinois, you must obtain a Class M driver's license or an "M" classification on your regular driver's license (Class "DM"). To legally operate a moped or a motor scooter with less than 150-cc engine displacement, you must have either

Figure 19–8 You must pass a special driving test or pass a state-approved rider safety course to get a motorcycle license or endorsement.

a valid driver's license of any classification or a special Class L license. The process of applying for a Class M or L license is identical to that of applying for a Class D license in terms of identification requirements and procedure. All applicants under the age of eighteen must have the consent of their parent or guardian to obtain an instruction permit or license and must have completed a driver education course.

You must be at least sixteen years old to apply for Class L instruction permit. Unless you are enrolled in the Motorcycle Rider Course (MRC) sponsored by the Illinois Department of Transportation, you will not be allowed to obtain a Class M instruction permit until you are eighteen. With this permit, you may not carry a passenger. You are restricted to riding only during daylight hours, and you must be under the direct supervision of a licensed motorcyclist who is twenty-one or older with at least one year of riding experience.

You must pass a vision screening, a written test that includes questions on motorcycle laws, and an Alternate Motorcycle Operator Skills Test (ALMOST). If you are applying for a Class L license, you will take the test on a moped or motor scooter with an engine displacement of 150-cc or less. Be aware that mobile Driver Services facilities are not equipped to perform motorcycle driving tests, so call beforehand to find out where you can take the ALMOST.

Before the driving test, the examiner will verify that your motorcycle, motor scooter, or moped is properly equipped, registered, and insured and that it is in good working order. Be aware that you will be required to wear eye protection if your vehicle is not equipped with a windscreen. You will be tested on your ability to balance and control the cycle, start and stop, shift gears, ride in a straight line at a slow speed, make sharp turns, weave through cones, make a U-turn, and swerve quickly to avoid an obstacle.

The MRC provides riders with 20 hours of professional instruction at training sites located throughout the state. This course, which is available to both beginners and experienced riders, is mandatory for all motorcycle instruction permit and license applicants who are minors and is designed to meet Motorcycle Safety Foundation standards. Graduates over the age of eighteen are exempt from the ALMOST as well as the written test if they submit their "Rider Education Course Completion Card" to Driver Services along with their license application and required fee.

ID Cards

Driver Services offers identification (ID) cards to those Illinois residents who do not have or do not want a driver's license. Similar in appearance to driver's licenses, ID cards are accepted as valid proof of age, identity, and signature, but they do not carry any driving privileges. You may not possess both an Illinois driver's license and an ID card. When applying for an ID card, you must pay a fee and provide

factoid

There are more than 200,000 licensed motorcyclists in Illinois.

the same forms of identification as required for a permit or license.

Illinois ID cards for most adults are good for five years and expire on your birthday. Senior and Disabled ID cards are good for ten years. ID cards issued to those between the ages of fifteen and twenty-one, like driver's licenses for the same age group, expire three months after the holder's twenty-first birthday. The procedures for correcting and replacing an ID card are also similar to those for a license.

19–2 REGISTERING AND INSURING YOUR VEHICLE

Just as the government monitors who is on the roadways by licensing drivers, it keeps track of vehicles to make sure that they are as safe as the people who operate them through a process known as **registration.** All motor vehicles, including mopeds, and trailers must be registered each year with the Secretary of State's **Vehicle Services Department** to make sure that they are in compliance with state laws. In exchange for providing accurate information about your vehicle and paying various fees and taxes, you are issued a registration certificate and license plates that allow you to operate it on public roadways.

All motor vehicles must have a valid **certificate of title** before they can be registered in Illinois. The title is a document that shows proof of ownership. It lists details about the vehicle such as make, year, model, and identification number (VIN). If the vehicle is financed, it also lists the "lienholder," the institution (usually a bank or credit agency) that has formal ownership of the vehicle until the loan ("lien") is paid off. Vehicle Services is responsible for processing all titles as well as registrations.

Acquired from a Private Party

If you buy a vehicle or acquire one as a gift from another person, he or she must transfer the original title to you. The prior owner must complete the top portion of the back of the document, indicating his or her name, the date of sale, and the vehicle's odometer reading, and sign it. You must then print your name and sign the back of the document in the spaces provided. Be aware that erasures or alterations on a certificate of title are illegal and will void the document.

If you "assume" the previous owner's loan, the seller will not have a title. In this case, you must contact the lienholder directly, who can provide you with a copy of the title, loan papers, or other documents verifying the transfer of legal ownership.

As soon as the title (or an acceptable substitute) has been transferred to you, you should notify your insurance company to make sure that it is covered by your policy. Be aware that the seller is obligated to remove the license plates and registration certificate from the vehicle before transferring ownership to you. You have a 24-hour grace period to operate the vehicle without li-

factoid

There are more than 9 million registered vehicles in Illinois.

cense plates or registration. This time is provided for you to drive the vehicle from the seller's property to your own, if necessary. While driving during this period, you must carry evidence of the transaction with you, such as a bill of sale or the title itself (carrying the title is recommended only if you are driving directly home).

Alternatively, you can obtain a **five-day permit.** This permit will allow you to temporarily drive the car until you get it registered. Simply present the prior owner's certificate of title signed over to you at any Driver Services or Vehicle Services facility, currency exchange, car dealer, or any business authorized by the Secretary of State to accept money from the public for vehicle title and registration fees.

Illinois law requires you to apply for title and registration on your vehicle within twenty days of the date of sale. Vehicle Services maintains four branch offices in metropolitan Chicago and several stand-alone customer service centers that can process your application. Most Driver Services facilities can also process your application. Whatever facility you choose will process both your registration and titling information.

In addition to the prior owner's certificate of title signed over to you, you must fill out an application and pay a fee and any taxes due. The tax is based on the model year of the vehicle if the selling price is less than $15,000. Taxes on vehicles with a selling price of $15,000 or more are calculated based on the purchase price. You usually will not be required to show proof of insurance, but be aware that your signature on the application form affirms that you do have insurance for the vehicle.

After all of your paperwork has been reviewed, you will, in most cases, receive a sixty-day **temporary registration permit.** This permit must be attached to the inside lower-left corner of your vehicle's rear window. It allows you to drive until you receive your registration certificate, two license plates (unless you are transferring your old plates to the

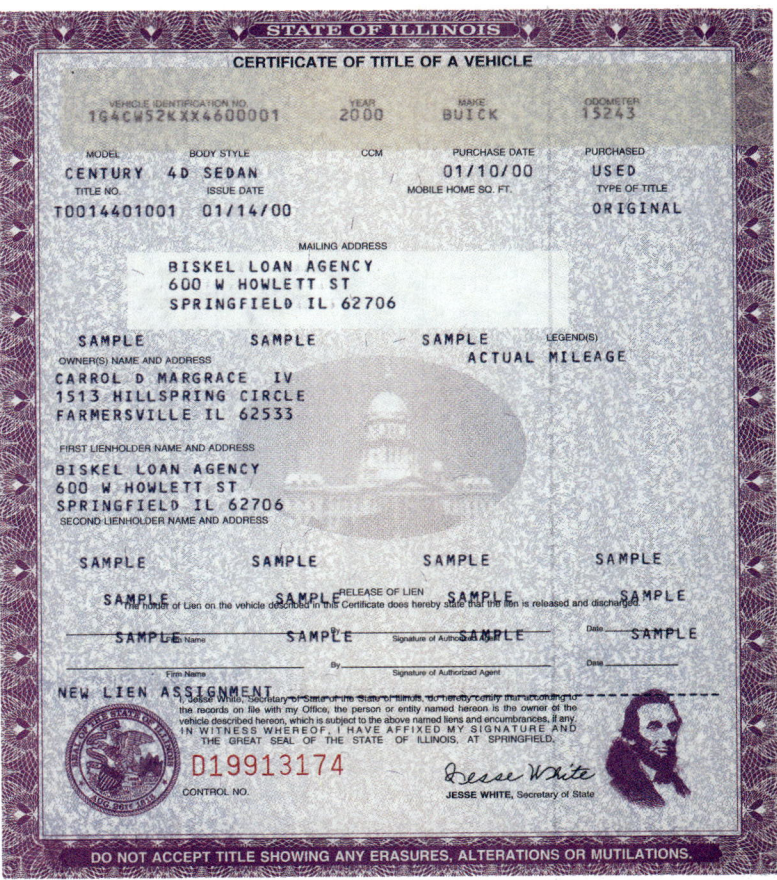

Figure 19–9 The certificate of title is your proof of vehicle ownership.

factoid

The VIN of all cars manufactured beginning in 1968 is embossed on a metal strip located on the lower-left corner of the dashboard.

License plates were standardized in 1957, when Congress declared that every plate issued in the United States be 6 × 12 inches (15 × 30 cm).

vehicle) for front and rear mounting, and a sticker for the rear license plate. These items are usually mailed out, although some Driver Services and Vehicle Services facilities are equipped to issue them to you on the spot. If you order personalized or "vanity" plates, which take longer to process, you can obtain a ninety-day permit. Be aware that if you have not received your plates and registration documents by the time your permit expires, *you* are responsible for applying for an extension in person.

Both license plates must be displayed on a passenger vehicle. If you are registering a motorcycle or trailer, you will get only one rear plate. The sticker, which bears the month of registration and year of expiration, should be immediately affixed to the lower-center portion of the rear plate. Motorcycle stickers go on the upper-right corner of the plate. A different color is assigned each year to stickers to make it easier for law-enforcement officers to quickly verify that your registration is current.

You must have your registration certificate with you at all times while driving, and you must be prepared to present it to any police officer who requests to see it or to other parties to a collision in which you may be involved. Failure to do so can result in a fine up to $1,000. If your registration is expired, your vehicle can be impounded until all due taxes and fees have been paid.

You will receive a new title in the mail in six to eight weeks. If you financed the purchase, the title will be sent to the lienholder. After the vehicle is paid off, the lienholder will then mail it to you. Once you have the title, make sure that it clearly indicates that the loan has been satisfied or you will not be able to sell the vehicle later. If the title is not accompanied by a letter on official stationery verifying that the loan is paid in full, request one as extra

DRIVING TIPS **Custom Plates**

Are you a member of the National Guard? A ham radio operator? Do you support the University of Illinois? Do you love the environment? Do you have an antique vehicle? Vehicle Services offers motorists a choice of numerous custom license plates. Nature lovers can choose a plate featuring the state bird, the cardinal, in a patch of prairie grass. Those who wish to support the state's violence prevention programs can order a "Prevent Violence" plate depicting a dove with an olive branch. Hunters can support the Wildlife Habitat Fund by choosing from six different designs of sporting series plates. Veterans can select from several types of plates that commemorate their service to the nation. Prices for most custom plates range from $15 to more than $50, plus annual taxes and fees. Prices are higher if you want to personalize the plate. Those who meet the state's eligibility standards can obtain a disabled plate for a nominal fee or parking placard at no cost.

protection. Once you receive the title, store it safely away. Do not keep it in your vehicle.

Purchased from a Dealer

If you buy a new or used vehicle from a dealer in Illinois, the dealer will complete the registration and titling paperwork for you and submit it Vehicle Services. If you are not financing the vehicle, the certificate of title will be mailed to you. If you purchase the vehicle with a loan, the original title will be sent to the lienholder. The dealer will give you a temporary registration permit, which will allow you to drive until your plates and registration documents arrive. In some cases, these items will be mailed to the dealer for you to pick up, but usually they are sent directly to you.

Transferring or Canceling a Registration

In Illinois, registration and plates are linked to the owner, not the vehicle. Whenever a vehicle is transferred from one party to another, the new owner must either apply for a new registration or transfer an existing registration to the new vehicle. The previous owner must either cancel his or her registration or transfer it to another vehicle. The owner on the registration cannot change, and both vehicles must be of the same class. For example, a registration cannot be transferred from a car to a motorcycle.

To transfer an existing registration to a newly acquired vehicle, you can take your registration certificate and license plates with you to the dealer where you bought the new vehicle and they can process the transfer for you. Otherwise, you can go to a Driver Services or Vehicle Services facility and follow the same procedure that you would to apply for a new registration. You must transfer the plates on the old vehicle to the new one. After you pay the registration fee, you will be issued a registration certificate with the new vehicle information. If you sell your vehicle and do not transfer the license plates to another vehicle, you must turn the plates in to a Driver Services or Vehicles Services facility or destroy them.

Out-of-State Vehicles

If you move to Illinois from another jurisdiction, you must title and register your vehicle(s) within

From 1912 to 1918, front license plates in Illinois had holes in them to allow air to pass through to the radiator.

REALITY CHECK — *Keep Your License Plates Visible*

License plates must be illuminated at night with a white light, even if they are reflectorized. This requirement helps law-enforcement officers and witnesses identify vehicles in the event of a collision or traffic violation. In Illinois, it is illegal to tint, cover, or otherwise obscure your plates, or to add punctuation to personalized plates, because they may hinder others' ability to read the letters and/or numbers.

factoid

Due to wartime material shortages, license plates issued in Illinois from 1943 to 1948 were made of soybean pulp, making them prime targets for hungry goats, cows, horses, pigs, and dogs.

thirty days of becoming a resident. To convert your out-of-state title and registration, you follow the same basic procedure as if you were applying for a new registration. If you financed your vehicle purchase and the loan has not yet been paid off, you may submit a copy of the title, bill of sale, out-of-state registration, manufacturer's certificate of origin (MCO), or loan papers in lieu of the title itself. If you paid taxes on the vehicle in the three months prior to the move, you are exempt from paying any taxes but you still must submit the proper tax forms.

If you purchase or lease a new or used car from an out-of-state dealer, you must submit the MCO and bill of sale. If you buy a used vehicle from a private party in another jurisdiction, you must submit the title that has been signed over to you. Along with the application form, fees and taxes due, you must surrender the out-of-state plates, if any.

Registration Renewal

Registrations are valid for one year. Passenger vehicle and light truck registrations expire at midnight on the last day of the month indicated on the license plate sticker. Motorcycle registrations expire on March 31, while recreational vehicle and trailer registrations expire on December 31. Vehicle Services will send a renewal notice to your last-known mailing address sixty to ninety days prior to the expiration date. Registration can be renewed in person at any Driver Services or Vehicle Services facility, by phone, or by mail. Your bank, savings and loan, or credit union may also be authorized to renew your registration for you.

If all the information printed on your registration certificate and renewal letter is still correct, you need return only the letter along with all required fees. If anything has changed, you will be required to produce evidence that you have transferred the plates to a new vehicle. Vehicle Services will mail you your new registration certificate and license plate sticker.

If you move within Illinois, you must report your change of address to Vehicle Services within ten days. The required change-of-address form is the same as that used for driver's licenses. Most people use the same form to update both registration and driver's license information at the same time. For any other changes to your registration or title, or if you need to replace a lost, destroyed, or damaged license plate or certificate of title, contact Vehicle Services to find out what forms and procedures are required. Note that if any changes must be made to your registration, you will be required to correct the title information also.

Vehicle Emissions Inspections

To help reduce the problem of air pollution, the Illinois Environmental Protection Agency (IEPA) administers an emissions testing program known as **Air Team** in all or part of eleven of the state's counties. Most gasoline-powered

Automobile emissions testing is required in more than 180 metropolitan areas nationwide.

vehicles in these counties, which embrace metropolitan Chicago and East St. Louis, are required to get an emissions test every other year. Vehicles manufactured before 1968, farm equipment, race or show cars, vehicles used exclusively in parades by organizations, diesel-powered vehicles, motorcycles/motor scooters/mopeds, and vehicles powered by alternative fuels (such as natural gas or electricity) are exempt from the test.

If you live in an area where emissions tests are required, the IEPA will assign your newly registered vehicle to a month in which you must get it tested. About one month before the beginning of your "inspection month," you will be mailed a notice containing instructions, testing locations and hours, and information about your vehicle provided to the IEPA by Vehicle Services. Review the test notice carefully to make sure that all of the vehicle data are correct.

You have four months from the assigned month to get the inspection or to notify the IEPA of incorrect information or a change in the status of your vehicle (a postcard is enclosed with each test notice for this purpose). If you have not complied within this time period, you will be informed in writing that failure to have your vehicle inspected within ninety days will result in the suspension of both your driver's license and vehicle registration. You will receive two or more warning letters, so there is no excuse for not getting tested.

Air Team emissions inspection stations are set up like automated car wash facilities. After pulling into a lane of the inspection station and taking a wait-time ticket, your vehicle will be driven through the testing station by a certified inspector after you and any passengers exit the vehicle. You may observe the testing procedure through large windows from a waiting area. The test takes about ten minutes and is offered on a drive-in basis only at no cost. A sign out front informs you of the wait time. Each test station also has a telephone hotline that will provide you with the current wait time.

The main portion of the test consists of placing a hose over the tailpipe that carries the exhaust to special sensors. The sensors in turn transmit data to a computer that analyzes the levels of hydrocarbon, carbon monoxide, and oxides of nitrogen in the emissions. If your vehicle is model year 1981 or newer, the test will be performed while your vehicle is placed on a treadmill-type device called a dynamometer. The dynamometer test is designed to be the equivalent of a 2-mile (3-km)

Figure 19–10 There are thirty-five state-licensed Air Team testing stations located in the Chicago and East St. Louis areas.

driving tips

Wait times at Air Team inspection stations are usually shortest on Wednesdays, Thursdays, and Fridays.

urban trip in which the engine will be operated from idle up to 58 mph (93 km/h).

Once the test is completed, the results will be electronically transmitted to the IEPA. You will also be given a vehicle inspection report (VIR). This report is your record of the test. If you pass the test, you will also be issued a compliance certificate, which must be kept in your vehicle at all times.

If you fail the emissions test, you must have the necessary repairs performed and get the vehicle reinspected. Testing stations are not authorized or equipped to make repairs. In addition to the VIR, you will be given a series of documents that will explain what repairs need to be made, where you can get your vehicle repaired, what must take place once repairs are completed, and what options are available if you fail the post-repair inspection. In certain cases, you may be able to obtain a waiver. Twelve-month "hardship" extensions are also available for qualified vehicle owners. A customer service representative is available at each testing facility to assist those who have failed the test.

If for some reason your license and registration are suspended for failure to comply with the Air Team testing requirements, you may, as a last resort, obtain a five-day trip permit (not to be confused with the five-day permit) from a Driver Services or Vehicle Services facility. This permit will allow you to take your vehicle to an inspection facility and get an emissions test. The five-day period should be sufficient time to get any needed repairs and inspection completed in addition to the test itself. Once you are in compliance, you should wait at least three days before driving again to give the Secretary of State's office a chance to clear the suspension actions from your records.

Financial Responsibility

To legally operate a motor vehicle in Illinois, you must be able to pay for any injuries, deaths, or damage that may be caused by your vehicle. Although you could deposit money with the State Treasurer or obtain what is called a "surety bond" from an authorized company, getting insurance from a licensed insurance company is the most common and convenient way to satisfy the state's **financial responsibility** requirement. When you register a vehicle, your signature on the registration application affirms that you are in compliance with the law.

Insurance is acquired by paying a **premium,** a specified amount of money for a period of time, usually six months or a year. In most cases, you have to make a down payment, and then have the option of paying the balance in monthly installments. The provisions of your insurance are detailed in a **policy,** a written contract between you and the insurance company that outlines your coverage and how much the insurance company is liable for in the event of a collision or other incident resulting in damage to your vehicle.

If you are involved in a collision and another party makes a

driving tips

To increase your chance of passing an emissions test, you should have the oil and air filter changed, and operate your vehicle at least 15 minutes before testing.

valid **claim,** or request for payment, against you, you must first pay a deductible. A **deductible** is a fixed amount of money that you must pay before your insurer begins paying out on the claim. The higher the deductible, the lower the premium. Once the deductible is satisfied, the insurance company will then pay the balance of the damages up to the limits specified in the policy.

Most types of insurance are tied to a particular vehicle. You may be required to specify in your policy which household members (such as a spouse or dependents) and any other persons are customarily permitted to operate your vehicle, but any licensed person to whom you give permission to drive is usually covered in the event of a collision. If you drive someone else's vehicle, make sure that the vehicle is insured and that the owner has authorized you to operate it. Before you let someone else drive your vehicle, make sure that he or she is covered by your policy.

Insurance companies will issue you an ID card that must be carried with you at all times when you are driving. Be aware that if you do not have *evidence* of financial responsibility at the time of a traffic stop or at the scene of a collision, you can be charged with not having insurance even if, in fact, you do. At the very least, you will have to waste a day in court proving that you were insured. As an added precaution, it is a good idea to keep a copy of your insurance card in both your wallet and in a handy place in your vehicle, such as the glove compartment.

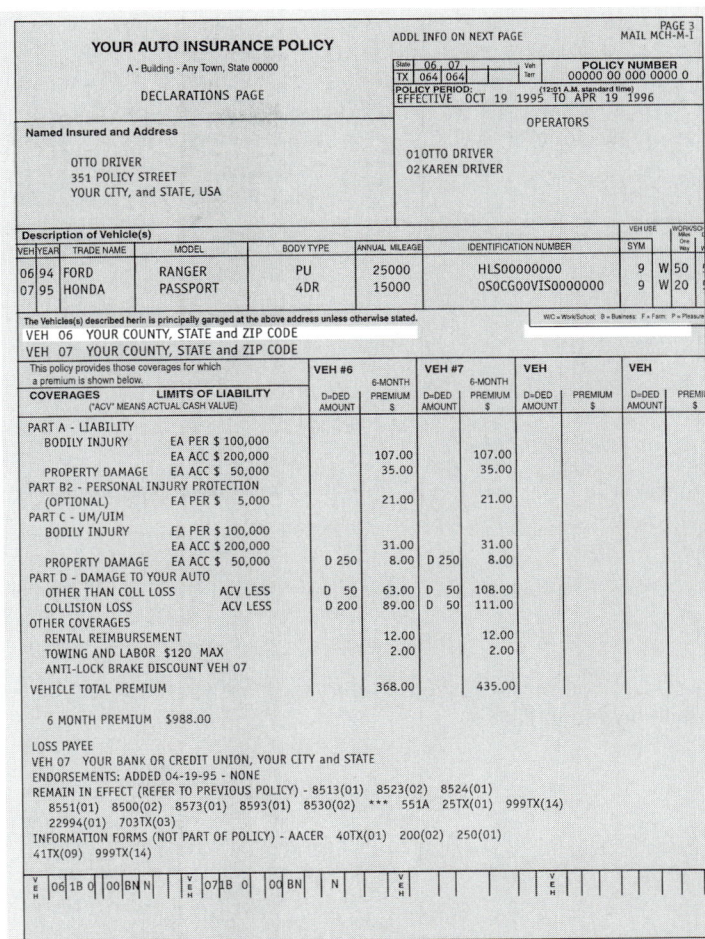

Figure 19–11 The conditions of insurance coverage are detailed in your policy.

Required Coverage

You will, as part of your policy, be required to carry **liability insurance** to cover both bodily injury and property damage caused by you or anyone authorized to drive your vehicle. *Bodily injury liability* pays for medical expenses and other damages, such as lost wages and "pain and suffering," to people injured or killed by you, family members, or anyone driving

factoid

The average annual premium for an auto insurance policy in Illinois is more than $700.

ILLINOIS INSURANCE CARD

The coverage provided by this policy meets the minimum liability limits prescribed by law.

NAME OF INSURED

POLICY NUMBER

EFFECTIVE DATE   EXPIRATION DATE

VEHICLE DESCRIPTION
YEAR   MAKE/MODEL   VEHICLE IDENTIFICATION NUMBER

ILLINOIS LAW REQUIRES
ID CARD TO BE CARRIED IN THE VEHICLE AT ALL TIMES.

USAA Building   San Antonio, Texas 78288

Figure 19–12 You must carry proof of financial responsibility at all times while driving in Illinois.

Figure 19–13 What was the name of this 1921 vehicle, assembled using sheets of solid copper and a rare form of ebony wood for trimwork, that would have cost even the best of drivers a sky-high automobile insurance premium?

your vehicle with your permission. In any one collision, your policy must pay up to $20,000 for bodily injury to, or the death of, one person and up to $40,000 for injuries to, or the deaths of, two or more persons. *Property damage liability* pays for losses caused by you, a family member, or someone authorized to drive your vehicle to another vehicle or other property. The minimum required limit for property damage liability is $15,000.

Uninsured motorist bodily injury (UM) coverage protects you, family members who live with you, and occupants of your vehicle from losses arising out of a collision with a "hit and run" driver or an at-fault uninsured driver. It pays for medical expenses that would normally have been covered by the other driver's insurance. Illinois law requires you to carry limits of at least $20,000 per person and $40,000 per occurrence of UM coverage.

Supplemental Coverage

Given the rise in medical and auto repair costs, a collision can be financially devastating if you have only the minimum coverage required by law. The money paid by your insurance company might pay only a fraction of the actual expenses associated with a crash you cause, and you can be sued in court for the difference. Also, it is important to realize that liability insurance does not cover the cost of injuries to you or your passengers, or damage to your vehicle, if you are at fault.

Insurance companies in Illinois offer many types of optional coverage, and it is strongly recommended that you buy extra protection. For starters, you should definitely purchase much higher limits of both bodily injury and property damage liability. The Illinois Department of Insurance recommends coverage limits of at least $100,000 for bodily injury to, or the death of, one person and up to $300,000 for injuries to, or the deaths of, two or more persons. The suggested limit for property damage liability is $50,000. The peace of mind that you will get by boosting your liability coverage is easily worth the higher premium you will pay.

You should also invest in insurance that protects you, family members who live with you, and occupants of your vehicle from losses arising out of a collision with an at-fault driver whose liability coverage is insufficient to pay for your bills. Illinois law requires that **underinsured motorist bodily injury (UIM) coverage** be purchased if you buy any amount of UM coverage in excess of the required minimum amount.

Two other important types of optional insurance, especially for newer vehicles, are collision coverage and comprehensive coverage. **Collision coverage** covers the cost of repairing or replacing your vehicle at its current market value if it is damaged after overturning or colliding with another vehicle or object. It pays for damage even if you are responsible or if your vehicle is damaged while parked in a lot or on the street. This type of coverage is often required if you have a loan or are leasing your vehicle. In some cases, the lender may purchase the protection and add the cost to your loan. This is known as "forced placement," and it will definitely cost you more than if you bought the coverage yourself.

If you do not elect to purchase collision coverage, your insurance provider may offer you the option of purchasing **uninsured motorist property damage (UMPD) coverage.** This insurance pays for damage to your vehicle caused by an identified, at-fault uninsured driver.

Comprehensive coverage protects you from the loss of your vehicle or accidental damages to your vehicle caused by anything except a collision, including fire, theft, vandalism, glass breakage, or a storm. It pays for repair or for replacement costs up to the current market value of the vehicle, less the deductible. Be aware that while some insurance companies treat contact with an animal as a collision claim, others will classify it as a comprehensive claim.

If you or someone else in your vehicle is hurt in a crash, you will be glad that you have **medical payments coverage.** This insurance protects you, your family members, and any passengers in your vehicle who are killed or injured in a collision regardless of who is at fault. It also pays for funeral and medical expenses if you or a family member is killed or injured as a pedestrian. Some insurance companies in Illinois may require you to purchase medical payments coverage along with liability insurance.

Check with your insurance company or, in some cases, a car dealer or financial institution about other types of optional insurance that are available, including "umbrella" coverage (which increases protection beyond the limits of your policy), "gap" coverage (for those leasing a vehicle), towing and labor coverage, rental reimbursement, and mechanical breakdown coverage.

Shopping for Insurance

Take the time to shop around for an insurance policy. There are about 300 licensed insurers in Illinois, and the costs of premiums as well as available types of coverage vary greatly. If you do not want to spend the time researching individual insurance carriers, you can

---

bumper sticker sightings

HIT ME . . . I NEED THE MONEY

---

factoid

A seventeen-year-old male may have to pay up to three times as much for insurance as a thirty-year-old male.

Insurance premiums are, on average, higher in Illinois than in all neighboring states.

purchase a policy through an agent licensed to sell insurance for numerous companies. In some cases, you can obtain insurance over the phone or by mail.

Insurance companies charge a premium determined by statistics that indicate the likelihood that certain groups of people will be involved in a crash. Factors used to determine the cost include driving record, age, miles driven annually, gender, marital status, residence, and type of vehicle. Having a bad credit rating or being uninsured for more than thirty days will make you a "high-risk" driver in the eyes of some insurers.

Insurance companies typically offer a variety of discounts to make coverage more affordable. For example, if you have more than one vehicle insured by that company, safety belts, factory-installed air bags or side bags, antilock brakes, an antitheft device installed on the vehicle, or good grades in school (usually 3.0 grade-point average or higher), you will generally pay less for insurance. Those with a clean driving record are generally eligible for a "good driver" discount. Drivers who are fifty-five years or older who complete a state-approved defensive driving course are entitled to a discount for the following three years.

If you are unable to find an insurance company that will sell you a policy with the minimum requirements, you must, as a last resort, apply to be in the **Illinois Automobile Insurance Plan.** Under this "assigned risk" program, you will be assigned to a participating insurer, who is obligated to insure you for three years. You will, of course, have to pay a higher-than-average premium. By maintaining a good driving record, however, you will significantly increase the chance of later finding an insurer who will take you on voluntarily.

The Questionnaire Sampling Program

To make sure that all drivers are in compliance with the state's financial responsibility requirements, Vehicle Services administers a **questionnaire sampling program.** Each week, verification forms are mailed to 5,000 randomly selected vehicle owners asking them to provide their insurance carrier's name and policy number.

If you receive a questionnaire and fail to respond within thirty days or if you provide information that cannot be verified, Vehicle Services will mail you a notice that they are about to suspend your registration. You will have forty-five days to provide current evidence of insurance such as a valid insurance card, the "Declarations" page of your policy, or a copy of your insurance application signed by you and your insurance provider. Failure to do so will result in the suspension of your registration. To reinstate your registration, you will have to show proof of insurance and pay a reinstatement fee.

Driving Without Insurance

If you are caught operating a vehicle without insurance or some other accepted form of financial responsibility by a law-enforcement officer, you will be charged with what is known as a "business offense." If convicted, your registration will be suspended and you will face a minimum of $500 and up to $1,000 in fines. If you are convicted of driving while your registration is suspended for lack of insurance, the minimum fine is increased to $1,000 and the maximum goes up to $2,000. If you are caught driving while under this type of suspension a second or subsequent time, or, if you produce a counterfeit insurance card when stopped by a law-enforcement officer, you will be charged with a misdemeanor, which may result in jail time as well as a fine.

If your registration is suspended for driving without insurance and you are placed under court supervision, you will have to pay a reinstatement fee and purchase a very expensive "SR-22 insurance" policy that you must maintain for one year. Any lapse in coverage will immediately subject you to the continued suspension of your registration.

If you are found to be at fault in a collision involving the death or injury of any party and were driving without insurance at the time, you could lose your driving privileges for some time. The Secretary of State's office is empowered to suspend your driver's license and registration until you pay the damages (if they exceeded $500) or demonstrate that you have set up some kind of payment plan with other parties who successfully sue you in court. You will also be required to maintain SR-22 insurance for at least three years.

19–3 YOUR DRIVING RECORD

Your **driving record** is a record maintained by Driver Services of all your motor-vehicle violations and reported collisions, including those that may have occurred in other jurisdictions. Law-enforcement officials, courts, insurance companies, and agencies in other jurisdictions share information with their counterparts in Illinois, so all of them have access to your record at any particular moment.

Penalties for Traffic Convictions

If you are ticketed for a traffic offense, you will be required to "post bond" with the officer who issued the citation. You may either pay the bond in cash or surrender your driver's license on the spot. If you surrender your license to the officer as bond, the yellow copy of the ticket acts as a temporary license until you pay the fine. For minor moving violations, you usually have the option of paying the fine

Figure 19–14 A law-enforcement officer who pulls you over has complete access to your driving record.

driving tips

For a small fee, you may obtain a copy of your driving record at any local Driver Services facility.

by mail. Otherwise, you must contact the designated courthouse by the date indicated on the ticket to pay the fine in person or schedule a hearing in court. The officer will indicate on the back of your ticket whether or not a court appearance is required.

Once in court, you have the option of pleading "guilty" or "not guilty" before a judge or magistrate (an attorney empowered to act as a judge). The officer who issued the citation may appear to give testimony against you, and in some cases you may even face a state's attorney. If the judge/magistrate finds you guilty, he or she may sentence you to jail or probation, suspend your license, or impose other penalties besides a fine depending on the circumstances of the case and your driving record. The penalties for breaking a traffic law generally depend on the severity of offense and the number of times you have been convicted.

If you pay a traffic ticket by mail (which is an admission of guilt), your driver's license will be mailed back to you. You will get your license back immediately if you go to court pay the fine in person. If you fail to appear in court or mail in the fine by the date indicated on your ticket, a new court date/mail-in deadline will be set. Failure to comply with this second notice will result in the loss of the bond, an additional fine, and the possible suspension of your driver's license. If a court appearance was required for the original offense, a warrant for your arrest may be issued.

Appearing in court to plead "guilty" to a traffic violation rather than paying by mail will not guarantee a reduced fine. The judge/magistrate may, however, take into account your willingness to take time off from work or school to appear in person and accept responsibility for your actions.

Court Supervision

Many jurisdictions in Illinois also allow you to request **court supervision.** In exchange for admitting guilt and agreeing to fulfill various requirements outlined by the judge/magistrate for a given pe-

---

REALITY CHECK — Out of Sight, But Not Out of Mind!

Illinois is a member of what is known as the Non-Resident Violator Compact. This agreement allows law-enforcement agencies of different states to cooperate in managing traffic violations committed by those traveling through their state. If you are cited for a traffic violation in another state that belongs to the compact, you have two options. You may stay in the state and argue the case or pay the fine, or you may sign an agreement to pay the fine once you get home. Keep in mind that offenses you may commit in other jurisdictions while driving with an Illinois license will go on your driving record. A judge or magistrate might also consider violations committed in other states when assigning a penalty for an offense committed in Illinois.

riod of time (anywhere from ninety days to two years), you will receive a lesser punishment than what would otherwise be assessed if you were convicted of a traffic offense.

One of the most common conditions issued under a court supervision order is attendance at a **defensive driving course.** Offered by the National Safety Council, this course varies in length from four to eight hours and is designed to teach defensive driving and collision prevention techniques. It also promotes an attitude of mutual understanding, courtesy, and respect on the road. Those between the ages of sixteen and twenty-four are offered a special curriculum oriented toward young drivers called "Alive at 25."

If you complete the specified conditions of your court supervision within the mandated period of time without further violations, the offense will not be entered into your driving record. Be aware that the policies regarding the use of court supervision are set by each local circuit court and can vary, not only from county to county, but from judge to judge.

License Suspension and Revocation

Driver Services can suspend or revoke your license at any time if you fail to drive safely. A **license suspension** leads to a temporary loss of the privilege to drive; your license is taken away and then returned to you after a certain period of time (never more than one year).

A **license revocation** is more serious. In this case, Driver Services cancels the legal permit given to you to drive. When the period of revocation expires (never less than one year), you must apply for a new license, which will be granted only if Driver Services has determined that you are fit to drive. You will be required to retake some or all of the tests. You must also carry an "SR-22 insurance" policy for the following three years.

The length of a license suspension or revocation generally depends on the seriousness of the offense. Driver Services can suspend your license if you are convicted of three traffic violations within any one-year period. Drivers under the age of twenty-one will have their license suspended if they commit two violations within two years. Your license may be suspended for a number of other offenses, including the following:

- Failure to report a collision
- Not having insurance at the time of a collision in which you were at fault and for which you owe damages
- Failure to stop for a school bus that was loading or unloading children
- Having ten or more unpaid parking tickets
- Failure to get a required vehicle emissions inspection
- Failure to appear in court or pay a fine for any offense
- Possessing, displaying, or using a fraudulent permit/license

reality check

If you are convicted of a traffic violation in another jurisdiction that would cause your license to be suspended or revoked in Illinois, Driver Services will suspend or revoke your license when the other state notifies Driver Services of the conviction.

factoid

Non-DUI license suspensions and revocations remain on your driving record for ten years, while DUI-related suspensions and revocations remain indefinitely.

Figure 19–15 The costs of owning a vehicle can add up quickly if you do not obey the law.

- Failure to make child support payments

Driver Services has the right to suspend your driving privileges at any time if it determines that you are physically or mentally incompetent to drive.

Your license will be revoked if you are convicted of driving under the influence, "hit and run," reckless driving three times within one year, reckless homicide or conduct, fleeing a law-enforcement officer, using a motor vehicle in the course of committing a felony, or lying on Driver Services or Vehicle Services paperwork. If your license has been revoked, your vehicle registration(s) may also be suspended.

Driver Services will notify you of any suspension or revocation action by certified mail and ask you to surrender your license if you have not already done so. You are entitled to a hearing with Driver Services, during which time you will have the opportunity to present evidence and explain why your license should not be suspended or revoked.

In certain cases, those who have had their license suspended or revoked may be eligible to receive limited driving privileges. If you are sixteen or older and need to drive to or from work or school, as part of your employment, or to obtain regularly required medical care or drug treatment for yourself or a family member, Driver Services may issue you a **restricted driving permit (RDP)**. This permit specifies exactly when and where you can drive, and is valid for only one year. If your license has been revoked, you must reapply each year for a new RDP.

If your license was suspended for getting three traffic convictions within a one-year period, Driver Services will mail you an application for a **probationary license** along with your "Notice of Suspension." If you successfully complete a defensive driving course before the expiration of your suspension, your instructor will forward your completed application to the Driver Services office in Springfield, which will in turn mail you the license. A probationary license allows unrestricted driving for a three-month period.

Be aware that if you are caught driving with a suspended or revoked license, you face up to $25,000 in fines and up to six years in prison. If your license has been suspended or revoked as a result of a conviction for Driving Under the Influence (DUI), "hit and run," or reckless homicide, you face a mandatory minimum sentence of seven consecutive days in jail or thirty days of community service. If you are caught driving with a license suspended or revoked due to an insurance-related violation, your vehicle may be impounded as well.

19–4 BUYING AND SELLING A VEHICLE

Before you decide to buy a vehicle, weigh your options and think hard about your transportation requirements for the near future. Many types of vehicles are available, some of which are obviously not suitable for you whereas others could meet your needs very well.

New or Used?

If you are ready to buy a car, you must first decide whether to purchase a new or used vehicle. A new vehicle will initially cost you more than a used car, but one of the biggest advantages of buying a new car is that you get a more comprehensive and longer-term warranty. A **warranty** is a written guarantee that the car manufacturer or dealer will repair or replace any defective parts or systems within a set amount of time or miles driven. With such protection, you should spend less on maintenance in the first years of driving the vehicle, and you will not be socked with a big-ticket expense like replacing the transmission. A new car is also more likely to hold up better under heavy use or long-distance traveling, and replacement parts are generally cheaper and easier to find. Finally, a new car is likely to have more safety features that could save your life.

On the other hand, a used car often has a lower purchase price. A good used car often represents a better dollar value because a previous owner has absorbed most of the cost of depreciation. **Depreciation** is the decline in a vehicle's resale value over time. It varies according to vehicle condition, age, mileage, and optional equipment. In most cases, used cars are also cheaper to insure.

Researching the Market

Once you have settled on a new or used car, you must decide what kind of car is right for you. Do you want something sporty, practical, or economical? How will you be using the vehicle? How many passengers will you carry? Will you be towing a trailer or carrying heavy loads? How many miles will you travel on average? Will you be going on long or short trips?

Once you settle on the type of vehicle that fits your needs, budget, and image, take the time to gather data on the various makes and models available. Bookstores, libraries, credit unions, banks, the Internet, insurance companies, mechanics, car dealers, and manufacturers are all good sources of

Among the automobiles manufactured in Chicago in the early 1920s were the Birch, Shaw, Hamlin-Holmes, Bradley, and Marshall.

A new car can depreciate as much as 50% in value during the first year that you own it.

TIRES

The tires that come as original equipment on your new car are warranted separately by the tire manufacturers (including the compact spare tire). A separate warranty statement is included.

If a tire is damaged during the warranty coverage period because of a vehicle defect in factory-supplied materials or workmanship, the tire will be replaced by the ACME Motor Company.

ROADSIDE SERVICE

Your vehicle is eligible for the ACME Roadside Assistance Program. This Program is separate from the New Vehicle Limited Warranty, but the program period is concurrent with the Full Warranty period. Under this program ACME will cover towing, spare tire mounting, fuel delivery, jump-starts, and lockout problems. For more information call 1-800-555-ACME.

Towing necessitated by warrantable failure beyond the Full Warranty is covered under the applicable warranty.

For daily rental units that must be towed because a covered part has failed during the Full Warranty period. ACME will cover towing to the nearest dealership.

WARRANTY SUMMARY

New Vehicle Limited Warranty		Emissions Warranty	
Full Warranty	4/50,000	Emissions Performance Warranty	3/50,000
Safety Restraint System	4/50,000	Emissions Defect Warranty	4/50,000
Rust Perforation	5/Unlimited	Short Term	4/50,000
		Long Term	7/70,000
Years/Miles In Service			Years/Miles In Service

Figure 19–16 The warranty offered by a manufacturer or car dealer can be a major factor in your final choice of a vehicle.

Figure 19–17 Do as much research as you can before selecting a vehicle.

up-to-date information. Narrow your choices by comparing prices, safety records, maintenance and insurance costs, fuel economy, available options, warranties, customer satisfaction, repair records, theft history, and other important factors.

Buying a New Car

If you are buying a new car, visit several auto dealerships to get a first-hand look at the vehicles you are interested in. If possible, take along someone who has bought a car before to benefit from his or her experience. Find out what is available, and determine what you are actually going to have to pay to drive one off the lot. Keep in mind that some car salespeople are paid by commission and tend to be aggressive in going after business. Ask to speak with the sales manager or fleet manager. Be polite but firm when negotiating prices and options.

When shopping for a new car, compare dealer prices carefully. A new car's **list price**, or "retail price," is the manufacturer's suggested retail price (abbreviated MSRP) for the car and all factory-installed options. The **"sticker price"** is the list price plus destination charges and additional options and services provided by the dealer. The "destination charge" is the cost of shipping the vehicle from the manufacturer or, for foreign cars, from the point of entry, to the dealership. This charge may vary by location. Dealers often add a "preparation charge" to prepare a vehicle for sale by tuning up the engine, checking the emissions, and performing other services. Because many manufacturers already pay dealers to "prep" a car, be wary of high preparation charges.

The actual selling price of a car is also affected by manufacturer or dealer incentives and the time of the year when you buy. Some manufacturers offer a **rebate**, a special payment to the customer,

Highway heroes

Around midnight on April 19, 1992, while responding to a report of a disturbance at an auto dealership on Illinois 38 in West Chicago, Officer Michael A. Browning was struck by a man attempting to steal a car from the showroom and died soon after.

Figure 19–18 Sports cars may be fun to drive, but they are generally less safe and more expensive both to maintain and insure.

to boost sales on certain models. Dealers may also offer discounts at the end of the year to get rid of their inventory and make room for newer models.

When discussing prices with a salesperson, it is a good idea to know the **invoice cost,** the amount the dealer paid for the car. You can get this information from various auto magazines, consumer groups, and pricing guides, but you may also want to consult an auto pricing service, which for a small fee can prepare an accurate report within a couple of days on a specific model's invoice cost.

Make sure that you also consider the warranty offered by various manufacturers and dealers, which are often negotiable. Car dealers are required to post a warranty statement in one of the windows of the vehicle (usually on the driver's side of the car) outlining which parts are guaranteed and for how long. Most new cars come with a standard warranty on major parts that lasts from three years or 36,000 miles (58,000 km) to five years or 60,000 miles (96,500 km). Dealers may also offer an "extended service contract" on all or part of the standard warranty.

Buying a Used Car

If you cannot afford a new vehicle, consider buying a used one from either a car dealer or a private seller. Selecting a used car is more time consuming and complicated than buying a new one because you have to do more detective work. Used cars can be great deals, but they can also be an endless source of regret if you find out too late that the vehicle you purchased was an auto-repair nightmare.

New-car dealers sell their best quality trade-ins and returned lease cars that have been reconditioned. These vehicles are generally in good condition, have low mileage, and include a limited warranty covered by the dealer's service department. On the downside, they are among the most

Figure 19–19

Car dealers typically make a bigger profit on a used car than on a new one.

Figure 19–20 You can negotiate for a lower price with most car dealers.

Figure 19–21 Searching classified ads for used cars or special deals may cost you less, but it also carries risks.

expensive on the market. Most sell their vehicles "as is," and the risk of acquiring a car with serious mechanical problems or defects is greater. It is always a good idea to check a used-car dealership out with the Better Business Bureau before you make a purchase there.

You stand the best chance of finding a bargain if you purchase a used car from a private seller rather than a dealer, but you also take a bigger risk. Unless the seller can demonstrate that the vehicle has been properly maintained and serviced over the years, you will know nothing about its history. It takes time to set up an interview with a person, and you are likely to come across many vehicles that do not live up to the glowing description you read in the newspaper classified advertisement. You must also handle all the registration paperwork by yourself.

The **wholesale price** is what a dealer is willing to pay for a used car. Wholesale prices for used cars vary depending on the model, year, mileage, and their general condition. Before buying a used car, you should consult several used-vehicle valuation guides on the market. These guides list the average used-car prices based on actual sales by dealers and wholesale auctions, average loan prices, and

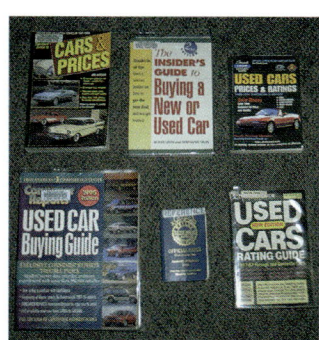

Figure 19–22 Valuation guides are invaluable resources if you are buying a used car.

Figure 19–23 Thoroughly inspect a used car yourself before spending the money to have a mechanic look at it.

mileage tables used to adjust prices. Check "on-line" as well.

Once you have settled on a used car, thoroughly examine it. You can be sure that the previous owner or dealer will do all that he or she can to make the vehicle appear problem free. When inspecting a used car, go in the daytime when you can see well and wear old clothes that you do not mind getting dirt, grease, or rust on. It is also a good idea to take along a friend for a second opinion and to help you test drive it.

The only way to get the feel of a used car, just like a new car, is to test drive it. If possible, test drive the car on both highways and city streets, on hills, and on different road surfaces. Watch for any warning lights on the instrument panel. Test the brakes to see how hard you need to push the pedals to come to a complete stop or whether the car pulls to one side or the other. When you accelerate, the speed should gradually and smoothly increase. Listen for unusual noises from the front, back, or under the hood. If you hear grinding, whirring, clicking, clunking, squealing, or hissing, odds are that only a day in the service bay is going to stop it. If the car wobbles, rattles, or shakes, rub your chin, tell the owner that you will "think about it," and look elsewhere.

Before you purchase a used car, always have a mechanic give it a thorough diagnostic check. For a moderate fee, a mechanic will provide you with a report listing the parts or systems that meet acceptable operating standards and those that fail inspection and need repairs. If you or your mechanic determine that repairs need to be made, either walk away from the deal or negotiate these costs into the final sale price.

If you are buying a vehicle from a private party, remember to have the registered owner complete the top portion of the back of the certificate of title, sign it, and give it to you. He or she should also remove the license plates. Keep in mind that, until the vehicle is registered in your name, you cannot drive the vehicle more than 24 hours after the time of purchase unless you obtain a five-day permit. You should arrange, if possible, to have the other party bring the vehicle to your residence.

Financing Your Vehicle

The **purchase price** of a vehicle is the amount of money that the dealer or owner is willing to accept. The least expensive way to buy a car is with cash because you will not have to add any interest

The Ford Model T, the first mass-produced car, sold for less than $300.

Figure 19–24 A car advertised as having "low monthly payments" usually requires a large down payment.

cost to the final purchase price. If like most people you cannot pay cash for a car, your other options are to finance the purchase with a loan or to lease the vehicle.

Auto Loans

The total cost of a car loan depends on the interest rate, the total amount of money borrowed, and the length of time required for repayment. Car loans generally run from two to five years in duration. Shop around for the best interest rate (often called the annual percentage rate, or APR) available. Most car loans are limited to 90% of the purchase price, but 100% financing is sometimes also available. Keep in mind, however, that the larger the down payment you can make, the less you will have to pay for the car in the long run.

When you finance a car, you must put up what is known as collateral in case the loan is unpaid. In most cases, the collateral is the vehicle itself. The lender will retain the certificate of title to the vehicle until the loan is fully paid. If you are a full-time student or lack a sufficient credit rating, the lender will require an adult to co-sign the loan. The adult who co-signs is responsible for the repayment of the loan.

Leasing a Car

Many people prefer to finance their car by leasing rather than purchasing it. *Leasing* is just a fancy word for renting. Instead of buying and owning the vehicle, you will be renting it. The main advantage of leasing is that monthly payments are up to 30% lower than payments on a car loan. Another advantage is that you often do not have to make a down payment. Leasing lets you avoid the risk of depreciation because you pay for the life of the car only while using it. Unlike buying a car, you are not stuck with it if it does not hold its resale value well.

Leases usually run from two to five years, with three years being the average term. The longer the term of the lease, the smaller your monthly payments will be but the higher the total cost of the vehicle once interest is added. You can lease from either a new-car dealer or an independent leasing company. The fees and penalties of leases vary widely, so always remember to compare the total cost of the lease, not just the monthly payments.

One of the largest potential costs of leasing is exceeding your annual **mileage allowance.** Most leases limit you to 12,000 to 15,000 miles (19,300 to 24,100 km) per year. If you drive more than the amount specified in your contract, you will have to pay a penalty for each mile over that limit. Over the course of the lease, this can add up to thousands of dollars if you drive much more than you anticipated.

Selling Your Vehicle

Buying a car is an investment. One of the reasons it is important to take care of your vehicle is that you will want it to be worth something when it comes time to sell it or trade it in for a new car.

Where am I?

Figure 19–25 The guiding philosophy of this home's architect, Mies van der Rohe, also expresses the truth about low down payments: "Less is more."

There are many reasons why you might want to get rid of an older car. Costly repair bills or low gas mileage might make your vehicle too expensive to operate for your current needs. Replacement parts may be harder to find. It may need considerable body work or mechanical repairs you are not willing to pay for. Maybe you just want a change or can afford to upgrade. Whatever the reason, selling a car—just like buying a car—requires research, paperwork, and a little "elbow grease."

Thoroughly clean the car inside and out. If any major work needs to be done, determine whether you can get your money back when you sell the car or if it would be better to sell the car as is and deduct those costs from the final price. Organize your records on gas mileage and routine maintenance so that you can show them to potential buyers. This openness with buyers will go a long way to convincing them of the value of your car.

Check some of the used-car valuation guides and websites for an estimate of what your car is worth based on its condition and mileage. Visit several dealers and compare what they would pay you for the car. Go through the classified ads to see what cars similar to yours are selling for. The price you can expect to get normally will be somewhere between the price charged by other private party sellers and the dealers' prices. Once you have a good idea of what you can expect, set your asking price and the minimum amount that you would be willing to accept as a compromise.

Advertise in the weekend editions of local newspapers when people will be more likely to shop for a car. Specialized auto classified ad publications, which usually pair photos with descriptions, are often the best way to show off your car. When advertising your vehicle, include the make, year, and model of the car. Describe selling points that will draw buyers such as "low mileage," "full maintenance history," or "original owner." If you advertise a firm price, you run the risk of scaring some people away. To increase your bargaining leverage, list a higher price than you are willing to settle for but say that it is negotiable. Providing a price range will spare you telephone calls from people who cannot afford the car.

Most people will want to test drive the car or have it inspected by a mechanic. Go with the potential buyer on the test drive not only to protect yourself against theft but to answer any questions he or she may have while driving. Make sure that the person has a valid license.

Once you have a buyer, you must transfer the certificate of title to him or her after completing the top portion on the back and signing it. Remember to remove the license plates and, unless you are transferring your registration to another vehicle, destroy them or turn them in to a Driver Services or Vehicle Services facility. Finally, notify your insurance company after the transfer to avoid a potential headache should the buyer get into a collision or in trouble with the law.

Figure 19–26 When advertising your vehicle, include as many selling points as you can.

You should only accept a certified check or cashier's check made payable to you as payment. If the buyer insists on paying with a personal check, go with the buyer to his or her bank to cash the check. *Never transfer the certificate of title until you are paid in full.*

If you do not want to sell your car yourself you can trade it in for a new or used car at most dealers. In general, trade-ins are a bad value. You pay a heavy price for the convenience of having the person who sells you a new car to take your old one off your hands. If you are determined not to sell the car yourself, take the time to make the vehicle as presentable as possible and get offers from several dealers. Find out what the estimated trade-in value of your vehicle is through a reputable pricing guide. Most importantly, never negotiate a trade-in with a dealer until after you have settled on a purchase price for the new car. Otherwise, you have no idea whether you are getting a good value for your old car.

DRIVING TIPS — Avoiding "Lemons"

Like most jurisdictions, Illinois has what is called a "lemon law" to protect consumers from getting stuck with a defective motor vehicle, commonly called a "lemon," purchased at or leased from a dealership. A lemon is defined as any vehicle other than a motorcycle that has been in the shop at least four times for the same problem or has been out of service for a cumulative total of thirty days. Only defects that substantially impair the use of the vehicle, decrease the market value, or affect the safety of the vehicle are covered. Consumers who buy or lease a new vehicle under warranty are entitled to a vehicle buyback, less the value of the miles driven, or a comparable replacement vehicle if the dealer or manufacturer is unable to fix it within one year or 12,000 miles (19,300 km), whichever occurs first. Be aware that claims under this law must be filed with the authorized manufacturer's representative, not the dealer, within one year of the date of purchase.

YOUR TURN

19–1 The Licensing Process
1. What is the purpose of a driver's license?
2. What is the purpose of an instruction permit?
3. What are the requirements for obtaining an Illinois driver's license?

19–2 Registering and Insuring Your Vehicle
4. What is financial responsibility?
5. What is a deductible?
6. What are different types of insurance coverage?

19–3 Your Driving Record
7. Why is it important to maintain a clean driving record?
8. What can lead to the loss of your driving privileges?

19–4 Buying and Selling a Vehicle
9. What are the relative advantages of buying a new or used car?
10. What precautions should you take before purchasing a used car?
11. What factors should you consider when getting an auto loan?
12. How should you prepare a vehicle for sale?

SELF-TEST

Multiple Choice
1. Which of the following is *not* used by automobile insurance companies to determine rates?
    a. your driving record
    b. your marital status
    c. your driver license number
    d. your age
2. Which of the following is *not* a valid use of your driver's license?
    a. proof of your ability to operate a motor vehicle
    b. proof of your age
    c. proof of your signature
    d. proof that you have liability insurance
3. Loaning your driver's license to another person:
    a. may result in a one-year jail sentence if you are convicted.
    b. will allow that person to drive legally if he or she is already properly licensed.
    c. is permitted only in emergencies.
    d. is allowed only if he or she is test driving your car.
4. A certificate of title is a document that shows:
    a. insurance coverage.
    b. driver's license restrictions.
    c. the interest rate of a car loan.
    d. proof of ownership.
5. Most standard warranties run from:
    a. one to two years.
    b. one to three years.
    c. three to five years.
    d. five to seven years.

Sentence Completion
1. License _____ allow you to drive only under certain conditions, such as with certain types of adaptive equipment or while wearing prescription eyewear.
2. _____ is the decline in a vehicle's resale value over time.
3. A license _____ by Driver Services leads to a temporary loss of driving privileges.
4. Acquiring insurance is the easiest and most convenient way to satisfy Illinois's _____ requirements.
5. Your _____ contains a history of all your motor vehicle violations and reported collisions.

Matching
Match the concepts in Column A with examples of the concepts in Column B.

Column A
1. __ Protection from "hit and run" driver
2. __ Graduated licensing
3. __ Premium
4. __ Air Team
5. __ Policy
6. __ Certificate of title
7. __ List price
8. __ Comprehensive insurance

Column B
a. MSRP
b. Protection from theft
c. Cost of an insurance policy
d. Contract between you and insurer
e. Curfew
f. VIN
g. Chicago and East St. Louis
h. UM

Short Answer
1. What is the purpose of the driving test?
2. What affects a car's depreciation?
3. How can you price a used car?
4. What is a restricted license?
5. What is a diagnostic check?

Critical Thinking
1. A friend has accumulated several unpaid parking tickets while at school in another state, but he is graduating later in the year and plans to return home. He tells you that the Secretary of State's office will never find out about the tickets when he re-registers his car and applies for a new license. What should you do?
2. After visiting a few used-car dealers and responding to several classified ads, you find two used vehicles of the same make and model that meet your requirements. One of the cars is for sale by a private seller. It needs some fixing up but it also has relatively low mileage. The other vehicle is for sale at a used-car dealer. It has 20,000 more miles (32,200 km) than the other car but comes with a limited thirty-day warranty. The cost of the car at the dealer is $2,000 more than the one offered by the private seller. What should you do?

PROJECTS
1. Call three different insurance companies for estimates of how much it would cost to insure your vehicle to meet Illinois's minimum liability requirements. Share the results with your class. Discuss how rates are affected by where you live, how far you drive, the market value of your vehicle, and other factors.
2. Go to the library and look up your vehicle in consumer magazines, valuation guides, or on the Internet to find out how it ranks in terms of safety, cost to insure, customer satisfaction, warranty coverage, and other indicators of value. Determine the same information for two other models from the same year that are similar in size and design but that are made by different manufacturers. Is your vehicle a good buy?

20 chapter

Vehicle Maintenance, Equipment, and Security

A vehicle is much like your own body. Take good care of it, and it will serve you well for a long time to come. Maintenance is the key to keeping your vehicle running smoothly and reliably. What also contributes to greater road safety are special kinds of equipment and accessories. If you are a disabled driver, your vehicle can be outfitted with special equipment to adapt it for your use. You will also have to decide whether you want to obtain devices to protect your vehicle against the possibility of theft.

Chapter Objectives

Upon completion of this chapter, you should be able to:

20–1 Preventive Maintenance

1. Understand what preventive maintenance is.
2. Describe how to check tires for inflation and wear.
3. Identify the fluids you should check as part of routine maintenance.
4. Understand the proper way to wash and wax a vehicle.

20–2 Selecting a Mechanic

5. Understand how to select a mechanic.
6. Understand how to avoid poor workmanship.

20–3 Vehicle Equipment and Accessories

7. Identify three types of safety-oriented equipment available.
8. Identify several of the devices used to adapt a regular vehicle for a disabled driver.

20–4 Vehicle Security

9. Describe three basic types of antitheft devices available.
10. Understand how a vehicle tracking device works.

KEY TERMS

preventive maintenance
adaptive equipment
antitheft devices
car alarms
kill switches
tracking device

20–1 PREVENTIVE MAINTENANCE

The routine care you give your vehicle to avoid more serious repairs is called **preventive maintenance.** Preventive maintenance includes day-to-day care, such as changing the oil, filling the tires with air, and checking levels of key fluids, as well as routine servicing. Because vehicle requirements may vary greatly, you should always check your vehicle's owner's manual to determine the manufacturer's recommended schedule of preventive maintenance. Following the schedule will save you from paying expensive repair bills later. Ignoring the recommended service intervals may also void the manufacturer's warranty on your vehicle.

Tire Maintenance

Tires are one of the easiest parts of your vehicle to monitor. However, they often get overlooked. Keep in mind that it is easier and safer to change a tire at home than on a busy street with traffic whizzing by. Take the time to research the best tires for your vehicle when it comes time to replace them. Spending a little extra on better tires can often save you time and trouble in the long run.

Make sure that the tires on your vehicle are the proper size. If you are not sure that they are or if the tires do not "look right," check with the manufacturer's recommendations in the owner's manual. Never mix different types of tires. Your tires should be of equivalent size, quality, and tread design.

Inflation

Properly inflating the tires ensures that they will grip the road evenly and wear at the same rate. Tires without sufficient air grip the road only at their outer edges, which causes the tires to squeal, makes steering more difficult, leads to heat buildup, decreases fuel economy, and wears the tires out faster. Underinflated tires are also more prone to blowouts. Overinflating the tires allows only the center of the tires to grip the road, which leads to a harder ride, a reduction in the depth of the tires' print, and faster wear.

Check tire pressure at least once a month. Make sure that you do so *only when the tires are cold.* The heat from friction with the road causes air in the tires to expand, so pressure readings taken right after driving are not accurate. If only one tire is low on air pressure, check carefully for a nail, which can cause a blowout if the pressure drops below 20 pounds per square inch. When you check your regular tires, remember also to check the spare. Station wagons and cars with engines in the rear may require higher pressure for rear tires than for the front, whereas some front-wheel drive cars require higher pressure for the front tires.

Wear

It is illegal, as well as dangerous, to drive on tires that are badly worn. Before getting into your vehicle, note the condition of all tires. "Bald" tires with little or no tread provide almost no traction

Driving on underinflated tires can reduce tread life by as much as 50%.

Before 1910, when it was discovered that adding carbon compound to tires increases abrasion resistance and tread life, all automobile tires were white.

Figure 20–1 Check tire pressure *before* you drive.

DRIVING TIPS — Tire Inflation in Cold Weather

Tires need more air in cold weather. A tire that may have lost pressure during the summer and fall driving season could easily become underinflated on a freezing day. Pay special attention to air pressure in your tires during the winter!

and reduce directional control. They are also more likely to get punctured or go flat. Bald tires will also double the risk of skidding and hydroplaning on rain-slick surfaces. You should acquire a special gauge to measure the minimum depth of tread possible before the tire goes bald. Most tires have tread-wear indicators that tell you when they should be replaced. If your tires do not have tread-wear indicators, you should replace them when the tread wears down to no less than $\frac{1}{8}$ inch (3 mm).

Look for bumps, bulges, tread or sidewall cracks, and embedded nails, glass, or metal. Plies or cords should never be exposed. A frequent loss of pressure in a tire suggests a slow leak, which can cause an emergency when you drive. For equal wear, tires should be rotated on a regular basis, which means switching them from front to rear and from one side to the other. Check your owner's manual for the recommended guidelines for your tires. When rotating the tires, make sure that they are balanced so that the weight is evenly distributed as the wheels turn. This will provide better steering control, a smoother ride, and a longer tire life.

Fluid Levels

Fluids are the lifeblood of your vehicle, so do not take chances. Check your fluid levels regularly, and occasionally check under your vehicle for signs of leaking fluids. The color of any puddle on the pavement indicates the type of fluid leaking out. To make leaks easier to see, leave a large white sheet of paper under the vehicle overnight. Repair any leaking system before you have trouble.

Oil

If you have any reason to suspect a problem with your oil or if the oil gauge/indicator light is acti-

Figure 20–2 Driving on bald tires increases the risk of getting a flat and skidding.

AUTO ACCESSORIES — Tool Sets

To help you perform routine maintenance on your car, invest in a standard automobile tool set. These tools can be carried in your trunk or stored in your garage. Remember to purchase correctly sized tools—either "standard American equivalent" (SAE) for most North American vehicles or "metric" for some North American vehicles and almost all imports.

vated, stop the vehicle immediately! If you keep driving, you may irreparably damage your vehicle's engine. Running an engine without oil is a sure death sentence for the engine. Check the oil level at least once a month, more frequently on older cars, and change the oil and oil filter as specified in your owner's manual—at the least, every six months or 3,000 miles (5,000 km). If you drive in hot weather or spend much of your time in rush-hour traffic, you should change it even more frequently.

The oil level should be checked when the engine is cool and not running. To check the oil level, remove the dipstick and wipe it off. Reinsert it and remove it once more. If the oil level registers below the line, add the proper amount of oil. *Do not rely solely on your oil gauge to tell you when to change the oil.* By the time the warning light flashes, the engine has been without oil long enough to harm the machinery.

Transmission Fluid

Keep an eye on your transmission fluid level. Replacing or refilling transmission fluid is easy and can save you costly repairs in the future. The color of the fluid should be a bright, cherry red. Have a mechanic examine the transmission if the color is dark and has a burnt smell. Transmission oil should be changed every 25,000 to 35,000 miles (40,000 to 55,000 km) depending on the type of vehicle.

Brake Fluid

Have the fluid level in the dual master cylinder checked every time you have the oil changed and replace it at least once a year.

A properly maintained vehicle will last up to 50% longer as well as stay "in tune" with the environment.

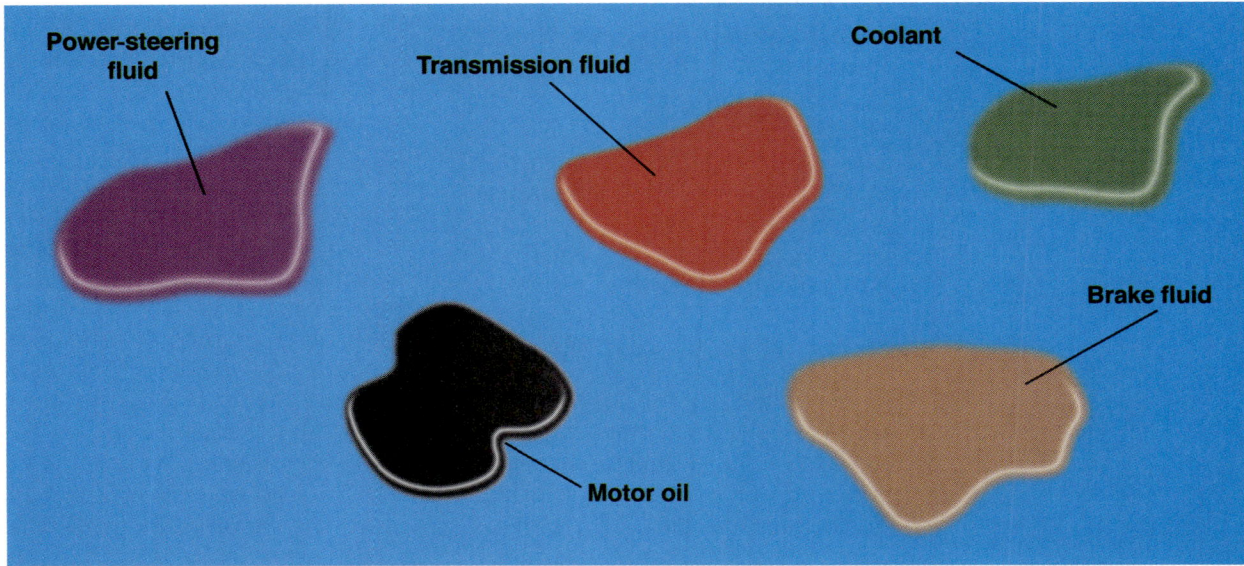

Figure 20–3 Learn to recognize what is leaking from your vehicle.

Radiator coolant reduces the temperature at which engine fuel burns from 4,500°F (2,480°C) to 275°F (135°C).

Brake fluid attracts water from condensation and humidity in the air, causing corrosion in the master and wheel cylinders and shortening their lives.

Coolant/Antifreeze

Check the level of coolant in the radiator overflow tank. Insufficient levels of coolant may result in overheating and damage to the engine. If the coolant level is low, start the engine and add more. *The engine should be running while the coolant is being added.* This prevents cracking of components from a sudden temperature change in the hot radiator. In warm weather, water alone can be used if antifreeze is unavailable. If you notice any puddles of coolant/antifreeze, which may be bright green, yellow, or pink, under your car, have your cooling system checked *immediately.*

Other Fluids

Make sure that you have the proper amount of battery fluid, if necessary, for your battery. Also be certain that you have enough windshield washer fluid. You do not want your wipers smearing mud or other foreign substances all over your windshield just because you forgot to refill the windshield washer fluid reservoir.

Belts, Hoses, and Wires

A loose belt in the engine may cause the electrical system to malfunction. Loose belts can also contribute to inefficient cooling of the engine and to problems with the power steering and air conditioning. Belts should feel tight when you push down on them. As a rule of thumb, have the belt tightened if you can push it down a distance equal to the thickness of the belt. Replace frayed or cracked belts immediately. Check hoses and hose connections for leaks. Wires that are loose, broken, or disconnected, as well as cracked insulation on wires, should be repaired. Check to see that battery cables are tightly connected and terminals are free of corrosion.

Brakes

The braking system is obviously one of the most important features governing both your safety and the safety of others on the road. Pay strict attention to its maintenance. Test brakes by stepping on the brake pedal. You should feel firm resistance and the vehicle should come to a smooth, direct stop. Also, the pedal should stay well above the floor.

Let a mechanic know as soon as possible if you must depress the brake pedal very low to get it to work, if the vehicle pulls to one side when you brake, if one tire locks when braking, or if you detect a "spongy" feeling in the brake pedal. Other problems include grabbing or uneven brake action, accompanied by squealing or chattering noises. You may have completely worn the brake pads down, resulting in metal braking on metal. Also, let your mechanic know if you need to pump the brakes or push the brake harder than usual to stop the car.

The Engine

The quickest way to diagnose problems with your engine is from your exhaust. If your exhaust changes to a blue or smoky white color, you can assume engine damage is occurring or has already occurred. Turn the motor off immediately to prevent any further damage!

If you notice a foul, sulfuric odor like rotten eggs, your vehicle might have a damaged catalytic converter. Defects in the exhaust system create noise and increase the risk of carbon monoxide poisoning for the vehicle's occupants. The exhaust system should be replaced when a part rusts or becomes damaged in some other way. To protect your catalytic converter, you are required by law to use only unleaded gasoline.

Always pay attention to the temperature in your engine. Overheating can result from the breakdown of your water pump or from a malfunctioning thermostat. If you hear a suspicious knocking noise, stop and turn off the engine.

Steering and Suspension

Be alert to problems with your vehicle's steering system, such as vibration of the steering wheel, excessive "play" in the wheel, or difficulty in steering even though tires are properly inflated. You may also have a mechanical problem in your steering system if you feel a wobbling or if the vehicle pulls to one side under normal driving conditions. If you experience considerable sway or bounce when driving over a bump or turning corners, or if you detect rattling, your vehicle could have faulty shock absorbers. Be sure to report these problems to a mechanic as soon as you detect them.

Routine Maintenance Schedule

Based on recommendations in the owner's manual of your vehicle, set up a routine maintenance schedule. Keep a record of when and what type of service is done, and whether you do the work yourself or have a mechanic do it. Here is a typical routine maintenance schedule:

- *Every week:* Check the levels of the vehicle's oil, windshield washer fluid, and radiator coolant/antifreeze.

- *Every month:* Check the condition of the tires, belts, and hoses. Inspect the automatic-transmission, brake, and power-steering fluid levels. Clean dead bugs out of your radiator grille. Wax the car to protect the paint.

- *Every six months:* Check and clean the battery cables and terminals. Check the condition of the brakes.

- *Once a year:* Check the air conditioner refrigerant and shock absorbers. Replace the air filter and windshield wiper blades. Flush out the radiator and refill it with coolant. Hose down the undercarriage to remove dirt

Figure 20–4

DRIVING TIPS: Improving Fuel Efficiency

Improvements in design and performance have made modern cars more fuel efficient and more mechanically reliable than older vehicles. However, you can take the following additional steps to save money on gasoline:

- Perform preventive maintenance as recommended in your vehicle's owner's manual. Failing to tune up your car regularly can reduce fuel efficiency by 30%.
- Do not exceed the posted speed limit. In addition to being both illegal and dangerous, driving at high speeds sharply reduces mileage.
- Avoid stop-and-go and nonessential driving. Travel when traffic is light, and combine daily errands into one trip.
- Avoid "jackrabbit" starts and unnecessary accelerating, lane changing, and downshifting. Most vehicles get optimum fuel economy when cruising at a steady speed of 35 to 45 miles per hour (55–70 km/h).
- Keep your tires properly inflated and aligned to make your engine work less hard.
- Avoid excessive idling. Turn off the engine while waiting for passengers.
- Remove excess cargo, and store items in the trunk rather than on the roof. The lighter your vehicle, and the less air drag it has, the better your mileage.
- If you are buying a new or replacement vehicle, avoid high-polluting "gas-guzzlers" that will eat away at your budget as well as at the ozone layer. Think twice about whether "riding high" in an sport utility vehicle is worth 50 cents extra per mile.

and salt after a cold driving season. Check the headlight alignment.

Your owner's manual will outline specific maintenance schedules for your vehicle based on the number of miles that you drive. In most cases, these procedures should be done by a certified mechanic.

Refueling

Pay close attention to your fuel gauge. You never want to let a vehicle run out of fuel because dirt left in the bottom of the tank will clog up the fuel filter. If the fuel filter becomes blocked, your engine will not receive the fuel it needs for combustion to take place in the cylinders. Always turn your engine off before refueling.

The types of fuel delivered at the fuel pump are indicated on the pump. If your vehicle requires unleaded or diesel fuel, a manufacturer's notice normally appears on the instrument panel near the fuel gauge, around the fuel cap, and in the owner's manual. The fuel octane rating necessary to prevent engine knocking or "pinging" during normal driving is also noted in the owner's manual. Be aware that most vehicles are meant to run on regular unleaded gas and will derive little, if any, benefit from high-octane fuel, so check your manual before you spend an extra 10 or 20 cents per gallon on "premium" or "supreme" grades.

7,500 miles (12,000 km) or at 6 months	22,500 miles (36,000 km) or at 18 months
☐ Change engine oil. ☐ Replace engine oil filter.	☐ Change engine oil. ☐ Replace engine oil filter. ☐ Inspect brake linings.
15,000 miles (24,000 km) or at 12 months	**30,000 miles (48,000 km) or at 24 months**
☐ Change engine oil. ☐ Replace engine oil filter. ☐ Lubricate steering linkage.	☐ Change engine oil. ☐ Replace engine oil filter. ☐ Replace air cleaner filter. ☐ Replace spark plugs. ☐ Inspect drive belt, adjust tension as needed. ☐ Lubricate steering linkage. ☐ Drain and refill automatic transmission fluid. ☐ Drain and refill transfer case fluid.

Figure 20–5 Each vehicle has its own recommended maintenance schedule.

If you elect to keep an emergency gas can in the trunk of your, make sure it is completely empty. Storing a container of gas in your car is a major fire hazard, and a rear-end collision or even extreme summer heat can cause an explosion. If you ever use the gas can to refuel your car, thoroughly wash it out with water before returning it to your trunk. If you have a pickup with a bedliner, always take the gas container out of the vehicle before filling it up to avoid an explosion caused by built-up static electricity. For the same reason, you should always turn off any electronic personal communication device such as a cellular phone or pager while refueling.

Protecting the Exterior

If you want to preserve the value of your vehicle for resale or if you just want it to look as nice as possible for as long as you own it, you will want to protect it from the damaging effects of the environment, including acid rain, tree sap, bugs, and corrosion from road salt and dust. Wash your vehicle frequently, especially if you live in areas where salt is used to

factoid

During the 1920s, when the first gasoline dispensers that did not require hand pumping appeared, it took about 8 minutes to fill a typical car's 5-gallon (19-l) tank.

Figure 20–6 Always turn off your engine before refueling.

factoid

In 1974, before the rise of self-service, only 6% of gas stations in the United States allowed customers to pump their own gas.

provide traction during snowstorms. You should also bear in mind that sunlight can prematurely damage a vehicle's exterior paint job, as well as fade and crack interiors exposed to sunlight through the windows.

If you wash your car by hand, wait until it is cool outside. Keep your vehicle out of direct sunlight as much as possible. Clean the tires and wheels first so that you will not splash water all over your newly cleaned vehicle. Spray the wheel wells and undercarriage to remove corrosive agents.

When washing your car, avoid household liquid soaps and laundry detergents, which can damage the luster and finish of the vehicle by stripping the protective wax coat. Flood the car's surface with water before applying suds. This will remove the surface dirt and soil that may otherwise get rubbed into the body and cause scratches. Using a sponge or mitt, wash the car from the top down, in straight lines and using overlapping strokes. When drying your car, open up all the doors and lids to enable partially exposed areas to dry completely. You may want to use a chamois ("shammy"), which is a piece of soft leather, to help dry the vehicle.

Wax your car at least six times a year, more often if it is exposed to salt air, road salt, or industrial air, or if you have to park it outside. If water does not bead up on the car's surface after a rain, waxing is needed. Because hot metal surfaces can cause wax chemicals to damage the finish, do not wax your car in direct sunlight. If you choose to wax your car on a really humid day, the wax may streak. Never wax large areas of the car at one time. Apply the wax to small areas with a foam applicator. Remove it with a shining cloth, rubbing in a circular fashion.

If you want the outside of your vehicle to really dazzle your friends and neighbors, you can find dozens of specialty cleaning items at any local auto supply store. Besides various types of polishes, waxes, and rubbing compounds you can buy paint protector/sealant, bumper patch, tire glaze, scratch remover, convertible top cleaner, and fiberglass resin. Before you use exotic chemicals on your vehicle, however, read the directions and your owner's manual carefully. If you bought a new car, check with the dealer to see if it has been treated with a protective finish, which can be damaged by many cleaning products.

No matter how often you wash and wax your vehicle, exposure to the elements will cause the

Figure 20–7 Use household detergents on your dishes and laundry, not your car.

bumper sticker sightings

DO NOT WASH—
THIS VEHICLE IS A
SCIENTIFIC EXPERIMENT

Figure 20–8 You should wax your car often to protect its finish.

exterior paint to oxidize and fade in color. The color of the roof, hood, and trunk will often fade more rapidly than the color on the sides of the car. Long-term exposure to the sun's ultraviolet (UV) rays can cause paint to fade after one to two years, to crack after three to four years, and to lift and peel after five to six years. To minimize the effects of environmental damage to your car, keep it covered with a tarp or in a garage or carport if you are not going to drive it for a while.

Protecting the Interior

It is just as important to keep the inside of your vehicle clean as it is to protect the outside. Vacuum your car regularly, and occasionally clean it out. Having a thick film of dust on your dashboard or fast-food wrappers on the floor is embarrassing when you have passengers. It can also be dangerous—a golf ball, cola can, or apple core that gets wedged under the accelerator while you are driving can cause a crash. A number of available products are available on the market to clean upholstery, leather, vinyl, and the inside of your windows. You can also buy products that will deodorize the interior or protect it from UV rays.

20–2 SELECTING A MECHANIC

Sometimes new-car warranties require you to have certain services on the car performed by the dealer to enjoy the full protection of the warranty. Other services may be performed elsewhere. After the warranty expires, you may

Figure 20–10 Always get an estimate before authorizing repairs on your vehicle.

Wild Wheels

Figure 20–9 What is the name of this car, introduced in 1972, that had most of its black Naugahyde upholstery on the *outside*?

factoid

The most popular color for a car is silver, followed by white, black, blue, green, and red.

Figure 20-11 Let only certified mechanics work on your vehicle.

factoid

Nearly 40% of selected plastic parts in automobiles are recycled from damaged or discarded cars.

have all services performed any place you choose, including at the dealership, at your local garage, or at a service station. You might also consider an independent mechanic. No matter who works on your vehicle, find someone you trust. Ask friends and relatives where they have had their vehicles serviced and if they were satisfied with the work.

There are many ways to protect yourself from poor workmanship at auto repair shops. One of the best ways to protect yourself is to shop around. Do not be content with merely one estimate from one mechanic regarding a repair. Get two or three estimates. And do not automatically fall for the lowest price. A golden rule of thumb is to eliminate the highest and lowest estimates. The mechanic with the highest estimate is probably charging too much, and the lowest may be cutting too many corners.

Check a garage's reputation before you authorize it to perform repairs. Contact your local Better Business Bureau and the Illinois Attorney General's Consumer Protection Division to see if any consumer complaints have been filed against the shop. Make sure that the mechanics are certified by the National Institute for Automotive Service Excellence (ASE).

Take the time to survey the facility. If the service bay is littered with junk or the employees seem unprofessional, get out of there. If you are satisfied, show some interest in the job being done on your vehicle. Without trying to diagnose the problem, be specific in describing what you see, hear, feel, or smell, where the problem seems to be located, and what your vehicle is doing when it occurs. You may be surprised to find out how nice a mechanic is when he or she knows that you are paying attention.

Before you have any work done, *ask for an estimate in writing.* You may request either an itemized or a nonitemized estimate indicating the total price for repairs performed. The itemized estimate should include a list of each part, whether it is new or used, and a description of all charges for parts, diagnostic testing, assembly, disassembly, and labor. It should also include a statement by the mechanic declaring whether the repairs are required or suggested. The date of the estimate and your vehicle's odometer reading should be indicated as well. As extra protection against unnecessary repairs, request that all parts (except warranty and exchange items) be returned to you for inspection.

According to Illinois law, a repair shop may not charge more than 10% above an itemized estimate or any amount above a nonitemized estimate without your permission. Also, it may not perform work you do not authorize either in writing or over the phone.

After the repairs are completed, obtain a repair order or invoice describing the work done. It should list each repair, the cost of each part used, labor charges, the vehicle's odometer reading when it entered the shop, the total cost, and any charges covered by warranty. Compare the repair order against the estimate to make sure that only authorized work was performed and that you were charged the proper price for the parts used. If you were later contacted by the shop because unforeseen circumstances caused the actual cost of repairs to exceed the estimate, inspect both the invoice and the estimate to see if the merchant recorded the date, time, phone number, and additional cost. Inspect each replaced part and, if necessary, ask the mechanic to point out what was wrong with it.

If your best efforts to protect yourself from a bad mechanic fail, first return your car to the repair facility that did the work. Provide a written list of the problems, and keep a copy of the list. It is also a good idea to get a written statement from another mechanic defining the problem and outlining how it may be fixed. Give the repair facility a reasonable opportunity to fix your car. Speak directly to the service manager and have him or her test drive the car to observe the problem. You may be pleasantly surprised to discover that he or she will want to keep your business by pleasing you.

If you continue to have problems, report the facility to the Better Business Bureau and the Consumer Protection Division. A call or letter from them may persuade the mechanic to take action. If you charged the repairs, send a certified letter to the credit card copy explaining the situation; the company may decide not to honor the charge. Mail copies of all your correspondence to the merchant, who will probably prefer mediation to going to court. If all else fails, bring a lawsuit against the mechanic in small claims court. If the disputed charges are high, it may be worthwhile consulting an attorney.

No matter what your problems are with the car, keep accurate records. Copies of your service invoices, bills you have paid, letters you have written to the manufacturer or the repair facility owner, and written repair estimates from an independent mechanic are essential in helping resolve your problems. Not only will these documents help you out legally, but they will also help lead to a quicker resolution of your complaint.

20–3 VEHICLE EQUIPMENT AND ACCESSORIES

Equipment and accessories exist to help you with emergencies, to assist disabled drivers, and to improve the comfort and appearance of your vehicle.

Figure 20–12 This home, once occupied by a prominent militia general, state supreme court judge, and U.S. senator, is now maintained by the White County Historical Society.

Where am I?

factoid

One in seven jobs in the United States is dependent on the automotive industry.

Figure 20–13 Keep copies of all records related to repair work on your car.

Safety-Oriented Equipment

Dozens of products are on the market that can make your vehicle safer or contribute to your becoming a safer driver. Interior, oversize rearview mirrors can be installed to provide the driver with a more extensive rear view and to reduce the size of your blind spots. Reinforced bumpers provide extra protection in the event of a head-on or rear-end collision. Door guards provide extra protection in tight parking situations.

Some of the most important safety equipment that you can acquire are items that help the driver's ability to see. Up to three fog lights or other auxiliary driving lights can be mounted on the front of your vehicle to improve your visibility at night and during bad weather. In addition, one spotlight can be mounted on the front of your vehicle to assist your sight in emergency situations and to read signs. To make your vehicle more visible, especially if your vehicle is an older model, consider installing daytime running lights, a center high-mounted rear brake light, clearance lamps and cab lights, and extra reflectors or reflector strips.

Vehicle Accessories

When it comes to vehicle accessories, the list is almost infinite. Certain accessories can "expand" the cargo area of your vehicle, such as ski racks mounted on the vehicle's roof or side; luggage webs, racks, or bubble cases mounted on the roof or on the rear of the vehicle; bike racks; and bungee cords, which are stretchable ropes with hooks on the ends that work wonders when you have to tie down cargo. You can increase the storage capacity of the interior of your vehicle by adding a customized console.

Other accessories allow you to enhance the exterior appearance of your vehicle, adding value, increasing life span, and preventing costly repairs down the road. Mobile tents or car covers can protect your vehicle from the elements if you have no garage. Window vent guards keep unwanted moisture from entering the passenger compartment. Exterior paint can be protected by installing a hood-mounted bug and stone deflector and/or a leather or vinyl protector for the front hood and bumper. You can also purchase covers for the exterior lights.

Some accessories, such as specialty wheel rims, license plate frames, airdams, air deflectors, flares, scoops, and spoilers may

Figure 20–14 Fog lights can greatly increase your vision as a driver.

Figure 20–15 Bicycles are hard to stow in a vehicle without a specialized rack.

Figure 20–16 Keeping your car covered when not in use will protect the paint from the harsh effects of the environment.

Figure 20–17 Replacing your existing wheels can be a less expensive way to individualize your vehicle.

make the vehicle more aerodynamic and/or stylish.

Some interior accessories are designed to protect the carpet and vinyl inside your vehicle from wear and sunlight, such as dashmats, floormats, seat covers, and steering wheel covers, whereas others merely add to your driving enjoyment and pride of ownership, such as car audio systems, racing pedals, shift knobs, and dashboard overlays.

Pickups, sport utility vehicles, and vans support additional accessories that can enhance appearance, contribute to driver safety, and help make these vehicles even more practical than they already are. Winches mounted on the front of the vehicle use either tough wire cord or chain to pull other vehicles and objects out of the way or out of dangerous situations. Brush guards and taillight covers protect the grille and lights from damage during off-road driving. Roll bars affixed to the chassis of pickups and sport utility vehicles add more structural support in the event of a rollover crash. Marker lights, which are small lights mounted along the sides of a vehicle, aid in making the vehicle more visible to other drivers.

Several different options are available when you want to either increase the cargo carrying capacity of your vehicle, protect the vehicle from cargo damage, or increase the amount of interior space. Exterior bed-mounted tool boxes for pickups are containers specially built to stow tools and other materials securely. Tonneaus cover the bed of a pickup, creating a secure, enclosed storage space. Cargo covers get around the problem in sport utility vehicles of not having a secure, locked trunk. These covers can be placed in the rear of a sport utility vehicle to hide stored items. Similarly, cargo vaults can be installed in pickup beds or in rear compartments of sport utility vehicles for safe and secure storage of belongings.

Pickup beds get a lot of rough treatment, and unsecured cargo tends to badly scratch and dent the bed. To protect your bed, you can purchase a tough plastic bedliner or rubber bed mat. You may

Figure 20–18 Installing a tonneau over the bed of the pickup creates a secure, enclosed cargo space.

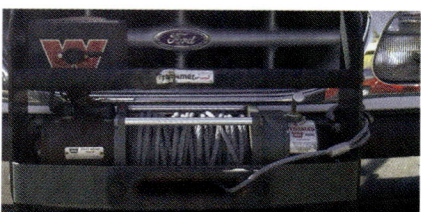

Figure 20–19 A winch can be helpful in emergencies.

Figure 20–20 This accessory encloses the bed and greatly increases its size.

want to consider spider netting lift gates and flexible bed covers that "mold" to your cargo and keep it from jostling around. If you want to increase the interior space of a pickup, you can get a camper shell, which comes in a variety of shapes and sizes. This will create a large enclosed space out of your bed. If you want to increase the space on a van, look into getting a van top, which extends the roof vertically and allows for more headroom for passengers. Installing door steps and running boards on the undercarriage of high-profile vehicles will assist drivers and passengers to safely get into and out of these vehicles with a minimum of effort.

Equipment for Disabled Drivers

Just because you are physically disabled does not mean that you cannot or should not learn how to drive. Given the advances in automotive **adaptive equipment** technology in recent years, it has become possible for people with many types of physical disability, including cerebral palsy, spina bifida, amputations, muscular dystrophy, head and spinal cord injuries, and multiple sclerosis to drive.

If you are physically disabled and you find that learning to drive in a standard driver education course is impractical or not working out, there are disabled driver education specialists in Illinois who can help you master the fundamentals of driving and work with you to obtain your license. Depending on your particular situation, it is possible to obtain financial help for this specialized training from various state and federal agencies, your insurance company, disease-specific organizations, and local civic/religious groups.

If you are a disabled driver, the process of adjusting a vehicle for your use centers around finding the appropriate mix of adaptations that will work specifically for your needs. Of course, not every disabled person is the same, and what works for one type of disability will often not work for another type of disability. Adaptive equipment is as unique as is the particular disability that it is being used to accommodate. Some of the more common devices used to assist with the steering function are spinner knobs, amputee spinners, foot steering mechanisms, and "tri-pins" that can be attached to the steering wheel to allow you to steer if you cannot grip the wheel. Adaptive devices for the various controls on the vehicle can be as varied as remote switches, extensions to existing controls, automatic clutches, pedal blocks, left foot accelerators, and hand controls that allow someone who cannot use their legs to control signals and pedals with their hands. Certain vehicles can also be outfitted with a lift on the side or rear to enable persons to drive while in their wheelchair.

Although the cost of adapting a vehicle can sometimes be prohibitively expensive, many vehicle manufacturers have disability programs whose main purpose is to assist physically disabled individuals acquire a new car, either via lease or purchase. These manufacturers may even refund disabled drivers for some or most of the cost of the adaptive equipment necessary to make a new car drivable for their particular disability.

20–4 VEHICLE SECURITY

Unfortunately, we live in a time when we all need to be concerned about the security of our vehicles. Because vehicles are valuable, mobile, and often parked in the open on public streets, it is no surprise that they are one of the most stolen possessions in society.

When and Why Vehicles Are Stolen

A significant percentage of auto thefts are committed during the early morning when the owners are asleep and their vehicles are unattended. According to recent studies, about one-quarter of all stolen vehicles had the ignition key in them and in plain view of the thief when they were stolen! About 40% of stolen vehicles had at least one unlocked door at the time they were taken. Only about 10% of the stolen cars had any type of antitheft device installed.

Contrary to popular belief, the most targeted vehicles among thieves are not fancy luxury cars or sports cars. The average stolen vehicle is an ordinary family car, sport utility vehicle, or minivan that is a few years old whose spare parts value has risen. In other words, because the model is getting older, the availability of spare parts from the manufacturer and other suppliers is decreasing. This increases the value of the parts of that model and makes those vehicles still on the road a more attractive target for thieves. Some vehicles are actually worth two to three times their value on the black market when disassembled into parts!

Preventing Vehicle Theft

The typical car thief is an opportunist, looking for an "easy" car to steal. The more time that it will take to break into your vehicle, the less likely he or she will bother to steal your car. A completely unprotected car can be stolen in less than 30 seconds. The vehicle owner can become an active part of the solution to car theft by doing a variety of things

factoid

More than 50,000 vehicles are stolen in metropolitan Chicago each year.

Know your neighbor

You are more likely to have your car stolen in Gary, Indiana, or Milwaukee, Wisconsin, than in Chicago.

to prevent his or her car from being stolen.

A large part of playing the antitheft game successfully is knowing where and how to park your vehicle. Always use the emergency brake when parking, even if you are on level ground, because this will delay a thief who intends to push your car away for a quick tow. Turning your wheels toward the curb when you park (which you should always do unless you are facing uphill) also makes it more difficult for a thief to tow the car. Make sure that you do this when the engine is still on because your steering column lock will prohibit you from fully turning the tires once the engine has been shut off.

Most thieves avoid doing their dirty deed in open view, so always park your car in well-lit areas with a lot of pedestrian traffic. If there is a parking lot security guard or attendant, try to park as close to where this person is stationed. If a secure garage or parking structure is available where the thief never even has a chance to see the vehicle in the first place, use it.

Before leaving your vehicle, always remember to take your keys out of the ignition, lock your car, and completely close the windows. Never leave your car running and unattended. Do not leave valuables in plain view to tempt thieves. Do not hide a spare key on the outside of the car because thieves can usually find these keys. Instead, keep a spare in your pocket or purse.

Do not let your guard down when leaving your vehicle with a mechanic for repair work. Investigate the mechanic before you entrust him or her with your vehicle. Remember to get a signed receipt from the repair shop that shows it has taken custody of your vehicle. Give a mechanic only your car keys, and clean out valuables from your vehicle before you drop it off.

Antitheft Devices and Systems

The next level of prevention is to install some type of antitheft device on your vehicle. **Antitheft devices** are any equipment installed on your vehicle that either deters, disrupts, or helps catch a car thief. Many automobile insurance companies will offer you a discount on your comprehensive insurance policy if your vehicle is equipped with one or more of these devices. There are three basic types of antitheft devices: warning devices, shut-off devices, and tracking devices.

Warning devices usually emit either a sound, signal, or a visual cue intended to scare a would-be car thief away. Audible **car alarms** are equipped with sensors that trigger a loud siren or a constant beeping of the vehicle's horn. Most people are so used to hearing these alarms go off by mistake, however, that they do not investigate when they hear them, and the owner is usually out of hearing range. "Silent" car alarms are normally more effective because they are inaudible to the thief but trigger an alarm with the owner or the police.

Steering wheel locks are metal bars with a hooking mechanism on both ends that are locked into place either on the steering wheel or between the steering wheel and the brake pedal. This makes it nearly impossible to steer or otherwise operate the vehicle, a major problem if you are a thief trying to make a quick getaway.

Steering column collars are metal collars that can be fitted around your steering column to prevent a thief from accessing the ignition or breaking into the ignition wires. They are particularly effective on older vehicles that are more easily "hot-wired."

Shut-off devices operate on the principle that thieves will often bypass your ignition and attempt to hot-wire the car to get it to start. A variety of products defend against this. The two most commonly used are smart keys and kill switches.

Smart keys are ignition keys that are computer-coded or radio-controlled to signal your vehicle to start. Without this exact key, the vehicle's motor will remain "shut off," even if the thief has hot-wired it. Kill switches are sometimes called starter disablers, fuse cut-offs, ignition disablers, or fuel disablers. They are secondary ignition switches concealed in an unlikely location inside the vehicle that work by inhibiting the flow of electricity or fuel to the engine. To start the vehicle, you must restore that flow by inserting a separate key or manually activating the switch. When a thief hot-wires a vehicle and it still does not start, he or she will suspect a kill switch is somewhere on the vehicle. Most thieves are too nervous to spend the time looking for the kill switch, especially if it is well concealed, so they will move on to another vehicle. A cheap alternative when you will be leaving the car parked for long periods is to remove the coil wire from the distributor cap. The vehicle will not start without this part.

bumper sticker sightings

IF YOU LOVE YOUR LIFE AS MUCH AS I LOVE MY CAR, DON'T STEAL IT

Figure 20–21 Using a high-voltage electrical shock, this steering wheel lock goes one step further than most to deter theft.

You can also equip your vehicle with a tracking device. With this system, if your vehicle is stolen, you report its disappearance as soon as possible to a central response center. That center will then activate a tracking device hidden somewhere on the car that emits a signal to the police or to the central response center, allowing them to find the exact location of the vehicle. This system has proven to be very successful in recovering stolen vehicles quickly, sometimes even before they are stripped of parts.

YOUR TURN

20–1 Preventive Maintenance

1. What is preventive maintenance?
2. How do you check tires for inflation and wear?
3. What fluids should you check as part of routine maintenance?
4. What is the proper way to wash and wax a vehicle?

20–2 Selecting a Mechanic

5. How do you select a mechanic?
6. How can you avoid poor workmanship?

20–3 Vehicle Equipment and Accessories

7. What are three types of safety-oriented equipment?
8. What devices are used to adapt a regular vehicle for a disabled driver?

20–4 Vehicle Security

9. What are three basic types of antitheft devices available?
10. How does a vehicle tracking device work?

SELF-TEST

Multiple Choice

1. You should check tire pressure:
   a. when the tires are cold.
   b. when the tires are hot.
   c. every time you change your oil.
   d. when the tires are "bald."

2. What routine maintenance should be performed on your vehicle every month?
    a. Replace the radiator cap.
    b. Replace the fuel filter.
    c. Check the exhaust system for leaks.
    d. Check the power-steering fluid.
3. The paint on the _____ of your car will oxidize less rapidly than the other parts.
    a. sides
    b. hood
    c. trunk
    d. roof
4. If you were a disabled driver and wanted to obtain a refund for a spinner knob you wanted to place into a new car you just purchased, where would you seek the refund?
    a. the Red Cross
    b. a motor-vehicle manufacturer
    c. the federal government
    d. your family physician
5. What is the typical type of motor vehicle stolen by car thieves?
    a. ordinary family car
    b. pickup
    c. sports car
    d. luxury four-door sedan

Sentence Completion

1. For equal wear, tires should be _____ on a regular basis.
2. The oil level in your car should always be checked when the engine is _____.
3. _____ lights on the sides of a vehicle greatly enhance its visibility.
4. A spinner knob is a common piece of _____ equipment used by disabled drivers.
5. Starter disablers and fuel disablers are two types of _____.

Matching

Match the concepts in Column A with examples of the concepts in Column B.

Column A
1. __ Preventive maintenance
2. __ Exterior appearance accessory
3. __ "Pinging" engine
4. __ Pedal block
5. __ Tire rotation
6. __ Increases interior space
7. __ Shut-off device
8. __ Itemized estimate

Column B
a. Description of parts and labor
b. Van top
c. Car bra
d. Disabled driving adaptation
e. Checking fluid levels
f. Octane rating
g. Front to rear
h. Ignition disabler

Short Answer

1. What documents should you have to help protect you in a dispute with a mechanic?
2. Why are bald tires dangerous?
3. What does a kill switch do?

4. What are some of the disabilities that can be accommodated by adaptive equipment in vehicles?
5. Why are so many vehicles stolen in the early morning?

Critical Thinking

1. After hearing your brakes make a grinding noise, you decide to take your car to a mechanic. You go to a garage that is offering a great deal on brake jobs. He writes you out an estimate to replace the brake pads, and you leave the car with him. When you return at the end of the day, he submits a bill to you that is four times the estimate. He explains that you needed new rotors, but he could not reach you at the telephone number you left him to discuss the problem. What should you do?

2. You just got a part-time job in a high-crime area of your city, and you must drive there every day. Your boss does not have enough parking spaces for all of the employees, so you are forced to find your own parking on the street, usually about two blocks away and out of hearing and viewing range. You are worried about the security of your vehicle but do not know which vehicle antitheft device is the appropriate one for you to purchase. What should you do?

PROJECTS

1. Call a local automobile repair shop and ask the owner or manager if you could spend a few hours observing what the mechanics do and how they do it. Ask to watch a "lube job," "brake job," tune-up, and other procedures. Take notes and share your experience with your class.

2. Go with a friend to a local automobile supply or accessory store and investigate all of the different antitheft devices available. Compare the prices, the claims of the manufacturers, and the warranties offered. Then go to the library or use the Internet at home to determine what the consumer press has said about each of these devices. How do they stack up against one another? Analyze all the information you have collected and come up with a recommendation for protecting your vehicle effectively and for a reasonable amount of money.

21 chapter

Driving Away from Home

Whether you are going on a short overnight trip to the beach, lake, or mountains or a long vacation far from home, you must be prepared to deal with unfamiliar roads, customs, weather conditions, and driving regulations. If you are hauling a boat, towing a trailer, moving furniture to a new home, driving a recreational vehicle, or renting a car, you must familiarize yourself with different instruments, controls, and maneuvering capabilities.

Chapter Objectives

Upon completion of this chapter, you should be able to:

21–1 Road Trips

1. Understand how to prepare for extended road trips.
2. Describe the basic features of a road map.
3. Understand how highways are numbered in North America.

21–2 Driving Unfamiliar Vehicles

4. Understand what precautions you should take when driving an RV or moving van.
5. Understand how to safely load and pull a trailer.
6. Understand what precautions you should take when driving an SUV.

KEY TERMS

scale
legend
mileage (distance) chart
loop
spur
recreational vehicles (RVs)
sport utility vehicles (SUVs)

21–1 ROAD TRIPS

Driving away from home can be fun and exciting, but without adequate preparation, your pleasant getaway could easily turn into a nightmare.

Planning Ahead

The key to enjoying any extended road trip away from home is to plan ahead. Most auto and travel clubs will provide you with maps, suggested routes, and travel guides at no extra charge. Some clubs can warn you about roads that are closed or under construction, tolls, and other potential problems you can expect to encounter on your route.

Maps and travel guides are also available from service stations; bookstores; local chambers of commerce; and state/provincial, regional, and city tourism bureaus. The Internet is a great resource for planning trips. Browse the World Wide Web for maps, directions, and detailed information on your destination. Computer software is available as well to help you plan trips.

Take seasonal weather conditions and traffic congestion into account when you plan the dates of your trip. Choose the simplest route that you can with the fewest number of hazards, even if it takes a little longer to get to your destination. Try to time your passage through cities and areas with heavy traffic at times other than rush hour. Nothing is more frustrating than spending your vacation stuck in the same traffic jams that you face on your way to work or school.

Figure 21–1 There are many resources for planning road trips.

Organize your trip so that you do not have to drive too far in one day. Change drivers frequently. Allow plenty of time to relax and rest overnight along the way. If you are taking your trip during a peak travel season, reserve lodging ahead of time so you do not get stuck without a place to sleep. To break up the trip, plan to explore some of the tourist attractions along your route.

Before You Leave

Before you begin a long road trip, always make sure that your vehicle is in the best possible working order. It is a good idea to have it serviced before leaving. If you are traveling far from home, especially through less populated areas, pack emergency supplies. Keep an extra set of car keys in your pocket or with a passenger.

Figure 21–2 Vacations and other extended road trips require planning and advance preparation.

Do not forget to take your auto club membership card or a roadside service telephone number.

Figure out a budget that includes expenses for lodging (hotels, motels, hostels, or campsites), meals, entertainment, and fuel. Make sure that you include in your budget any additional expenses such as entrance fees, parking fees, equipment rental, ski-lift tickets, fishing permits, and souvenirs. Take along extra funds in case of an emergency or a vehicle breakdown. If possible, use traveler's checks and credit cards to cover your spending, and carry only a minimum of cash. In case they get lost or stolen, leave a copy of your traveler's check receipts and credit card information with a friend or relative. Put coins in a location accessible to the driver for toll roads, bridges, and tunnels.

Always tell a family member, friend, or neighbor what your intended route is and places where you can be reached along the way if necessary. If you are stuck in a remote area or are the victim of a crime, this information will make it easier for authorities to locate you. It is always a good idea to regularly check in with somebody during your trip to let him or her know all is well. If no one will be at home while you are away, arrange with someone to look after your pets and collect your mail. You should also postpone your newspaper subscriptions. Uncollected mail and newspapers are open invitations to thieves. Set timers to turn your home lights on and off so the house or apartment will not look empty while you are away.

Using Road Maps

Maps are invaluable to the traveler to find out where you are and how to get to your destination.

factoid

The first "motel," combining "motor" and "hotel," opened in California in 1924.

Figure 21–3 If you do not book lodging in advance, you may have difficulty finding a room.

Figure 21–4 Keep an assortment of coins in your car to pay tolls on road trips.

index. For example, a city listed as "F-24" would be found within the space on the map where imaginary columns extending from "F" and "24" intersect.

The **legend** explains the colors and symbols used on the map to represent different types of roadways and common landmarks such as airports, rest areas, campgrounds, and ports of entry. On detailed city maps, the names of schools, shopping centers, parks, hospitals, places of worship, monuments, and other important sites usually appear on the map next to the symbol.

Many maps have a **mileage** or **distance chart,** a table that lists distances in miles/kilometers between key points on roadways. Depending on the scale, distances and approximate travel times may also be indicated directly on the map next to the roadway, usually using a system of colored arrowheads.

Besides displaying the various roadways and ferry routes in a given area, they highlight points of interest, types of terrain, climate differences, time zones, and other useful information. The **scale** of the map determines its level of detail. State and provincial maps usually have an index of towns, cities, and recreation areas, as well as small inset maps of major cities and downtown areas. City maps generally have a street index to enable you to locate a particular address.

Of course, a map is useful only if you know how to read it. On most maps, the top border corresponds to the direction north. A compass or arrow marked somewhere on the map indicates the actual orientation. The letters and numbers that appear at intervals on the edges of the map correspond to the letter–number codes used in the

Highway and Freeway Numbering

When you are traveling away from home, it is useful to know the numbering system for highways and freeways. These identification numbers are posted on signs at regular intervals along the roadway. On United States and Canadian freeways and highways, routes that run in a general north–south direction have odd numbers. Even numbers are assigned to routes running in a general east–west direction. In the United States, the greater the even number, the farther north the road is. The greater

Figure 21–5 Maps are essential for any road trip to an unfamiliar area.

© GeoSystems Global Corp.

DRIVING TIPS — Maps Are Not Always Reliable

Just because a road or highway is on a map does not mean that you will be able to use it. Some maps clearly mark roads that are unpaved, under construction, or closed during certain times of the year, but the information may be inaccurate or out of date, depending on how old the map is. To determine the actual condition of a questionable roadway, contact local highway officials to make sure it is open and safe to use for the type of vehicle that you are driving.

the odd number, the farther east the road is.

Most routes are one- and two-digit numbers. Alternative routes are generally designated with three-digit numbers, with the last two digits representing the main route. If the first digit is even, the alternative route is a **loop** that goes through or around a city. If the first digit is odd, the alternative route is a **spur** that leads directly into a city. Business loops and spurs that go to or around business sections of cities have the same one- or two-digit number as the main highway.

Crossing the United States–Canadian Border

Traveling between the United States and Canada is almost as easy as driving from one state or province to another. The primary difference is that Canada uses the metric system, whereas the United States does not. In several Canadian provinces, signs also appear in French or in both French and English.

If you are a citizen of either the United States or Canada, you need a valid driver's license, a passport or other proof of citizenship, and your vehicle's title certificate or registration to cross the border. You must keep these with you at all times while driving. If you plan to stay longer than six months in either country, contact immigration officials to determine whether you need a visa.

In general, penalties for driving under the influence, licensing requirements, and insurance laws are more strict in Canada than in the United States. Each state and province has its own laws regarding when visitors must obtain a license and register their vehicle in that jurisdiction, but generally this applies only if you become a resident in that state or province. In most cases, exchange students are exempted from these requirements. Drivers are not required to show proof of insurance at the border, but they must have it if they are involved in a collision or are pulled over by the police.

21–2 DRIVING UNFAMILIAR VEHICLES

At one time or another, you may end up driving an unfamiliar vehicle such as a rental car, recreational vehicle, sport utility vehicle, or moving van, or you may

Figure 21–6 In which general direction does this route run?

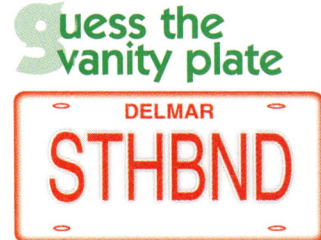

Figure 21–7

Figure 21–8 United States drivers in Canada must be prepared to tackle the metric system and, in some provinces, signage in French.

have to tow a trailer. Before taking to the road, it is important for you to familiarize yourself with the limitations and capabilities of each of these vehicles.

Rental Cars

When renting a car, take a moment to learn how it differs from your own vehicle. Before you get in, walk around the vehicle to get a sense of how long and wide it is. Thoroughly inspect it for dents or scratches and report them to the rental agency. Make sure that the tires are properly inflated and that a flat tire kit and chains, if necessary, are stored in the trunk. Look at the layout of the dashboard. Note the location of warning lights, gauges, and hazard lights. Activate all switches, controls, and locks to make sure that everything works. Locate the emergency brake and adjust the seat and all mirrors. Take the car for a trial run so that you get used to the response of the steering wheel, brakes, gearshift, and accelerator. If you feel uncomfortable with the vehicle, request another one more similar to your own.

Most rental car companies require you to return the car with the same amount of fuel as when you picked it up. If you do not, they will charge you high rates to refill the tank themselves. Allow yourself plenty of extra time to stop and refuel before you are due to return the car.

Recreational Vehicles

Recreational vehicles, often called **RVs,** include motor homes, pickup campers, some kinds of vans, and trailer combinations of various types and sizes. They are generally much harder to handle than cars because of their size, weight distribution, and awkward shape. RVs share the same blind-

Figure 21-9 Take time to inspect a rental car before you take it off the lot.

spot problems as trucks, but they have less power and are less stable. Their high profile makes them vulnerable to gusting winds, and most lack the aerodynamic design and durable construction of commercial rigs. Many campers have heavy rear sections that are prone to veer from side to side. The high center of gravity of some RVs increases the risk of rollover on turns.

RVs accelerate, brake, and maneuver much more slowly than cars. Always maintain an extra-long following distance when driving an RV. When possible, try to avoid backing up because most RVs have very limited rear visibility. Instead, drive around the block and reach your destination head-on. If you have no other choice, first get out and walk around the vehicle to check for any obstacles that you cannot see from the driver's seat. Have another person stand outside the RV to help guide you.

When driving in windy conditions, reduce your speed and keep a firm grip with both hands on the steering wheel. In extreme winds, stop in a sheltered area until things quiet down. Be prepared to feel the effects of wind gusts when driving out of a protected area such as a tunnel or when passing another large vehi-

The first rental car, the Hertz "Drivurself," was introduced in 1925.

DRIVING TIPS — Rental Car Insurance

Rental car companies normally offer all renters extra insurance on their vehicles. Some of this coverage may duplicate the coverage of your own insurance policy. Although in many cases your insurance company will cover you in a rented car just as if you were driving your own vehicle, it often will not cover damages for revenue lost when a rental vehicle is being repaired. Before deciding whether to purchase additional insurance on a rental car, check with your insurer first.

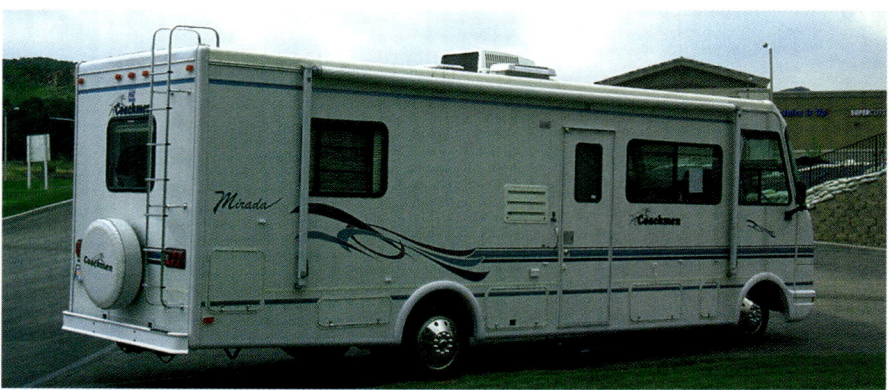

Figure 21–10 Driving a recreational vehicle is very different from driving a car.

cle such as a tractor-trailer truck. Because driving an RV can be mentally and physically exhausting, plan frequent rest stops and switch drivers more often than you would in a car.

Being conscious of the large size of your RV will prevent you from hitting overhead objects such as overpasses or service station canopies. If your RV is rented, the total height of the vehicle may be marked on the outside, near the driver's door. If not, check with the rental agency or in the vehicle manual to find out the correct height. As a reminder, it is a good idea to mark this height inside the vehicle within view of the driver. Most overpasses and tunnels will be marked with a clearance height. If your vehicle is too tall, or very close to the clearance height, or the passage seems very narrow, *do not proceed!* Turn off the roadway and find another route that can accommodate your vehicle such as a local truck route. Local regulations on extremely tall or wide vehicles may even restrict you to such a route.

Moving Vans

The challenges of driving a moving van are similar to those of an RV. Before passing under bridges or overpasses, know the van's height so that you can judge your clearance. Increase your following distance and give yourself additional time to speed up, brake, and stop. Make your turns slow and wide. When steering, compensate for wind gusts. Always load the heaviest items first, over the axles, and try to distribute the weight of the load evenly.

Most moving van rental agencies will inspect their trucks adequately before turning over the keys. However, you should still take a walk around the vehicle to check its condition. If you see any major damage, check that the rental agency has noted it in its records. Make sure that all lights and locks are working and that the tires have good tread and are

Wild Wheels

Figure 21–11 Which company, famous for making tractors and other kinds of farm equipment, built this early RV for use in African expeditions in the 1930s?

Figure 21–12 Make sure that the truck you rent is appropriate for the size and weight of the load you carry.

properly inflated. When you leave the rental location, take a drive around the block to check that the brakes feel and sound good. If you are not satisfied with the condition of the van, return to the agency and ask for another vehicle in better condition.

Towing a Trailer

Vehicles towing trailers are even less stable than RVs and moving vans. Most jurisdictions require towed trailers to be driven in the far-right lane of traffic except when passing, preparing to turn, entering or leaving a highway, and driving on a highway with four or more lanes in each direction, in which case you may use the two right-hand lanes.

Pulling a trailer puts a strain on your vehicle and its engine. If you are going to hitch a trailer to your vehicle, especially for long distances, study your owner's manual for specific procedures to prepare your vehicle. Besides checking engine fluid levels and filters, it is generally recommended that you increase the air pressure in your vehicle's rear tires to the maximum recommended limit. You will also need safety chains to connect the trailer to the car in case the hitch breaks, and special oversize side-view mirrors on both sides to allow you to see the full length of the trailer. Remember to remove these mirrors from your tow vehicle after you have unhitched the trailer.

Figure 21–13 Be sure to properly connect the brake lights when attaching a trailer.

DRIVING TIPS — Towing at Night

If you will be towing a trailer at night, check the alignment of your headlight beams *after* you have hooked up the trailer. The added weight of a trailer often tilts the headlights of a car or pickup truck upward into the eyes of oncoming drivers, which is dangerous and illegal.

Do not try to hook up a trailer by yourself if you have never done it before. If you are renting the trailer, have the rental agent show you how to hitch and unhitch it and to connect the brake lights so that other roadway users to the rear of the trailer will know that you are braking. Most vehicles with a trailer hitch are equipped with a special socket into which you can plug the brake-light cord extending from the trailer tongue. If your vehicle is not equipped with this socket, you can purchase special light bulbs that screw into your brake-light sockets and have electric cords extending from their sides that attach to the trailer brake lights. These light bulbs are available at trailer rental agencies and auto supply stores.

Always hitch the trailer to the tow vehicle *first,* and then load the trailer. When you are loading a trailer with loose cargo such as luggage, furniture, or camping equipment, always put the heaviest items in first over the trailer's axle. Try to distribute the weight of all other items evenly from left to right and front to back. Secure the load with a rope, net, or tarpaulin as necessary. *Never carry passengers in a trailer.* This is unsafe and is illegal in most jurisdictions. It is also illegal to tow more than one trailer or other vehicle at a time in a "trainlike" fashion.

Keep in mind that it takes much longer to stop, accelerate, and turn when towing a trailer. Increase your following distance, especially going downhill, and take your turns slow and wide. Because your vehicle is now much longer than it is without the trailer, keep passing and lane changes to an absolute minimum. Use lower gears on hills for added control and to reduce the load on your brakes. Keep an eye out for special posted trailer speed

Figure 21–14 Some roadways have lower speed limits posted for vehicles pulling trailers.

limits on highways and on roads with steep downgrades.

If you are towing something large and heavy, such as a boat or a camper, you will have to be prepared to adjust your steering to deal with gusting conditions and crosswinds created by vehicles as they pass you or as you pass them. If at any point your trailer starts to sway or fishtail, steer toward the center of your lane and gradually reduce your speed. If the trailer has a separate set of brakes, use them first while accelerating slightly. When the trailer stops swaying, use your vehicle's brakes carefully to slow down. If necessary, pull off the roadway to see if the load has shifted and whether it needs any adjustments.

Sport Utility Vehicles

Four-wheel drive vehicles, designed to operate both on- and off-road, are often called **sport utility vehicles (SUVs).** These vehicles differ from average cars in significant ways. They are usually taller, so you sit higher relative to the road. Because you can see traffic and the roadway ahead more easily, you may be tempted to follow the vehicle ahead of you too closely. Take care not to tailgate, a common bad habit of SUV drivers.

Because SUVs have a high center of gravity, they are more likely to "lean" as they go around a curve. They also have a rollover rate two to three times that of regular cars. Therefore, it is very important when driving an SUV to watch your speed and slow adequately for turns, curves, and even lane changes. *Never carry passengers in the open cargo area of an SUV.*

If you plan on doing frequent off-road driving in an SUV, you should have an experienced off-road driver show you some basic and advanced techniques. Off-road driving can be dangerous if you have little experience or understanding of how to maneuver a vehicle around difficult obstacles. Every off-road vehicle has different capabilities and limitations. Some general rules to follow when driving off-road are as follows:

- Drive directly up and down hills because traveling diagonally may result in a sideways slide.

- Cross ditches or logs at an angle so that one wheel at a time goes over the obstacle.

- Accelerate gently to avoid spinning your wheels. If a spin does occur, back off the pedal to regain traction.

- Do not wrap your thumbs around the steering wheel. They could break if the wheel kicks around suddenly when it hits rocks or other obstacles.

- Know your vehicle's performance potential.

- Vehicle speed should be reasonable, and not so fast as to endanger any passengers, pedestrians, or animals.

- Cross any two-way highways at a 90-degree angle.

SUVs have the highest occupant ejection rate of any vehicle.

- Drive in a manner that respects wildlife, land, and vegetation on public property. Stay in areas or on trails specially designated for off-road use.

In many areas of the United States and Canada, you must obtain a permit to "off-road" on public lands. It is your responsibility to comply with all regulations.

YOUR TURN

21–1 Road Trips
1. How do you prepare for extended road trips?
2. What are the basic features of a road map?
3. How are highways numbered in North America?

21–2 Driving Unfamiliar Vehicles
4. What precautions should you take when driving an RV or moving van?
5. How do you safely load and pull a trailer?
6. What precautions should you take when driving an SUV?

SELF-TEST

Multiple Choice

1. Odd-numbered highways in North America:
   a. run in a general north–south direction.
   b. run in a general east–west direction.
   c. are loops.
   d. are spurs.
2. The legend of a map:
   a. explains what the colors and symbols mean.
   b. determines its level of detail.
   c. lists the distances between key points.
   d. displays the map coordinates of major cities.
3. The scale of a map:
   a. shows where you can find tourism information.
   b. lists the fees of tollways.
   c. determines its level of detail.
   d. gives you average seasonal temperatures in major cities.

4. When preparing to tow a trailer:
   a. load the trailer before attaching it to your vehicle.
   b. hitch the trailer to your vehicle before loading it.
   c. place the heaviest items in the front.
   d. decrease the tire pressure in your rear tires.
5. A spur route:
   a. starts with an odd number and leads directly into a city.
   b. starts with an even number and goes to rural areas.
   c. can be an even- or odd-numbered route.
   d. is always connected to loop routes.

Sentence Completion

1. The _____ is a table that lists distances in miles or kilometers between key points.
2. Carrying _____ in a trailer is unsafe and illegal in many jurisdictions.
3. Pickup campers and motor homes are examples of _____ vehicles.
4. A _____ goes through or around a city.
5. Alternate highways have _____ numbers.

Matching

Match the concepts in Column A with examples of the concepts in Column B.

Column A	Column B
1. __ Scale	a. Miles per inch
2. __ Trailer hitch	b. Travel time between cities
3. __ Legend	c. Airports, rest areas, parks
4. __ SUV	d. Interstate 5
5. __ Spur	e. Off-road vehicle
6. __ North–south roadway	f. Highway 179
7. __ Mileage chart	g. Interstate 40
8. __ East–west roadway	h. Safety chains

Short Answer

1. What should you bring with you if you plan to drive a vehicle across the United States–Canadian border?
2. What should you do before driving a rental car off the lot?
3. What are some of the hazards of driving a recreational vehicle?
4. What are some ways that you can prevent fatigue when driving on long trips?
5. What should you do to prepare your vehicle for an extended road trip?

Critical Thinking

1. You are towing a trailer up a mountain on a two-lane highway. Just as you exit a tunnel, a large gust of wind causes the trailer to sway toward the oncoming traffic lane. At the same time, you observe an oncoming big rig in the other lane. What should you do?

2. You have just arrived in your favorite city for a two-week vacation. You decide to go on a road trip to visit some of the outlying towns. You visit a rental car agency, and the agent sets you up with a car with a manual transmission. As you pull out of the lot onto a steep incline, your car stalls. A truck is right behind you, but you are having a hard time coordinating the gearshift and clutch to start the car and move up the hill. What should you do?

PROJECTS

1. Using a map and a travel guide, plan a trip from one end of the country to the other. Make allowances for rest stops, tourist attractions, lodging, and places to eat along the way. Highlight your route on the map and indicate where you intend to stop. Calculate about how long it would take and how far you would travel.

2. Call a few rental car companies and get a copy of their standard insurance agreements available to people who are renting a vehicle. Check with your own insurance company to find out what is and what is not covered in your policy. Determine what, if any, additional coverage you will need to rent a car.

Answers to "Guess the Vanity Plate," "Wild Wheels," and "Where Am I?"

Chapter	Guess the Vanity Plate	Wild Wheels	Where Am I?
1	Ticket free (Figure 1–10)	Since 1936 (Figure 1–6)	
2	Don't be too near (Figure 2–7)	Convair Model 118 (Figure 2–8)	
3	Mad about you (Figure 3–8)	1907 Thomas Flyer (Figure 3–6)	
4	Stop on a dime (Figure 4–14)	1911 Marmon Wasp (Figure 4–8)	
5	Straighten up (Figure 5–9)	1903 Duryea (Figure 5–10)	
6	Pump it up (Figure 6–12)	Chrysler Airflow (Figure 6–7)	Fermi National Accelerator Laboratory, Batavia (Figure 6-16)
7	Bike to work (Figure 7–27)		Paramount Arts Center, Aurora (Figure 7–16)
8	Yield to me (Figure 8–10)	35 miles per hour (Figure 8–21)	Jarrot Mansion State Historic Site (Figure 8–6)
9	I hate lefts (Figure 9–6)	Roths' Beatnik Bandit (Figure 9–10)	
10	Hate the city (Figure 10–9)	"Articulator" bus (Figure 10–4)	Ulysses S. Grant Home State Historic Site, Galena (Figure 10-13)
11	I'm so slow (Figure 11–12)	1913 "Stanley Steamer" 810 (Figure 11–14)	John Deere Historic Site, Grand Detour (Figure 11-9)
12	Double trouble (Figure 12–16)	Lamborghini Diablo Roadster (Figure 12–6)	Graue Mill and Museum, Oak Brook (Figure 12-15)
13	Hog happy (Figure 13–13)	Unit Rig "Lectra Haul" (Figure 13–19)	Wheels Through Time Museum, Mount Vernon (Figure 13-14)
14	Dimwit (Figure 14–4)	Hummer (Figure 14–7)	Clark Bridge, Alton (Figure 14-13)
15	Gustbuster (Figure 15–10)	The Boeing Company (Figure 15–14)	Vandalia Statehouse State Historic Site, Vandalia (Figure 15-13)
16	I buckle up (Figure 16–12)	Isetta (Figure 16–10)	First Division Museum, Wheaton (Figure 16-4)
17	Skid row (Figure 17–11)	Reeves Octo-Auto (Figure 17–3)	Oliver P. Parks Telephone Museum, Springfield (Figure 17-6)
18	Designated driver (Figure 18–8)	Grandpa Munster of *The Munsters* (Figure 18–10)	Al Capone's Hideaway and Steak House, St. Charles (Figure 18-6)
19	Too few miles per gallon (Figure 19–19)	1921 Rolls-Royce Silver Ghost (Figure 19–13)	The Farnsworth House, Plano (Figure 19-25)
20	Steer clear (Figure 20–4)	Mohs Safarikar (Figure 20–9)	[John McCracken] Robinson-Stewart House, Carmi (Figure 20-12)
21	Southbound (Figure 21–7)	International Harvester Co. (Figure 21–11)	

Glossary

A

ABS: *See* antilock braking system.

acceleration: the rate of change of an object's speed.

acceleration lane: a temporary freeway lane that is an extension of the on-ramp and is designed to allow entering vehicles to match freeway driving speeds before merging into traffic.

accelerator: the far-right foot pedal used to control the amount of fuel fed to the engine. Also called the gas pedal.

adaptive equipment: specialized equipment that adapts vehicle controls to allow a physically disabled person to drive.

air bags: cushions located in the steering wheel and, in many vehicles, the passenger-side dashboard that inflate in a front-end collision and distribute the force of impact over a wider surface of the torso. On some vehicles, air bags located to the sides of vehicle occupants deploy in side-impact collisions.

Air Team: a program administered by the Illinois Environmental Protection Agency that requires most gasoline-powered vehicles in the Chicago and East St. Louis areas to have their emissions tested every other year.

alcohol/drug counseling: mandatory penalty for all DUI offenders in Illinois.

all-terrain-vehicles: off-highway motorized vehicles less than 50 inches (125 cm) wide, weighing 600 pounds (270 kg) or less, riding on three or more low-pressure tires, and equipped with a seat or saddle and handlebars or a steering wheel.

alternator: a generator that produces electricity to power a vehicle's electrical system.

angled parking: parking in a space that's angled to the curb, usually about 30 degrees.

antilock braking system: a braking system in which sensors are used to adjust hydraulic pressure to each wheel of a motor vehicle to keep the wheels from locking, thus preventing the vehicle from skidding.

antitheft devices: equipment installed on a vehicle that deters, disrupts, or helps catch a car thief.

assess: to predict potential threats on the roadway before they happen by accurately anticipating what others are going to do in dangerous situations. Part of the SAFE method of defensive driving.

ATVs: *See* all-terrain vehicles.

B

BAC: *See* blood alcohol concentration.

back-up lights: white or amber lights on the back of a vehicle that are activated when the transmission is set to REVERSE, indicating the driver's intention to back up.

banked road: a road that dips down in one direction, so that either the left or right side is higher than the other.

basic speed law: a law that states that it is illegal to operate a motor vehicle at a speed greater than is reasonable and proper for existing conditions, regardless of posted limits.

bicycle: any two-wheeled, human-powered vehicle propelled through a belt or chain.

bicycle lanes: lanes, usually on the far-right side of roadways, designated for bicycles and right-turning traffic only.

blind spots: those areas not reflected in either the rearview or sideview mirrors of a motor vehicle.

blood alcohol concentration: a percentage measurement of the level of alcohol in one's bloodstream.

blood test: the most accurate chemical test for BAC, typically administered to those taken to a medical facility following a crash in which alcohol is considered a factor.

blowout: a sudden loss of air pressure in a tire.

"blue slip": *See* certificate of completion.

brake fade: a condition in which brakes overheat after being applied continuously over time, such as when descending a long, steep mountain road.

"brake and hold": a reflex action to press the brake pedal and hold it down in an impending collision.

brake lights: red lights on the rear of a vehicle that are activated when the driver presses the brake pedal. They are designed to warn others that the vehicle is slowing down or stopping.

braking distance: the distance traveled by a vehicle from the time the brakes have been applied until it makes a complete stop.

braking skid: a skid caused by one or more of a vehicle's brakes locking up. It occurs most frequently in vehicles without an antilock braking system.

breath test: the most common chemical test for intoxication used by law enforcement. It consists of having a person blow into a tube attached to a breath-analyzing device, or "breathalyzer," that checks for the presence of alcohol using infrared light and/or "fuel cell" technology. The results of this test are admissible in court.

bus lanes: special lanes on the far-right side of the road, usually identified by a diamond symbol and other pavement markings as well as regulatory signs, reserved at certain times of the day for buses and right-turning traffic.

C _____

car alarms: optional warning devices installed on vehicles that emit a sound, signal, or visual cue to deter car thieves.

car jacking: an assault on a driver to steal his or her vehicle or personal property.

carbon monoxide: a colorless, odorless, tasteless gas found in the exhaust fumes of gasoline engines that under certain circumstances can seep into a vehicle's passenger compartment.

center of gravity: the point about which an object's weight is centered.

center turn lane: *See* two-way left-turn lane.

central vision: an area about 3 degrees wide directly ahead of a driver where everything can be seen clearly.

centrifugal force: the outward force or "pull" experienced when traveling in a circular path.

certificate of completion: a certificate issued to those who complete driver education. To qualify, you must have maintained a passing grade in at least eight classes for the previous two semesters of school. Minors applying for a regular driver's license must present this certificate along with their 25-hour certification.

certificate of title: a document that shows proof of vehicle ownership. All registered vehicles must have a certificate of title.

chain reaction: a series of collisions involving vehicles that impact one after another, usually front to rear.

child safety seat: a special seat for infants and young children that must be purchased separately and installed on top of the car seat.

choke: a device on older vehicles used to control the amount of air entering the carburetor while starting the engine.

claim: a demand to recover, from an insurer, losses covered by an insurance policy.

Class D license: the basic Illinois driver's license, which allows you to operate any motor vehicle (except motorcycles with an engine displacement over 150 cc) weighing up to 16,000 pounds (7,265 kg). It also permits you to tow any single vehicle, boat trailer, travel trailer, or trailer coach not in excess of 10,000 pounds (4,540 kg). Except in certain hardship cases, you must be at least sixteen years old to possess a Class D license.

clutch: a pedal on manual-transmission vehicles located to the left of the brake pedal that the driver must press while shifting gears.

collision coverage: optional insurance that covers the cost of repairing or replacing your vehicle when it is damaged after overturning or colliding with another vehicle or object.

color blindness: the inability to differentiate between certain colors, such as red and green. Color-blind drivers must interpret traffic controls by memorizing their shape and meaning.

color vision: the ability to see color. Color vision helps drivers determine the meaning of various signs and symbols.

comprehensive coverage: optional insurance that protects you from the loss of your vehicle or accidental damages to your vehicle caused by anything except a collision.

construction signs: orange signs, similar in appearance to warning signs, that alert you that you are in or about to enter a construction or maintenance zone.

controlled intersections: intersections in which some form of sign or signal is used to direct the flow of traffic.

controlled railroad crossings: railroad crossings that include some combination of crossing gates, flashing warning lights, traffic signals, and pavement markings.

cooperative driver testing certificate: a certificate issued to those who receive a final grade of A or B in their driver education course as well as pass the driving test when administered by their instructor rather than by Driver Services. Presentation of this certificate with a license application usually results in the driving test being waived.

cornering skid: a skid caused by losing traction on a curve.

court supervision: an option offered to those accused of violating a traffic law in which, in exchange for admitting guilt and complying with requirements outlined by a judge or magistrate for a given period of time, they are not formally convicted of the offense.

covering the brake: a way to be prepared to stop or slow suddenly that involves taking your right foot off the accelerator and holding it over the brake pedal as you cruise forward on your car's momentum.

crossbuck: a crossed white regulatory sign used at railroad crossings that is equivalent to a YIELD sign.

crossovers: special areas on divided highways where vehicles can turn around to go in the opposite direction. Unpaved crossovers are generally restricted to emergency vehicles.

crowned road: a road that is built higher in the center than on the sides.

cruise control: a device that allows you to regulate and maintain a vehicle's speed without using the accelerator.

curb parking: parking parallel to a curb with no surrounding vehicles.

D

deceleration lanes: temporary freeway lanes that allow drivers exiting a freeway to adjust to slower-speed conditions without blocking traffic in the through lanes.

deductible: a specified amount of money that an insured person must pay before his or her insurance company begins paying on a claim.

defensive driving: a method of driving that emphasizes anticipating and avoiding danger on the roadway.

defensive driving course: a course offered by the National Safety Council, varying in length from four to eight hours, that is designed to teach defensive driving and collision prevention techniques.

depreciation: the decline in a vehicle's resale value over time.

depth perception: the ability to judge the distance between two objects.

designated driver: a nondrinking person who volunteers in advance to drive others who have been drinking alcohol.

divided highways: multilane highways in which opposing directions of travel are separated by a fixed barrier or median strip.

double-merge lanes: lanes at the end of on-ramps in which drivers in adjoining lanes on the on-ramp must merge before they can merge onto the freeway.

double parking: parking parallel to another parked or stopped vehicle, blocking its movement as well as through traffic.

double vision: an uncontrollable rapid vibration of the eye that makes it virtually impossible to see.

downshift: to shift from higher to lower gears.

downshifting skid: a skid caused by downshifting too quickly.

driver education: instruction designed to help unlicensed and inexperienced drivers become familiar with the basics of vehicle control and rules of the road.

Driver Services Department: the state agency that grants and withdraws driving privileges and maintains driving records. It is part of the Illinois Secretary of State's office.

driving record: a record maintained by Driver Services of all of your reported collisions and motor vehicle violations, including those that may have occurred in other jurisdictions.

driving test: a test administered by Driver Services that evaluates license applicants' ability to safely operate a motor vehicle. You must be at least sixteen years and have practiced driving at least three months with an instruction permit to take the driving test.

Driving Under the Influence: the crime of operating a motor vehicle while under the influence of alcohol or other drugs. In Illinois, you are automatically presumed to be DUI if you have a BAC of 0.08% or higher if you are twenty-one or older. If you are under the legal drinking age, you can be convicted of DUI for having a BAC above 0.00%.

drop-offs: areas where the terrain literally drops off from the edge of the roadway without any shoulder.

DUI: *See* driving under the influence.

E

emergency flashers: *See* hazard lights.

emissions inspection: *See* Air Team.

execute: to carry out a decision to avoid an upcoming conflict on the roadway. Part of the SAFE method of defensive driving.

express lanes: special lanes found on some freeways, including the Kennedy Expressway and the Dan Ryan Expressway in Chicago, that are either reversible or allow drivers to bypass congestion caused by vehicles leaving the freeway at local exits. Express lanes are separated from other freeway lanes by a physical barrier.

expressways: *See* freeways.

F

fatigue: weariness or tiredness due to labor, boredom, eyestrain, and other factors that can severely affect a driver's reaction time and decision-making abilities.

field sobriety tests: "on the spot" roadside evaluations of a person's level of intoxication administered by a law-enforcement officer and consisting of a series of tests to detect physical or mental impairment caused by alcohol or other drugs.

field of vision: the area that you can clearly see directly in front of you, and to the sides, when looking straight ahead.

financial responsibility: the ability to pay for any injuries, deaths, or damage that might be caused by your vehicle.

find: to identify an escape route or "out" that allows you the best means of avoiding a conflict on the roadway. Part of the SAFE method of defensive driving.

five-day permit: a permit issued by Driver Services or Vehicle Services that allows you to temporarily drive a newly purchased vehicle until you can get it registered. Without such a permit, you may not drive a newly acquired vehicle without registration and license plates for more than 24 hours.

flooded engine: a condition that occurs when too much gasoline is in the engine. It is caused by depressing the accelerator too many times before starting the car.

following distance: the amount of space between your vehicle and the vehicle directly ahead of you.

force of impact: the force of a collision as determined by the magnitude of the kinetic energy of the objects that collide and the distance over which the kinetic energy is lost.

freeways: divided roadways with at least two lanes going in each direction that have controlled and limited access.

"fresh" green light: a traffic light that has just turned from red to green.

friction point: the position of the clutch at which the engine begins to engage the transmission.

fully protected turn: a protected turn in which you are permitted to turn only when you have a green arrow.

G

gauge: a scale with an indicator needle or numerical marker that keeps track of changing conditions like fuel level or speed.

gearshift: a device located on the console or right side of the steering column and attached to the transmission that allows the driver to change gears.

graduated licensing law: a recently enacted law that imposes special driving restrictions, including a prohibition on driving during curfew hours, and criminal penalties on drivers under the age of twenty-one. Restrictions are gradually withdrawn and sanctions are gradually loosened over the course of three stages: the permit stage, the initial licensing stage, and the full licensing stage.

gravity: an invisible force that objects exert on other objects.

gridlock: a situation in which traffic becomes so heavy that it comes to a virtual stop.

guide signs: green signs that tell you where you are, where you are going, or how to get somewhere.

H

"hairpin" turns: *See* switchbacks.

hand-over-hand method: a method of steering in which one hand pushes the steering wheel around and down as the other crosses over to pull the wheel even further down.

hand signals: signals using hand and arm motions that communicate a motorist's or cyclist's intention to turn, slow down, or stop. Hand signals are also used by law-enforcement officers, construction workers, and school crossing guards to direct traffic in special situations.

hazard lights: a signaling device that flashes all four turn signals at once to warn other drivers that a vehicle is moving slowly, stopped, or not functioning properly.

headlights: lights on the front of a vehicle that give off white light to help the driver see in conditions of low visibility. Most headlights have both low-beam and high-beam settings.

head-on collision: a front-to-front collision between two vehicles.

head restraints: front-seat cushions, sometimes adjustable, designed to support the head and neck of the driver and any front-seat passenger in the event of a rear-end collision.

highway: a main public roadway designed to carry traffic for long, uninterrupted periods at medium to high speeds.

highway hypnosis: an effect of open highway driving in which you literally become "hypnotized" by the road for several seconds and afterwards feel as if you just awakened from a dream.

Highway Transportation System: the network of roadways in North America designed to move people and goods in a safe, efficient, and timely manner across state, provincial, and international boundaries with a minimum of bother.

"hit and run": the crime of leaving the scene of a crash in which you are involved.

HTS: *See* Highway Transportation System.

hydroplaning: skating across a wet road after the tires lose contact with the road surface.

I

ignition: a switch, usually located on the right side of the steering column, activated by a key to start a vehicle's engine.

ignition interlock device (IID): a device installed on vehicles as a penalty for certain DUI offenses that requires you to breathe into a tube and register a BAC below 0.025% before the ignition can be activated.

Illinois Automobile Insurance Plan: an insurance program in which high-risk drivers unable to obtain a policy on their own are assigned to a participating insurer for three years.

implied consent law: a law that states that any person operating a motor vehicle is "deemed to have given consent" to have his or her breath, blood, or urine tested for the presence of alcohol or other drugs. According to this law, you are legally required to submit to a chemical test if a law-enforcement officer has "probable cause" to believe that you are under the influence while in actual physical control of a vehicle.

inertia: the tendency of an object at rest to remain at rest or an object in motion to continue in motion in a straight line until acted upon by a force.

inhibitions: the elements of your personality that stop you from behaving without regard to possible consequences.

instruction permit: a "pre–driver's license" available to qualified applicants who are at least fifteen years old (seventeen years, nine months if not enrolled in driver education). It allows you to practice driving with any parent, guardian, or responsible licensed driver approved by your parent or guardian in the front seat with you who is twenty-one years or older and has been driving for at least one year.

interchanges: grade-separated intersections connected by one-way ramps.

intersections: places where different roadways meet or cross.

intoxication: the condition of being affected by alcohol. The level of intoxication is determined by one's BAC reading.

invoice cost: the amount a car dealer pays the manufacturer for a vehicle.

I-Pass: an electronic toll collection system used on all five tollways in Illinois. Those who purchase I-Pass have their toll deducted automatically, thereby making it unnecessary to stop, roll down their window, and handle money at toll facilities.

J

jaywalking: crossing a street as a pedestrian without regard to traffic rules or signals.

judicial driving permit (JDP): a license with restricted driving privileges issued to qualified first-time DUI offenders who are eighteen or older. It is granted after the recipient serves at least thirty days of his or her statutory summary suspension.

jump-start: to recharge a car's battery by drawing electricity from another vehicle's battery using jumper cables.

K

kill switches: secondary ignition switches concealed inside vehicles to inhibit the flow of electricity or fuel to the engine.

kinetic energy: the energy that an object in motion has.

L

lane-use signals: overhead traffic signals used in tunnels, on bridges, at toll plazas, and above municipal streets with reversible lanes to indicate whether or not lanes are open to traffic moving in your direction.

law enforcement: agencies that enforce the vehicle codes and traffic laws governing the use of the HTS.

legend: a guide on a map that explains the colors and symbols used to represent different types of roadways and common landmarks.

liability insurance: required insurance coverage that pays for both bodily injury and property damage caused by you or anyone authorized to drive your vehicle.

license revocation: the termination, by Driver Services, of a person's driving privileges for a year or more. A person whose license has been revoked must reapply for a new license at the end of the revocation period.

license suspension: the temporary cancellation, by Driver Services, of a person's driving privileges for up to a year. A person whose license has been suspended has their license returned to them at the end of the suspension period.

limit line: a white or yellow stop line painted across the roadway at an intersection behind which vehicles must stop. In intersections with a crosswalk, the nearer crosswalk line serves as the limit line.

list price: the manufacturer's suggested retail price of a vehicle.

loading: adding weight to a vehicle's weight as measured when it is empty.

locking the brakes: an emergency braking technique in which the driver firmly presses and holds the brake pedal until the vehicle comes to a complete stop.

loop: an alternative route that goes through or around a city.

M

medical payments coverage: optional insurance that provides extra protection to you, your family members, and any passengers in your vehicle who are killed or injured in a collision regardless of who is at fault. It also pays for funeral and medical expenses if you or a family member is killed or injured as a pedestrian.

median strip: an area of space that separates opposing directions of travel on a divided highway.

merging area: the space where an acceleration lane merges with the freeway.

mileage allowance: the annual limit on the number of miles a person can drive a leased vehicle without paying a penalty.

mileage chart: a table on a map that lists distances in miles between key points on roadways.

milepost: a special type of guide sign that appears at 1-mile (1.6-km) intervals along sections of interstate highways to indicate your location on the roadway.

minimum speed law: a law that prohibits drivers from moving at such a slow speed as to impede or block the normal and reasonable movement of traffic, except when reduced speed is necessary for safe operation or in compliance with the law.

misdemeanors: criminal motor-vehicle offenses.

moped: any two- or three-wheeled motor vehicle that is incapable of exceeding 30 miles per hour (50 km/h) and which has an engine displacement of 50 cc or less. In Illinois, a moped is classified as a "Class L" motorcycle.

motorcycle: any two- or three-wheeled motor vehicle having a seat or saddle for riders, excluding tractors.

motor scooters: two- or three-wheeled vehicles that have a platform for the rider's feet instead of pedals and an enclosed engine, and that are capable of reaching speeds of 40 miles per hour (65 km/h) or more. In Illinois, most motor scooters are classified as "Class L" motorcycles. Those with an engine displacement of 150 cc or more are considered "Class M" motorcycles.

N

"no-zones": the blind spots of large vehicles such as tractor-trailer trucks and buses.

O

odometer: a gauge that displays the total number of miles that a vehicle has been driven since it was manufactured.

off-highway motorcycle: any two-wheeled motorized vehicle equipped with a seat or saddle and designed to travel off-road.

off-ramp: a one-way exit ramp from a freeway.

"off-tracking": the tendency of truck drivers to swing wide on a turn, especially a right turn, to avoid riding up over the curb, hitting parked vehicles or the median, or veering into oncoming traffic on the cross street.

one-way streets: roadways in which all lanes of traffic move in the same direction.

on-ramps: one-way entrance ramps used to gain access to a freeway.

open container law: a law requiring any alcoholic beverage transported in the passenger compartment of a motor vehicle, including the glove compartment, to be full, sealed, and unopened. It also prohibits drinking while driving.

orderly visual search (OVS): a scanning technique developed in the 1950s by Howard L. Smith that requires using selective glances in a constantly repeating pattern to monitor all the different areas around one's vehicle.

overdriving your headlights: traveling at speeds that prevent you from stopping within the distance illuminated by your headlights.

oversteering: turning too sharply when making a turn or rounding a curve.

OVS: *See* orderly visual search.

oxidation: the process by which the liver rids the body of alcohol by turning it into oxygen and carbon dioxide.

P

parallel parking: parking alongside a curb between two parked vehicles.

parking brake: a manually set brake that keeps a vehicle in place when parked and serves as an emergency brake if the standard brakes fail.

parking lights: lights that are white or amber-colored on the front of the vehicle and red on the rear of the vehicle. They are normally activated by the first setting on the headlight switch and are designed to be used only when parking.

pass: to overtake a vehicle directly ahead by entering an adjoining lane, gaining a speed advantage, and re-entering the original lane in front of the other vehicle.

passive restraint: a passenger restraint, such as an air bag or shoulder belt on some vehicles, that operates automatically.

PBT: *See* preliminary breath test.

pedestrian: any person who uses or crosses a roadway on foot or by means of a self-propelled device other than a bicycle.

pedestrian signals: special signals, sometimes activated by pushbuttons, that are mounted below traffic signals and used to control pedestrian traffic.

peer pressure: the influence that others of your own age have on you.

peripheral vision: the unfocused areas extending 180 degrees to the sides of your central vision.

perpendicular parking: parking at a 90-degree angle from the curb with the front or rear of the vehicle aligned with and parallel to the curb.

policy: a written contract between you and an insurance company that outlines your coverage and how much the company is liable for in the event of a collision or other incident resulting in damage to your vehicle.

power brakes: power-assisted brakes that require less foot pressure to operate than standard brakes.

power skid: a skid caused by accelerating too quickly, especially on a slick road surface.

preliminary breath test: a field sobriety test that calls for you to blow into a tube attached to a handheld device that registers an accurate estimate of your BAC. The results of this test are not admissible in court, but a high reading would be considered reasonable grounds for the law-enforcement officer who administers it to arrest you for DUI.

premium: a specified amount of money a person pays for insurance coverage.

preventive maintenance: day-to-day care and routine service given to a vehicle to avoid more serious repairs.

probationary license: a license with unrestricted driving privileges, valid for three months, issued to qualified applicants whose license has been suspended for getting three traffic convictions within a one-year period and who have completed a defensive driving course.

protected turn: a turn made from a turn lane posted with signs, road marked with arrows, and accompanied by a traffic signal arrow.

pullouts: widened off-road areas or extra-wide shoulders located at intervals on rural roads and highways to allow drivers to pull out of the way of faster-moving traffic or to turn around.

pumping the brakes: an emergency braking technique in which the driver alternately applies the brakes completely until they lock and then releases them in rapid succession.

purchase price: the amount of money that a car dealer or owner is willing to accept for a vehicle.

push-pull method: a method of steering in which one hand pushes the steering wheel up while the other pulls it down.

Q

questionnaire sampling program: a program administered by Vehicle Services that checks whether randomly selected drivers are in compliance with Illinois's financial responsibility requirements. Under this program, those who fail to respond to an insurance verification questionnaire have their registration suspended.

R

ramp signals: special signals found at the end of some on-ramps that use alternating red and green lights to allow one vehicle at a time to proceed onto the freeway.

RDP: *See* restricted driving permit.

reaction distance: the distance traveled by your vehicle during the time it takes you to identify the need to stop and react to a braking situation.

rear-end collision: a collision in which one vehicle hits another from behind.

rearview mirror: a wide, rectangular mirror either suspended from the roof or attached to the windshield of a vehicle that allows a driver to see what's directly behind him or her.

rebate: a special payment to customers offered by vehicle manufacturers designed to increase sales of certain models.

recreation signs: brown signs that tell you about nearby places of cultural interest and public recreation like historic sites, museums, and parks.

recreational vehicles: vehicles, such as motor homes, pickup campers, or trailer combinations, used for travel and recreational activities.

reflective pavement markers: square- or rectangular-shaped reflectors, colored yellow or white to match pavement markings, spaced at intervals between lane-line dashes and, on freeway ramps, along the edge of the roadway next to the lines to help drivers stay within lanes.

registration: a process requiring drivers to pay various fees and taxes and to provide information about a vehicle to Vehicle Services before they can legally operate it on the roadway. It is required annually for all motor vehicles and trailers.

regulatory signs: signs that tell you what you can or cannot do at certain times and places.

restricted driving permit: a license with restricted driving privileges, valid for one year, issued to qualified applicants over the age of sixteen who have had their license suspended or revoked.

restrictions: limitations placed on one's driving privileges by Driver Services, such as driving only with adaptive equipment or while wearing prescription eyewear.

reversible lanes: special lanes, separated from regular traffic lanes by dashed double-yellow lines and/or channeling devices, open to traffic moving in one direction at certain times of the day and in the opposite direction at other times of the day. They are used on high-density commuter roadways in some major cities.

revoked license: *See* license revocation.

riding the brake: resting the right foot on the brake pedal while driving. Because this constantly activates the brake lights, it confuses drivers behind you as to your intentions.

right-hand rule: a general rule that at uncontrolled intersections, four-way stops, and intersections with flashing red or broken signal lights, drivers on the left should always give the right-of-way to drivers on the right.

right-of-way: the right to use a certain part of a roadway when two or more users of the roadway want to use it at the same time.

road rage: a term used to describe aggressive driving, including deliberate tailgating, yelling at other drivers, making obscene gestures, purposely blocking other drivers' paths, and in extreme cases, assault.

"rock out": a technique to get out of mud, deep sand, or snow by moving forward and backward in rapid succession until the stuck tire or tires are freed.

rollover: a result of a collision in which a vehicle is flipped upside down or literally rolls over one or more times.

rotaries: *See* traffic circles.

route markers: signs used to identify numbered roadways or roadways designated for special purposes.

rubbernecking: a term used to describe the common driver habit of slowing down to observe a crash site.

rumble strips: narrow grooves or wide depressions cut into the pavement or along the shoulder to alert drivers to slow down and pay attention, or that they are straying off the roadway.

running lights: lights located on the front of a vehicle that automatically illuminate when the engine is started. They may be a dimmer setting of the headlights or a separate set of lights located next to the headlights.

RVs: *See* recreational vehicles.

S

SAFE: a defensive-driving strategy to evade potential danger on the roadway. SAFE stands for *scan, assess, find,* and *execute*.

safety belts: automatic or manually fastened belts designed to restrain vehicle occupants in the event of a collision or quick braking action.

scale: the level of detail of a map, as indicated by a line representing a larger unit of measure (for example, 1 inch equals 1 mile).

scan: to look ahead down the road to gather information about the complete driving scene. Part of the SAFE method of defensive driving.

semicontrolled intersections: intersections in which signs or signals are used to control one or several approaches, but not every approach.

semiprotected turn: a protected turn made from a turn lane but not accompanied by a traffic signal that directs your turn with a green arrow.

service signs: blue signs that inform you of nearby services such as hospitals, rest stops, and gasoline stations.

shoulder: a continuation of pavement or other stable surface that goes beyond the road boundary line.

side-impact collision: a side-to-side, side-to-front, or side-to-rear collision between two vehicles.

sideview mirrors: exterior mirrors mounted on the vehicle's doors that allow the driver to see along the sides of the car and neighboring lanes of traffic.

skid: a loss of directional control of a vehicle that occurs when a vehicle loses traction.

slow-moving vehicles (SMVs): vehicles not designed for high-speed travel, such as wide-load trucks, bulldozers, or agricultural harvesters. Most slow-moving vehicles are required to have a special SMV sign attached to the rear.

snowmobiles: self-propelled vehicles designed for operation on snow or ice or natural terrain, steered by skis or runners, and supported in part by skis, belts, or cleats.

sobriety checkpoints: special roadside safety checks in which police officers stop drivers to determine whether any are impaired by alcohol or other drugs.

soft shoulders: shoulders that slope downward or do not provide effective traction.

solo collision: a collision involving only one vehicle, such as with a tree or concrete barricade.

space cushion: an empty space between you and the cars and other objects on the roadway around you.

speed: the rate at which an object changes its position over time.

speed bumps: raised slabs of pavement used to force drivers to stop or slow to a near stop.

speed limits: the maximum legal speeds for operating a vehicle on a roadway under ideal driving conditions.

speedometer: a gauge that indicates how fast a vehicle is traveling in miles and/or kilometers per hour.

sport utility vehicles: four-wheel-drive vehicles designed to operate both on and off the roadway.

spur: an alternative route that leads directly into a city.

"stale" green light: a traffic light that has been green for a long time or since you first noticed it.

statutory summary suspension law: a law that mandates the immediate suspension of your driver's license if you have a BAC of 0.08% or higher. Those under the legal drinking age have their license suspended if they test positive for any alcohol.

steer: to control a vehicle's direction of travel.

steering wheel: a device that allows the driver to turn the front wheels of the vehicle.

"sticker price": the list price of a vehicle plus destination charges and additional options and services provided by the dealer.

stopping distance: the total distance required to stop from the time you first recognize the need to brake to the time your vehicle is no longer moving. The sum of reaction distance and braking distance.

suspended license: See license suspension.

SUVs: See sport utility vehicles.

switchbacks: curves, typically found on mountain roads, that are so sharp that they reverse direction.

synergistic effect: the enhancement of a drug's side effects caused by mixing it with alcohol and certain other drugs.

T

tachometer: a gauge that measures an engine's revolutions per minute (rpm).

tailgate: to follow other cars too closely while driving.

taillights: red lights located on the rear of a vehicle that are illuminated when the headlights are turned on.

temporary registration permit: a permit issued by Driver Services or Vehicle Services and mounted on the inside rear window of a newly registered vehicle that allows you to drive the vehicle until you receive your registration certificate, license plates, and license plate sticker in the mail. Most temporary registration permits are valid for sixty days.

three-point turn: a way to reverse direction on a narrow street in which a driver turns sharply to the left and stops as close as possible to the opposite edge of the roadway, and then backs up before moving forward in the opposite direction.

3-second rule: a test to check your following distance during low-speed driving. It calls for picking a fixed object directly ahead of you and counting off three full seconds. If you reach the object before you reach the count of 3, you are driving too closely to the vehicle ahead.

threshold braking: an emergency braking technique in which the driver uses a full and firm application of the brake pedal up to the point where the brakes lock and cause the vehicle to enter a skid.

T-intersections: intersections in which a side road joins a main road with through traffic at a right angle.

toll plaza: a line of tollbooths located at intervals along tollways and on toll bridges where the required fee to use the toll facility is collected from motorists, either in the form of cash payment or electronically.

tollways: controlled-access highways that require a fee to use.

torque: the ability of a force to cause an object to rotate.

tracking: a method of steering that allows you to keep your vehicle on the intended path of travel by looking toward the center of the lane ahead and making only slight movements with the steering wheel.

tracking device: a system that uses a device hidden on a vehicle to emit a signal to the police or a central response center to locate the vehicle if it is stolen.

traction: the friction between a vehicle's wheels and the surface of the road that makes it possible for drivers to move, change direction, and stop.

trade-ins: vehicles that people sell to a dealers in exchange for credit on a new or another used vehicle.

traffic circles: circular roadways that allow traffic from many different directions to intersect.

traffic laws: state or provincial regulations covering the licensing of drivers, registering and titling of vehicles, financial responsibility of drivers, minimum safety requirements of vehicles, the rules of the road, and infractions and penalties.

traffic signals: electronic traffic controls, primarily found at intersections and other areas where traffic is heavy or there is a high level of risk of a collision, that use red, yellow, and green lights.

tunnel vision: a 70% reduction in one's field of vision, caused by blurred peripheral vision and fixation on a narrow field ahead.

turn lanes: lanes added near an intersection to separate left- or right-turning traffic from through traffic.

turn-signal lights: flashing lights used to indicate right or left turns that are white or amber-colored on the front of the vehicle and red on the rear of the vehicle.

25-hour certification: a certificate, submitted to Driver Services by minors applying for a regular license, verifying that the applicant has received at least 25 hours of behind-the-wheel practice with an instruction permit.

two-point turn: a way to reverse direction in which a driver either backs into a driveway or side street on the right side of the street and then drives forward in the opposite direction ("reverse, then forward") or turns into a driveway or side street on the opposite side of the street and then backs up into the street before moving forward in the opposite direction ("forward, then reverse").

two-way left-turn lane: a special lane between two opposing directions of traffic designated for left turns only from either direction. It is marked by parallel solid and dashed yellow lines, sometimes accompanied by white arrows on the pavement that alternately point left and right.

U

UIM coverage: *See* underinsured motorist bodily injury coverage.

UM coverage: *See* uninsured motorist bodily injury coverage.

UMPD coverage: *See* uninsured motorist property damage coverage.

uncontrolled intersections: intersections without any form of traffic control device or roadway marking.

uncontrolled railroad crossing: a railroad crossing without any signs, signals, or roadway markings.

underinsured motorist bodily injury coverage: insurance that protects you, family members who live with you, and occupants of your vehicle from losses arising out of a collision with an at-fault driver whose liability coverage is insufficient to pay for your bills. It is optional in Illinois unless you purchase UM coverage in excess of the required minimum amount.

understeering: turning insufficiently when making a turn or rounding a curve.

uninsured motorist bodily injury coverage: required insurance that protects you, family members who live with you, and occupants of your vehicle from losses arising out of a collision with a "hit and run" driver or an at-fault uninsured driver.

uninsured motorist property damage coverage: optional insurance that pays for damage to your vehicle caused by an identified, at-fault uninsured driver. It is usually offered to those who do not elect to get collision coverage.

unprotected turn: a turn not made from a turn lane at an intersection and where there are no arrows—whether designated by a signpost or road markings, or signaled by a traffic control—to guide your turn.

urine test: the least accurate chemical test for BAC, typically used when drugs other than alcohol are suspected.

U-turn: a way to reverse direction in which the driver traces a path in the shape of a large "U."

V

vapor lock: a condition in which gasoline in a vehicle's fuel system becomes so hot that it boils and turns to vapor, causing the engine to "lock," or stop running. It occurs only in older vehicles that have fuel pumps in the fuel tank rather than on the engine.

variable message signs: electronic "signs" posted on some roadways or set up by work crews to provide drivers with up-to-date information about road conditions.

Vehicle Services Department: the state agency that registers vehicles and verifies that they are insured. It is part of the Illinois Secretary of State's office.

velocitation: an effect of open highway driving in which you unconsciously find yourself driving much faster than you intended.

victim impact panels: meetings with a panel of volunteers who relate how their lives have been changed by an alcohol-impaired driver. They are required of many DUI offenders in Illinois.

vision screening: a test administered by Driver Services that checks the visual acuity and peripheral vision of all permit and license applicants.

visual acuity: how sharply you can see an object at a specified distance.

W

warning lights: red or yellow indicators on a vehicle's dashboard that indicate serious problems or safety concerns.

warning signs: yellow signs that alert you to possible hazards ahead, a change in road conditions, or an upcoming intersection or pedestrian crossing.

warranty: a written guarantee that a car manufacturer or dealer will repair or replace any defective parts or systems within a set amount of time or for a certain number of miles/kilometers driven.

"weave" lanes: shared acceleration and deceleration lanes on a freeway in which vehicles entering and exiting the freeway share the right-of-way.

wholesale price: the amount of money a dealer is willing to pay for a used vehicle.

written test: a test administered by Driver Services that evaluates permit and license applicants' understanding of Illinois's traffic laws; signs, signals, and roadway markings; and vehicle equipment requirements.

Y

Y-intersections: intersections in which roads meet at unusual angles.

Z

zero tolerance law: a law mandating more severe criminal penalties for DUI to offenders under the age of twenty-one.

Index

ABS. *See* Antilock braking systems
Acceleration, 121
  lanes, 281–82
Accelerators, 83
  stuck, 401–402
Accessories, 498–500
  interior, 499
Age, of driver, 49
Air bags, 388–90
Air Team, 464–65
Alcohol. *See also* Driving Under the Influence; Drunk driving
  beverage content, *423*
  chemical tests for, 434–37
  judgment and, 427
  limits on consumption of, 429–30
  physiological effects of, 426–27
  ridding body of, 425–26
  types of, 423
All-terrain cycles (ATCs), 262
All-terrain vehicles (ATVs), 262
Alternate Motorcycle Operator Skills Test (ALMOST), 459
Altitude, effects of high, 342
Anger, and driving, 55–57
Animals
  freezing in tracks, 265
  wild, 264–65
Antifreeze, 490
Antilock braking systems (ABS), 131
Antitheft devices, 502–504
Anxiety, and driving, 58
Army Trail Road, 289
Arrows
  lagging left-turn, 232
  red, yellow, and green, 153–55
ASE. *See* Automotive Service Excellence
ATCs. *See* All-terrain cycles
ATVs. *See* All-terrain vehicles
Auto accessories, for emergencies, 412
Auto loans, 480
Automotive Service Excellence (ASE), 496

BAC. *See* Blood alcohol concentration
Backing up, 102–104, *104*

Batteries
  dead, 402–405
  fluid for, 490
  jump-starting, 403–404
Belts, checking, 490
Better Business Bureau, 496
Bicycles, 302–306
  driver responsibilities to, 303–304
Bicyclists
  don'ts for, 304
  responsibilities of, 305–306
  right-turning, 302
  in urban environments, 227
  use of horn toward, 303
Big rigs, 127, 281, 313
Blind spots, 79–80, *79*
Blockers, following, 210
Blood alcohol concentration (BAC), 423–24
  determining, 424–25
  for permit holders, 454
Blood test, 435
Blowouts, 397
Blue slip, 449
Bodily injury liability, 467
Brake pads, 400
Brake(s), 84–85, 490
  burning, 130, *130*
  covering, 228–29
  failure of, 399–401
  feel of, 130
  fluid for, 489–90
  light, *74*
  overheated, 401
  riding, 228
Braking, 129–34
  antilock systems, 131
  distance, 133–34
  emergency, 131–32
  heel on floor when, *130*
  method, 130–31
  skid, 407
  systems, 131, 400
  weather conditions and, *350*
Breakdowns, on freeway, 402
Breath test, 435
Bridges, 337–38
"Burning rubber," 123
Buses
  commercial, 317–18
  pedestrians around, *317*
  school, 318–19

Busey, Gary, 309
Business districts, 227

Canada, crossing border into, 513
Car alarms, 502
Carbon monoxide poisoning, 53–54
Car fires, 409, 411–12
Cargo carrying capacity, 499
Car jacking, 412–14
Carpool lanes, 161
CB. *See* Citizen's band radio
Cellular phones, 60
Center lines, yellow, *159*
Center of gravity, 127
Certificate of completion, 449
Certificate of title, 460, 482
  transfer of, 460–61
Chain reactions, *375*
Chamois, 494
Chemical tests, for alcohol, 434–37
Chicago
  driving in, 235
  freeways in, 280
Chicago Skyway, 289
Children, as passengers, 60–61
Child safety seats, 390–91
Citizen's band (CB) radio, 354
City driving, 227–31
Class D license, 447, 451, 459
Class L license, 459
Class M license, 459
Clearance, chassis and, 333
Cloverleaf, 271–72
Clutch, 86
Collector roads, 273
Collisions
  calling for help, 380
  causes of, 373
  exchanging information, 380–82
  force of impact and, *125*
  formal report, *382*
  giving aid at, 377–79
  head-on, 385
  insurance coverage, 469
  preventing, 383–86
  property damage, 382
  rear-end, 27, 385–86
  reporting, 383
  set-up, 376, 413–14
  side-impact, 385
  sketch of, *381*
  statistics on, 373

Numbers in *italics* refer to figures.

stopping at scene of, 376–77
techniques to avoid, 383–386
types of, 373–75
unavoidable, 386
warning others of, 379–80
with wild animals, 264
Color blindness, 48–49
Color vision, 48–49
Communication, of intentions, 30–32
Confrontations, between drivers, 56
Construction zones, 336
Controlled intersections, 171, 172
Coolant, 490
Cooperative Driver Testing Program, 452
Coordination, 49
   alcohol and, 427
Cornering skid, 407
Courtesy, toward others, 6
Court supervision, 472
Crashes. *See* Collisions
Crosswinds, 315
Cruise control, 86
Curb feelers, 106
Curbs, painted, 163–64
Curfew hours, 453
Curves
   entering, 128–29
   negotiating, 125–29
   speed and, *129*
Cutting corners, 208

Dangers
   ahead, 210–11
   avoiding, *23*, 39
   escaping, 38–39
   forms of, 37
   of highway and rural driving, 260–65
   on left side, 208–10
   potential, 36–39
   on right side, 208
   scanning for on freeways, 273–74
Dan Ryan Expressway, 289
Dean, James, 7
Debris, on roadway, 409–10
Deceleration lanes, 285
Deep water, escape from, 410–11
Defensive driving
   anticipation and, 24
   awareness and, 24
   good judgment and, 24–25
   knowledge and, 23
   practicing, 25–26
   safety preparation for, 24
   skills, 23–26
Department of Natural Resources, 263
Depressants, 428
Depression, driving with, 58

Depth perception, 48
Detours, 238
Diamond interchange, 271
Dips, 334–35
Direct connection interchange, 273
Direction
   reversing, 211–17
   ways to reverse, 216–17
Dirt bikes, 262
Disability, chronic, 50
Disabled drivers, equipment for, 500–501
Distractions, 58–59
   in unfamiliar areas, 62
Distributor roads, 273
Divided highways, 249–50. *See also* Freeways; Highways
   crossing, *252*
   crossovers, 255
   entering or crossing, 251–53
   entry and exit problems, 255
   line of cars on, 256
   roadside stands, 255–56
   turning left off, 253–54
   turning left onto, 251–53
Dividing lines, creeping over, 233
Double-merge lanes, 286
Downshifting, 97–98, 339
Driver education, 3–4
Driver intention, 201
Drivers
   mental condition of, 54–62
   physical condition of, 47–54
Driver Services Department, 448
Driveways, unmarked, 254–55
Driving
   challenges of, 454
   environment, 37–38
   readiness for, 8
   record, 471–74
   test, 451–52
Driving Under the Influence (DUI), 432–42. *See also* Alcohol; Drunk Driving
   administrative penalties for, 437
   conviction under, 474
   costs of, 439–40
   criminal penalties for, 437–39
Drop-offs, 335
Drugs, OTC, 428–29
Drunk drivers, avoiding, 431–32
Drunk driving. *See also* Alcohol; Driving Under the Influence
   alternatives to, 430–31
   preventing, 429–32
DUI. *See* Driving Under the Influence
Dynamometer, 465

East-West Tollway, 288–89
Edens Expressway, 288
Electrical shock, 411

Electrical system, monitoring of, *73*
Emergencies, causes of, 395
Emergency traffic patrols, 407
Engines, 491
   stalled, 401
   starting, 93–94
Equipment, safety-oriented, 498
Exit(s)
   lanes, *283*
   missed, 284
Experience, 26
Express lanes, 286–87

Fatigue, 50–53
Ferries, 290
Financial obligations, 5–6
Fishtailing, 519
Five-day permit, 461
Five-light signals, 153
Flares, 410
Flat tires, 397–99
Fluid levels, 488–90
Folded diamond, 271
Following distance, 26–29
   increasing, 257–58
   in low-speed situations, 28–29
   testing, *28*
Force of impact, 124–25
Fort Madison Bridge, 291
Four C's, 206
Freeways
   access to, *279*
   breakdowns on, 402
   dangers on, 286–91
   driving on, 271–78
   entering, 278–83, 281
   exiting, 283–86
   fear of, 273
   numbering of, 511–13
   passing on, 276–78
   potholes on, 274
Fuel
   additives, 230
   efficiency, 492
Full cloverleaf, 272
Funeral processions, 186–87

Gabor, Zsa Zsa, 448
Gauges, 71, *72*
Gearshifts, 85–86
Givens, Edward, 374
Glare, from sunshine, 332–33
Glare guards, 331
Gore points, 157
Graduated licensing, 453–54
Grammer, Kelsey, 100
Gravel roads, 333
Gravity, 121–22
   driving against force of, 339
Green arrows, 154–55

Green lights, 153, 174–75
  fresh, 174
  stale, 174
Gridlock, on city streets, 231

Hairpin turns, 340
Hallucinogens, 428
Hand signals, 31–32
Hashish, 429
Hazard lights, 361
Headlights
  alignment of, 330
  burned-out, *75*
  failure of, 402
  looking beyond, 327–28
  overdriving, 328–30
Head-on collision, 374
Head restraints, 387
Head trauma, 378
Hearing, 49
  impairment of, *50*
Heat shields, 365
Heat waves, 363
Hedges, Michael, 258
Height, driver–car compatibility, 49–50
Helmets, for motorcyclists, 313
Hennepin Canal Parkway State Park, 262
Herd instinct, 236
High-occupancy vehicles (HOV) lanes, 161
Highways
  defensive driving on, 256–60
  hypnosis on, 260
  numbering of, 511–13
Highway transportation system (HTS), 9–17
  breakdowns in, 17
  building, 15–16
  components of, *10*
  designing, 15–16
  maintaining, 15–16
  in North America, *12*
  order in, 14–15
  people using, 13–14
  relating to, 11–13
  sharing, *14*
Hills, 338–39
  starting on, 98
Hood latch failure, 405–406
Horns, 78
  using, 31
Horseback riders, 264
Horse-drawn vehicles, 264
Hoses, 490
Hot-wiring, 503
House signals, 153
HOV. *See* High-occupancy vehicles
HTS. *See* Highway transportation system
Hubbard's Cave, 338
Hydroplaning, 351–52

Ice cream trucks, 316
ID. *See* Identification cards
Identification (ID) cards, 459–60
IDOT. *See* Illinois Department of Transportation
IEPA. *See* Illinois Environmental Protection Agency
Ignition, 82–83
Ignition interlock devices (IID), 441
Illinois Attorney General's consumer Protection Division, 496
Illinois Automobile Insurance Plan, 470
Illinois Department of Natural Resources, 263
Illinois Department of Transportation (IDOT), 407
Illinois Environmental Protection Agency (IEPA), 464–65
Illness, 53
  chronic, 50
Implied consent law, 434
Inertia, 123–24
  moving passenger, *124*
  at rest, *123*
  straight line force, *126*
Injury, 53
Inspection month, 465
Instruction permit, 447–48
  driving with, 450
Instrumentation, 71–74
Instrument panel, *71*
Insurance
  driving without, 470–71
  for rental cars, 515
  required coverage, 467–68
  responsibility for, 466–67
  shopping for, 469–70
  supplemental coverage, 468–69
Interchanges, 271–73
Intersections, 171–80
  alley, 177–78
  at alleys, driveways, and private roads, *183*
  approaching uncontrolled, 176–77
  congestion at urban, *233*
  right-of-way at, 181–83
  rural, 250–51
  semicontrolled and signal-controlled, *181*
  squaring, 180
  with STOP signs, *182*
  tips for, 176
  types of, 177–80
  uncontrolled, 171
  Y, 251
Intoxication, BAC and, *424*
Invoice cost, 477
I-Pass, 290–91
  EXPRESS lane, 291

Jack-knifing, 313
Jackrabbit starts, 492

Jaywalking, 301–302
Jitters, behind-the-wheel, 9
Judicial driving permit (JDP), 440
Jumper cables, using, 404
Jump-starting a dead battery, *405*

Kennedy Expressway, 286
Kill switches, 503
Kinetic energy, 124–25
Kinison, Sam, 327

Lane-changers
  constant, 237
  near stop light, *237*
Lanes
  bicycle, 163
  bus, 163
  changing, 235–36
  changing on freeways, 275–76
  choosing, 231–32, 274–75
  old markings, 158
  positioning in, 231–40
  reversible, 162
  special, 160–63
  spill over into, 234
  temporary, 162
Laws
  ignorance of, 14
  physical affecting driving, 121–25
  types of, 5
Leaks, recognizing, *489*
Leasing, 480
Left-turners, hesitant, 206
Left turns, *202–203*
  dangers of, 203
  lagging arrows for, 232
  lanes for, 161, 201–204
  stealing, 204–205, 206
  three rights for, 205
  on two-way streets, 201
  types of, 201
Legal duties, 5
Lemon law, 482
License plates, 463
  custom, 462
Licenses
  applying for, 449–50, 452–53
  classes of, 447, 451, 459
  driving with fake, 5–6
  for motorcycles, 312
  restricted, 440–41
Licensing, 447–60
  correcting and replacing, 458
  for motorcycles and mopeds, 458–59
  for new residents, 456
  for nonresidents, 456
  refusal of, 456
  renewing, 457
  restrictions on, 455
  suspension and revocation of, 473–74
  violations, 458

Lien, 460
Light rail crossings, 191
Light(s), 74–76
  beating, 173
  dirty, 329
  red, yellow, and green, 153
Limit lines, 160
List price, 476
Livestock crossing areas, 263–64
Loading, 127
Loops, 273
Low mileage, 481

Mailbox bashing, 254
Maintenance
  history, 481
  routine schedule, 491–92
  schedule, *493*
Manufacturer's certificate of origin (MCO), 464
Manufacturer's suggested retail price (MSRP), 476
Marijuana, 429
McKinley Bridge, 291
MCO. *See* Manufacturer's certificate of origin
Mechanical failures, 399–406
Mechanics, selecting, 495–97
Medical payments coverage, 469
Merging area, 282–83
Michigan Canal State Trail, 262
Minutemen, 407
Mirrors, 78, *79*
  round-the-corner, 178
Misdemeanor, involving collision, 380
Mississippi River, 291
Mitchell, Margaret, 230
Mopeds, 312
  licensing of, 458–59
Motorcades, 186–87
Motorcycle Rider Course (MRC), 459
Motorcycles, 306–13
  blind spots and, *308*
  dangerous spots for, 310
  driver responsibilities to, 306–308
  fatality rate, *307*
  headlights and, *308*
  helmets for, 313
  licensing of, 458–59
  safety equipment for, 309, 389
Motorcyclists
  helmets for, 313
  responsibilities of, 308–11
Motor scooters, 312
Mountain roads
  driving on, 340–41
  passing on, 341–42
  U-turns on, 341
Mouth-to-mouth resuscitation, 378
Moving vans, 516–17
MRC. *See* Motorcycle Rider Course

MSRP. *See* Manufacturer's suggested retail price
Mud guards, 250

Narcotics, 428
National Oceanic and Atmospheric Weather Administration (NOAA), 362
New Harmony Bridge, 291
Nicholson, Jack, 56
Non-Resident Violator Compact, 472
No-passing zones, *259*
North-South Tollway, 289
"No zones," of trucks, *312, 313*
Nozzles, 72

Odometer, 71
Off-ramps, 273, 285–86
Off-road driving, 519–20
Off-roaders, 263
Off-tracking, 316–17, *317*
O'Hare International airport, 338
Oil, 488–89
  pressure monitoring, *73*
On-ramps, 273, 278–81
  with double-merge lanes, *281*
  merging from, *280*
Open container law, 441–42
Operating devices, 74–81, *75*
Orderly visual search (OVS), 35–36
Organ donation, 455–56
Overheating, 365–67
OVS. *See* Orderly visual search

Pacing, 136
Pakula, Alan, 410
Partial cloverleaf, 271
Parked cars, 229–30
  exiting, 110
Parking, 104–13
  at airports, 113
  angles, 110–11
  curb, 106
  entering and exiting spaces, 106
  exiting parallel space, 109–10
  on hills, 111–12
  parallel, 110
  perpendicular, 110–11
  prohibited, 105
  on right side of street, 107
Parking garages, 112
Parking lots, 112
  alleys of, 217
  large, 179–80
Parking spaces, on city streets, 230
Passengers, 60–61
  protecting, 6–7
Passing
  by fast drivers, 242
  general rules for, 240
  on highways, 258–60
  steps to, 241–43

when not to, 241
who not to, 239
Patton, Jr., General George S., 204
Pavement markers, 164–65
PBT. *See* Preliminary breath test
PDPS. *See* Problem Driver Pointer System
Pedestrians
  don'ts for, 300
  driver responsibilities to, 297–99
  intersections and, *299*
  responsibilities of, 299–301
  signals for, 157
  unseen, 298
  in urban environments, 227
Peer pressure, resisting, 429
Permits, applying for, 449–50, 452–53
Pets, in vehicles, 61
Police
  activity of, 414
  pulled over by, 436
Pollock, Jackson, 126
Potholes, 334
Power lines, downed, 411
Power skid, 406
Preliminary breath test (PBT), 434
Preventive maintenance, 487–95
Probationary license, 474
Problem Driver Pointer System (PDPS), 456
Property damage liability, 468

Questionnaire sampling program, 470
"Quick starters, right-hand," 175

Radar detectors, 136
Radiators, adding water to, *365*
Railroad crossings, 186–91
  approaching, *189*
  controlled, 188–90
  signs at, 146, *149*
  tips for, 188
Ramps, 273
  signals on, 156
RDP. *See* Restricted driving permit
Reaction distance, 133
Rear-end collision, 374
Rearview mirrors, 498
Recreational vehicles (RVs), 514–16
Red arrows, 153
Red lights, 153, 172–73
Refueling, 492–93
Registration
  of out-of-state vehicles, 463–64
  renewal of, 464
  transferring or canceling, 463
Rental cars, 514
Repairs, estimates for, 496
Residential streets, 228
Responsibility, of driving, 5
Restricted driving permit (RDP), 474
Retail price, 476

Riding high, 492
Right-of-way, 180–87
    emergency vehicles and, 185–86
    funeral processions and, 186–87
    merging lanes and, 184
    motorcades and, 186–87
    passing and, 183
    pedestrians and, 185
    roadway parking and, 184
Right turns, 207–11
    bike lanes and, *210*
    pedestrians and, *211*
    on red light, 211
    two-point, 212–14, *212–13*
Road maps, 510–11
    reliability of, 513
Road rage, 55–57
Roads
    conditions of, 333–38
    defensive driving and, 24–25
    gravel and unpaved, 333
    level, banked, or crowned, *128*
    multilane, 233–35
    narrow, 336–37
    not pedestrian friendly, 301
    two-lane, 232–33
Roadside safety checkpoints, 438
Road slopes, 127–28
Road trips, 509–13
    planning, 509
    preparation for, 509–10
Roadways
    changing conditions on, 38
    construction and maintenance of, *15*
    law enforcement and, *16*
    markings on, 157–65
    types of, *13*
Rock Island Centennial Bridge, 291
Rolling through, 173
Rollins, Jr., Howard E., 434
Rollovers, *375*
Route markers, 148–49
Rubbernecking, 61, *62*
Rumble strips, 164
Rural driving, 251–56
Rural intersections, 250–51
Rural roads, 250
    visibility on, *256*
Rush-hour traffic, 287–88
RVs. *See* Recreational vehicles

SAFE process, 33–39
    for driving in wind, 361
    for entering freeways, 281
    in large parking lots, 179
    for left turns, 206
    for passing, 240
    for passing trucks, 314
    on rural roads, 259
Safety belts, *7,* 386–88

Safety cushions, *125*
Safety seats, child, 390–91
Saturation patrols, 438
Scanning, 33–36
    OVS technique, *35*
School buses, 318–19
    passing, 319
Security, 501–504
Semicontrolled intersections, 171
Semis, 313
Shifting
    down, 97–98
    to higher gears, 97
Shoulders, 335
Side-impact collision, 374
Signaling, 77
    solo, 257
Signals
    automated, 175
    hand traffic, 157
    house, 153
    lane-use, 154–55
    pedestrian, 157
    ramp, 156
    timed, 177
    warning, 155–56
Signs
    colors of, 146–47
    construction, 150–51
    crossed white, 146
    DO NOT ENTER, off-ramp sign, *279*
    guide, 150
    local traffic, 152
    overhead on freeway, 287
    parking, 146–47
    recreation, 151–52
    regulatory, 145–47
    service, 151
    shapes of, *146*
    STOP, 172–73
    symbols on, 147–48
    traffic, 145–52
    TURNOUT, 341
    variable message, 156–57
    warning, 147–48
    YIELD, 173–74
    yellow, *148*
Silva, Jr., Trinidad, 173
Size, driver–car compatibility, 49–50
Skids, 406
    responding to, 408–409
    types of, 406–408
Slow-moving vehicles (SMVs), 260–61, 281
Smart keys, 503
Smith, Bessie, 27
Smoke, 360–61
SMVs. *See* Slow-moving vehicles
Snowmobiles, 262
Snowmobile Safety Certificate, 263
Snowplows, 358–59

Snow tires, 358
Sobriety tests, 433–34
Social Security Number (SSN), 449
Solo collision, 374
Space
    cushion, 29–30, *29,* 274
    managing, 26–33
Speed, 121
    visibility and, *34*
Speed bumps, 165
Speeding, saving time and, 133
Speed law
    basic, 135
    minimum, 135–36
Speed limits, 134–36
    posted, 135–35
Speedometer, 71
Spinal trauma, 378
"Spinning out," 409
Splash and spray, 349
Sport utility vehicles (SUVs), 519–20
Sprewell, Latrell, 275
SR-22 insurance, 473
SSN. *See* Social Security Number
Steering, 98–102, 491
    dry, 99
    failure, 401
    hand-over-hand method, 101–102, *101*
    push-pull method, 102, *102*
    straightening out, 102
    tracking, 99–100
    for turns, 100–102
Steering column collars, 503
Steering wheels, 81–82
    grip, 207
    hand position, 99
    locks, 503
    tilt-wheel, *82*
Stereos, maximum volume of, 59–60
Sticker price, 476
Stopping distance, *133,* 134
    speed and, *133*
Streets, one-way, 236–38
Stress, and driving, 57–58
Summer driving, 366
Suspension, 491
SUVs. *See* Sport utility vehicles
Switchbacks, 340

Tachometer, 71
Tailgating, 30, *242*
Temperature monitoring, *73*
Temporary permit, 461, 463
Terminology, 525–36
Theft, preventing, 501–502
Thomas, Derrick, 357
Threats. *See* Dangers
Time, managing, 26–33

Tires
  chains for, 358
  inflation, *122*, 487–488
  maintenance, 487
  repair kit, *399*
  wear, 487–88
Tolerance law, 440
Toll facilities, 288–91
Tomjanovich, Rudy, 160
Tool sets, 488
Towing
  at night, 518
  trailers, 517–19
Traction, 122–23
Traffic
  controls, 34
  disabled vehicle in, 403
  turning left in heavy, 205–207
  circles, 178–79, *183*
  collisions with pedestrians, *302*
  conditions, defensive driving and, 25
  convictions, penalties for, 471–72
  flow of, 239
  laws. *See* Traffic laws. *See also* Laws; Legal duties
  lines, 157–60
  signals, 152–57
Traffic laws. *See also* Laws; Legal duties
  enforcement of, 16–17
  understanding, *23*
Trailers, towing, 517–19
Transmissions
  automatic, *83, 93,* 95
  fluid for, 489
  manual, *85, 94,* 95–98, *96*
Tri-State Tollway, 288
Trucks, 313–17
  driving behind, 313–14
  merging, 315
  oncoming, 315–16
  passing, 314
Trumpet interchange, 273
Tunnels, 338
Turning. *See also* Left turns; Right turns; Turns
  basics, 197–99
  lane for, 198
Turn lanes, 160–62
  center, 204
Turns
  fully protected, 197
  left, 199–207
  semiprotected, 197
  signaling for, 199
  three-point, 216

  two point, 212–14
  U, 214–15
  unprotected, 198
Turn signals, 76–77

Uncontrolled intersections, 171
Underinsured motorist bodily injury (UIM) coverage, 469
Underpasses, 338
Uninsured motorist bodily injury (UM) coverage, 468
Uninsured motorist property damage (UMPD) coverage, 469
Unpaved roads, 333
U-turns, 214–15, 341

Vandalism
  mailbox bashing, 254
  of signs, 146
Vans, wheelchair-modified, *51*
Vapor lock, 365
Variable message signs, 156–57
Vehicle identification number (VIN), 460
Vehicle inspection report (VIR), 466
Vehicle(s)
  all-terrain, 262
  buying new, 476–79
  buying used, 477–79
  condition of and defensive driving, 25
  controls of, 81–86
  driving unfamiliar, 513–20
  emissions inspections, 464–66
  equipment and accessories for, 497–501
  exterior protection, 493–95
  financing, 479–80
  interior protection, 495
  knowledge of, 7–8
  off-road, 261–63
  registering and insuring, 460–71
  restraint systems, 386–91
  runaway, 85
  selling, 480–82
  slow-moving, 260–61
  squeaking by, 231–32
  stolen, 501
  test driving, 479
  types of, 13
  valuation guides for used, 478
  washing, 494
Velocitation, 260
VIN. *See* Vehicle identification number
VIR. *See* Vehicle inspection report
Visibility, 327–33
  defensive driving and, 25
  field of vision, 30, 47–48

  high beams and, 330
  increasing, 32
  managing, 26–33
  rain and, 349
  reduced, 256–57
  speed and, *34*
Vision, 328
  alcohol and, 426
  central, *48*
  field of, 30, 47–48
  screening, 450
Visual acuity, 47–49

Wabash Memorial Bridge, 291
Warranty, 475, 477
Waxing, 494
Weather, driving conditions and
  dawn, 331–32
  defensive driving and, 25
  dusk, 331–32
  dust storms, 361–62
  fog, 360–361
  hail, 362
  hot and cold, 363–67
  hurricanes, 362–63
  ice, 353–59
  leaves, wet, 351
  lightning storms, 362
  mud, 333–34
  rain, 349–51
  sand, 333–34
  snow, 356–58
  tornadoes, 362–63
  water, deep, 352–53
  wind, 315
Weave lanes, 286
Wholesale price, 478
Wide-load vehicles, 260–61
Windows
  defrosting, *364*
  dirty, 329
  frosted up, 364
Windshields
  washer fluid, 490
  water spray on, 78
  wipers, 77–78, 405
Winter driving, tips for, 355
"Winterizing" your vehicle, 354
Wires, 490

Yellow arrows, 153–55
Yellow lights, 153, 173–74
  flashing, *155*
  long or short, 174
  running, 174
Y-intersections, 251
Youthful Intoxicated Driver's Visitation Program, 440

Figure Credits

1–4	Courtesy of California Department of Transportation
2–8	San Diego Aerospace Museum, San Diego, CA
3–2	Sonic Care Hearing Associates
3–6	On display at the National Automobile Museum (The Harrah Collection), Reno, NV
3–14	Keith D. Cullom, M.F.P.
4–8	Courtesy of the Detroit Public Library, National Automotive History Collection, Detroit, MI
5–10	On display at the National Automobile Museum (The Harrah Collection), Reno, NV
6–6	Keith D. Cullom, M.F.P.
6–7	On display at the National Automobile Museum (The Harrah Collection), Reno, NV
6–16	Fermi National Accelerator Laboratory, Batavia
7–2	Illinois Secretary of State
7–9	Illinois Department of Transportation, Springfield
7–14	Illinois Department of Transportation, Springfield
7–15	Illinois Department of Transportation, Springfield
7–16	Courtesy of Aurora Area Convention and Visitors Bureau, Aurora
7–21	NYS-DOT photo
7–26	Illinois Department of Transportation, Springfield
8–6	Illinois Historic Preservation Agency, Cahokia
9–10	On display at the National Automobile Museum (The Harrah Collection), Reno, NV
10–4	Southern California Metropolitan Transportation Authority
10–13	Galena State Historic Sites, Galena
11–9	John Deere Historic Site, Dixon
11–14	On display at the National Automobile Museum (The Harrah Collection), Reno, NV
11–15	Courtesy of Yellowstone National Park, WY
12–1	Illinois Department of Transportation, Springfield
12–2	Illinois Department of Transportation, Springfield
12–3	Federal Highway Administration, Washington, D.C.
12–6	Exclusive, Inc. U.S. Distributor for Automobili Lamborghini, Santa Ana, CA
12–12	Photo courtesy of USS Massachusetts Memorial Committee
12–15	Grave Mill and Museum, Oak Brook
12–18	Illinois Department of Transportation, Springfield
12–19	Illinois State Tollway Authority, Downers Grove
13–8	Courtesy of John S. Riley and Associates, the Motorcycle Lawyers, Pasadena, CA
13–10	© NYS Department of Economic Development
13–14	Photo by Robert Albright/owned by Dale Walksler, Wheels Through Time Museum, Mt. Vernon
13–19	Courtesy of Eric Orlemann, owner, ECO Industrial Communications, Decatur, IL
14–7	Courtesy of AM General Corporation, South Bend, IN
14–13	Illinois Department of Transportation, Springfield
15–8	Courtesy of California Department of Transportation
15–13	Illinois Historic Preservation Agency, Vandalia Statehouse State Historic Site, IL. oldest existing Capitol Building.
15–15	Courtesy of the National Aeronautics and Space Administration, Houston, TX
16–1	Photo compliments of Arizona Department of Public Safety, Phoenix, AZ
16–2	Photo compliments of Arizona Department of Public Safety, Phoenix, AZ
16–3	Photo compliments of Arizona Department of Public Safety, Phoenix, AZ
16–4	© Cantigny First Division Foundation, Wheaton
16–5	Courtesy of Keith D. Cullom, M.F.P.
16–9	Illinois Department of Transportation, Springfield
16–11	Courtesy of the Detroit Public Library, National Automotive History Collection, Detroit, MI
16–14	TRW, Inc., Washington, MI
17–3	Courtesy of the Detroit Public Library, National Automobile History Collection, Detroit, MI
17–6	Photo courtesy of Ameritech Corporation Archives, Chicago
17–12	Illinois Department of Transportation, Springfield
Chapter 18 opener	Photo compliments of Arizona Department of Public Safety, Phoenix, AZ
18–6	Al Capone's Speakeasy & Steakhouse circa 1920, owner W.G. Brooks III, St. Charles
18–10	Courtesy of George Barris, Los Angeles, CA
18–11	Fullerton Police Department, Fullerton, CA
18–12	Nevada Highway Patrol
19–1	Illinois Secretary of State
19–2	Illinois Secretary of State
19–6	Illinois Secretary of State
19–7	Illinois Secretary of State
19–9	Illinois Secretary of State
19–11	Reprinted with permission of the USAA Educational Foundation, a non-profit organization
19–12	Reprinted with permission of the USAA Educational Foundation, a non-profit organization
19–13	On display at the National Automobile Museum (The Harrah Collection), Reno, NV
19–25	Photo courtesy of the Farnsworth House, Plano
20–5	Courtesy of Hammond Castle Museum
20–10	Courtesy of the Detroit Public Library, National Automotive History Collection, Detroit, MI
20–12	White County Historical Society, Carmi
21–5	Copyright by and courtesy of GeoSystems Global Corporation
21–11	Courtesy of Navistar International Transportation Corporation, Chicago, IL

Maps of Illinois

Map content provided by GeoSystems Global Corporation, publishers of the National Geographic Road Atlas, MapQuest, and other National Geographic map products.

To order your very own copy of the National Geographic Road Atlas, or to find thousands of other kinds of maps and atlases, visit **www.mapstore.com**

For door-to-door trip directions, visit **www.mapquest.com**

544 ◆ *Maps of Illinois*

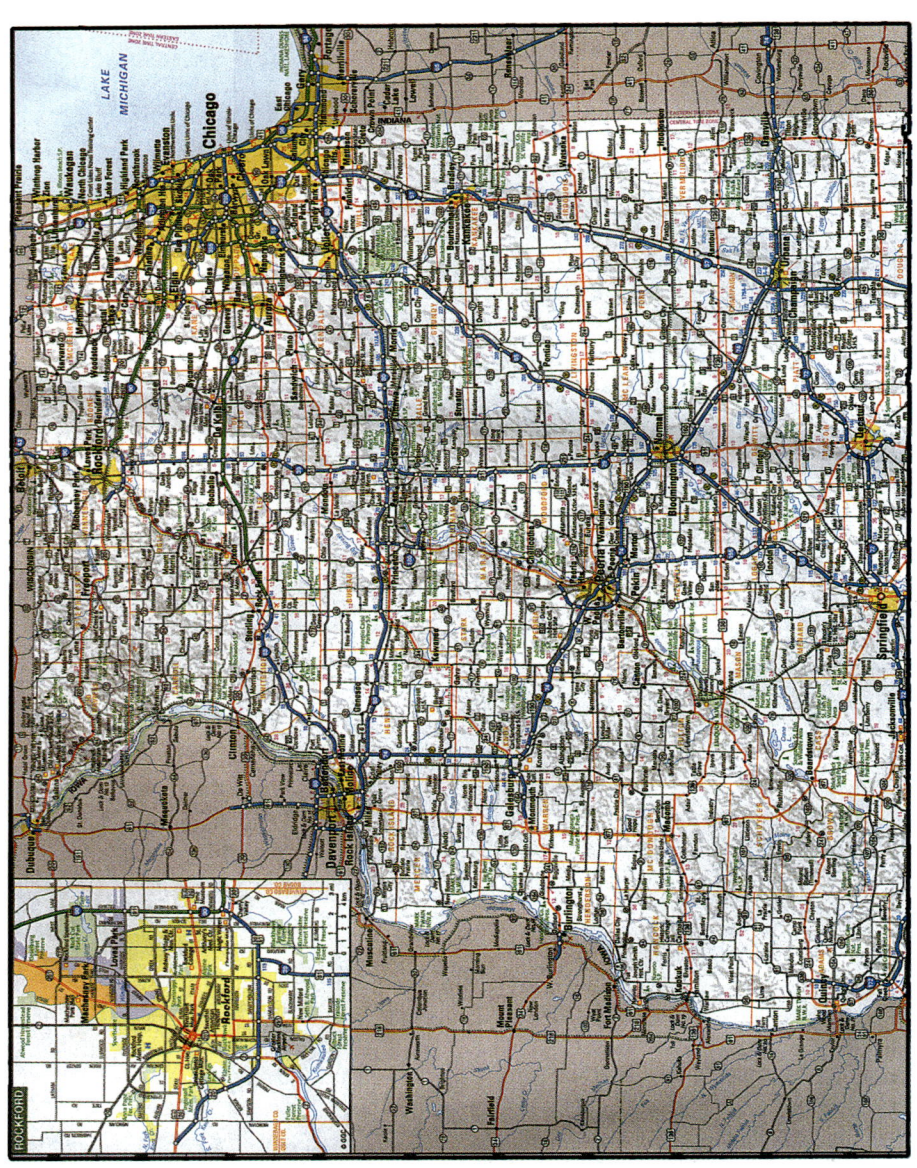

Maps of Illinois ◆ **545**

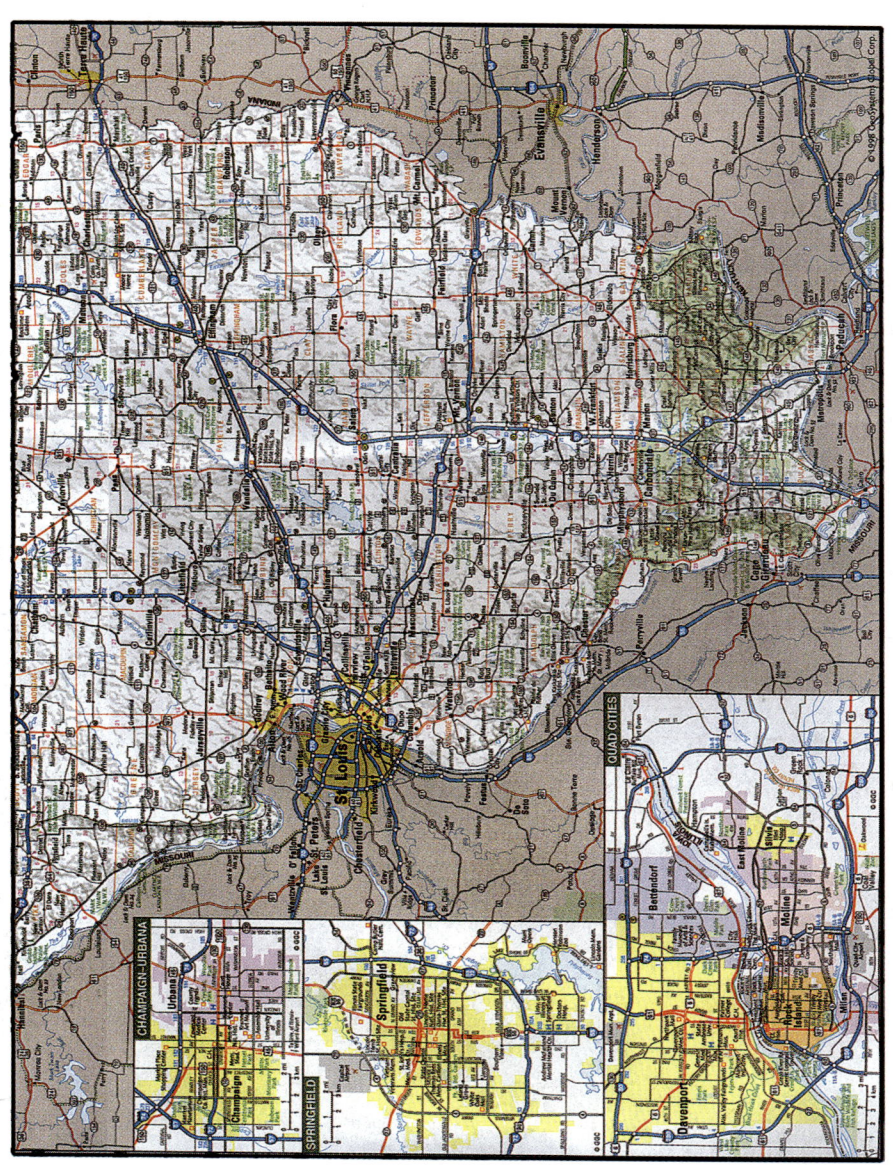

546 ◆ Maps of Illinois